Springer Series in
ADVANCED MICROELECTRONICS 6

Springer
Berlin
Heidelberg
New York
Barcelona
Hong Kong
London
Milan
Paris
Tokyo

Springer Series in
ADVANCED MICROELECTRONICS

Series editors: K. Itoh   T. Lee   T. Sakurai   D. Schmitt-Landsiedel

The Springer Series in Advanced Microelectronics provides systematic information on all the topics relevant for the design, processing, and manufacturing of microelectronic devices. The books, each prepared by leading researchers or engineers in their fields, cover the basic and advanced aspects of topics such as wafer processing, materials, device design, device technologies, circuit design, VLSI implementation, and sub-system technology. The series forms a bridge between physics and engineering, therefore the volumes will appeal to practicing engineers as well as research scientists.

1 **Cellular Neural Networks**
Chaos, Complexity, and VLSI Processing
By G. Manganaro, P. Arena, and L. Fortuna
2 **Technology of Integrated Circuits**
By D. Widmann, H. Mader, and H. Friedrich
3 **Ferroelectric Memories**
By J. F. Scott
4 **Microwave Resonators and Filters for Wireless Communication**
Theory, Design and Application
By M. Makimoto and S. Yamashita
5 **VLSI Memory Chip Design**
By K. Itoh
6 **Smart Power ICs**
Technologies and Application
By B. Murari, F. Bertotti, G. Vignola
7 **Noise in Semiconductor Devices**
Modeling and Simulation
By F. Bonani and G. Ghione
8 **Logic Synthesis for Asynchronous Controllers and Interfaces**
By J. Cortadella, M. Kishinevsky, A. Kondratyev, L. Lavagno, and A. Yakovlev
9 **Low Dielectric Constant Materials for IC Applications**
By P. S. Ho, W. W. Lee

Series homepage – http://www.springer.de/phys/books/ssam/

Reihen-Nr. 4076

ERRATA

Dear readers,

This is not the 2nd Edition of this book.
It is a corrected printing of the 1st Edition 1996.

We apologize for this inconvenience.

Smart Power IC's
Edited by B. Murari, F. Bertotti, G.A. Vignola
ISBN 3-540-43238-8
Springer-Verlag Berlin Heidelberg New York 2002

B. Murari · F. Bertotti · G. A. Vignola (Eds.)

Smart Power ICs

Technologies and Applications

2nd Edition

Contributions by

Andreini A., Borghi M. R., Botti E., Cognetti C., Contiero C.
Diazzi C., Fichtner W., Galbiati P., Gariboldi R., Kobayashi E.
Lonzi L., Maggioni M., Morelli M., Mukherjee S., Piccoli M.
Poletto V., Pedrazzini G., Rossi D., Speroni F., Stefani F., Storti S.
Sugawara Y., Tihanyi J.

With 480 Figures

Springer

Editors:

Dr. Bruno Murari
SGS-Thomson Microelectronics
Via Tolomeo, 1
20010 Cornaredo (MI)
Italy

Dr. Franco Bertotti
SGS-Thomson Microelectronics
Via Tolomeo, 1
20010 Cornaredo (MI)
Italy

Dr.-Ing. Giovanni A. Vignola
Via Don Luigi Sturzo, 15
20093 Cologno Monzese (MI)
Italy

Series Editor:

Dr. Kiyoo Itoh
Hitachi Ltd., Central Research Laboratory
1-280 Higashi-Koigakubo
Kokubunji-shi
Tokyo 185-8601
Japan

Professor Takayasu Sakurai
Center for Collaborative Research
University of Tokyo
7-22-1 Roppongi, Minato-ku,
Tokyo 106-8558
Japan

Library of Congress Catalog Card Number 61642

ISSN 1437-0387
ISBN 3-540-43238-8 2nd ed. Springer-Verlag Berlin Heidelberg New York
ISBN 3-540-60332-8 1st ed. Springer-Verlag Berlin Heidelberg New York

Springer-Verlag Berlin Heidelberg New York
a member of BertelsmannSpringer Science+Business Media GmbH

http://www.springer.de

Typesetting: Fotosatz-Service Köhler GmbH, Würzburg
Cover design: *design & production* GmbH, Heidelberg

Printed on acid-free paper SPIN: 10866393 62/3020 hu – 5 4 3 2 1 0 –

Foreword

Once a specialty market, Power Integrated Circuit Technology has grown in recent years to become a major business worth more than a billion dollars. The growth rate has also become more constant, too, as the market matures and the opportunistic "cherry picking" method of choosing applications gives way to a more systematic approach to the market.

At the same time, however, there has been a revolutionary change in the way that Power IC Technology is applied. Every system consists of a central core of processing elements isolated from the outside world by an umbrella of interface functions. The classic approach to partitioning is to divide these functions among a number of integrated circuits, each realized by a single-function technology like CMOS or bipolar. In smaller systems, however, it is necessary to combine different functions on one chip to minimize the component count. Moreover, since most systems involve some kind of interaction with actuators there is usually some sort of power element on the chip. As a result these circuits are best realized using mixed power technologies.

Mixed power technologies combining bipolar, CMOS and DMOS evolved in the mid-eighties, initially driven by the need to overcome second breakdown. At first mixed technology circuits simply replaced Bipolar power ICs, reducing dissipation. By 1990, designers were also taking advantage of the remarkable increase in complexity that CMOS logic and low dissipation DMOS allowed.

While single-function technologies evolve towards finer lithographies and higher speeds, there is a corresponding evolution in mixed technologies, as more functions are included to make more versatile technologies. The trend, in fact, is towards a modular approach where basic technology steps are designed to be compatible so that additional functions can be integrated when they are needed, at the expense of a few more mask steps.

Today mixed technologies are capable of integrating CMOS macrocells as complex as microcomputer cores. They are also able to integrate non volatile memory. With this capability many small systems can now be integrated into a single chip; others can be integrated into two chips where a delicate low voltage processing circuit is protected by a mixed technology circuit that integrates the rest of the system.

The new system oriented technologies are now applied to realize integrated circuits that integrate a smart battery charger and a chip that integrates almost all of the circuits of a camera. The software programmability of these "Super Smart Power" circuits brings new levels of functionality and versatility.

Another important trend is the move towards distributed "intelligence". By reducing functions to one chip Smart Power IC Technology can eliminate circuit boards. This is already the case in car alternator voltage regulators, and will also be true in automotive "smart connectors", where a control function is housed inside a cable connector. The decentralized approach will also be useful to reduce the amount of data to be transmitted. In an airbag system, for example, by adding intelligence to the sensor the raw data can be processed at high speed locally, so that the link to the control unit need only carry "control" data at a lower rate.

All of this is possible with the technology available today. Looking five years ahead to the millennium we can predict that many other important applications will be reduced to one chip. One crucial example is the home automation node – a combination of communications, control and high voltage. Analog telephones, too, will probably shrink to one chip.

In general, the mixed technologies evolving from smart power will be in the front line in the war against connections. Eliminating external connections to dice reduces cost and increases reliability at the expense of added complexity on the die. But as mixed system technologies shrink to the half micron level the cost of complexity diminishes rapidly.

Clearly what was once Power IC Technology is moving to a more central role, becoming much smarter and broadening its scope to include the integration of systems. Power Technology is thus no longer of interest only to power system designers; today it also interests digital designers and the architects of systems. To explore this new world they will need a guide, and the intention of the authors of this book is to fulfill this need.

One point, however, remains to be decided. What is "Smart Power"? In the industry a "Power IC" is usually defined by some arbitrary limit, while a "Smart Power IC" is a Power IC with some digital content. Perhaps it would be better to say that a Power IC is one where the customer pays more to have more power. On this question there is still much debate. One thing is clear, however: such is the pace of development in this field that we are unlikely to have a stable definition for some years to come. In the meantime, instead of worrying about the name of the rose, perhaps we should marvel at the way it blossoms and multiplies as we watch it.

B. Murari

Preface

As said by Mr. Murari in his Foreword this book is not only intended for Power System Design specialists as it makes a survey of the actual status and future trends of the very important family of Smart Power ICs but, due to their increasing variety of applications in many different fields ranging from telecommunications to automotive, it was also written for all those designers concerned with the development of digital circuits and with system architecture.

As the adjective "Smart" implies, Smart Power ICs are a new family of devices in which "intelligence" and "force" have been integrated into one chip.

This very important result was obtained as a consequence of a continuous research effort in the technological field. The availability of these new devices has promoted many new interesting applications.

New packing techniques and a more specific approach for reliability evaluation and product qualification have also been developed to cope with the new characteristics and improved performances of the new IC family.

The sequence of the book by which the chapters are presented follows this logical line and the book can ideally be divided into three parts.

The first chapters (Chap. 1 to Chap. 4) describe in detail the technological aspects.

The various technologies applied in design and production of Smart Power ICs are examined as concern their feasibility, advantages and limitations.

Chapter 5 opens the second part of the book, dedicated to a survey of the most important applications of Smart Power ICs, as it surveys all the commonest driving configurations applied in their design related to the different fields of application.

The other chapters of this second part, Chap. 6 (Motion control), Chap. 7 (Switching Regulators), Chap. 8 (Off-line ICs), Chap. 9 (Automotive Electronics) and Chap. 10 (Audio Amplifiers) describe the most important types of applications in various fields of interest.

In Chap. 11, which closes the second part of the book, some high complexity Smart Power ICs are described and an outline of future developments is also given.

In the last three chapters (Chap. 12, Chap. 13 and Chap. 14) which constitute the third part of the book, the Computer Aided Techniques available for Modelling, Design and Simulation of Smart Power ICs, the new package types and the new approach to Reliability evaluation and Product Qualification procedures as requested by this new family of devices are described.

This book is the result of a joint effort made by leading experts in the field from Europe, the United States and Japan. The editors want to extend their gratitude to all the authors that have made the editing of this book possible.

Most of the contributions were written by SGS-Thomson experts, as this Company is generally considered a leader in the field of Smart Power ICs.

The other authors belong to the following Companies: Philips USA, Siemens Germany, Hitachi Japan and Sanyo Japan.

The chapter on CAD Modelling, Design and Simulation was written by Professor Fichtner of Zurich Polytechnic Switzerland.

The book contains many circuit schematics, die photos, plots and tables and intends to be a valuable reference and source of updated information of the Smart Power ICs family for all those designers involved in industrial research and application laboratories.

The editors would like to thank Mr. P. Pistorio, President and CEO of SGS-Microelectronics, Mr. R. Paletto, President of SGS-Thomson Microelectronics s.r.l. and Mr. A. Romano, Corporate Vice President, DPG General Manager and Managing Director of SGS-Thomson Microelectronic s.r.l., for their decisive support and encouragement for sponsoring the initiative to edit this book.

The complexity of the book, due to the high number of figures, plots, tables, photographs and formulae and also to the different locations of some authors required a considerable effort for typing, correcting, and managing the preparation of the contributions in their final forms.

This book could not have been edited without the dedicated work of many person that cooperated to its realisation.

Therefore, we want to remember and thank all of them for the enthusiastic help given.

Miss A. Paciocco who reviewed the entire book as concerns the language; Mr. M. Casareto and Mr. C. Menescardi who prepared most of the figures, photos and graphs; Miss S. Petrelli for typing many contributions and last but not least Mrs. R. Desogus for the unvaluable secretarial work, coordinating all our collaborators and assuring the connection with authors and the editing publisher.

Special thanks are also due to Mrs. E. Raufelder, Springer-Verlag Technical Engineering Department for her kind and constructive help in the editing process of this book.

F. Bertotti
G. A. Vignola

List of Contributors

Andreini A.
SGS-THOMSON Microelectronics
Via Tolomeo, 1
20010 Cornaredo – Milano – Italy

Borghi M. R.
SGS-THOMSON Microelectronics
Via Tolomeo, 1
20010 Cornaredo – Milano – Italy

Botti E.
SGS-THOMSON Microelectronics
Via Tolomeo, 1
20010 Cornaredo – Milano – Italy

Cognetti C.
SGS-THOMSON Microelectronics
Via Olivetti, 2
20041 Agrate Brianza – Milano – Italy

Contiero C.
SGS-THOMSON Microelectronics
Via Tolomeo, 1
20010 Cornaredo – Milano – Italy

Diazzi C.
SGS-THOMSON Microelectronics
Via Tolomeo, 1
20010 Cornaredo – Milano – Italy

Fichtner W.
Institut für Integrierte Systeme
ETH-Zentrum, ETZ
CH-8092 Zürich

Galbiati P.
SGS-THOMSON Microelectronics
Via Tolomeo, 1
20010 Cornaredo – Milano – Italy

Gariboldi R.
SGS-THOMSON Microelectronics
Via Tolomeo, 1
20010 Cornaredo – Milano – Italy

Kobayashi E.
SANYO Company Ltd
LSI Division
180, Sakata – Oizum – Machi
Ora-gun, Gunma – Japan

Lonzi L.
SGS-THOMSON Microelectronics
Via Tolomeo, 1
20010 Cornaredo – Milano – Italy

Maggioni G.
SGS-THOMSON Microelectronics
Via Tolomeo, 1
20010 Cornaredo – Milano – Italy

Morelli M.
SGS-THOMSON Microelectronics
Via Tolomeo, 1
20010 Cornaredo – Milano – Italy

Mukherjee S.
PHILIPS Laboratories
345 Scarborough Road
Briarcliff Manor, NY 10510

Pedrazzini G.
SGS-THOMSON Microelectronics
Via Tolomeo, 1
20010 Cornaredo – Milano – Italy

Piccoli M.
SGS-THOMSON Microelectronics
Via Tolomeo, 1
20010 Cornaredo – Milano – Italy

Poletto V.
SGS-THOMSON Microelectronics
Via Tolomeo, 1
20010 Cornaredo – Milano – Italy

Rossi D.
SGS-THOMSON Microelectronics
Via Tolomeo, 1
20010 Cornaredo – Milano – Italy

Speroni F.
SGS-THOMSON Microelectronics
Via Tolomeo, 1
20010 Cornaredo – Milano – Italy

Stefani F.
SGS-THOMSON Microelectronics
Via Tolomeo, 1
20010 Cornaredo – Milano – Italy

Storti S.
SGS-THOMSON Microelectronics
Via Tolomeo, 1
20010 Cornaredo – Milano – Italy

Sugawara Y.
Chief Researcher of Hitachi Lab Ltd
Guest Chief Researcher of Kansai
Electric Power Company
Amagasaki – shi – Japan

Tihanyi J.
SIEMENS
Balanstr. 73
81541 München

Acronym List

BCD	Bipolar – CMOS – DMOS
BESOI	Bond and Etch – Back Silicon On Insulator
BGA	Ball Grid Array
CAD	Computer Aided Design
CDM	Charge Device Model
CMOS	Complementary MOS
CPU	Central Processing Unit
DI	Dielectric Isolation
DMOS	Double diffused MOS
EOS	Electrical Overstress
ESD	Electrostatic Discharge
FRR	Field Reduction Region
GTO	Gate Turn-Off Transistor
HBM	Human Body Model
HITFET	Highly Integrated Temperature Protected FET
IGBT	Insulated Gate Bipolar Transistor
IMS	Isolated Metal Substrates
JI	Junction Isolation
LCD	Liquid Cristal Display
LDMOS	Lateral DMOS
LOCMOS	Local Oxidation CMOS
LOCOS	Local Oxidation of Silicon
MCT	MOS Controlled Thyristor
PACE	Plasma Assisted Chemical Etching
PCB	Printed Circuit Board
PLCC	Plastic Leaded Chip Carrier
PQFP	Plastic Quad Flat Package
RESURF	Reduced Surface Field
SADDL	Self Aligned Double Diffused Lateral Transistor
SDDs	Smart Discrete Devices
SI	Self Isolation

SIMOX	Separation by Implanted Oxygen
SINFET	Schottky Injection FET
SITH	Static Induction Thyristor
SMD	Surface Mounted Devices
SMT	Surface Mounting Technology
SOI	Silicon On Insulator
SOP	Small Outline Package
SOS	Silicon On Sapphire
SPSDB	Single Poly-Si Direct Bonding
TCAD	Technology Computer Aided Design
TEMPFET	Temperature Protected FET
THT	Through Hole Technology
TOPFET	Temperature and Over Voltage Protected FET
TSOP	Thin Small Outline Package
VIP	Vertical Intelligent Power
VLSI	Very Large Scale Integration
VMOS	Vertical MOS
ZMR	Zone Melting Recristallization

Important Physical Constants and Some Properties of Silicon

Boltzmann's constant, K	1.38×10^{-23} J/K
Electron charge, q	1.6×10^{-19}C
Electronvolt, eV	1.6×10^{-19}J
Electron rest mass, m_o	9.11×10^{-28}g
Permittivity of free space, ε_o	8.86×10^{-14} F/cm
Permeability of free space, μ_o	1.256×10^{-8} H/cm
Thermal voltage at, V_T °	25.9 mV
Dielectric constant, K_{si}	11.7
Intrinsic mobility: *, °	
- Electron, μ_n	1350 $cm^2/(V.s)$
- Hole, μ_p	480 $cm^2/(V.s)$
Limiting velocity in bulk Si for electron and hole, V_{max}	1×10^7 cm/s
Diffusivity for minority carriers: (Si) *, °	
- Electron, D_n	29 cm^2/s
- Hole, D_p	11 cm^2/s
Thermal conductivity, σth °	1.5 W/K · cm

° At T = 300 K

* For background dopant concentration = 1×10^{16} cm^{-3}

Contents

CHAPTER 1

BCD Technology for Smart Power ICs

A. Andreini · C. Contiero · P. Galbiati

1.1 Introduction

The name BCD has been created in the mid-eighties to classify that family of mixed processes that allow to integrate into a single chip Bipolar, CMOS and DMOS transistors, forming a new power IC class called SMART POWER ICs [1–7].

Up till the mid-eighties Power ICs were realised using pure bipolar technologies previously developed together with their main application field, the audio amplifier market segment.

For the analog function required in these applications, the bipolar transistors were the best choice due to their amplification and matching properties.

The other major application field was the motor control one where the needed logic functions where implemented by bipolar I^2L structures.

With the growing requests of logic functions the bipolar I^2L approach resulted limited due to their design complexity, their high power consumption and their limited shrinkability with the lithography.

As these limits were not present in the CMOS components, at least at low frequency operation, to increase the smartness in the Power ICs it appeared evident that the only way was to make available CMOS in the place of I^2L.

During the Bipolar IC evolution the power that had to be supplied to the load increased, too. This increase reached the limit fixed by the allowed power dissipation and by the admitted consumptions.

The DMOS power transistor permitted to overcome all these limitations because it has no driving DC current requirement and it efficiently works in fast switching conditions.

The previous topics have been realized with the idea to merge the discrete DMOS technology, rising in the first eighties, with the junction-isolation bipolar technology used for many years to construct Power Integrated Circuits.

The realisation of this idea led to the birth of the Multipower-BCD [8].

1.2 Smart Power Technology Classification

Several different technological approaches to smart power have been explored, but all such technologies share the same basic concept of merging different

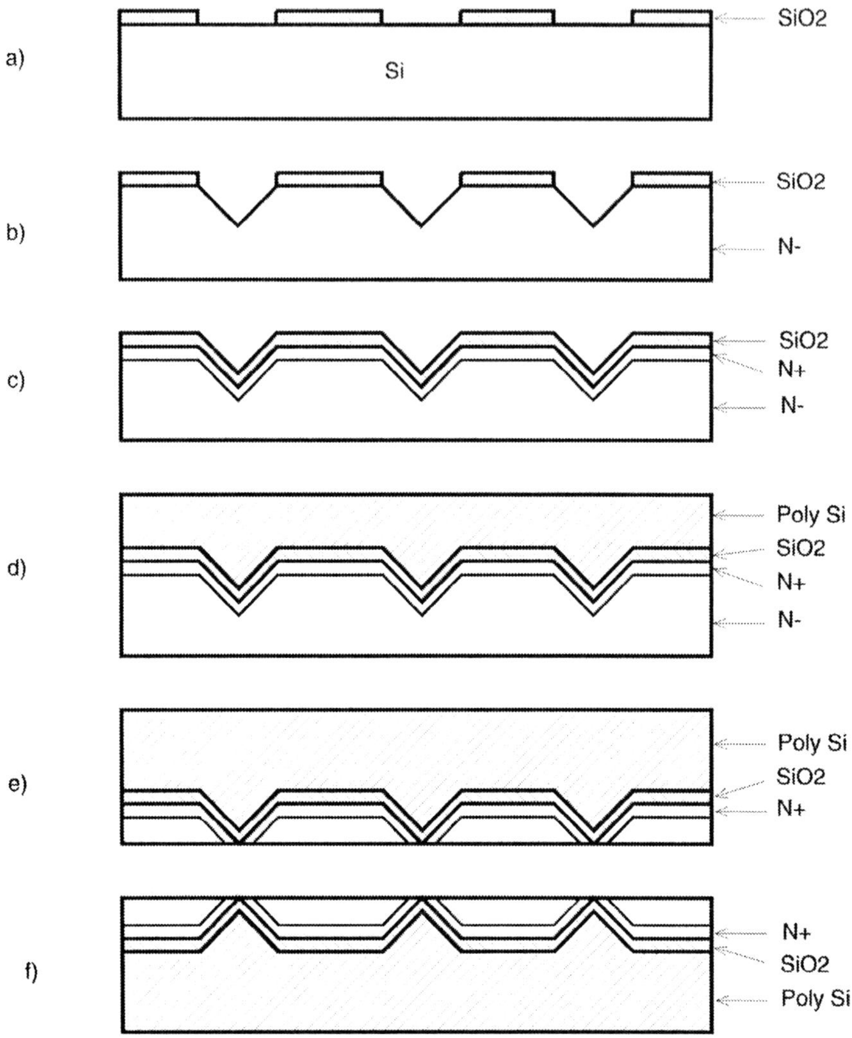

Fig. 1.1 Conventional Dielectric Isolation manufacturing process

structures into the same chip, taking advantage of the similarities in processing techniques.

Smart power technologies share many characteristics and may be conveniently classified according to the isolation technique (dielectric, self or junction isolation), by the power device structure (DMOS, bipolar, IGBT or SCR) and by the position of the drain/collector contact (on the surface or the back of the die).

1.2.1 Isolation Techniques

Dielectric Isolation (DI)

Dielectric isolation allows the realisation of silicon islands completely surrounded by oxide with advantages such as the low parasitic capacitance of components to substrate, the absence of leakage current and a reduction in the size of high voltage components. However, this method requires a more complex and costly manufacturing process (Fig. 1.1).

Moreover, since oxide is a poor conductor of heat it limits the integration of power devices.

Alongside the conventional construction methods developed more than 25 years ago [9], new techniques are being studied, such as Silicon Direct Bonding [10], lateral epitaxial growth over oxide [11] with different recrystalization techniques, and SIMOX [12–13]. These new techniques, however, are still at an experimental stage and will still require a little time before they are established. In addition, with these techniques it is difficult to realise vertical structures, even though a few examples have been reported [14–15].

Today the most promising DI approach is based on the SOI (Silicon On Insulator) substrate obtained by the Silicon Direct Bonding technique with isolation plugs realised by trench etches (Fig. 1.2) [16].

Self-Isolation

Self-isolation is used when the integrated devices do not need to be completely isolated because some terminals are shared.

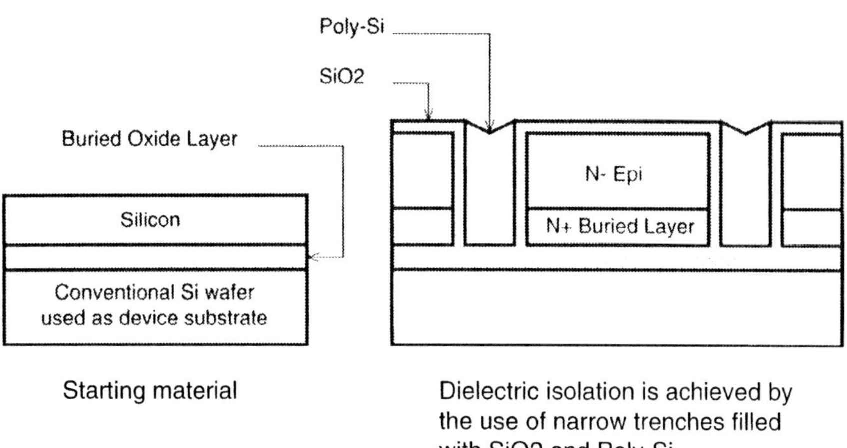

Fig. 1.2 Dielectric Isolation approach using Silicon Direct Bonding technique plus narrow trenches filled with SiO_2 and polysilicon

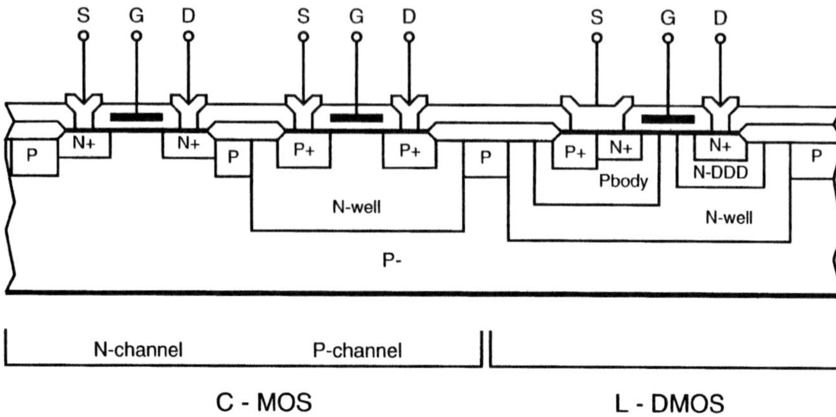

Fig. 1.3 Self isolation approach on P-type substrate: this example allows to integrate CMOS and Lateral DMOS components with drain contact on top chip surface and limited source follower capability

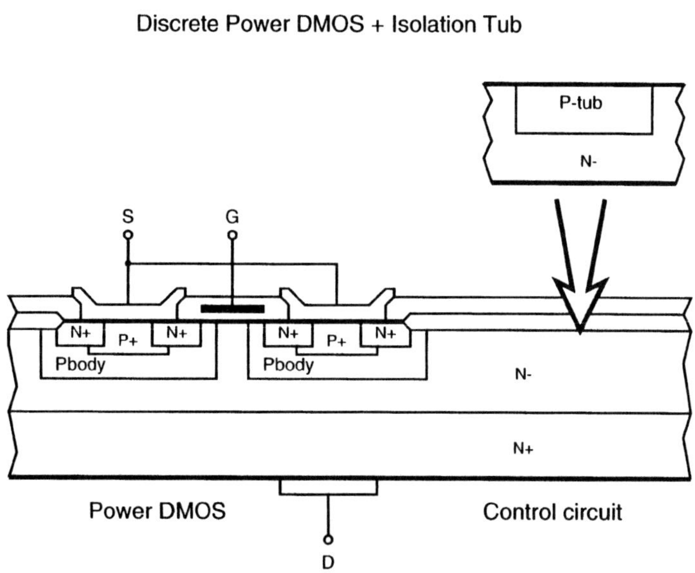

Fig. 1.4 Self isolation approach on N^-/N^+ substrate: this example allows to integrate NMOS and Vertical DMOS components with drain contact placed on the back side of the die

In this case the process is very simple and generally consists of a basic flowchart, for example CMOS or DMOS, to which a few extra steps are added to integrate additional structures. On the other side, this technique is not very flexible and is more susceptible to parasitic effects connected with biasing and supply.

Self-isolation lends itself to the integration both of high voltage and of high current devices. In high voltage devices the most common approach is carried out with the drain/collector contact on the top of the chip (Fig. 1.3). Typical applications are display drivers [17] and PWM controllers [18]. In high current devices the drain/collector contact is usually positioned on the rear of the chip (Fig. 1.4) [19].

Junction Isolation

With the junction-isolation technique the separation of the silicon islands where the various components are integrated is obtained through reverse-biased junctions realised by diffusing P regions through the entire depth of the N-type epitaxial layer grown on a P-type substrate (Fig. 1.5 a). This technique is the most widely used because it offers the best compromise between cost and versatility. With various methodologies it is possible to extend the applicability of this technique to both lateral and vertical structures, high voltage and high current. With the up and down isolation approach (Fig. 1.5 b), it is straightforward to produce integrated 250 V vertical DMOS devices [20–21] and the feasibility of devices up to 450 V has been demonstrated [22]. Voltages up to 1200 V have been reached applying the REduced SURface Field (RESURF) technique [23] to lateral DMOS devices [24]. Higher currents are obtained in junction isolation processes where the power device is integrated following the vertical scheme [19, 25, 26].

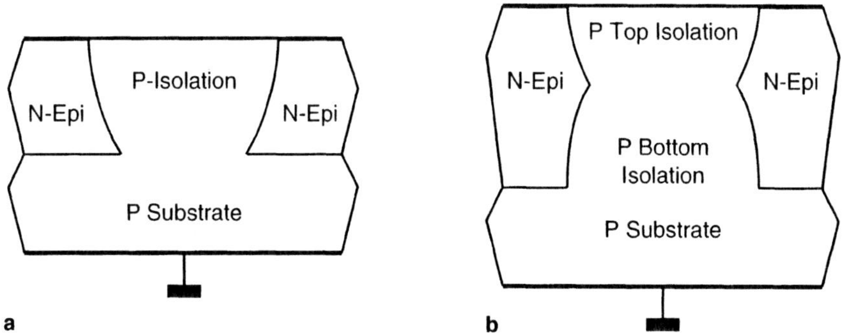

Fig. 1.5 Junction Isolation approach: **a** down P diffusion in thin epitaxial layer; **b** up and down P diffusion in thick epitaxial layer

Table 1.1 Comparison of MOS and Bipolar power transistors

MOS	BIPOLARS
Main differences	
– *Majority carrier device*	– *Minority carrier device*
no charge storage effects high switching speed drift current	charge stored in the base and collector low switching speed diffusion current
– *Voltage driven*	– *Current driven*
high input impedance no dc driving current consumption simple drive circuitry	low input impedance dc driving current consumption complex drive circuitry
– *Negative temperature coefficient of drain current for V_{GS} slightly higher than V_{TH}*	– *Positive temperature coefficient of collector current*
no thermal runaway no second breakdown limitation	thermal runaway effect susceptible to second breakdown
– *Low transconductance*	– *High transconductance*
– *Drain current is proportional to channel perimeter over length ratio*	– *Collector current is proportional to emitter area*

1.2.2 Integrated Power Structures

The most common power device used in a Smart Power Integrated Circuit is the DMOS [27] transistor for the well-known advantages that it offers compared to bipolar transistors (Table 1.1).

Several kinds of new power devices like IGBTs [28], IEGTs [29], ESTs [30], MCTs [31], just to mention the major ones, have been developed to compete with the performances of DMOS in high voltage and high current applications, but all these approaches are only aimed at market niches. Besides, as these are SCRs or thyristor-based structures, to overcome the problems coming from the current injection into the substrate, that can be a limiting factor to make them suitable as power elements in an integrated circuit, more recently the majority of these developments have been realised on SOI substrates.

Vertical & Horizontal Structures

As far as the geometry of power device integration is concerned, the positioning of the drain/collector contact on the surface or the rear of the chip is determined not only by the technology [25] but also by the needs of the application.

In a vertical structure the current flows through the device from the top to the bottom, across the substrate and the die attach area to the package itself (Fig. 1.6). This scheme, with the substrate above ground potential, is like that of

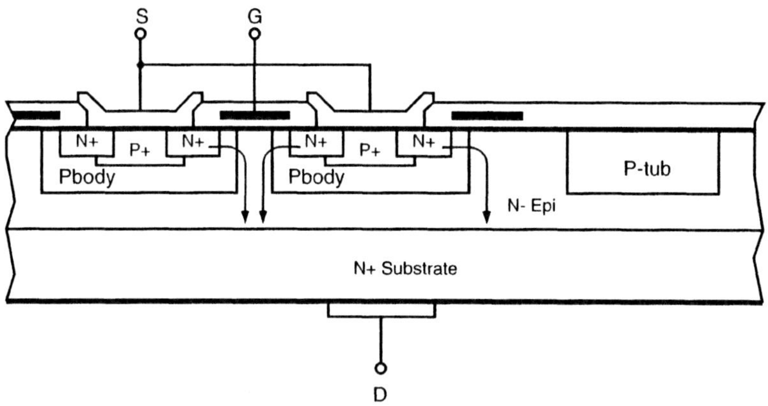

Fig. 1.6 Current flow in a Vertical structure

discrete power devices with the addition of an isolated structure containing the control circuitry. The vertical structure allows very high current densities but there is the limitation that in one chip it is only possible to integrate one power transistor, or several with the collectors or drains shared.

In a horizontal structure the current enters and leaves the chip through the upper surface of the chip. In this case, too, an isolated structure can be added that contains the control circuitry. This structure is derived from that adopted in standard power integrated circuits where the substrate is simply a mechanical support and a heat conductor. In the horizontal structure several variations can be distinguished on the basis of the current and voltage required.

Low Voltage High Current

Using DMOS as the power element the best structure in terms of specific ON-resistance is the lateral structure for voltage operation of few tens of volts (Fig. 1.7 a).

High Voltage Low Current

The horizontal lateral structure that exploits the RESURF technique to reduce the surface electric field, is the most suitable for applications above 250–300 V (Fig. 1.7 b).

In the junction-isolation approach this solution is generally limited to electrical configurations where the source floating capability is limited to few tens of volts.

This limitation can be overcome using a SOI approach.

Medium Voltage Medium Current

In mixed horizontal/vertical structures the current flow is vertical in the power device structure but the current is brought to the surface through buried layers

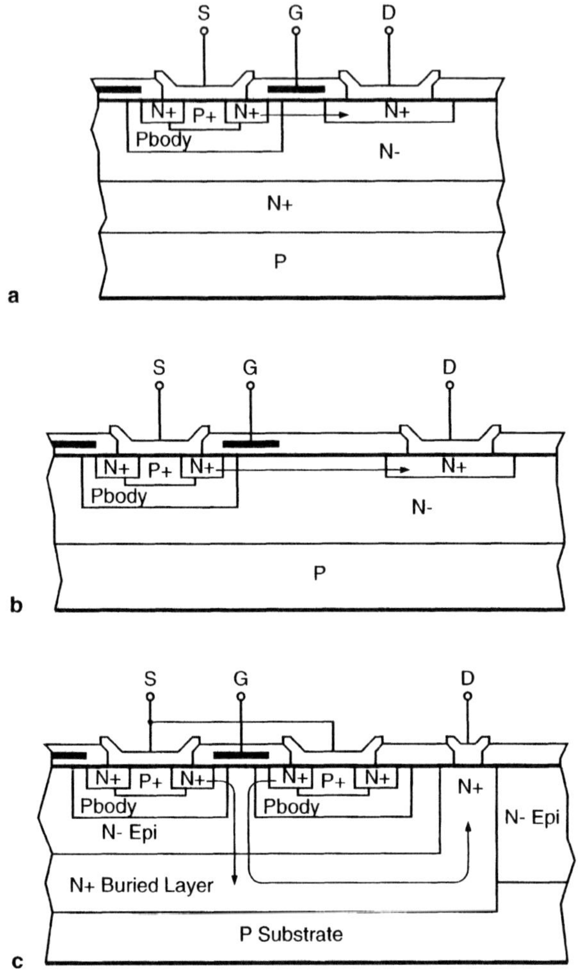

Fig. 1.7 Current flows in Horizontal structures

and sinkers (Fig. 1.7c). This approach allows the integration of any number of power devices connected in any way; in addition it offers medium currents and high voltages.

1.3 Multipower-BCD

The name Multipower originates from the possibility to integrate into one chip any number of DMOS power transistors interconnected in any way thanks to the use of the junction-isolation approach. Thus it becomes possible to integrate

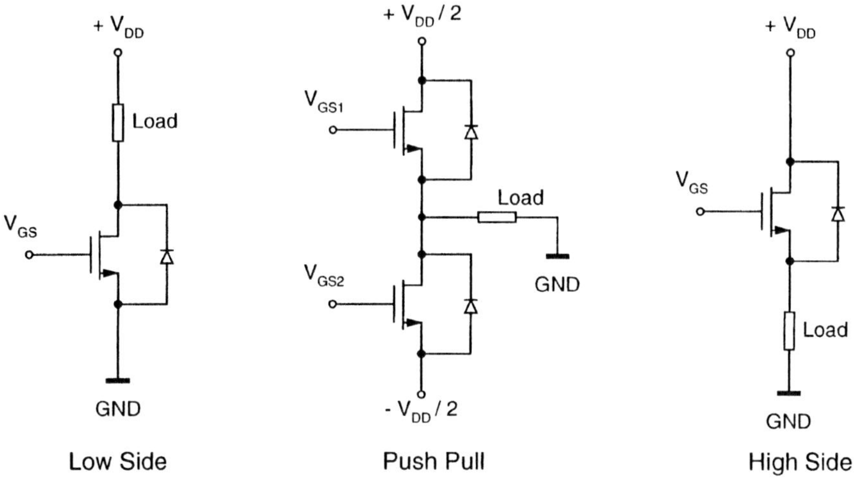

Fig. 1.8 Flexibility in the output stages

High-Side Drivers, Low-Side Drivers, Half bridges, full H-bridges and even multiple bridges (Fig. 1.8).

This feature distinguishes Multipower-BCD from the other Power IC family centred on a single power DMOS transistor, built like a discrete component, with the addition of control circuits.

This difference in flexibility makes Multipower-BCD usable in a broader application field than the other family named "Smart Discrete" that is dealt with in Chapter 3.

1.4 BCD History and Evolution

In Fig. 1.9 the genesis and evolution of the whole BCD process family are shown.

The given dates refer to the availability of the first working silicon.

The BCD process evolution has developed not only in the voltage direction (from 16 to 700 V), but also towards the reduction of the minimum lithography that is today scaled down to 0.6 µm in BCD-V.

In fact, another important advantage of the BCD is the dramatic increase of complexity induced in the Power ICs. This has been possible because in the BCD technology the power component is an MOS transistor where the current density depends on the geometrical ratio W/L (channel width/channel length) that can be improved simply by reducing the line width. Differently, in the power bipolar components the current density depends on the emitter area and therefore progress in microlithography does not bring any appreciable improvement in density.

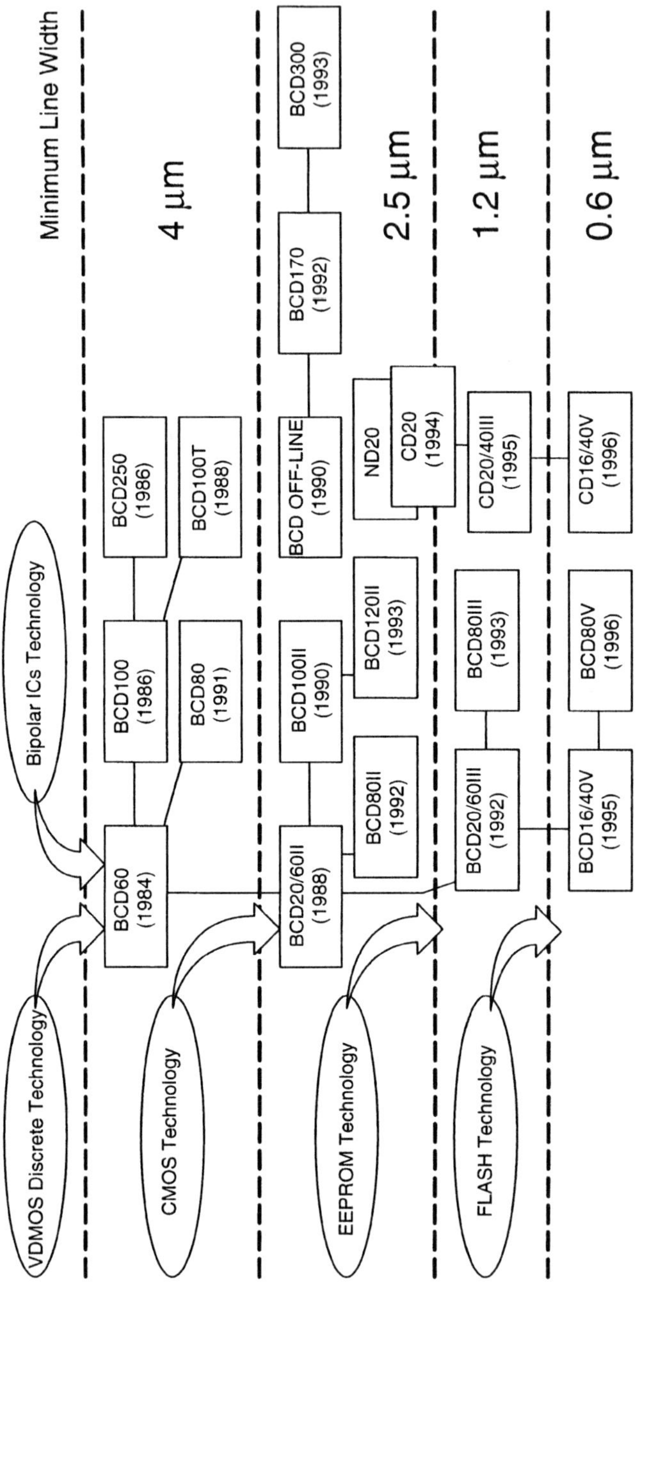

Fig. 1.9 Genesis and evolution of the BCD process family

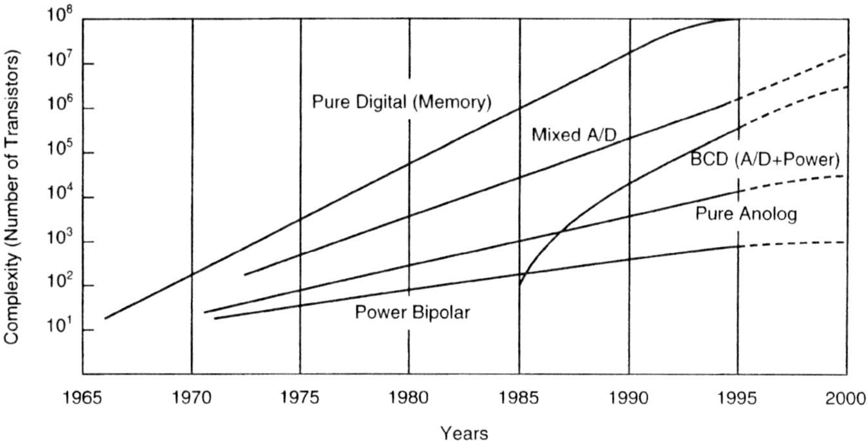

Fig. 1.10 Integrable transistor number versus time for different technology families

Combining the MOS compatibility of the BCD process with the gain obtained by moving from analog to digital signal processing, gives a complexity curve for BCD that is almost as steep as the digital memory curve (Fig. 1.10) and much steeper than the one for bipolar Power ICs.

1.4.1 First Generation (BCD-I)

The architecture of the first generation BCD process was based on the silicon gate self-aligned Vertical DMOS structure (VDMOS). Bipolar and CMOS components were added to this by applying the same doping profiles and polysilicon layer in a standard junction-isolation technology (Fig. 1.11); 4 μm was the minimum lithography used and 60 V, 100 V, 250 V were the maximum operating voltage classes developed.

The first commercially available product appeared in 1985 and it was a general purpose H-bridge motor driver IC designed in BCD60 process (Fig. 1.12) [32].

This IC demonstrated immediately the advantage of DMOS in this application because the DIP package version delivered 1.5 A continuous output current (the DMOS ON-resistance is 0.3 Ω) outperforming a similar bipolar bridge in a more expensive power package.

Products developed subsequently integrated more complex control circuits, such as a complete 10 A switch-mode regulator [33] or a stepper motor controller/driver.

However, the complexity of these circuits was limited not by the technology but because the full exploitation of the capability of the technology had to wait until system designers began to change their approach to partitioning.

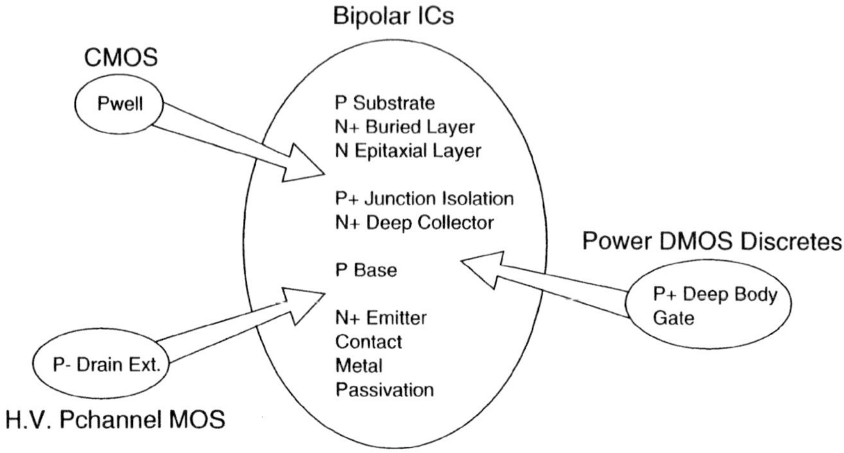

Fig. 1.11 Process architecture of First BCD generation

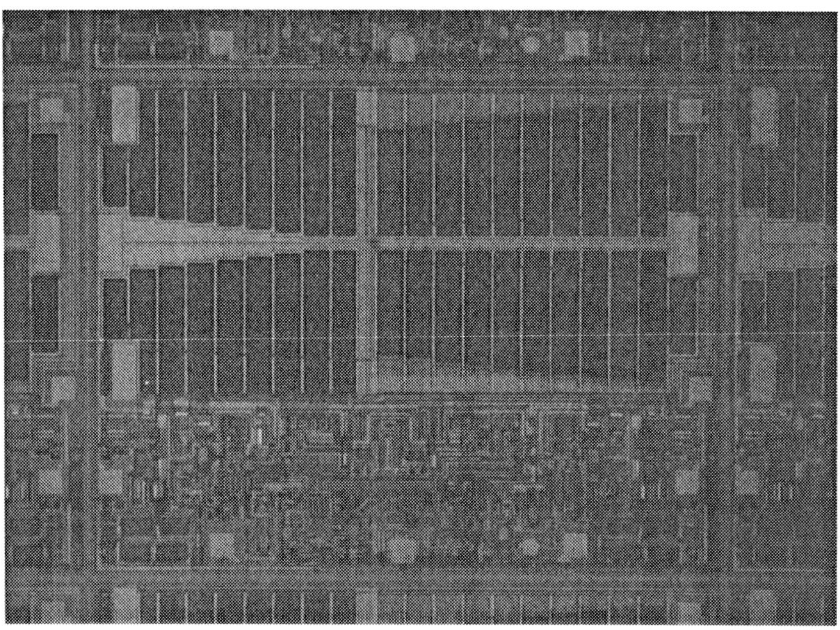

Fig. 1.12 Chip photograph of the first commercial BCD product: an H-bridge motor driver

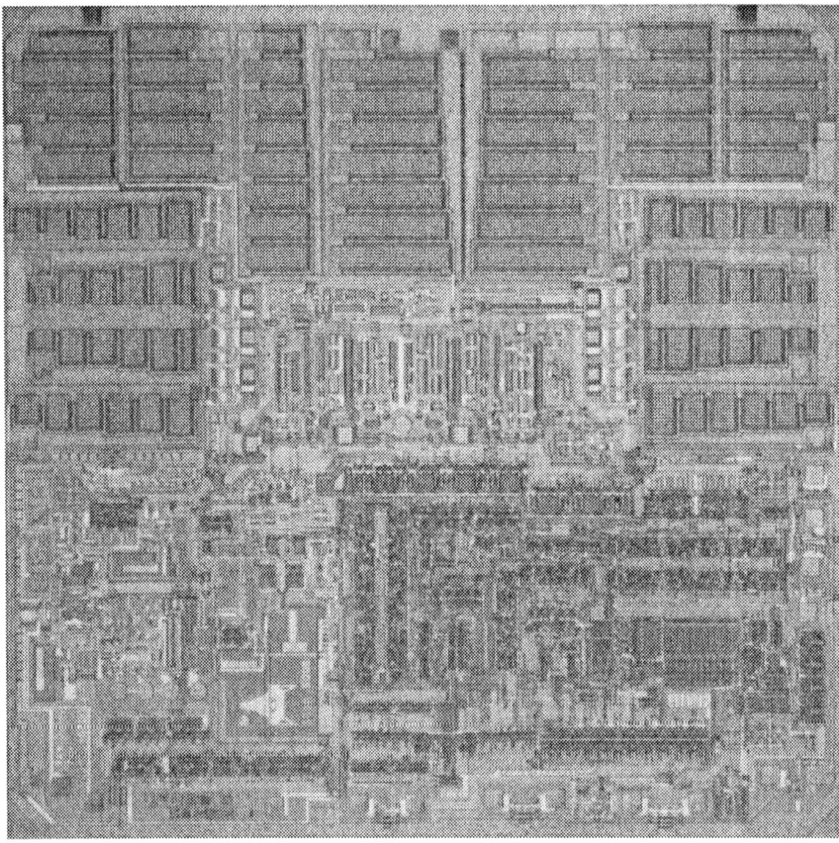

Fig. 1.13 Photograph of a single chip solution for electronic typewriter including three motor drivers and a solenoid driver

This was demonstrated in 1988 when a single chip power system was produced for a typewriter application (Fig. 1.13) [34].

Though this type of solution was possible using the first BCD process, the die size necessary for such levels of complexity made it too costly for many applications.

Therefore to make chips of this complexity truly cost effective it was essential to shrink the technology from the original 4 μm of the first generation to 2.5 μm of the second generation.

1.4.2 Second Generation (BCD-II)

In this new generation the core of the process was centred on CMOS to improve circuit design complexity (Fig. 1.14).

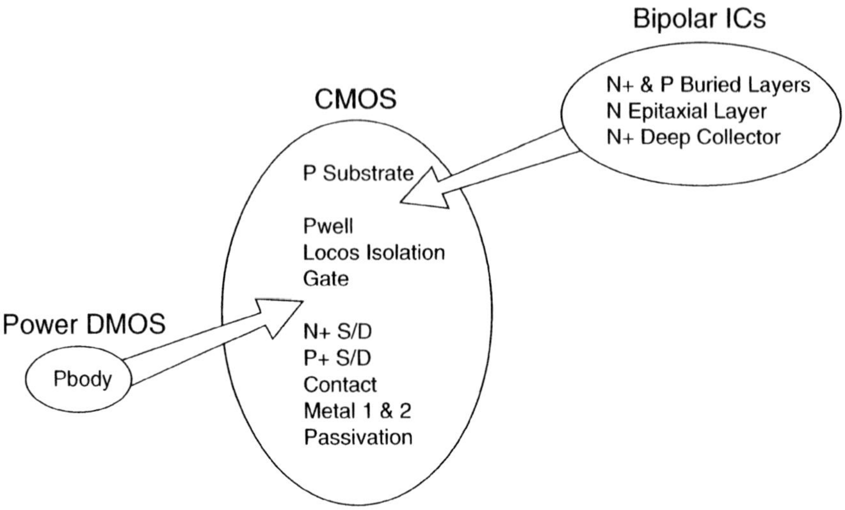

Fig. 1.14 Process architecture of Second BCD generation

This evolution approximately doubled both the signal section component density and the power section current density.

The availability of the second generation has widened the application fields and therefore many versions have been developed where the power devices are rated at 16 V, 20 V, 30 V, 40 V, 60 V, 80 V, 100 V and 120 V.

One of the first examples of a BCD-II product is the one shown in Fig. 1.15 which integrates into a single chip all the circuits needed to control and drive both the spindle motor (a three-phase brushless DC motor) and the head positioner (a "voice coil" actuator) in a hard disk drive. This chip includes 12 power DMOS transistors plus about 10 000 other transistors, though even this is still a long way from the limit of the technology.

To cover applications that require to be supplied directly from the mains, like the lamp ballast, in the second generation a version named BCD OFF-LINE has been developed, where the DMOS transistors are able to withstand reverse voltages as high as 700 V exploiting the RESURF approach. Similar to the other second generation BCD processes, also the BCD OFF-LINE architecture is based on a CMOS process [35].

An additional polysilicon level is introduced to build up the high voltage components and a twin-well structure is used (Fig. 1.16).

From this technology, designed primarily for industrial and off-line applications, it is possible to derive versions operating at lower voltages (170 V – 300 V) that, with some additional structures, will also find applications in the display driving fields.

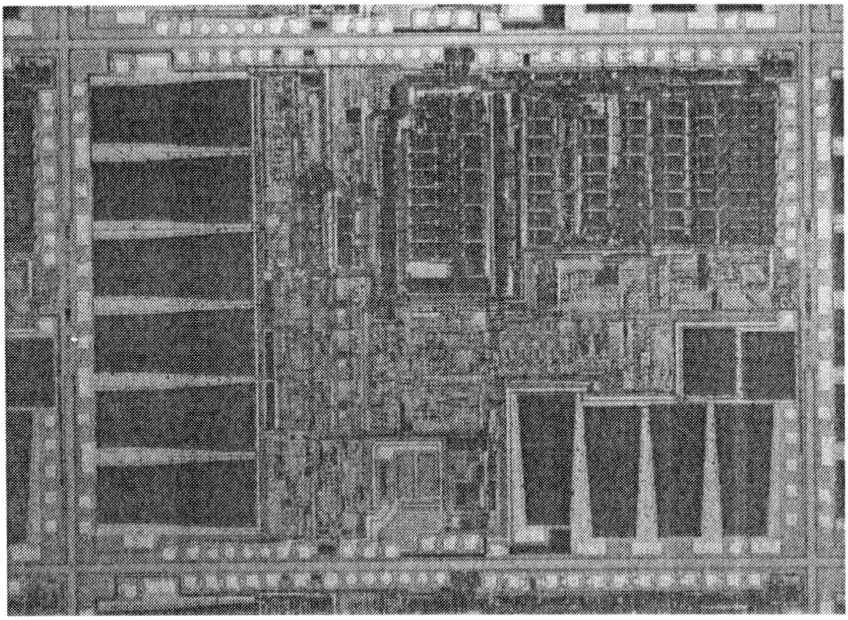

Fig. 1.15 Photograph of a single chip combo circuit for hard disk drive applications realised in BCD-II

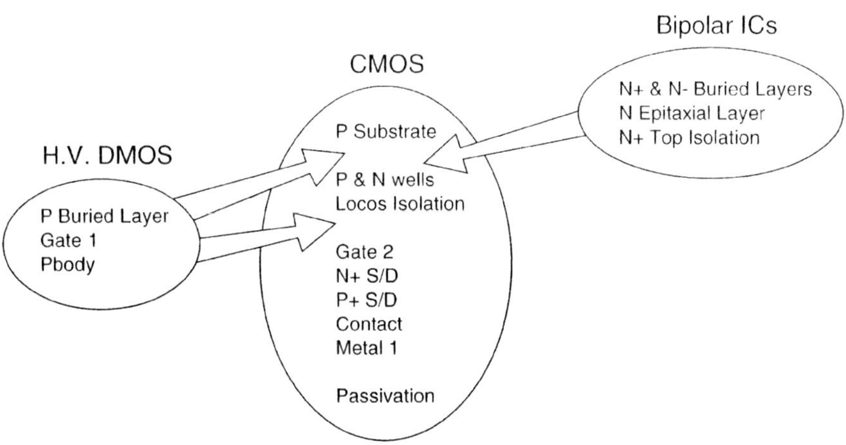

Fig. 1.16 Process architecture of BCD OFF-LINE for high voltage applications

1.4.3
System Oriented Generations

With the appearance, in the early nineties, of the third generation at 1.2 μm (BCD-III) where non-volatile memories (EPROM and EEPROM) are integrable too, the complexity level reached in the BCD processes has allowed to generate a new complete class of Power ICs, named "System Oriented" [36].

This name comes from the capability of these circuits to integrate into a single monolithic solution entire systems, previously realised with more chips made with single-function technologies, like pure CMOS, bipolars and non-volatile memories.

In applications where space is limited this approach is very useful because it reduces complex systems to a minimum number of chips, just one in some cases. The trend towards portable equipment brings a need for space reduction but at the same time reduces power requirements, encouraging the move to solutions of this type.

The architecture of the third generation is based on the concept of modularity where additional process blocks for optional structures and components can be added to a basic process flow-chart without creating any disturbance. The core of the process is based on a twin-well CMOS technology used to produce microprocessors, memories, gate arrays and other logic devices (Fig. 1.17) [37].

Moreover to reconcile the conflicting demands of packing density and current capacity there are three levels of metallization: two for high density

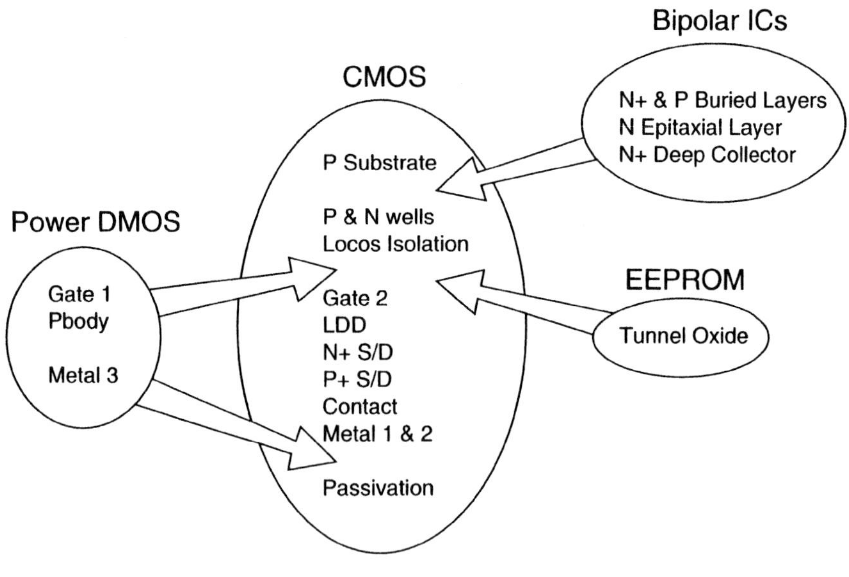

Fig. 1.17 Process architecture of Third BCD generation

interconnections in the signal section and a thicker layer for low resistance connections in the power sections.

With respect to the previous generation the current density in power devices is doubled and the density in signal components is almost tripled.

The third generation BCD is the first real example that allows designers to take advantage of existing CMOS libraries and even to include standard microcomputer cores and EEPROM memories to realize "Software programmable Smart Power" chips (SSP = Super Smart Power).

An interesting example of a complete system realised in a single chip solution using BCD-III is the Super Smart Battery Charger, that is a monolithic PWM current controller driving a 60 V, 0.1 Ω VDMOS with embedded a standard ST6 (8 bit) microcontroller with 2 Kbytes of internal EPROM and 64 bytes of internal RAM (see Chapter 11).

In the mid 90s, a 0.6 μm generation (BCD-V) has been introduced giving the possibility to integrate Flash memories, too.

At this level of lithography much more complex circuits will become feasible integrating other micro and DSP cores widening the filed of application for system oriented technologies.

In automotive, for example, they can be used to make single-chip smart switches for multiplex wiring systems. They can also be used in the computer field to make hard disk drives, in telecom to produce a one-chip cellular phone, in home automation to make a one-chip network node, in cameras to control all the motors and display functions, in industrial control and in many other fields. Other possible applications include medical implants, display drivers, flat-panel displays and even one-chip products with built-in sensors.

1.5 Process Architecture and Device Technology

In this section, the major steps involved in the BCD process construction are comprehensively listed and described to reach a complete understanding of the process architectures and of the function of each step putting them in relation with the integrable components.

In all BCD processes the construction sequence starts from a <100> P-type substrate whose doping level is chosen according to the maximum required voltage capability.

The key processing steps are

- N^+ buried layers Antimony or Arsenic doped are implanted to create low resistive paths for collector or drain current, to avoid punch-through limitation of floating P regions towards the P substrate and to kill he gain of the parasitic vertical PNP with the P substrate acting as a collector.
- N buried layers Phosphorous doped are added in high voltage processes both to increase the breakdown voltage of N^+ buried layers versus substrate and to implement source follower capability in RESURF type high voltage devices.

- P buried layers are introduced with isolation function: a "seed" of dopant is implanted before epitaxial growth and in subsequent thermal treatments this diffuses upwards to meet a downward diffusion from the surface. This approach is followed to save silicon area. In fact, a simple diffusion of a single-junction isolating layer from the top would require prolonged thermal treatments and consequently a larger lateral diffusion (Fig. 1.5 b).
  The same P buried layer can be used as buried anode of a diode featuring very low current losses towards the substrate because the gain of the parasitic substrate PNP associated to this diode is almost negligible (Fig. 1.18).
  In the RESURF approach P buried layers are introduced and dimensioned mainly to work together with the substrate in sustaining high reverse voltages (Fig. 1.19).
- The N epitaxial layer grown at this step, is dimensioned to obtain the best trade-off between maximum (close to ideal) breakdown voltage and minimum epitaxial bulk resistance contribution to ON-resistance when VDMOS is used as a power component. In BCD processes in which VDMOS is not implemented the epitaxial layer is only added to allow the formation of buried layers and the doping level is determined mostly by implanted N- and P-well regions.
- The junction-isolation structure is completed by the introduction of the top isolation. In the first BCD generation this is a heavily doped layer obtained by a solid source as in the traditional junction-isolation bipolar Power ICs and the P-well regions to realize NMOS transistors for CMOS circuits are introduced by ion implantation at a specific step before the top isolation deposition.

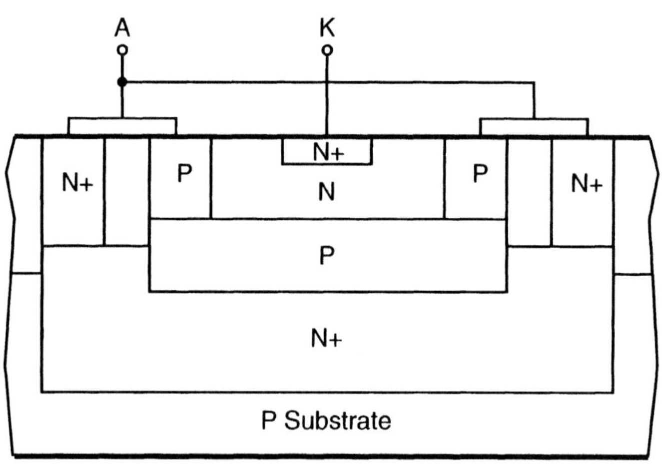

Fig. 1.18 Cross-section of a Low Leakage Diode (LLD) structure with P buried layer used as buried anode

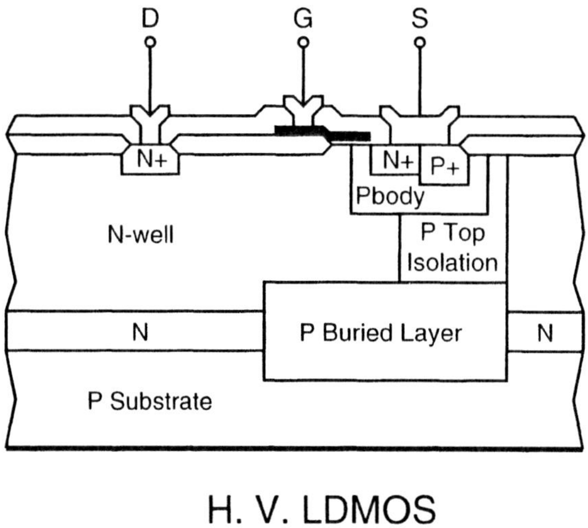

Fig. 1.19 Cross-section of RESURF type high voltage Lateral DMOS with P buried layer working as electric field shaping structure by enhancing depletion layer near source region

From the second generation onwards the top isolation function is realized by the P-well itself saving one mask and heavy process steps.
A separated step is still adopted in thick epi-layer processes and when power LIGBT (Lateral IGBT) devices are implemented.

- N-well regions are inserted when high density CMOS circuits are obtainable like in the third BCD generation at 1.2 μm.
  N-well regions are also introduced in high voltage RESURF type processes to better control the total charge in the drift regions of the RESURF acting structures (Fig. 1.19).
- Highly doped sinker plugs are realized by a Phosphorous deposition to contact the N^+ buried layer in order to bring up the current to the die surface in the vertical power devices. In the thick epi BCD processes, these sinker plugs are used in conjunction with buried sinker plugs to limit the area occupancy due to large side diffusions (Fig. 1.20).
- Highly doped P layers are implanted to form contact regions, lateral PNP emitter and collector and in the CMOS N-channel FET as a channel stopper on the P-well surface (BCD-I). The same layer is used also in the body of the DMOS component to contact the body itself and to reduce the value of the pinched resistance under the N^+ source region, that is, the base resistance of the parasitic intrinsic bipolar NPN transistor. As a consequence, the parasitic bipolar latch-back limit increases, making the DMOS more robust.
  The position of this step at this point in the process flow is peculiar to the first generation, while from the second generation on it has been moved to the

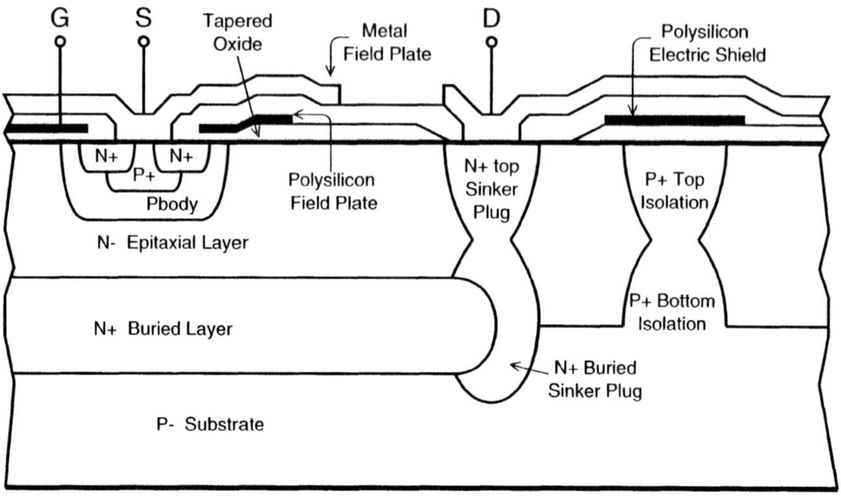

Fig. 1.20 Cross-section of high voltage Vertical DMOS in thick epitaxial layer in which deep drain contact to N^+ buried layer is obtained via an up-and-down scheme for N^+ sinker diffusion to reduce the area occupance

final part of the process flow, like in a pure CMOS process architecture to make source/drain regions for PMOS, in order to increase the CMOS density.

In this way, once obtained the platform to allocate the different integrable components, the BCD process flow continues with more specific construction steps aiming at the realization of the active part of the components and at the achievement of their performances.

- The active areas are defined in different ways depending on the process generation.
  In BCD-I, that is centred on the DMOS component, a thick oxide (field oxide) is patterned and etched after a high dose, low energy implant to obtain tapered oxide profiles with a very low angle. The slope of this oxide step is very important to realize high efficiency polysilicon field plates used to withstand the voltage in DMOS structures (Fig. 1.20). Furthermore, the smoothed oxide edge improves the metal step coverage even with very thick oxides.
  To encounter the requests of much denser CMOS logic, starting from BCD-II, the LOCOS technique has been adopted (Fig. 1.21a). This technique is based on the selective growth of a thick oxide (field oxide) on a silicon surface masked by a patterned silicon nitride.
  The nitride protects the underlying silicon from oxidizing, while allowing a thermal SiO_2 layer to grow on the exposed silicon: as a result the field oxide is semi-recessed under the starting surface. Therefore, the width of the base of the field oxide step defining the active area regions is reduced with respect to the tapered approach where this step is obtained by a prolonged wet etch.

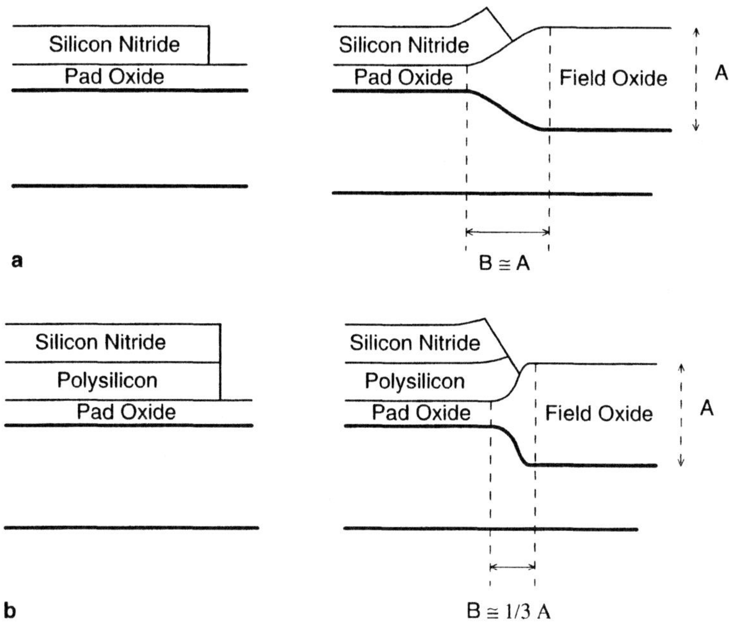

Fig. 1.21 Field oxide isolation techniques: **a** LOCOS; **b** Poly-Buffered LOCOS (PBL). The PBL technique allows to reduce the "bird beak" length compared to standard LOCOS

With the migration of the minimum lithography to values lower than 0.8 μm (BCD-V), the Poly-Buffered LOCOS (PBL) technique is used to further increase the transistor density (Fig. 1.21b).

- With the introduction of the LOCOS technique to form the component active areas in the BCD processes, it is also possible to obtain self-aligned channel stoppers on the CMOS body regions. These channel stoppers are formed by N- and P-type regions, named N field and P field, implanted before the field oxide growth and are masked by the nitride left to inhibit the surface oxidation. Therefore, these layers can be positioned only under the thick oxide and never inside an active area (Fig. 1.22a). The result is the increase of the surface concentration of P well, N well, N epi and consequently of the parasitic threshold voltage of polysilicon gate on field oxide.
  The P-field layer is also used as a drain extension in the P-channel MOS transistor to increase its voltage capability up the maximum value requested by the process rating (Fig. 1.22b).
- The gate oxide is grown inside the active area and it is used in all the MOS components. The required CMOS transistor density of the third generation BCD process forces to reduce the gate-oxide thickness to values not compatible with the high voltage performances needed in the vertical DMOS and high voltage P-channel MOS (H.V. PMOS) transistors. In BCD-III there-

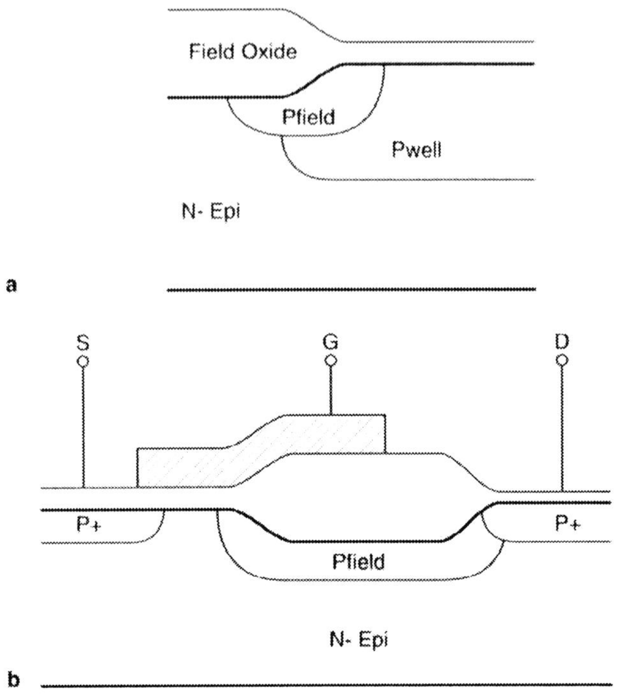

Fig. 1.22 P-field layer main functions: **a** self-aligned channel stopper to increase parasitic threshold voltage on P-well regions; **b** drain extension region in high voltage P-channel MOSFETs

fore there is a first oxidation step used as gate-oxide growth for DMOS and H.V. PMOS transistors only, while the gate-oxide growth for the CMOS components is shifted after the first polysilicon layer deposition and DMOS body diffusion step. In this case, a second polysilicon layer is also added and N^+ and P^+ source and drain implants are following. This approach has been adopted also in the BCD OFF-LINE version due to the very high voltage target for the DMOS components.
- At this point optional implants can be added to the process flow to adjust the MOS threshold voltages.
- The first polysilicon gate electrode, deposited and N doped at this step, is also used as field plate, channel stopper and as an interconnection level.

According to the modular approach, that is peculiar to the more advanced BCD process generations, additional steps can be inserted in the process flow before the polysilicon deposition in order to implement the EPROM an EEPROM cells.

- The control gate is constructed by a dedicated high dose N implant performed after the removal of the nitride layer used to define the active areas.

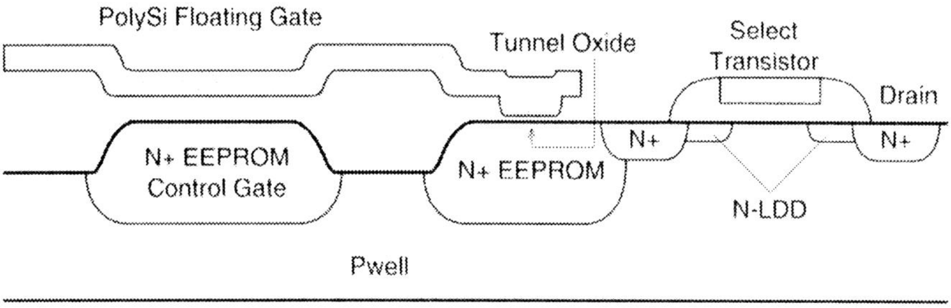

Fig. 1.23 Cross-section of a single-Poly structure EEPROM cell

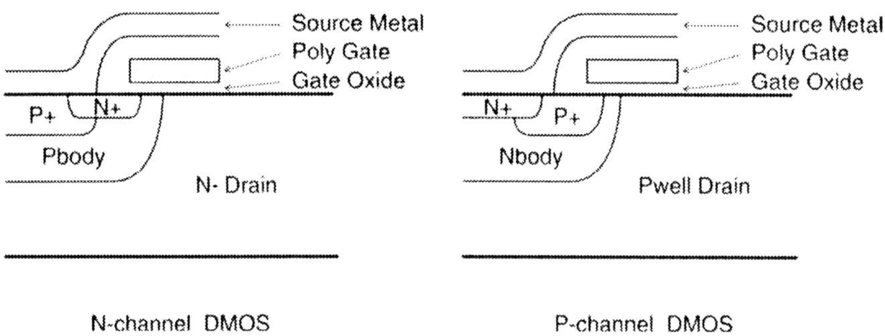

Fig. 1.24 N-channel and P-channel DMOS structure cross-sections

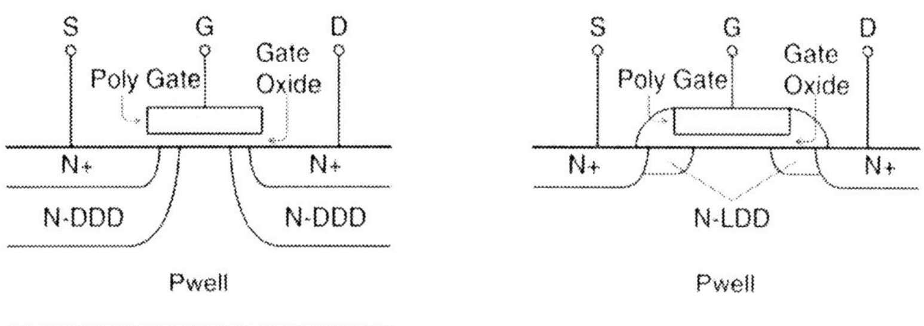

Fig. 1.25 Cross-section of DDD and LDD NMOS structures. The DDD or LDD layers are introduced to improve NMOS ruggedness against hot electron degradation

- In the EEPROM structure an extra mask is used to strip off the gate oxide from the areas where the tunnel oxide must be grown (Fig. 1.23).
- Self-aligned N and P implants are performed and diffused to form the double-diffused channel regions of P-channel and N-channel DMOS transistors (Fig. 1.24).

  The process parameters of these DMOS body implant and diffusion steps are key factors in the trade-off between the off-state breakdown and the on-state

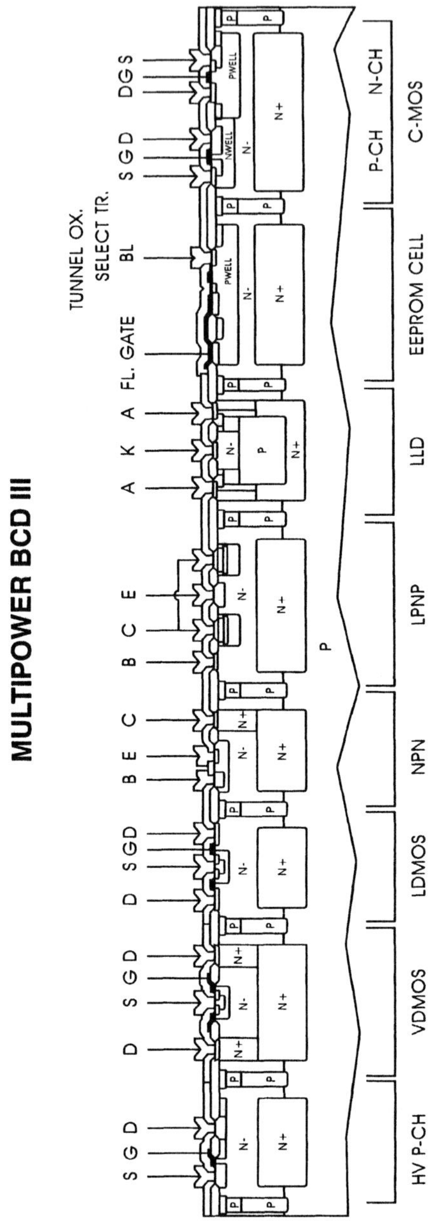

Fig. 1.26 Cross-section of BCD-III process through first metal deposition

performance that must be stated in the high voltage power DMOS components. The P-body layer has a wide use as the base of NPN transistors and as P-channel MOS source and drain regions.

- Standard DDD (Double Doped Drain) and LDD (Lightly Doped Drain) implant and spacer formation steps are carried out to build up source and drain extension regions of N-channel and P-channel MOS transistors to be used in high performance CMOS structures (Fig. 1.25).
  In the same way, the availability of the spacer technique gives benefit in increasing the drain to source maximum voltage capability of compact Lateral DMOS (LDMOS) components.
- After the N^+ and P^+ implants used to form source, drain and contact regions, a standard Boron Phosphorous Silicon Glass (BPSG) layer is deposited and the whole structure is provided with the conventional next contact opening step and with the first thin metal deposition.
- The following adopted metallization scheme strongly depends on the BCD process generation and can be distinguished in two major families: the single level, double thickness metal process and the multiple level, multiple thickness metal process. The former, used in the first and partially in the second generations, is obtained stacking a second thick metal on the first metal layer to increase the overall metal thickness in order to obtain a low resistance path for the current to be handled in the power sections and to improve the ruggedness against electromigration phenomena.
  The latter has been imported in the BCD processes to make the CMOS part compatible also in the interconnection scheme with pure CMOS processes.
  In this way, the portability of complex logic circuits is guaranteed and the advantage of the VLSI design can be fully exploited.
  Anyway, this scheme has been adapted to the Smart Power IC requirements, dimensioning the last metal level thickness in order to be used in the power components.

As an example of the BCD process architecture, in Fig. 1.26 the BCD-III cross-section is shown, with the main integrable components.

1.6 Integrable Devices: DMOS Transistors

Both bipolar and DMOS power transistors are widely used in Smart Power ICs. The choice of the power elements for application in switching conditions is based, in addition to the physical comparisons given in Table 1, on the efficiency in silicon area consumption as shown in Fig. 1.27.

For high voltage applications (i.e. >100 V) the DMOS is better than the bipolar even in processes with a relatively large minimum lithography (4 μm). The DMOS advantage is due to the fact that the its drain-to-source breakdown voltage fixes the device voltage capability that is determined by reaching the critical electric field at the body/drain junction.

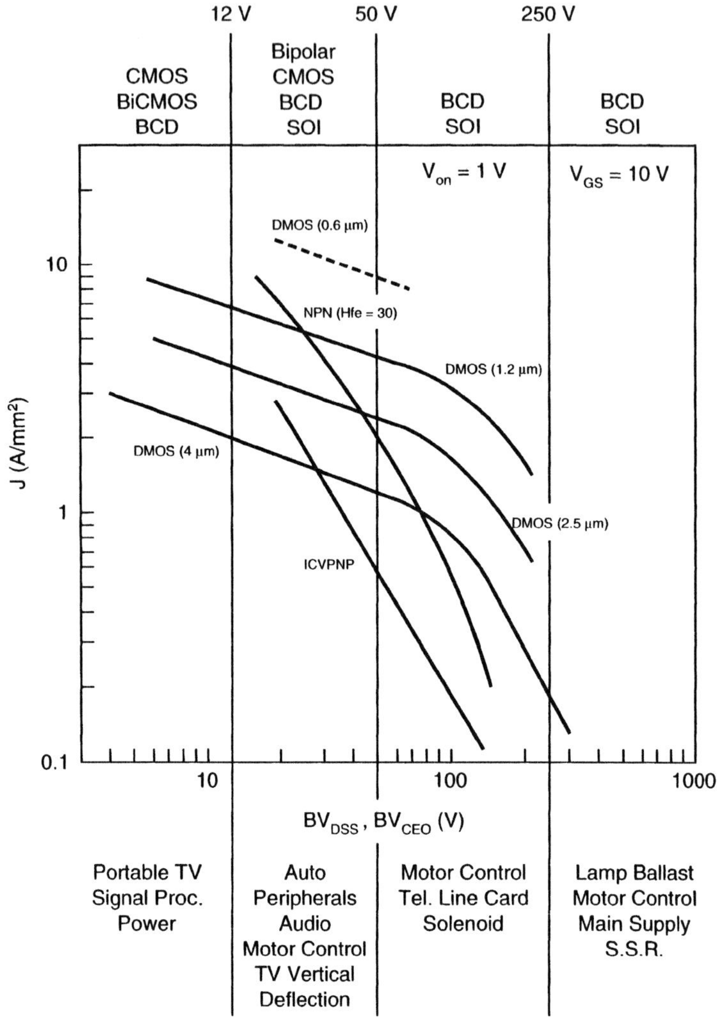

Fig. 1.27 Comparison between power bipolar and power DMOS transistors in terms of current density versus voltage capability for different process generations

In the bipolar device the transistor current gain lowers the breakdown voltage of the base/collector junction. Therefore to guarantee the same blocking voltage capability the bipolar transistor needs a thicker and more resistive epitaxial layer that penalizes its on-state performance.

Moving on to reduced minimum lithography, the DMOS transistor has taken advantage both in vertical configuration and particularly if realized in lateral arrangement. As a consequence, the breakeven point is shifted down to 20 V.

In this section, after a brief description of the DMOS structure, the optimization of its ON-resistance, that is the key parameter for the DMOS transistor operating in switching applications, will be deeply treated together with the other two parameters Vth and BVdss that are fundamental in the device construction design.

Furthermore, the DMOS performances in the saturation region will be analyzed, too, as this component is becoming of interest also for the realisation of amplification stages in audio and telecom application fields.

DMOS structure

In the DMOS device, the channel region is obtained by the difference in lateral diffusion of two different impurity profiles introduced through the same opening in the polysilicon layer. The obtainable DMOS channel length is therefore very short and it does not depend on lithography accuracy.

Two different ways to build up the device exist depending on the drain position that determines the current flow direction.

In the vertical approach the power DMOS device consists of a number of transistors with parallel source and gate electrodes integrated into a single isolated area with a common drain. The integration of this component requires bringing the drain contact to the surface via a N^+ buried layer and a N^+ sinker diffusion.

In this case, the current flows through the sinker to the buried layer and then up to the surface (Fig. 1.7c).

On the contrary, in the lateral approach the drain, positioned in front of each source, forces the current to flow horizontally (Fig. 1.7a, 1.7b).

1.6.1 Off-State Performance

To obtain the best trade-off between maximum (close to ideal) breakdown voltage and minimum epitaxial bulk resistance, in the vertical DMOS design considerations and investigations concentrate in the edge termination structure of the P body/N epitaxial junction.

In almost all BCD process generations an efficient and reliable solution consists of a polysilicon field plate overlaying the gate and field oxides forming a standard biplanar structure [38]. The best dimensioning of the polysilicon field plate is determined using a bi-dimensional device simulator. This operation allows to highlight the high electric field critical points of this solution and it allows to fix the proper epi-layer thickness and resistivity specification to achieve the best trade-off between the minimum contribution on the on-state resistance and the required breakdown voltage.

One of the most critical points of the biplanar structure is the transition region between the thin gate oxide and the thick field oxide as pointed out in Fig. 1.28.

In this figure, the case of the first generation vertical DMOS edge termination is depicted showing the influence of the level angle of the transition region on

Fig. 1.28 Electric field at point Q versus the level angle of the filed oxide with a constant applied voltage

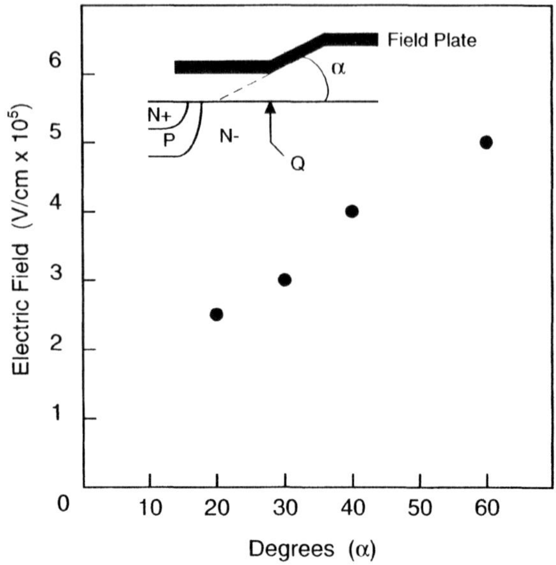

the surface electric field under that point. In fact, the electric field arising in that point when the angle is wide, causes electron injection into the oxide giving lowered and walking-out breakdown voltage values. The use of the LOCOS technique to define the active areas, from the second BCD process generation onwards, allows to reduce the electric field in the above mentioned critical region by adding a floating P-type ring, self-aligned to the active area. This layer helps to make a more uniform voltage partitioning (Fig. 1.29).

This ring can be formed by the P-field diffusion, already used in CMOS to increase the parasitic threshold voltage of polysilicon interconnections on field oxide over the P well, without requiring any additional mask and any extra ion implantation step.

To integrate the 250 V vertical DMOS device a triplanar structure must be adopted: it is realized extending the source metal beyond the polysilicon field plate edge over a third oxide level with an increased thickness (Fig. 1.20). The third field plate level serves to reduce the surface electric field at the edge of the polysilicon field plate that, otherwise, limits the component voltage capability.

In the vertical DMOS component the breakdown voltage condition can also occur between the inside cells. In this region, the spacing between two neighbouring cells plays a fundamental role in fixing the maximum electric field value as is shown in fig. 1.30. When the spacing becomes too large, the positive combined effect of the adjacent P-body depletion layers and of the gate field plate disappears. Each cell works like an individual P body/epi junction protected by a planar polysilicon field plate placed on a thin gate oxide. This is a less efficient structure than the biplanar one.

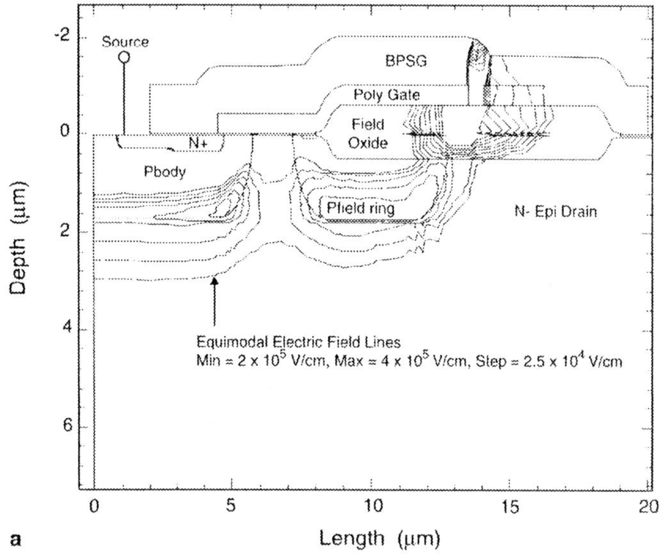

a

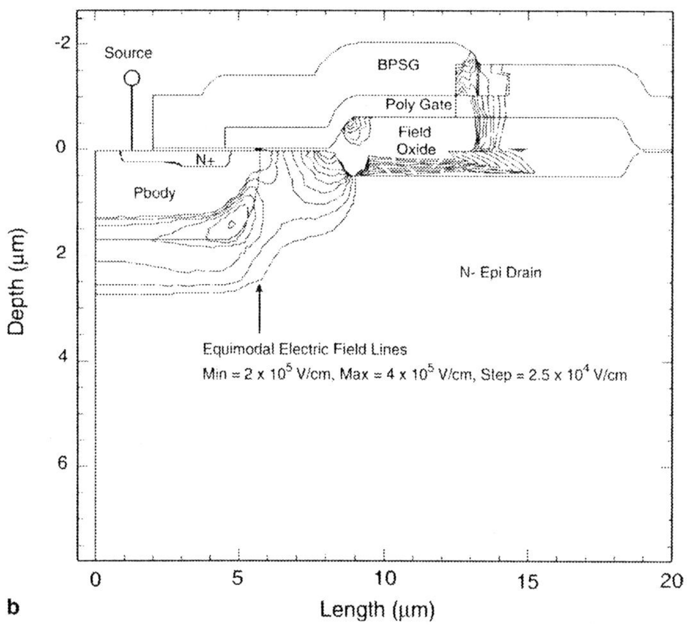

b

Fig. 1.29 Cross-section of VDMOS edge termination using LOCOS showing equimodal electric field lines. A self-aligned P-field floating ring (structure **a**) avoids the arising of a high electric field at the transition point between gate oxide and field oxide, present instead in the structure **b**) without P-field ring

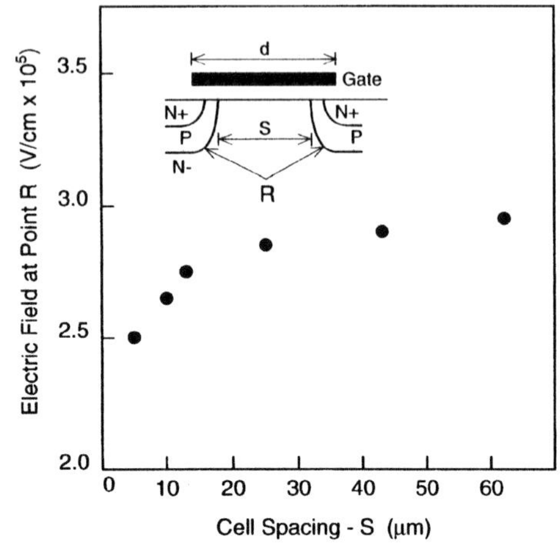

Fig. 1.30 Electric field at the inside cells as a function of the cell spacing in the case of 250 V BCD Vertical DMOS

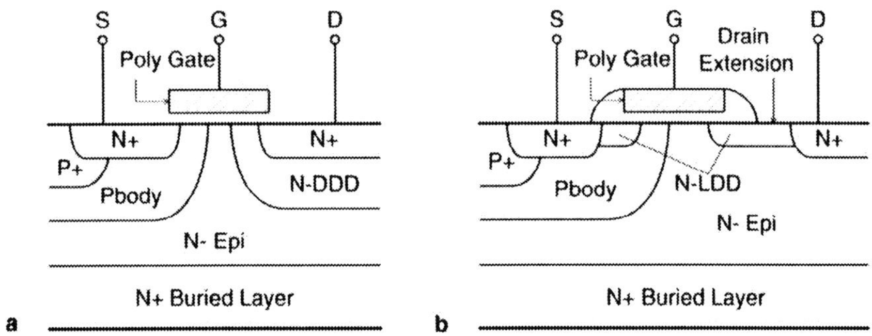

Fig. 1.31 Cross-sections of low voltage (from 16 V to about 30 V) Lateral DMOS using (a) N^- DDD or (b) N^- LDD layers in the drain region to reduce high electric field across the gate oxide

For very high off-state breakdown voltage values (> 250 – 300 V) the vertical structure is not suitable because it would require an epi layer so thick to become unpractical to create isolated islands by using the standard up and down junction isolation plugs.

To overcome this limitation, a completely different approach is followed: the RESURF principle (see Chapter 2). Exploiting this technique a lateral DMOS component can be placed in a thin epitaxial layer, withstanding voltages as high as 700 V.

When the lateral DMOS structure is used for low voltage applications, if drain and source are diffused in the same active area opening, the breakdown is typically located at the drain edge under the polysilicon gate. To reduce the electric field at that critical point, an extended drain structure is realized using different ingredients (DDD, LDD) depending on the BCD process generation, as shown in Fig. 1.31.

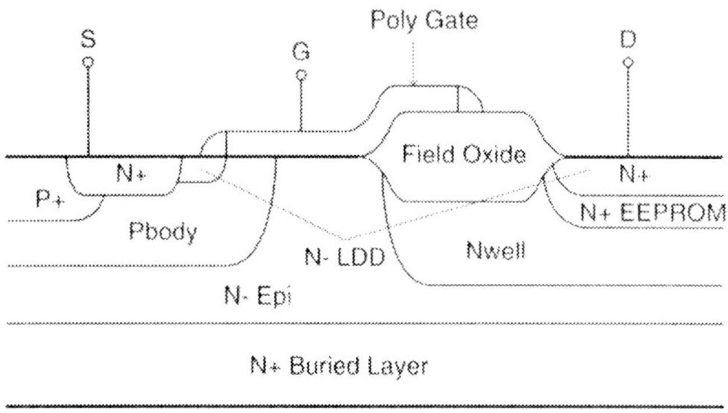

Fig. 1.32 Cross-section of medium voltage Lateral DMOS. The example shown is the 40 V device from BCD-III

With the reduction of the minimum lithography and of the LOCOS minimum size, a different lateral DMOS structure can be adopted that merges the advantages of the biplanar polysilicon field plate to improve the off-state voltage capability with the advantage of the lateral structure with respect to the vertical one for the on-state characteristics (Fig. 1.32).

1.6.2 Threshold Voltage

The value of the threshold voltage is mainly related to the thickness of the gate oxide and to the peak concentration of the laterally diffused body in the region included between source and drain.

The doping profile of the body region is multifunctional in an integrated circuit and consequently many trade-offs must be satisfied. Also when used in the DMOS transistor, the P body layer characteristics must match with conflicting requirements:

- the DMOS channel length must be as short as possible and its threshold voltage must be reasonably low;
- the channel punch-through voltage must be higher than the maximum voltage capability,

Table 1.2 DMOS channel lengths and threshold voltages through different BCD generations

BCD GENERATIONS	CHANNEL LENGTH (μm)	THRESHOLD VOLTAGE (V)
BCD-I	1.50	3.0
BCD-II	1.25	2.1
BCD-III	1.00	1.7
BCD-V	0.5	1.2

- the gain of the parasitic NPN, intrinsic to the DMOS structure, must be kept as low as possible in order to increase the DMOS ruggedness.

Table 1.2 gives the scaling of channel length and threshold voltage for DMOS components in different BCD generations.

1.6.3
ON-Resistance

Vertical DMOS Transistor

The layout of an integrated vertical DMOS device consists of alternated source and drain fingers where the source fingers consist of arrays of elementary cells.

The R_{ON} is strictly dependent on the topological layout that is shape and size (packing density) of the unit source cell and cell arrangement used to fill the silicon surface.

In the Multipower-BCD the source cells are square-shaped and they are positioned on a square matrix: it has been shown in different works [39, 40] that this is one of the most efficient ways to maximize the channel width/area ratio.

The corners of the square cell are rounded or smoothed in order to avoid excessively high electric fields close to the MOS channel area when high drain to source voltages are applied. In the integrated vertical DMOS transistor the position of the drain electrode on the top must be also considered in the R_{ON} computation because it adds an extra contribution with respect to the case of the discrete device, that depends on the N^+ buried layer drift region, the N^+ sinker contact region and the layout choice for drain and source interdigitation.

The layout configuration is identified using the following mathematical model that provides the product $R_{ON} \times$ Area as a function of the physical and geometrical parameters involved in the structure (Fig. 1.33):

$$R_{ON} \times Area = \left\{ \frac{Rc}{N} + \frac{1}{12} \times R_{bl} \times N + \frac{1}{2} \times R_{bl} \times \frac{2Y}{l} + \frac{Rsk}{l \times X_{sk}} \right\} \times l \times (N \times l + X_{sk} + 2Y) \tag{1.1}$$

where:

- R_C is the ON-resistance of the DMOS cell and consists of four different components:
  - R_{ch} is the channel resistance,
  - R_{acc} is the accumulation layer resistance,
  - R_{JFET} is the parasitic JFET ON resistance,
  - R_{epi} is the epitaxial layer drift resistance.

  The expression used for R_C relates to the well-known model of the DMOS structure [39].
- R_{bl} is the N^+ buried layer sheet resistance;
- R_{sk} is the N^+ sinker resistance per unit area;

- **l** is the cell pitch = X_{cell} + d, where X_{cell} is the DMOS cell window and d is the spacing between two DMOS cells;
- $\mathbf{X_{sk}}$ is the N^+ sinker window width;
- **N** is the number of cells between two sinker drain contacts;
- **Y** is the distance between the edge of the sinker window and half spacing of the last cell (this value includes the extension of the junction edge termination region).

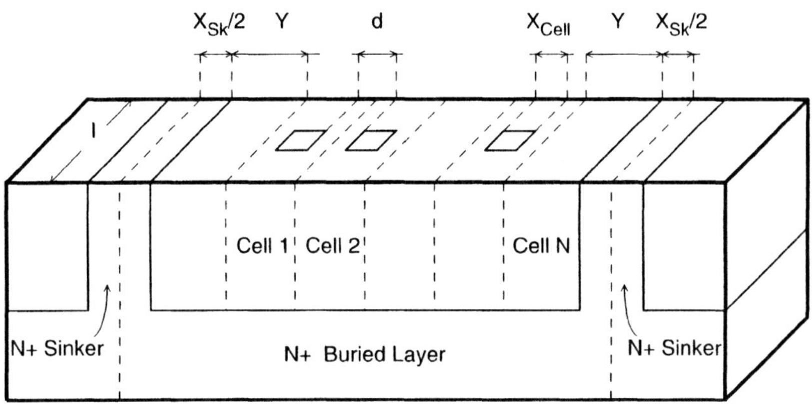

Fig. 1.33 Model for the $R_{on} \times$ Area calculation of the vertical DMOS

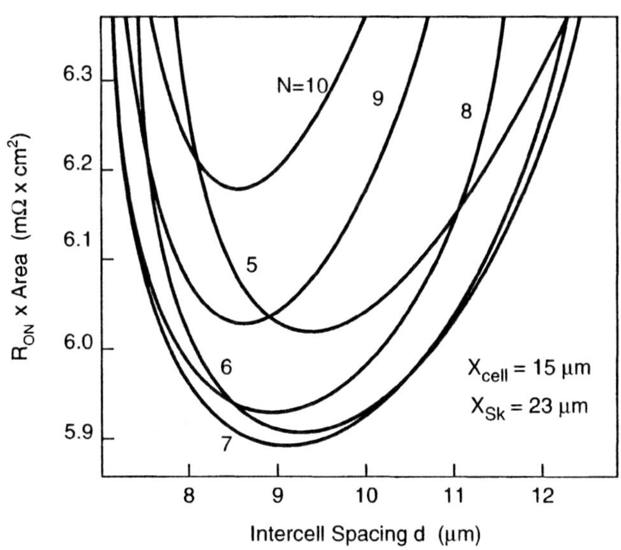

Fig. 1.34 $R_{ON} \times$ Area versus intercell spacing d (parameter N: number of cells between two drain sinker contacts)

Varying the parameters X_{cell}, d, N and X_{sk} it is possible to fix the geometrical configuration that minimizes the $R_{ON} \times$ Area product that provides a measure of the efficiency of the DMOS design.

In Fig. 1.34, the $R_{ON} \times$ Area functional dependence on the intercell spacing d can be observed having as a parameter the number of cells N between two drain sinker contacts. In the reported case, that refers to the first BCD generation, the minimum value is obtained at a value of d in the range from 9 to 10 µm and at N = 7 [8].

This model takes into account only the silicon contributions to the ON-resistance.

With the reduction of the minimum lithography in the BCD process generations, the silicon $R_{ON} \times$ Area decreases and the metal contribution can no longer be neglected.

A new model is therefore developed that represents the DMOS device by a resistive network in which the silicon and metallization resistances are connected together in a complex way related to the power layout arrangement [41]. The equivalent resistive network for each array and for the whole device are shown in Figs. 1.35 and 1.36.

Using these models it is possible to evaluate, for a given power layout, the relative weight of the different ON-resistance terms and to identify how much they are affected by the process and layout parameters [42].

In the high voltage case (250 V), the epitaxial layer thickness and resistivity are the only parameters that influence the ON-resistance for a wide range of device size.

Lowering the voltage range (60 V and 100 V) the metal and the MOS channel contributions become comparable to the epitaxial layer term.

Evolving towards finer lithography generations and therefore increasing the DMOS channel perimeter density, the silicon contribution to the total ON-resistance reduces to levels that make the metal interconnection series resistance

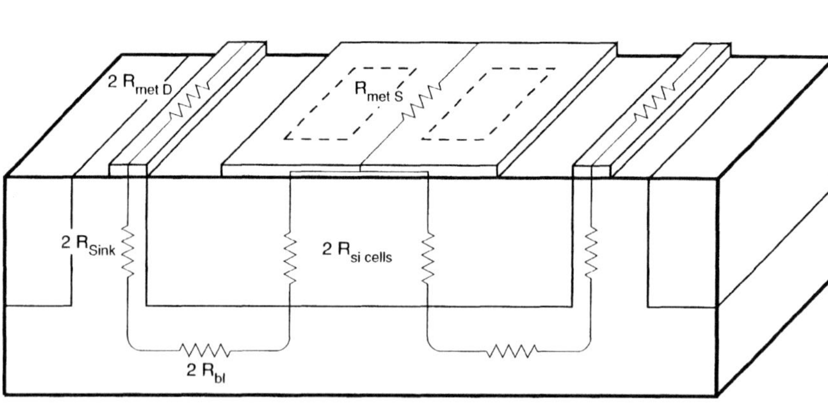

Fig. 1.35 Equivalent resistive network for a VDMOS basic finger element showing a simplified case of two cells

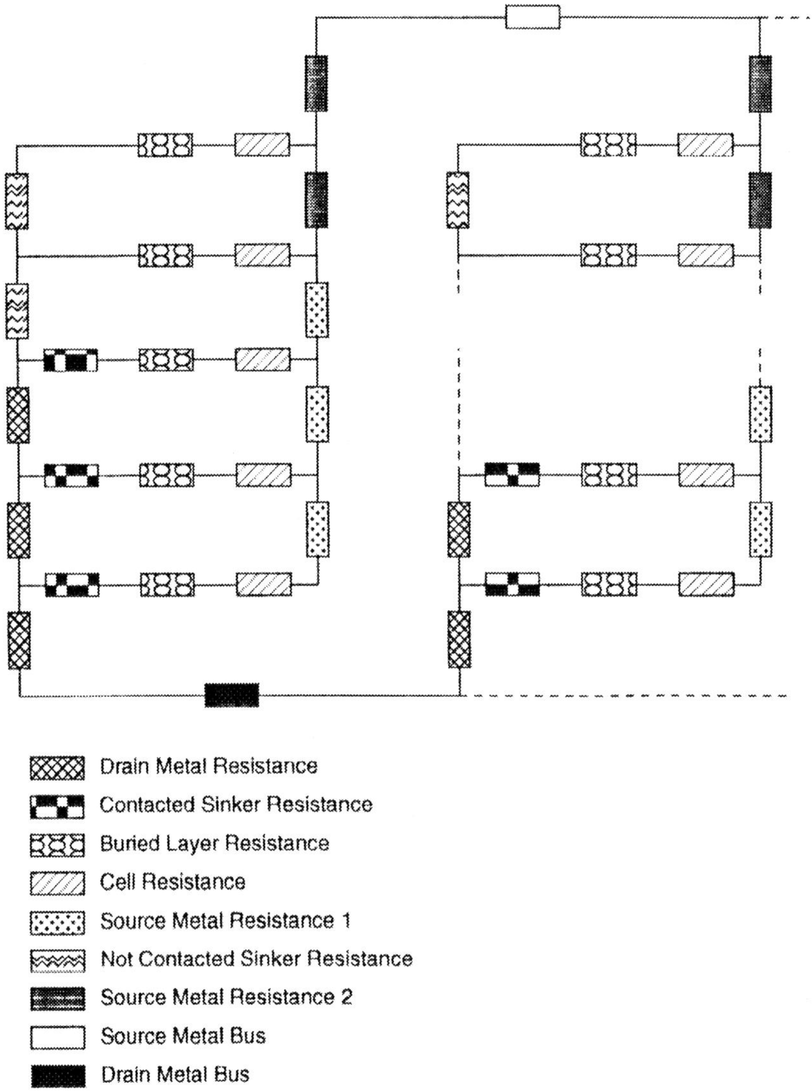

Fig. 1.36 Equivalent resistive network of an integrated VDMOS power device

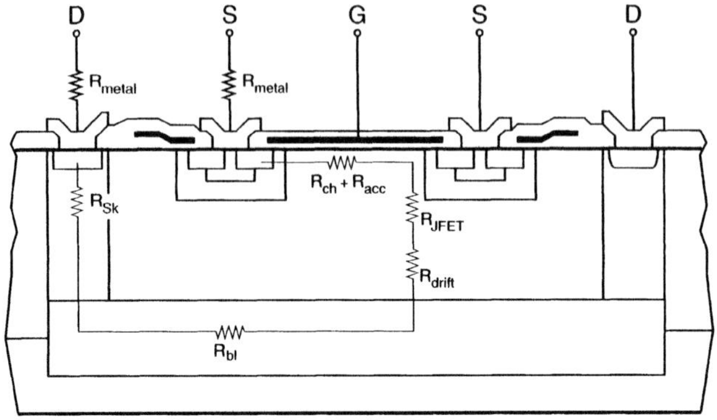

Vertical DMOS

- High voltage capability
- Low ON resistance

$$R_{ON} = R_{ch} + R_{acc} + R_{JFET} + R_{drift} + R_{bl} + R_{Sk} + R_{metal}$$

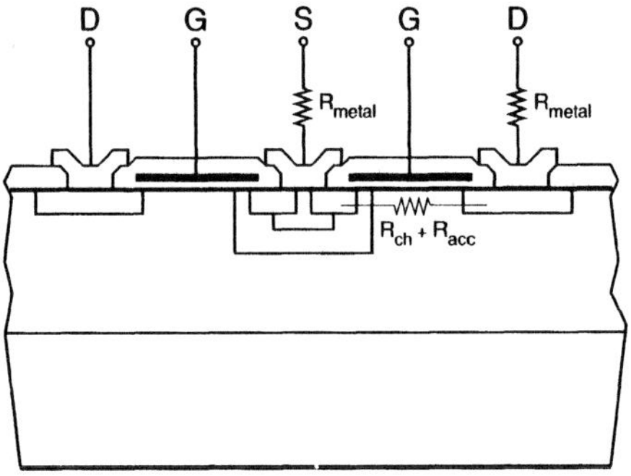

Lateral DMOS

- Low voltage capability
- Very low ON resistance

$$R_{ON} = R_{ch} + R_{acc} + R_{metal}$$

Fig. 1.37 Contributions to the total ON-resistance in vertical and lateral DMOS transistors

relevant. The metal term reduces adopting a multiple thickness and multiple level metal process instead of a double thickness single level one used in the first BCD generation.

Lateral DMOS Transistor

As far as the power LDMOS is concerned, the R_{ON} contributions are those due the MOS channel, accumulation region, drain drift region and metal layer.

The relative weight of the drain drift region depends on the maximum reverse voltage that must be sustained by the component: it varies from a negligible value to about 90% ranging from 20 V to 700 V.

In low voltage applications the lateral DMOS is advantageous compared to the vertical component because of the absence of different silicon contributions to the total ON-resistance (Fig. 1.37).

In this case, the percent contribution introduced by the metal interconnections becomes more relevant and therefore it becomes mandatory to introduce a multi-level metal technology in order to optimize the source and drain metallization layout scheme.

For example, in the third BCD process generation a 1 mm^2 N-channel power LDMOS device performs R_{ON} equal to 0.25, 0.12, 0.08 Ω if one, two or three metal levels are respectively used.

Fig. 1.38 shows the specific ON-resistance versus the drain-to-source breakdown voltage for the different BCD process generations.

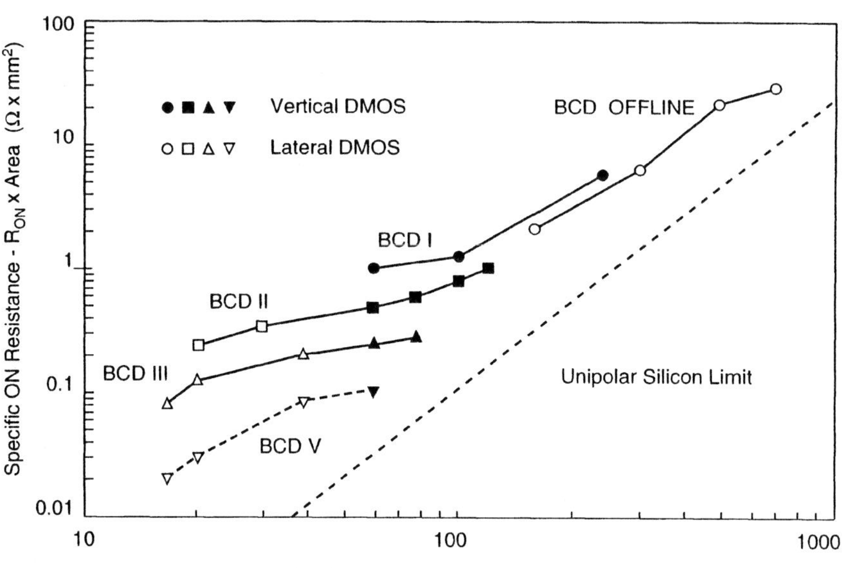

Fig. 1.38 Specific ON-resistance versus drain to source breakdown voltage for different BCD process generation

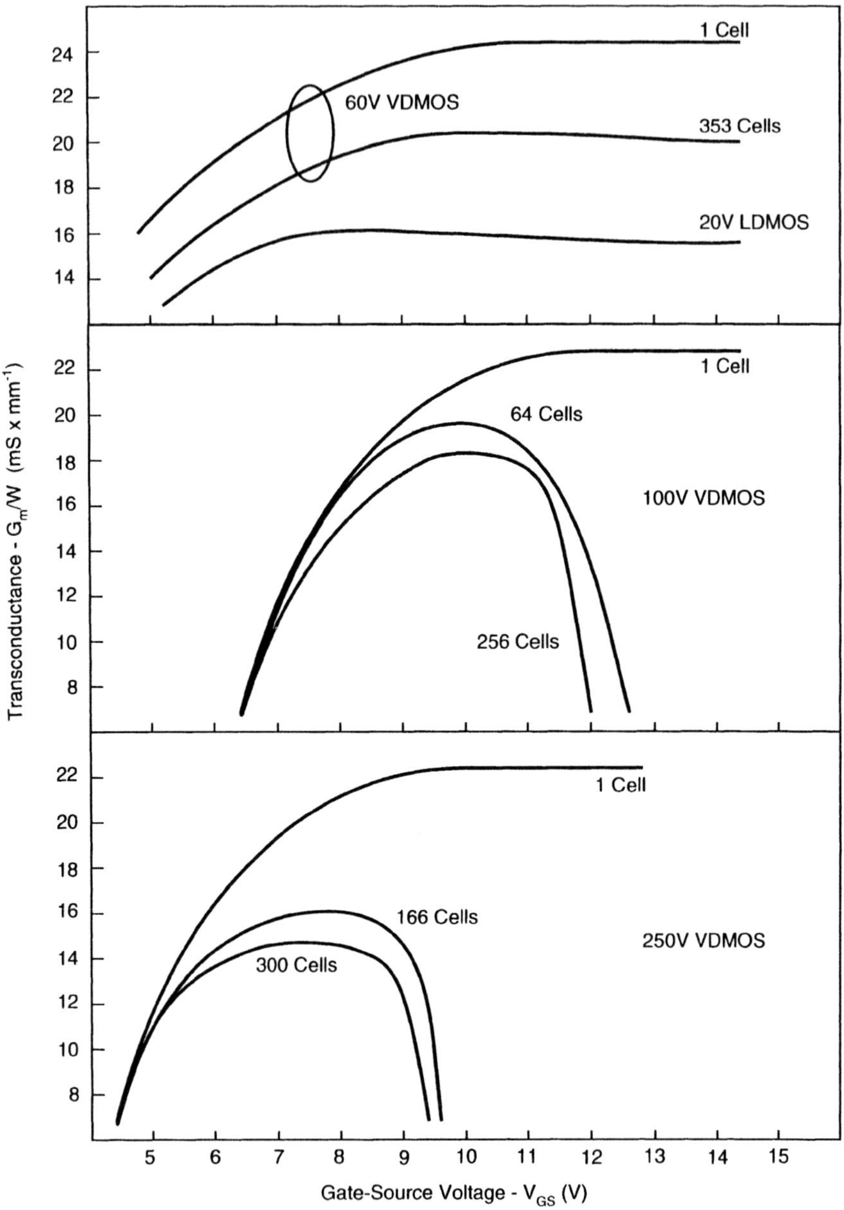

Fig. 1.39 Transconductance versus gate-to-source voltage for vertical and lateral DMOS in different voltage classes of BCD-I

1.6.4
Transconductance

The behaviour of transconductance versus gate-source voltage is reported in Fig. 1.39 for the Vertical DMOS transistor in three different processes.

The maximum transconductance value is limited by thermal effects while the quasi saturation effect explains the drop at high gate voltages. The quasi-saturation effect is caused by the carrier velocity saturation in the drift region between the P-body cells because of the high electric field, the low impurity concentration in the drift layer and the narrow cell spacing that acts as a channel of a parasitic JFET [43].

In the lateral DMOS device the quasi-saturation effect is not present because the alternated drain and source layout avoids the JFET formation and the transconductance is only temperature limited (Fig. 1.39).

1.7
Other Integrable Devices

1.7.1
High Voltage P-channel MOS (H.V. PMOS) Transistor

To satisfy the driving requirements of the high voltage DMOS device when used in high-side driver configuration, it is necessary to have a component of inverse polarity sustaining a supply voltage exceeding the DMOS maximum voltage by a value equal to its driving voltage.

The high voltage capability is achieved by an extended drain lateral structure realized by a properly doped region that allows the depletion region to spread mainly into the lightly doped drain instead of into the channel region. The drain extension also gives less Early effect. Its extension length and doping level control the maximum electric field at the drain but, at the same time, introduce a series resistance into the device.

This layer is obtained by a specific Boron implant step, in the first generation BCD processes (Fig. 1.40), while from the second generation onwards, the P-field

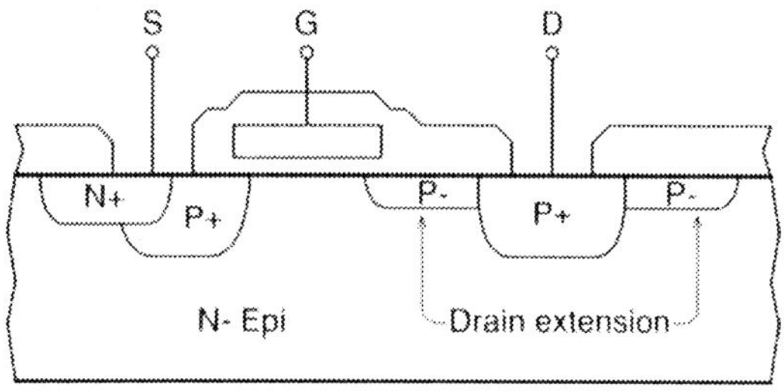

Fig. 1.40 High Voltage P-channel MOS transistor cross-section in BCD-I

layer used in the CMOS structure to increase the parasitic threshold voltage value, has such a doping level as not to require a specific processing step to get the drain extension function (Fig. 1.22b).

1.7.2 Bipolar Transistors

To obtain bipolar transistors no extra doping layers are required because the elements used to realize CMOS and DMOS structures can be combined in different ways to build up lateral PNPs with different base width and two kinds of NPN structures.

The electrical characteristics of the lateral PNP transistors are connected to the epitaxial layer resistivity and to the spacing between emitter and collector.

In BCD-I, the emitter and collector regions are realized simultaneously with the highly doped P layer used as body contact in the DMOS cells (Fig. 1.41a). In BCD-II and BCD-III the LOCOS field oxide scheme and the shallower P^+ junction depth, force to introduce into the collector region both the P-body and the P-well layers that, having higher junction depths, improve the efficiency in collecting the holes injected by the emitter (Fig. 1.41b).

The BV_{CEO} capability is related to the base width and to the collector construction.

When full voltage capability is needed, the same technique used in the high voltage P-channel MOS transistor can be implemented to increase the base to collector breakdown voltage.

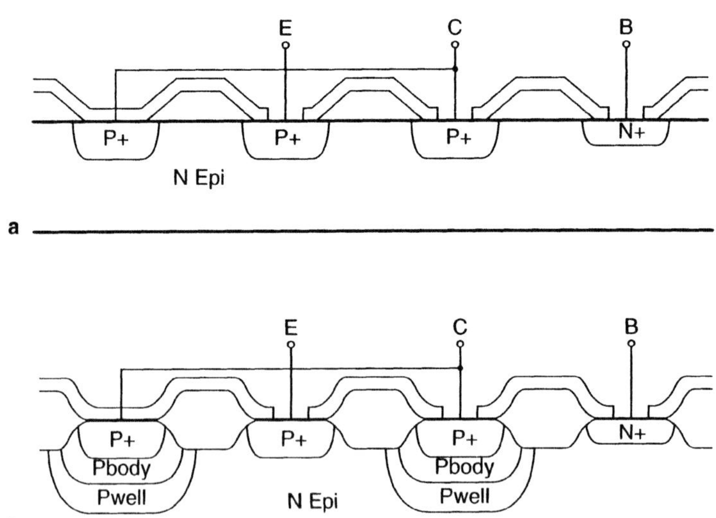

Fig. 1.41 Cross-section of the Lateral PNP transistor: **a** BCD-I; **b** BCD-II and BCD-III

Fig. 1.42 Cross-section of different types of NPN transistors

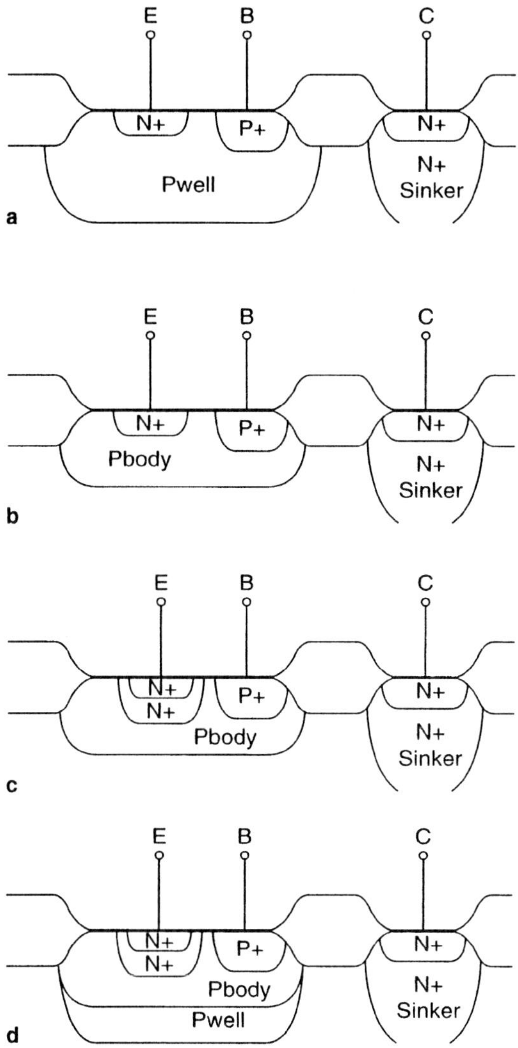

A first kind of NPN transistor can be made up by using as a base the CMOS P-well layer and as an emitter the N^+ source/drain contact (Fig. 1.42a). Due to the low doping profile of the P-well layer, coupled with the shallow emitter, the component exhibits high gain, but, owing to the high base width, the cut-off frequency is low.

A second one replaces the P well as a base profile with the P-body layer of the DMOS transistor (Fig. 1.42b): in this case the gain is lower but very constant (over more than two decades of collector current) and the cut-off frequency is increased.

From BCD-III onwards, the introduction of the single-Poly EPROM and EEPROM allows to exploit the N^+ control gate as a high efficiency emitter. Because its junction depth is deeper than the N^+ source/drain contact, if the P-body layer is used as a base (Fig. 1.42c), the BV_{CEO} is lower (> 16 V) but a cut-off frequency value higher than 1 Ghz can be reached. When a higher BV_{CEO} is required, the same structure is embedded in a P-well region (Fig. 1.42d), allowing to achieve 30 V.

While the BV_{CEO} values are fixed by the doping profiles involved in the process, the BV_{CBO} values can be increased to the maximum voltage capability of the process by protecting the base/collector junction with field plates.

To reduce the collector contribution to the saturation resistance a sinker plug is normally used in the collector region of all the NPN transistors.

1.7.3
CMOS

The CMOS part represents the structure that mostly benefits of the lithographic reduction in terms of density and speed, therefore the evolution of the BCD processes has been mainly driven by the increasing demand to integrate more and more complex CMOS circuits. In fact, while in BCD-I the CMOS is obtained just adding a P-well layer to the DMOS process, from BCD-II onwards the architecture of the processes is conceived to optimize CMOS features.

The main advantage in size reduction going from BCD-I to BCD-II comes from the introduction of the LOCOS field isolation and from the use of specific implants to realize PMOS source/drain regions. In BCD-I, in fact, P^+ rings surround each NMOS active areas to avoid P-well surface inversion performing the field isolation and the DMOS P-body layer is used to form source and drain for PMOS.

In both process generations the CMOS gate oxide an polysilicon gate electrode are common to the DMOS structure and the epitaxial layer is the PMOS body. Moreover, when matched threshold voltages are required, an optional Boron implant is added to the PMOS channel region.

In BCD-III a dedicated gate oxide and polysilicon layer are introduced together with a self-aligned twin-well structure. The minimum gate lengths are equal to 1.2 μm therefore to avoid limitation to the maximum operating voltage induced by punch-through and hot electron effects, a LDD (Lightly Doped Drain) solution is adopted on the NMOS device using spacers (Fig. 1.25b).

In BCD-V the gate lengths are reduced to 0.8 μm and the LDD solution becomes necessary for the PMOS transistor, too.

As regards immunity to latch-up, the CMOS takes advantage of the presence of the low shunt resistance of the N^+ buried layer positioned under the whole CMOS structure and of the availability of the sinker plug to reduce the series resistance to the PMOS body contact.

In the circuit design it is useful to have not only 5 V CMOS but also higher operating voltage capabilities to directly interface different supplies and to drive the DMOS power stage. To satisfy these requirements drain extension solutions

Table 1.3 5 V CMOS characteristics in different BCD process generations

	L-gate (μm)		Thoxgate (nm)	V_{TH} (V)	K_{LIN} (μA/V²)		CMOS density (tr./mm²)
	NMOS	PMOS			NMOS	PMOS	
BCD-I	4.00	7.00	85	1.30	27	10	650
BCD-II	3.00	4.00	60	1.10	40	15	1600
BCD-III	1.20	1.20	28	0.90	60	21	4000
BCD-V	0.8	0.9	20	0.85	85	28	15000

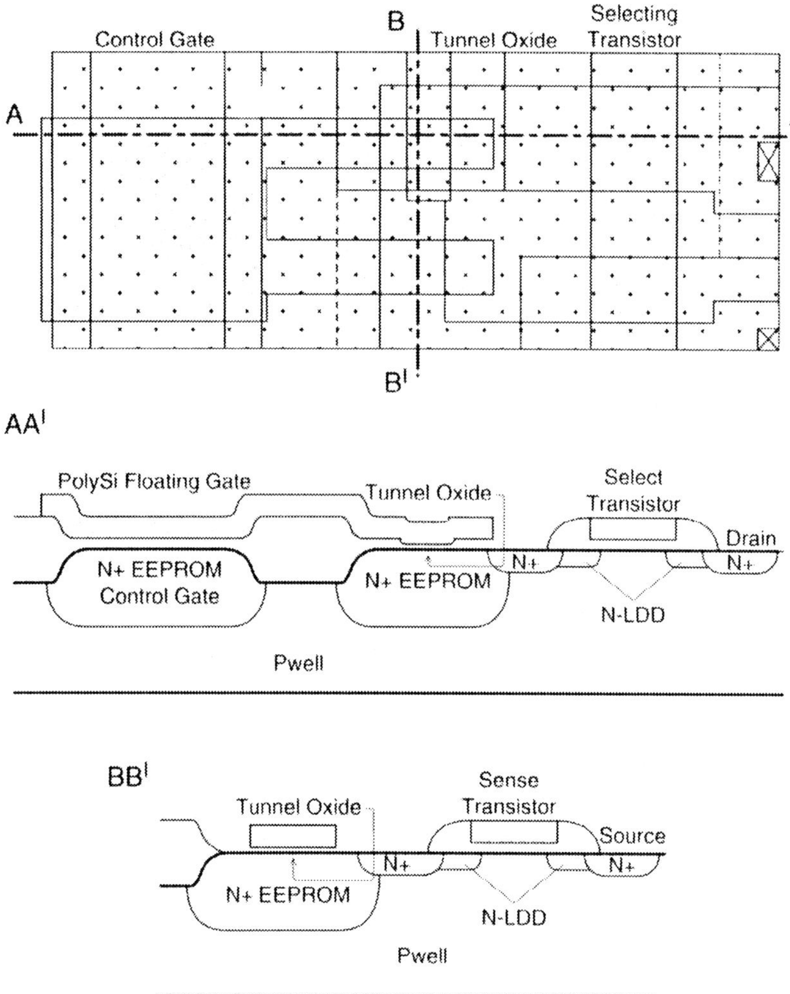

Fig. 1.43 Cross-section and layout of a single-poly EEPROM cell integrated in BCD-III

are adopted and maximum operating voltages up to 12–15 V are normally achieved.

In Table 1.3, the main CMOS geometrical and electrical features are summarized for the different BCD process generations.

1.7.4 Non-Volatile Memories

Non-volatile memories are introduced in the BCD processes from the third generation onwards.

The two-polysilicon layer approach used in BCD-III is primarily adopted to separate the high thermal-budget steps, like the DMOS P-body drive-in done after poly 1, from the low thermal-budget steps made after poly 2. This allows the growth of a 90 Å tunnel oxide, used in the EEPROM cell, without affecting its quality that is not compatible with high temperature treatments.

EPROM and EEPROM cells are single-poly structures (Fig. 1.43) in order to contain the process complexity. The control gate function is realized by an N^+ layer diffused together with the DMOS P-body layer [44].

To program the memory cells, high voltage (14–17 V) NMOS transistors are required. This high voltage capability is achieved using extended drain structures.

1.7.5 Low Leakage Diodes

Two different high voltage diode structures are integrable in the BCD processes with the possibility of limiting the current injected into the substrate by the parasitic PNP switched on by the anode/cathode forward-biased junction.

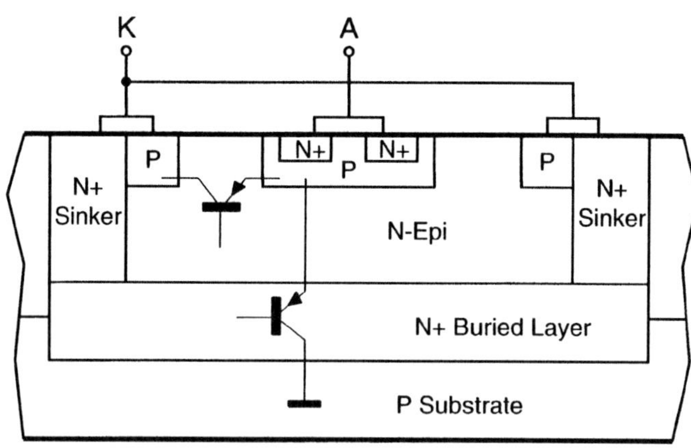

Fig. 1.44 Cross-section of the low leakage base/collector diode

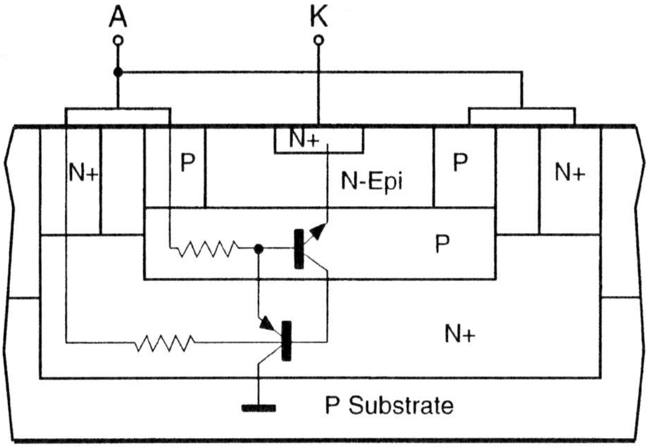

Fig. 1.45 Cross-section of the Low Leakage Diode (LLD) showing the associated parasitic substrate PNP and NPN transistors

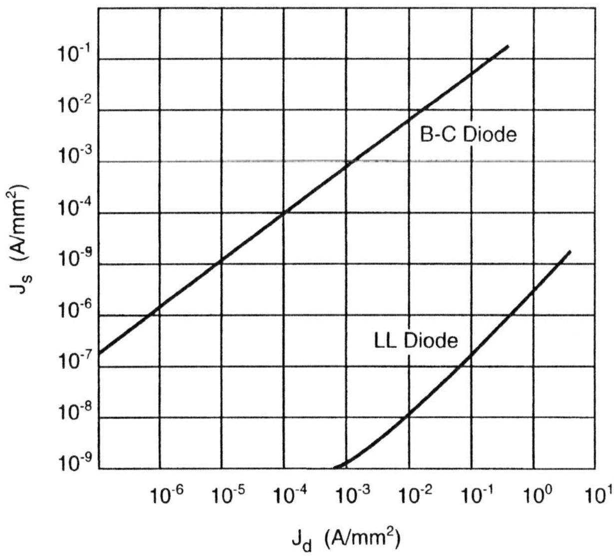

Fig. 1.46 Parasitic current density J_s injected into the substrate versus forward current density J_d in low leakage base/collector diode and in LLD

The simpler one is a NPN base/collector diode where the base is surrounded by a N^+ sinker ring short-circuited to a P ring acting as a lateral PNP extra collector (Fig. 1.44).

To further reduce the current loss, the emitter efficiency of the parasitic PNP is decreased adding the N^+ source/drain layer in the anode region [45].

In this kind of diode a current in the range of 1% of the diode current is injected into the substrate.

In the BCD processes that use an up-down junction-isolation approach, another kind of more efficient low leakage diode can be implemented [46].

In Fig. 1.45, the cross-section of this structure is shown together with the associated parasitic elements.

The buried anode is embedded in the N^+ buried layer and the deep anode contact is surrounded by a N^+ sinker ring. The deep anode and the N^+ sinker ring are short-circuited.

In this way, the base/emitter junction of the associated parasitic PNP can not be forward biased and the current loss into the substrate becomes negligible (Fig. 1.46).

The embedding of the anode in an N-type region isolates the diode from the substrate and makes it possible to use this component also tied to the high voltage rail.

1.8 Parasitic Structures

In a mixed process using the junction-isolation technique like Multipower-BCD, the large variety of different active components that can be integrated into the same chip cohabits also with parasitic structures of two species: surface and bulk.

1.8.1 Surface Parasitic Elements

The surface parasitic elements are MOS structures induced by polysilicon and metal interconnections, plastic packages or more in general by undesired external ionic contamination and by any charging effects due to high electric fields that, acting as parasitic gates, create surface inversion connecting two silicon regions of the same polarity.

In order to prevent surface inversion, channel stopper structures have to be inserted across the possible inversion path.

- At first, one way can be achieved by increasing the surface doping concentration in the channel region by inserting a heavily doped layer normally available in a mixed process (Fig. 1.47).
  To be effective the channel stopper width and surface concentration must be properly dimensioned according to the required maximum voltage drop between silicon and parasitic gates.

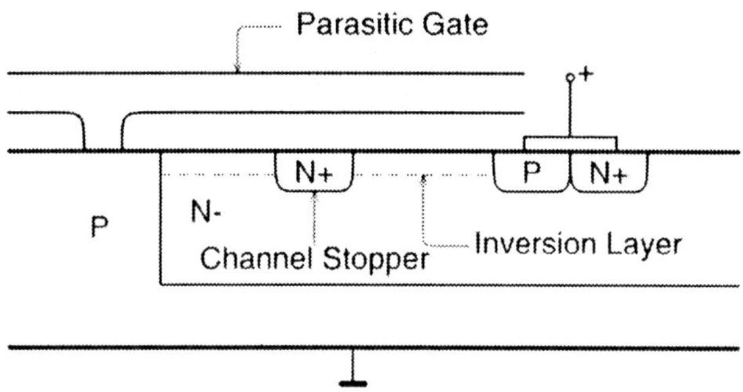

Fig. 1.47 Example of a possible inversion path stopped by a heavily doped region

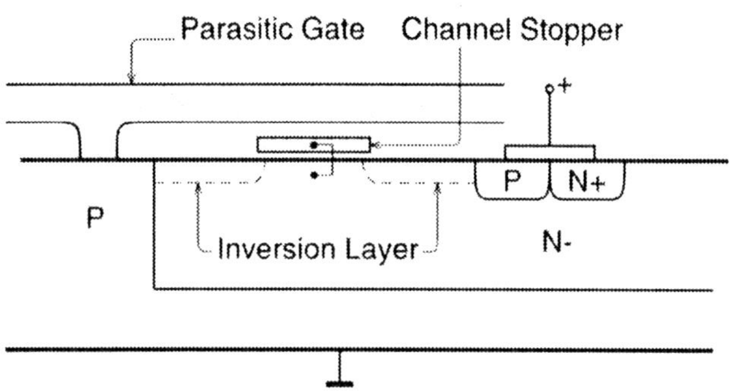

Fig. 1.48 Example of a possible inversion path stopped by a polysilicon electrostatic shield

- A second way of realization is to use polysilicon or metal interconnections as electrostatic shields tied to the same voltage of the silicon regions that have to be protected. In this case, the effectiveness depends on the channel stopper width that is also related to the underlying dielectric thickness (Fig. 1.48).

1.8.2
Bulk Parasitic Elements

The two main bulk parasitic elements are bipolar structures that can be activated in the following ways.

- When an epitaxial pocket bias is forced below the substrate potential, the base/emitter junction of a parasitic NPN is switched on (Fig. 1.49). Any other epitaxial pocket acts as a collector of this parasitic lateral device. This induces a higher power dissipation in the chip and can cause loss of driving control.
- When a P-type diffusion is biased above the supply level in an epitaxial pocket tied to the supply voltage, a parasitic vertical PNP transistor is turned on

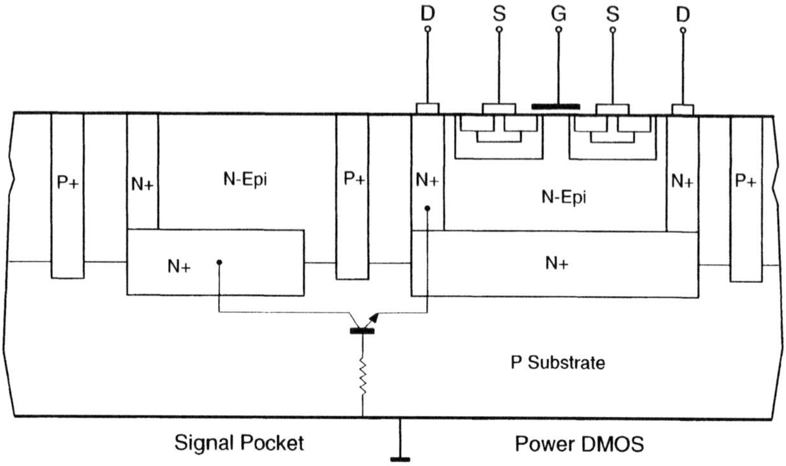

Fig. 1.49 Parasitic lateral NPN transistor between two epitaxial pockets

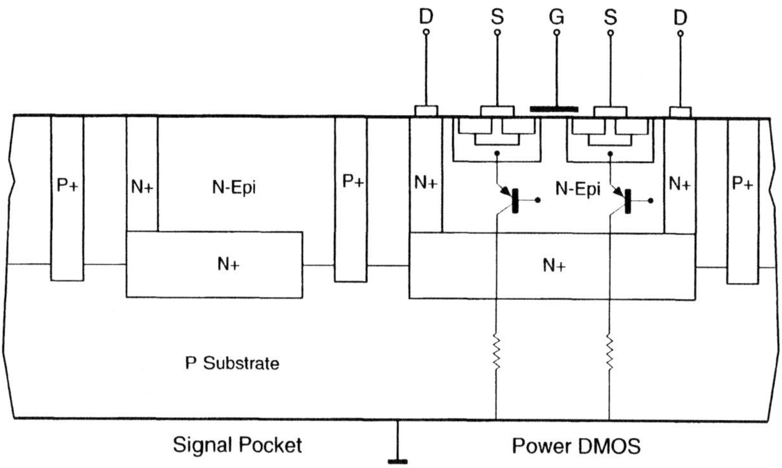

Fig. 1.50 Parasitic vertical PNP transistor

having the substrate acting as a collector. The PNP collector current increases locally the substrate potential; neighbouring epitaxial pockets that are biased to an equally or weakly positive potential compared to the substrate, act as emitters of parasitic lateral NPN transistors (Fig. 1.50).

The combined effect of parasitic lateral NPN and vertical PNP transistors builds up a SCR structure that can be regenerative and induces device burn-out.

In a junction-isolation approach the bulk parasitic elements cannot be completely suppressed and there are several techniques that limit their negative influence on the active components. These techniques mainly consist in the introduction of extra collectors for the parasitic lateral NPNs implemented via dummy epitaxial pockets, in the use of a diode (like the DMOS body/drain diode) in parallel to the epi/substrate junction, in order to sink part of the current flowing in the parasitic lateral NPN base/emitter junction, and in careful positioning of the critical components and of the substrate contacts in the IC layout.

Sometimes another approach to the problem is the reduction of substrate resistivity to limit the substrate voltage rise during parasitic PNP activation. However, low resistivity substrates enhance the parasitic NPN action via a reduced base resistance, therefore a careful choice of the substrate resistivity is often important in the control of the junction-isolation parasitics [47].

1.9 Electrostatic Discharge (ESD) Protection Structures

In the integrated circuits created with BCD processes, the large variety of possible applications, the multi-supply IC environment and the wide number of different basic components that can be integrated, require the adoption of unconventional ESD protection strategies and structures.

While the pure CMOS part can be protected via conventional ESD protection structures, like lateral NPN transistors or grounded gate field oxide MOS transistors [48–49], the mixed and high voltage parts of the ICs often require appropriate combination of the above structures to match the voltage requirements of the circuit or specific structures which may be realised as an active clamp implemented via a macro circuit. The active clamp is designed to provide a current path from the supply lines to ground, while all the pins that are related to those lines can be connected to the active clamp by plain diodes. An ESD network is therefore enforced which guarantees pin-to-pin protection against the ESD discharge.

This approach is very flexible as the active clamp can be designed with reference to the different existing ESD standards [50–53].

1.10 Future Perspectives

After more than ten years of evolution of BCD processes, a remarkable progress has been made in the increase of the voltage capability spectrum, of the variety of the integrable component arsenal and in the reduction of the lithography features. This progress has been pursued favouring the ability to provide flexibility in order to satisfy the wide range of Smart Power applications instead of running after the leading edge silicon technologies. This explains why BCD is now and will always be two or one generations behind BiCMOS and pure CMOS (Fig. 1.51).

Today the first products designed in BCD-III at 1.2 μm are going to be commercially introduced while the fifth generation designed at 0.6 μm is under

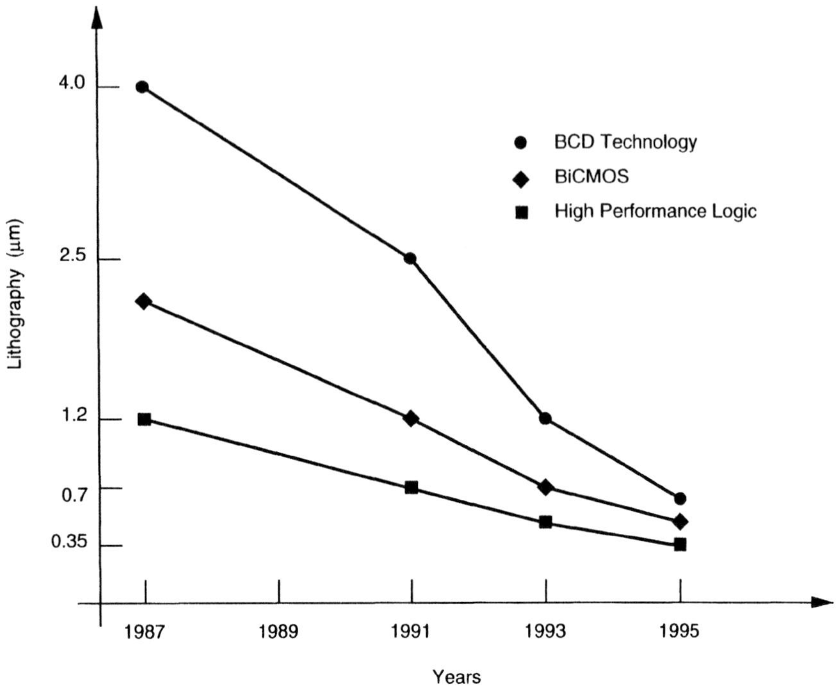

Fig. 1.51 Lithography trend in different process families

development. This last process version, which will include not only EEPROM but will also be compatible with Flash memory if asked by the market, will allow to integrate extremely complex systems on a single chip. A unique dedicated VLSI circuit will replace at a competitive price many systems realised today using a multi-chip approach.

In the near future, in the BCD development road map it will not be possible to ignore the Dielectric Isolation approach using SOI + trench. In fact, the integration of the active components in fully dielectrically isolated tubs will permit not only a further increase of the packing density, but also to suppress the bulk parasitics eliminating the interactions between the power stages and the more and more sophisticated and sensitive signal part.

Besides, an improvement in the frequency operation and a reduction in the current substrate leakage will be also achieved. Finally, it has to be mentioned that, together with the usage of the Dielectric Isolation approach, a worse thermal conductivity will have to be dealt with. To overcome the induced problems in power dissipation, new substrate materials with higher thermal conductivity will have to be developed.

References

1. Fischetti MA (1985) Solid State, IEEE Spectrum, p 60
2. Intelligent Power Devices, Electronic Engineering, p 84, January 1985
3. Travis B, (1985) Mixed-process, CMOS-Driven Power ICs Come Slowly but Inexorably to Market, EDN, p 51
4. Bierman H (1985) Smart Power ICs Move from Promise to Production, Electronics, p 40
5. Cini C, Contiero C, Murari B, Intelligent Power Technology: A Reality in the 1990s, Proceedings of 1988 International Symposium on Power Semiconductor Devices, Tokyo, p 88
6. Baliga BJ (1990) Smart Power Technology: An Elephantine Opportunity, IEDM Tech Dig
7. Rumennik V (1985) Power Devices Are in the Chip, IEEE Spectrum, p 42
8. Andreini A, Contiero C, Galbiati P (1986) A new Integrated Silicon Gate Technology Combining Bipolar Linear, CMOS Logic, and DMOS Power Parts, IEEE Trans Electron devices, vol ED-33, n 12, p 2025
9. Bean KE, Runyan WR (1977) Journal of Electrochemical Society, p 124, 5 C
10. Ohasi H, Ohura J, Tsukakoshi T and Simbo M (1986) IEDM Tech Dig, pp 210–213
11. Caller GK and Trimble LE (1987) Proc ESC Symp on High Voltage and Smart Power Devices
12. Izumi K (1978) Electron Device Letters, vol 14, p 593
13. Nakashima S, Maedd Y and Akiya M (1986) IEEE Trans Electron Devices, vol ED-33, p 126
14. Gabriel R (1991) An Intrinsic Save Smart Power IGBT, EPE Firenze, p 179
15. Weyers J, Vorgt H (1992) A 50 V Smart Power Process with Dielectric Isolation by Simox IEDM Tech Dig, p 225
16. Goodenough F (1994) SLIC Ejects Relays from PABX Line Cards, Electronic Design, p 55
17. Heftman G (1986) Display Driver ICs Light the Way for Flat-Panel Displays, Electronic Design, p 29
18. Rumennik V (1992) A 1200 V BiCMOS Technology and its Applications, Proceedings of International Symposium on Power Semiconductor Devices & ICs, Tokyo, pp 322–327
19. Wrathall RS (1985) The Design of a High Power Solid State Automative Switch in CMOS-VDMOS Technology, 16th PESC Tech Dig
20. Ludikhuize AW (1986) A Versatile 250/300-V IC Process for Analog and Switching Applications, IEEE Trans Electron Devices, vol ED-33, n 12
21. Contiero C, Andreini A, Galbiati P, Lombardi C (1987) Experimental and Numerical Analysis of the High Voltage Structures Implemented in the New Mixed Process Multipower BCD 250 V, Proc ECS Symp on High Voltage and Smart Power Devices, pp 31–40
22. Mastromatteo U, Villa F (1986) 16th ESSDERC Conf Digest, pp 204–205
23. Appels JA and Vaes HMJ (1979) High-Voltage Thin-Layer Devices (RESURF Devices), IEDM Tech Dig, pp 238–241
24. Chang MF, Piefer G, Yilmaz H, Wildi EJ, Hodgins RG, Owyang K and Adler MS (1986) IEEE Trans Electron Devices vol ED-33, pp 1992–2001
25. Contiero C, Andreini A, Galbiati P, Storti S (1987) Design of a High-Side Driver in Multipower-BCD and Vipower Technologies, IEDM Tech Dig, p 766
26. Zambrano R, Ferla G, Musumeci S & Paparo M (1985) 17th ESSDERC Conf Digest, pp 653–656
27. Grant DA an Gowon J (1989) Power MOSFET Theory and Applications, John Wiley Interscience
28. Baliga BJ, Adler MS, Gray PV, Love RP (1982) The Insulated Gate Rectifier (IGR): a New Power Switching Device, IEDM Atech Dig, p 264
29. Kitagawa M, Omura I, Hasegawa S, Inoue T and Nakagawa A (1993) A 4500 V Injection Enhanced Insulated Gate Bipolar Transistor (IEGT) Operating in a Mode Similar to a Thyristor, IEDM Tech Dig, p 679
30. Shekawr MS, Baliga BJ and Nandakumar M (1991) Experimental Demonstration of the Emitter Switched Thyristor, Proceedings of ISPSD, p 128
31. Temple VA (1984) MOS Controlled Thyristors (MCT's), IEDM Tech Dig, p 282

32. Cini C, Diazzi C, Rossi D (1985) A High Efficiency Mixed Technology Motor Driver IC, PCI Proceedings
33. Cini C, Diazzi C, Rossi D (1986) High Frequency, High Current Monolithic Switching Regulator in Bipolar/CMOS/DMOS Mixed Technology, Power Conversion International Conference, Munich
34. Pietrobon G, Rossi D (1989) A Reconfigurable DMOS Control Chip for an Electronic Typewriter, Proceedings of ISSCC, pp 200–201
35. Mukherjee S, Amato M, Wacyk I and Rumenik V (1987) LDMOS and LIGT's in CMOS Atechnology for Power Integrated Circuits, IEDM Tech Dig, p 778
36. Murari B (1993) Mixed Power IC Technologies and Small System Integration, invited paper presented at the 19th European Solid-State Circuits Conference, Sevilla, Spain
37. Contiero C (1995) BCD Technologies for VLSI Smart Power ICs, 25th ESSDERC
38. Conti F and Conti M (1972) Surface Breakdown in Silicon Planar Diodes Equipped with Field Plate, Solid-State Electron, vol 15, pp 93–105
39. Sun SC and Plummer JD (1980) Modeling of the ON-Resistance of LDMOS, VDMOS and VMOS Power Transistor, IEEE Trans Electron Devices, vol ED-27, n 2
40. Contiero C, Galbiati P and Gandolfi L (1983) Layout Analysis of a Power MOS Unitary Cell To Properly Design the Value of the ON-Resistance, in Proc 11th MIEL (Zagreb)
41. Galbiati P, Grandi L, Zullino L and Contiero C (1989) LDMOS and VDMOS Electrical Characteristics in Multipower – BCD Technologies: Modeling and Layout Optimization, Proc ECS Symp on High Voltage and Smart Power Devices
42. Fourneris G, Bekkara N, Benkoski J, Zullino L, Spatafora D, Martino G (1993) Demosthenes. A Technology-Independent Power DMOS Layout Generator, EURO-DAC, pp 178–183
43. Darwish MN (1986) Study of the Quasi-Saturation Effect in VDMOS Transistors, IEEE Trans Electron Devices, vol ED-33, n 11
44. Cacharelis PJ, Hart MJ, Manley MH, Frake SO and Knecht MW (1988) A Modular 1 μm CMOS Single Polysilicon EPROM PLD Technology, IEDM Tech Dig
45. Pacor B (1978) Diode Formed in Integrated-Circuit Structure, US Patent n 4,117,507
46. Eklund KH (1985) IC Diode with Reduced Substrate Leakage – Has Annular, Heavily Doped Region Surrounding Layer and Countacting Second, Heavily Doped Bottom Diffusion, EP Patent n 86210-B
47. Widlar RJ (1991) Controlling Substrate Currents in Junction-Isolated IC's, IEEE Journal Solid-State Circuits, vol 26, n 8, p 1090
48. Terletzki H, Rish L (1989) Electrostatic Discharge Test Structures for CMOS Circuits, Proc IEEE 1989 Int Conf on Microelectronic Test Structures, vol 2, n 1
49. Fujishin E, Garrett K, Levis MP, Motta RF, Hartranft M (1985) Optimized ESD Protection Circuits for High Speed MOS/VLSI, IEEE Journal Solid-State Circuits
50. Electrostatic Discharge Sensitivity Classification, MIL-STD-883C, Notice 8, Method 3015.7, March 22 1989
51. Draft Standard for ESD Sensitivity Testing – CDM Component Testing, EOS/ESD-DS5.3-1993, EOS/ESD Association Ing
52. Van Roozendaal L et al (1990) Standard ESD Testing of Integrated Circuits, Proc EOS/ESD Symposium
53. Electromagnetic Compatibility for Industrial-Process Measurements and Control Equipment – Part 2: ESD Requirements, IEC 801-2, 2nd ed, International Electrotechnical Commission, April 1991

CHAPTER 2

Technologies for High Voltage ICs

Dr. S. Mukherjee

2.1 Introduction

With the advent of the microprocessor, electronic controls have found their way into a wide range of applications, from simple household appliances to large-scale and complex industrial controls. This rapid growth, coupled with advances made in power semiconductor devices, has spurred the semiconductor community to focus more on integrating the power conditioning sections of these systems. What has typically remained beyond the scope of VLSI/ULSI is the power conditioning sections of the systems and direct interface to transducers. This has led to the need for Power Integrated Circuits (PICs) or High Voltage ICs [1], defined as ICs combining high voltage and/or high current components monolithically with low voltage/low current control components. Integration of the power conditioners promises to provide improved performance, increased functionality, enhanced reliability and compact solutions. The first applications of Power Integrated Circuits (PIC), nearly 20 years ago, were in voltage regulators and audio amplifiers [2, 3, 4]. Today PICs have been developed for such applications as power converters, motor controllers, fluorescent lighting ballasts, automotive electronics, video amplifiers, bridge circuit drivers and display drivers.

In spite of the large application potential of PICs, the growth during the last two decades has been much slower than expected. Several reasons can be found to explain this. First, the cost of PICs, until recently, has been 5 to 10 times higher than the corresponding discrete or electromechanical solutions. Second, with the advances in device physics and technologies, the cost of discrete power devices has decreased rapidly thereby posing a severe competition to PICs. Third, only a limited number of special purpose applications have been explored and general purpose high volume standard products such as memory chips in conventional ICs has not emerged. However, with the increasing thrust for miniaturization new applications are emerging and advances in power devices and technologies are bringing the cost down.

During the past two decades several approaches have been explored in developing technologies for high voltage ICs. Typically two approaches have been prominent. One approach involves starting with an optimized high voltage device structure and building low voltage control devices around it. The other approach involves starting with a low voltage CMOS technology and incorporating high voltage devices in it. The former approach is more popular than the latter. The low voltage devices are typically bipolar and CMOS, and design considerations include minimizing the total mask count, process complexity and resulting cost. The consequence of the approach followed is reflected in the performance and cost trade offs in the resulting technologies.

Device isolation and reduction of unwanted interaction between circuit components are key elements to reliable and predictable operation in high voltage ICs. Therefore, a scheme for the classification of high voltage IC technologies is in terms of the type of isolation technique employed. Three types of isolation schemes are encountered, namely: junction isolation (JI) (as in bipolar processes); self-isolation (SI) (as in MOS processes); and dielectric isolation including Silicon-On-Insulator (SOI). Of these, the junction and self-isolation technologies which have received the most attention in the industry. Examples of these technologies are seen in Bipolar CMOS and DMOS (BCD) processes ranging in maximum voltage rating of 200 volts to 700 volts. Other approaches include thin epi RESURF concept based high voltage lateral DMOS transistors and CMOS and bipolar control devices. Generally, four voltage ranges have been considered. These are: < 50 volts; 50 – 100 volts; 100 – 500 volts; and above 500 volts. JI and SI technologies have addressed all these ranges and are commonly employed in the industry where a limited number of high voltage devices are required on the same chip. Applications of these technologies include switch mode power supplies, electronic ballasts for fluorescent lamps, motor controls, automotive electronics, and display drivers. Unlike junction isolation, dielectric isolation technologies provide superior performance but usually at a higher cost, determined primarily by the starting material. Therefore these technologies have been applied where cost pressures are less severe. The first and prominent applications of DI technologies have been in telecommunications ICs where multiple high voltage devices are integrated into the same chip. More recently, the advent of the direct wafer bonding technique has given rise to increased interest in Silicon-On-Insulator, which promises a lower cost solution to a large class of high voltage ICs.

This chapter describes the technical elements of some of the major high voltage IC technologies that are in existence today and discuss their applicability and limitations. High voltage devices are described in relation to these technologies. New emerging technologies, either in research labs or in early industrialization phases, are described.

2.2 Process Technologies

Broadly, three classes of technologies, based on the techniques for isolating the high and low voltage components, have been developed for PICs. These are: Junction-Isolated (JI), Self-Isolated (SI) and Dielectrically-Isolated (DI) technologies. In junction-isolated processes devices are electrically isolated from each other by reverse biased p-n junctions as in bipolar processes. In self-isolation, common in MOS technologies, thick field oxides on the surface between active devices coupled with adequate doping below the oxides prevent field inversion at the maximum voltage and thereby provide the required isolation. JI and SI processes have been mostly used in bulk silicon starting material and can therefore also be termed bulk technologies as opposed to DI technologies. Dielectric isolation provides oxide layers around each active device to isolate individual components. Each of these technologies has found its way into applications that result in optimal performance per unit cost. While significant developments have occurred in isolation technologies, technical problems still remain, requiring continued research effort.

Within the bulk technology, further classification occurs based on whether the power devices are implemented vertically, as in discrete devices, or laterally. The vertical technologies have also been termed smart power as opposed to high voltage IC or power IC which refers primarily to lateral technologies. This terminology reflects the fact that vertical technologies are primarily based on discrete power device processes with added elements to provide control functions. The term smart discrete is also found in the literature for variants in this category. It should be pointed out that the existing terminologies are very often used interchangeably.

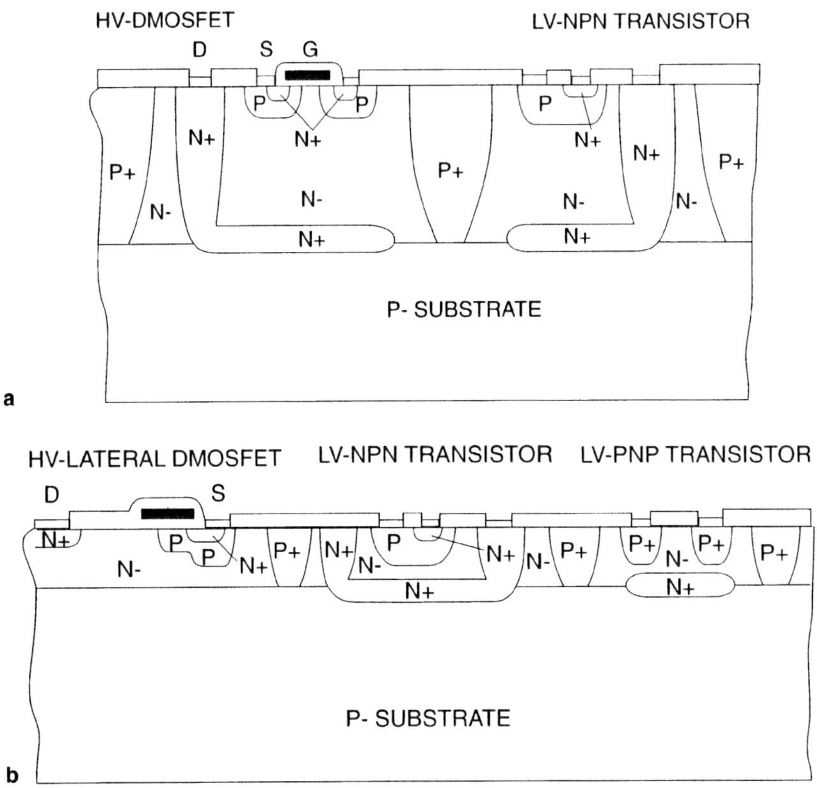

Fig. 2.1 Cross-section of two planar high voltage IC processes based on **a** vertical high voltage DMOS and **b** lateral high voltage DMOS

2.2.1
Bulk Technologies: Junction Isolation (JI) and Self-Isolation (SI)

The earliest processes for PICs employed junction isolation similar to that used in bipolar technologies. Several variants of this approach have been reported over the years for different applications. With the advent of MOS power devices, self -isolation using thick oxides, has been combined with junction isolation. Of the different technologies that have emerged in this area, two classes of processes can be distinguished. One class consists of integrated high voltage and low voltage devices (CMOS and Bipolar) with all the current conducting active device terminals on the same surface of the wafer. BCD (Bipolar CMOS and DMOS) processes [5] ranging from 60 volts up to 650 volts fall in this category. Figure 2.1 shows two examples of high voltage IC processes, one based on a high voltage vertical DMOS device (Fig. 2.1a) and the other based on a high voltage lateral DMOS (Fig. 2.1b). A broad range of devices (low and high voltage) is available in these processes for a large class of applications. Bulk processes with breakdown voltages up to 1200 volts [6,37,38] have been reported. In some cases, a thick epitaxial layer is used for the vertical DMOS power devices with up-down diffusion to obtain drain contact on the surface. For breakdown voltages above 250 volts, such processes require thick epitaxial layers (> 25 microns) and deep diffusions through these epitaxial layers. Controllability of such processes in manufacturing requires special attention.

2.2.1.1
Lateral High Voltage IC Technology

Thin epitaxial layer based lateral devices with dense CMOS and bipolar control have received attention in the industry [7]. The thin epitaxial layer (4–5 microns) and associated processes (implanted drift region of the lateral power devices) provide more compact solutions in many applications.

One of the techniques for achieving increased packing density in high voltage IC is the reduced surface field RESURF concept [8] utilized in lateral high voltage devices such as the lateral double diffused MOS transistor (LDMOS). A simple schematic of an LDMOS is shown in Figure 2.2. The device consists of a double diffused channel region and an extended drift region of an n-type doped epitaxial layer on a p-type substrate. It has been shown that by proper choice of the epitaxial layer thickness, doping (Nd) and substrate doping (Nsub), optimal on-resistance can be achieved for a given maximum voltage requirement. The basic phenomenon employed in this device is depletion of the entire drift region due to the substrate/epi layer junction at the maximum applied voltage. The breakdown voltage is a sensitive function of the drift region doping and reaches a maximum at an integrated doping level of 1 to $2\times10^{12}\,cm^{-3}$. Once the drift region and substrate doping concentrations are fixed, the breakdown voltage can be designed by choosing the appropriate drift region length (Ld) up to the maximum voltage determined by the substrate doping. Under these conditions, devices with lower than maximum breakdown voltage and correspondingly

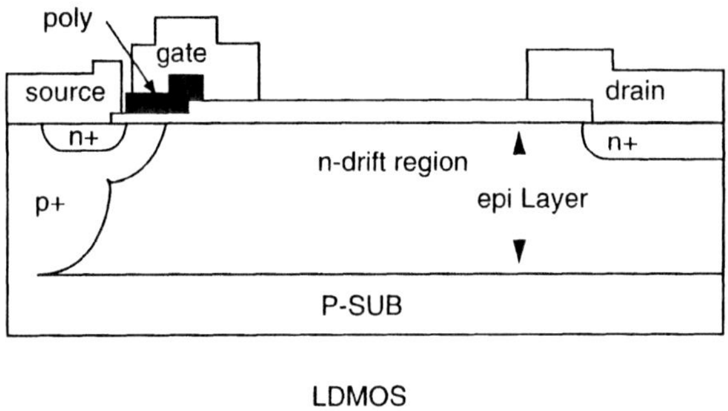

Fig. 2.2 Cross-section of a basic lateral DMOS (LDMOS) transistor

smaller drift region lengths are not optimal in terms of on-resistance. However, for an integrated process design, the flexibility offered by this structure can be employed effectively to address a wide variety of applications that require different maximum voltage capability.

Figure 2.3 shows the cross-section of a high voltage IC process designed using the RESURF concept. In this process, the low voltage section consists of dense CMOS and bipolar components and forms the basis of the overall process architecture. The fabrication process is based on a dual well CMOS process with self aligned n and p wells fabricated in an n-type epitaxial layer grown on a high resistivity p-type substrate. Prior to growing the epitaxial layer, p and n-type buried layers are defined on the substrate. The p-type buried layer is used for field shaping in the LDMOS devices to obtain improved RESURF effect and consequently reduced specific on-resistance [9]. It is also used in the low voltage sections to reduce p-well resistance and increase immunity to CMOS latch-up together with the n-type buried layer in the n-wells. Two layers of poly-silicon are used in this process. The first polysilicon layer is used to form the gates of the high voltage devices (LDMOS and lateral insulated gate bipolar transistors, LIGBTs) with a 600 A gate oxide. The second layer of poly-silicon is used for self-aligned low voltage n and p-channel transistors. This allows independent optimization of the high and low voltage sections and can be used to provide high density CMOS without sacrificing the high voltage device performance. The two layers of polysilicon also allow voltage-independent capacitors, which are used frequently in analog integrated circuits. The isolation between low voltage components is primarily thick field oxide grown using the standard LOCOS isolation technique in conjunction with field threshold adjust implant to provide an adequate barrier against field inversion. In addition to this, isolation between high and low voltage components is achieved by providing deep p-type isolation diffusions that extends all the way from the surface to the p-type substrate.

Some of the devices available in this technology are listed in Table 1 with selected electrical parameters.

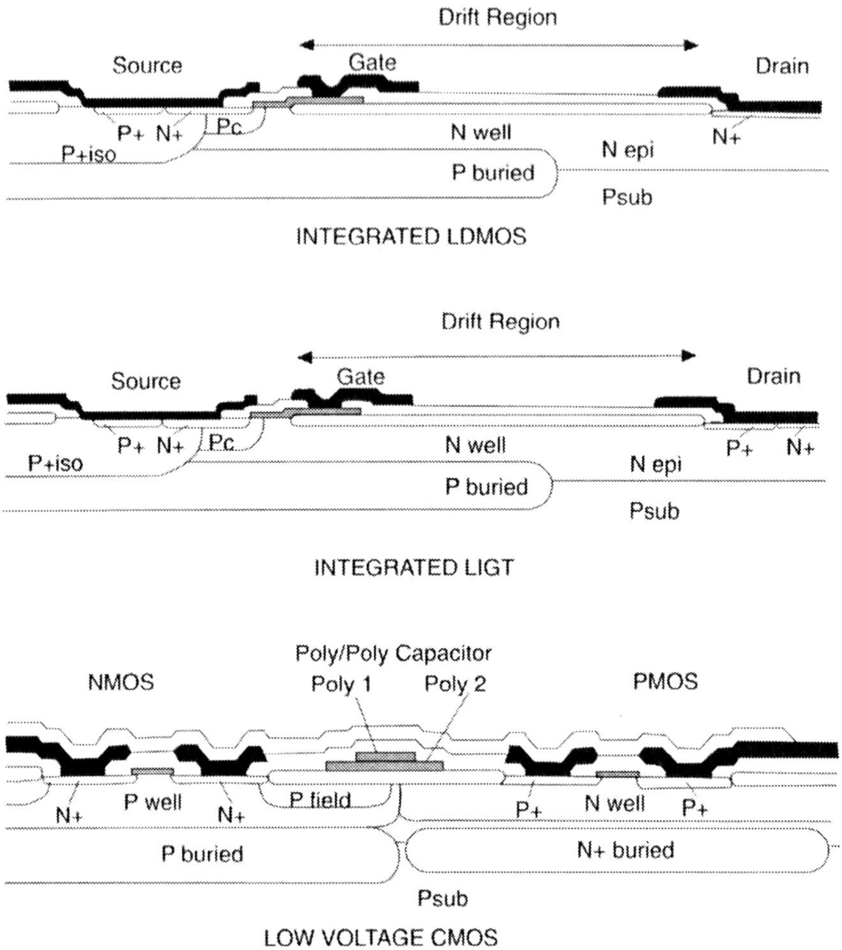

Fig. 2.3 Process cross-sections in a thin epi high voltage bulk IC process

The specific on-resistance of the high voltage LDMOS device in the process is shown in Figure 2.4 as a function of breakdown voltage. Also indicated in this figure is the minimum on-resistance calculated from the intrinsic resistance of the drift region. Here specific on-resistance is defined as the product of the on-resistance at gate bias of 12 volts and the active area of the device.

It should be noted that the LDMOS device in this process is designed for use primarily with the source terminal connected to the substrate. It is possible to design structures with source independent of the substrate, and under such conditions the maximum voltage between the source and substrate is typically limited to less than 100 volts. For applications requiring higher source voltage

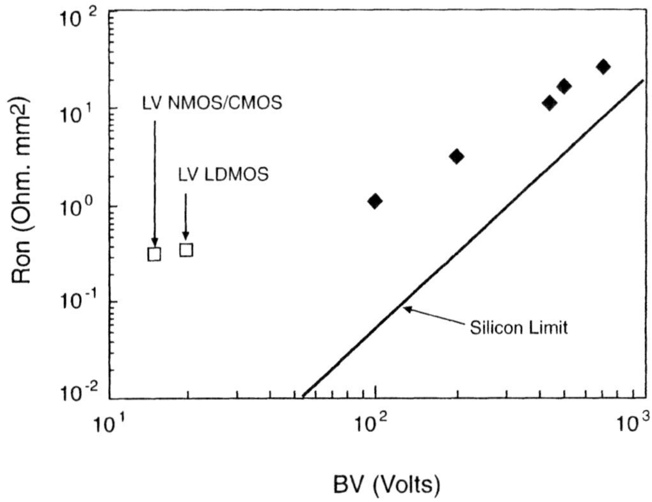

Fig. 2.4 Specific on-resistance vs breakdown voltage of LDMOS transistors

Table 2.1 List of devices in the high voltage IC process shown in figure 3

Device name	Electrical parameter	Comments
LDMOS	Breakdown Voltage up to 750 volts, Vt = 2.5 volts	On resistance optimized for 650 volt device.
LIGBT	Breakdown Voltage up to 750 volts, Vt = 2.5 volts	On resistance is voltage dependent and up to 8 × lower than corresponding LDMOS.
High Voltage cross-over	Breakdown Voltage up to 800 volts	
High Voltage JFET.	Breakdown Voltage up to 750 volts	
Medium voltage DMOS	Breakdown Voltage around 30 volts	BV adjustable to higher values.
Low voltage CMOS	Breakdown Voltage up to 17 volts, n and pmos Vt = +/− 1 volt.	
Vertical Bipolar npn.	Beta ~ 40.	

capability, such as in half or full bridge power stages, it is usually more economical to use discrete power devices. However for certain applications with limited load power it is possible to use other bulk power IC processes with thick epitaxial layers [6] that allow source high operation up to 750 volts or higher.

In addition to the LDMOS transistor, lateral IGBTs (LIGBT) are available in the technology as listed in the Table 2.1. The operation of the LIGBTs is based on minority carrier injection from the drain junction in the on state [7, 10, 11]. The breakdown voltages of the LIGBTs are comparable to those of the LDMOS counterparts and are usually derived from optimized LDMOS designs with modifications in the drain region to obtain minority carrier injection. The earliest implementations consisted of replacing the n+ drain diffusion by a p+ diffusion to obtain maximum injection and consequent conductivity modulation. A consequence of this is increased turn-off time during switching. To optimize the turn-off time, modification to the drain region to control minority carrier injection has been suggested [10, 39]. These modifications consist of providing a shorting n+ diffusion in conjunction with the p+ drain diffusion. Depending on the design geometries (that is, ratio of n+ to p+ diffusion and

a

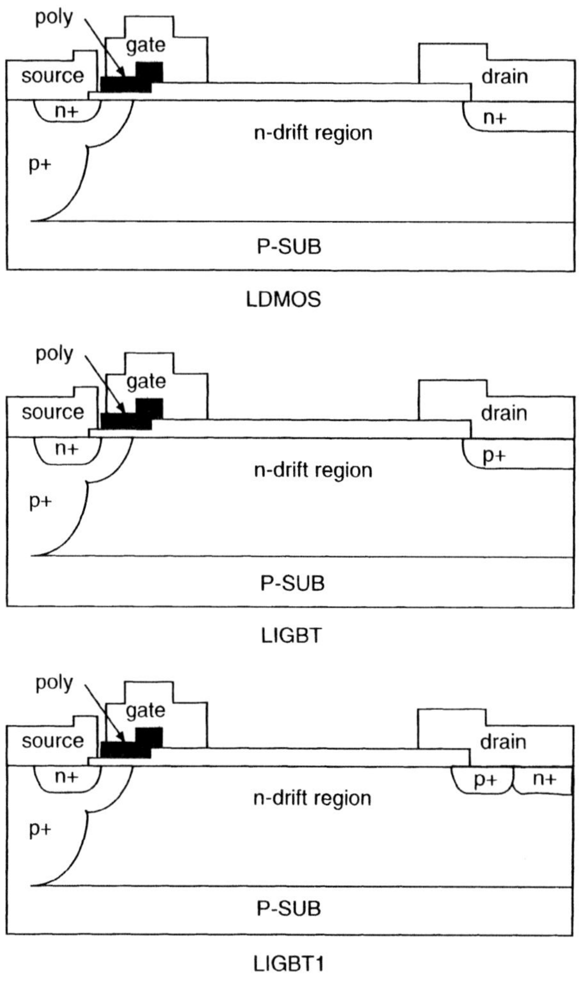

Fig. 2.5 **a** Cross-section of LDMOS and LIGBTs with different drain designs. **b** On-state characteristics of LIGBTs in comparison with LDMOS. **c** Switching turn-off times of corresponding LIGBTs shown in **a**

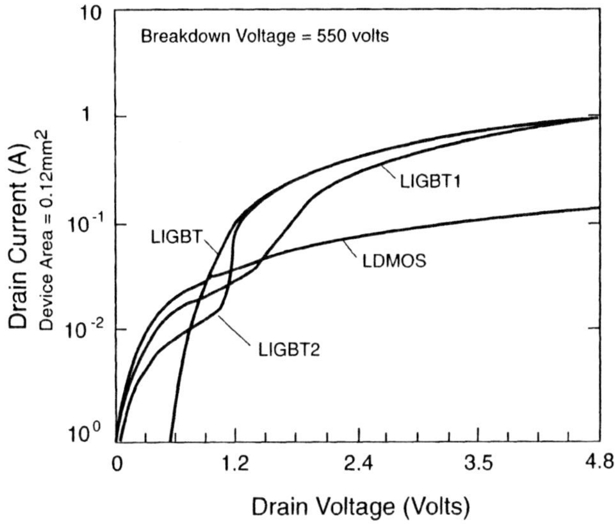

Fig. 2.5 b

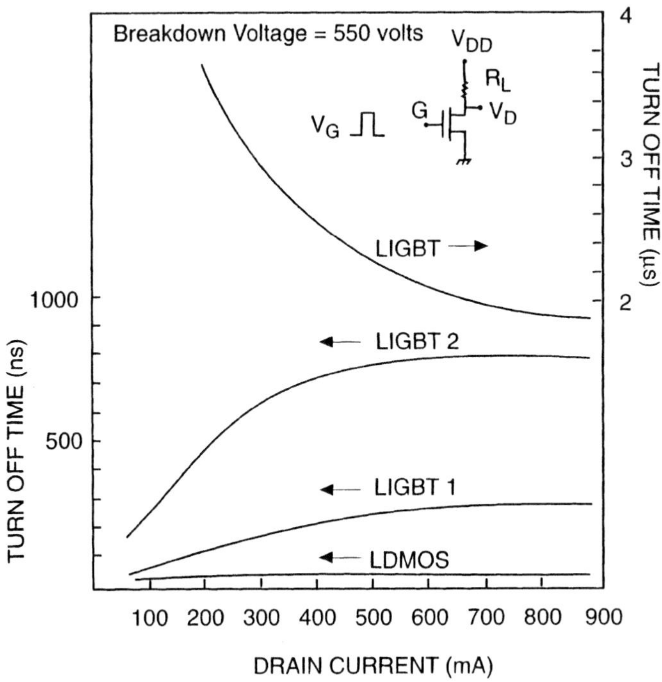

Fig. 2.5 c

lengths of the regions in the current conducting path) the on state and switching characteristics can be tailored to meet the requirements of different applications. The on-state and switching characteristics of a number of drain design variations are shown in Figure 2.5 a, b and c together with the equivalent LDMOS characteristics. From this it is seen that for increasing conductivity modulation and consequent increase in drain current for a given forward bias, the turn-off time is increased. This is due to increasing minority carrier storage time in the device.

An important consideration in the design of LIGBTs is the current density at which the device exhibits latch-up due to the turn-on of the parasitic silicon controlled rectifier (SCR). The latch-up current density determines the maximum current density that can be handled. Furthermore latch-up current density decreases with temperature thereby limiting the usable current range in an application. A number of approaches to increase the latch-up current density have been investigated [11]. All of these involve reducing the shunting resistance of the parasitic lateral npn bipolar emitter base junction. An approach that can provide immunity to latch-up involves interruption of the n+ source diffusion at certain intervals by p+ diffusion regions. This also results in a penalty in maximum current density under forward bias conditions.

Other high voltage devices available in the thin-epi technology include the SINFET [12] that incorporates a Schottky junction at the drain. This structure results in limited minority carrier injection in the on-state and fast turn-off time during switching. This device requires optimized Schottky barrier technology and typically exhibits increased forward voltage drop compared to standard LIGBTs.

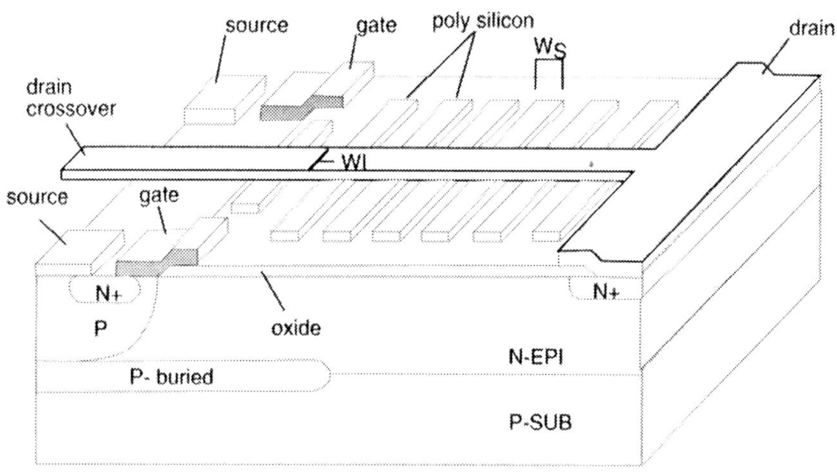

Fig. 2.6 High voltage cross-over structure on thin epi layer high voltage IC process. This structure employs a series of interpoly- capacitors on the drift region of a LDMOS structure and the drain metal is allowed to cross over the source

Figure 2.6 shows an example of a high voltage crossover structure available in this technology [13]. This structure employs a series of interpoly capacitors on the drift region of a LDMOS structure, and the drain metal is allowed to cross over the source junction for interconnection. By proper design of the capacitor stack and the drain metal, the drain voltage is dropped in the oxide between the drain metal and the poly-capacitors thereby maintaining a low field in the underlying silicon.

Another important element in most high voltage processes is the final passivation of the devices and the packaging methods utilized in the technology. It is important to minimize the total charge in the dielectric layers over the high voltage device drift regions and provide sufficient gettering to avoid mobile charge. Typically the total charge in the dielectric layers can be compensated by appropriate doping in the drift regions, however, mobile charge can cause unwanted drift in breakdown voltage.

2.2.1.2
Vertical Power IC Technologies

The other class of processes (so-called the vertical processes) [14] that have emerged are based on vertical DMOS devices as the basic power device. Here current conduction occurs vertically through the silicon wafer with the substrate acting as the drain terminal of the power device. Other processes employ vertical bipolar transistors as the power switch. These processes, shown in Figure 2.7, are typically restricted to a breakdown voltage of 60 to 80 V but with current capability of 10 A or more. These are predominantly used in applications requiring high current handling and relatively low voltages, such as small motor drives, automotive electronics, and printer driver.

A specific implementation of a high current PIC process is shown in figure 2.8 [15]. This process is based on a vertical DMOS power switch fabricated on an n-epi on an n+ substrate. The epitaxial layer doping and thickness are optimized for the power switch with a maximum voltage capability of 60 volts. The control section consists of a standard p-well based CMOS architecture with special high voltage (60 volts) p and n-channel transistors designed with the process steps used in the low voltage section.These devices employ extended drain or lightly doped drain concepts (for example, the drain region of the high voltage n-mos is formed by n-field threshold implant, and the source region of the p-mos is formed by a p-well diffusion). To allow independent optimization of the control circuit and of the power switch, two layers of poly-silicon are used. The first polysilicon layer is used for the gates of the VDMOS transistor and the second polysilicon layer is used for the control CMOS devices. The body of the double diffused VDMOS transistor is defined separately from the p-well diffusion of the control CMOS. A total of 11 masking layers is used in the process and self isolation is utilized in the CMOS section. The VDMOS transistor in this process has a specific on-resistance of 140 $m\Omega\ mm^2$.

An improvement on the VDMOS device, in terms of specific on-resistance, is the trench-DMOS or the UMOS [16] transistor structure shown in Figure 2.9. In

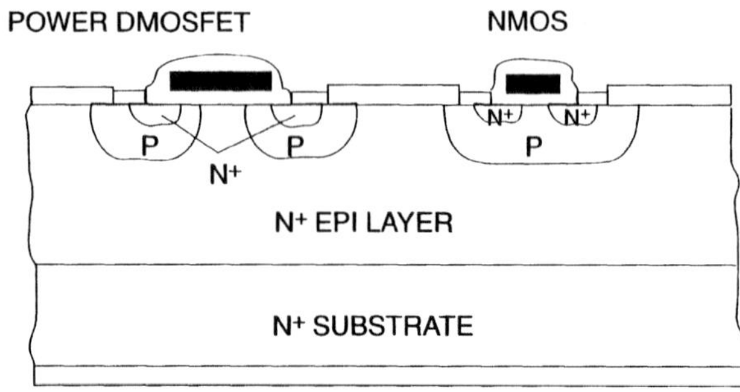

a

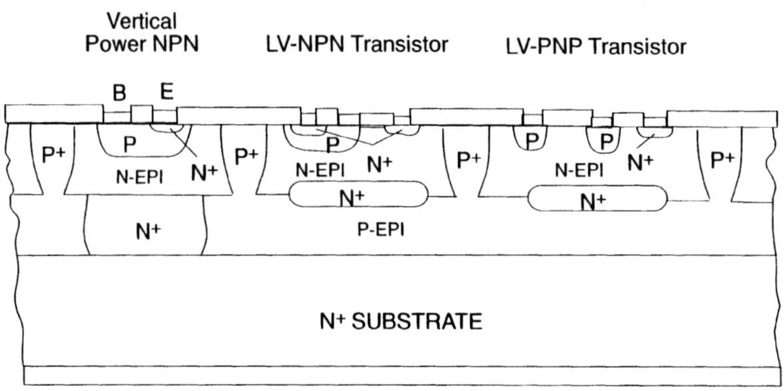

b

Fig. 2.7 High current PIC processes with vertical Power transistors. **a** Using VDMOS as power switch. **b** Using veritcal bipolar as power switch

this figure the different components of the on-resistance of the two structures are indicated. The major differences in the two structures is in the cell pitch (smaller for the trench-DMOS due to vertical trench sidewall channel) and the absence of R_{pinch} component in the trench-DMOS structure. The combination of these results in a substantial reduction in the specific on-resistance for the trench-DMOS in comparison with VDMOS (typically by a factor of 2). Also, by proper design of the trench depth, epi layer thickness and doping and edge termination, the breakdown voltage of the trench-DMOS can be close to that of the planar junction breakdown. These advantages have led to the integration of the trench-DMOS structure in a high current power IC process [17].

Figure 2.10 shows the process cross-section of a high current PIC process incorporating trench DMOS as the power switch. Compared to the VDMOS based PIC, this process employs one extra masking layer to define the trench regions. The key process steps are as follows:

1. Starting material n-epi (doping $\sim 1 \times 10^{16}\,\mathrm{cm}^{-3}$, thickness $\cong 4$ microns) on n++ substrate.
2. P-well formation.

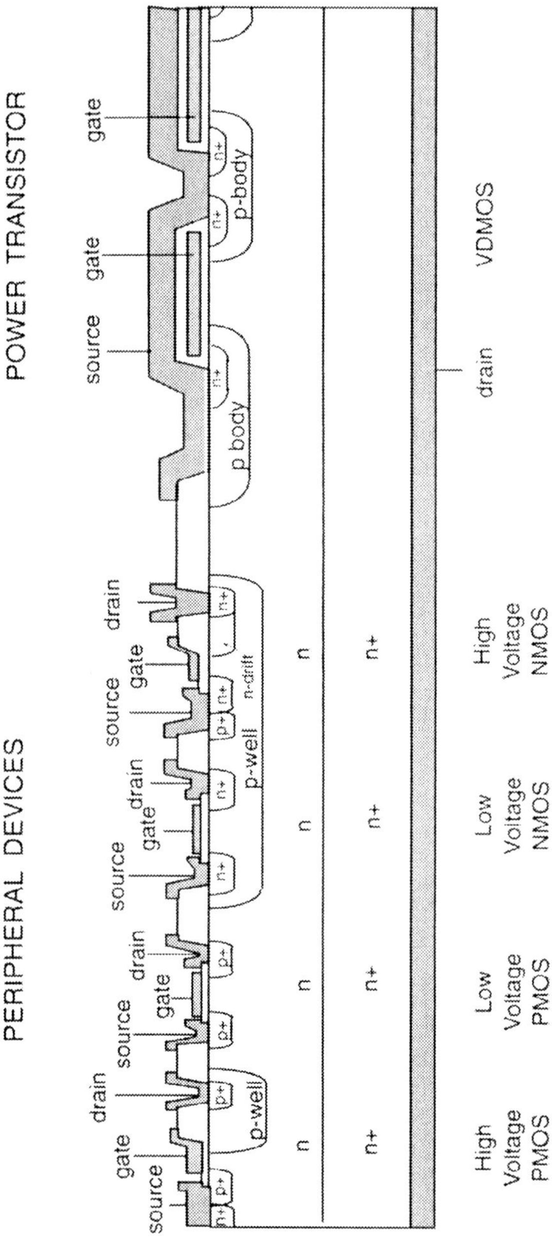

Fig. 2.8 A high current PIC process with dense CMOS and power switch

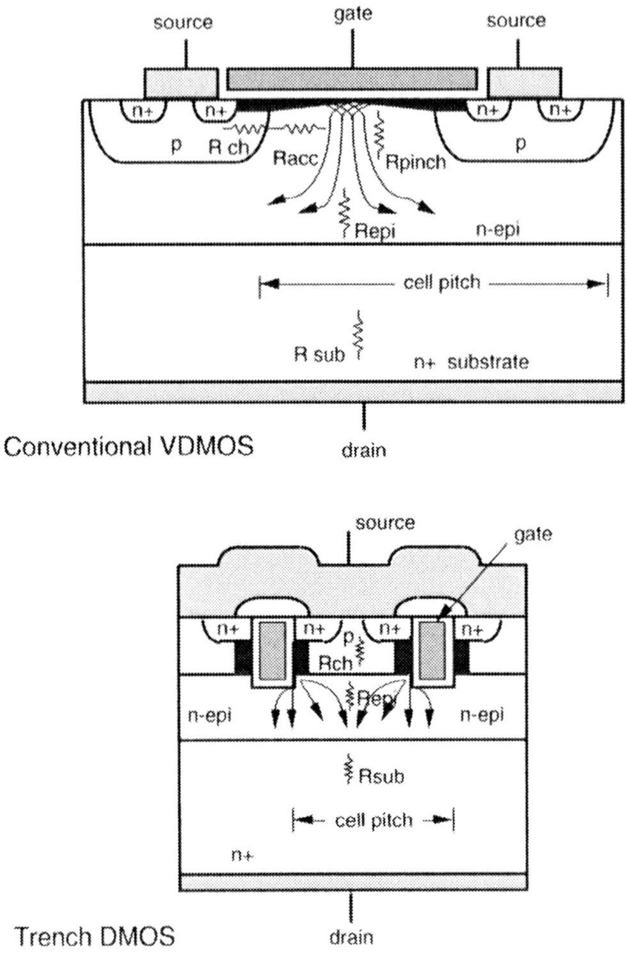

Fig. 2.9 Comparison of cross-sections of VDMOS and trench-DMOS transistors showing different components of on-resistance in each case

3. N and P field implantations.
4. LOCOS oxidation.
5. P-body of trench-DMOS formation.
6. Trench-DMOS formation using first poly-silicon layer. (Trench etching, oxidation, poly fill and planarization)
7. CMOS formation using second poly-silicon layer.

To allow independent optimization of the trench-DMOS and the control CMOS, two separate poly-silicon layers are used and the CMOS p-well is separated from the p-body diffusion of the trench-DMOS. The trench-DMOS p-body diffusion

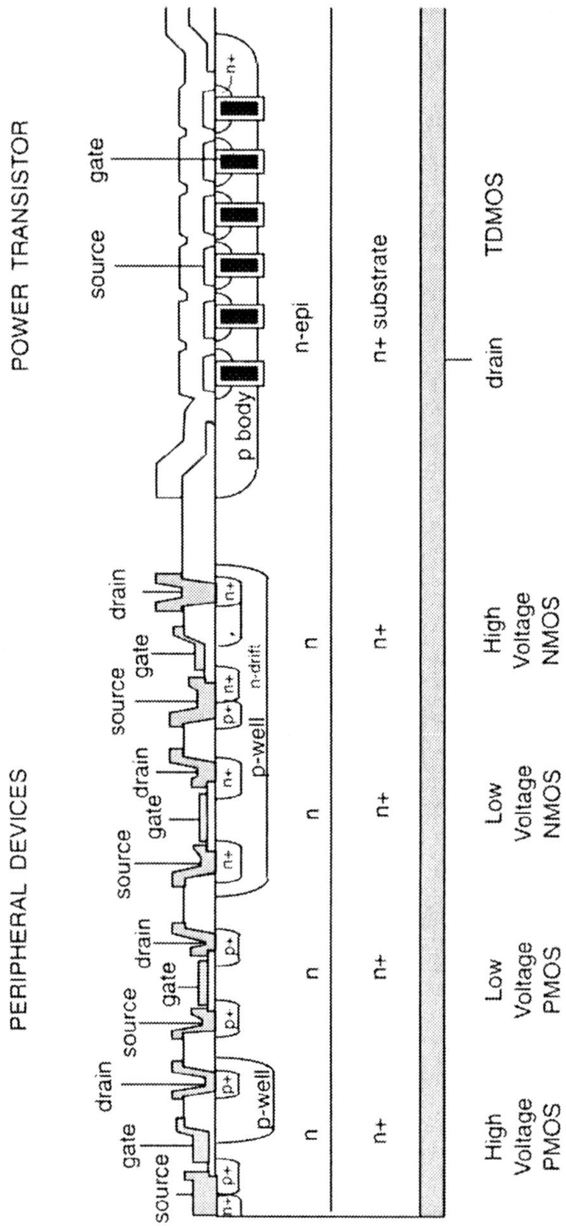

Fig. 2.10 Cross-section of a high current PIC process incorporating trench DMOS as the power switch

is adjusted to obtain a threshold voltage of around 2 volts and the doping profile is tailored to minimize resistance across the n+ source and p-body junction for increased ruggedness. It should be pointed out that the trench-DMOS due to its vertical construction can be inherently more rugged (less susceptible to bipolar turn on during avalanche breakdown) compared to the VDMOS. Specific on-resistance of the trench-DMOS devices are around 80 mΩ mm^2 for 60 volts breakdown voltage.

2.2.1.3 *Applications of Bulk PIC Technologies*

Several factors have contributed to the slow growth of PICs compared to what was expected 20 years ago. Unlike VLSI, PIC applications have relied on replacement of existing discrete device or mechanical switch based implementations. This has resulted in severe cost competition. Furthermore, the absence of standard PIC components such as memory devices in VLSI technologies has limited the widespread production and marketing of PICs. Today most PICs can be considered as custom ASIC designs. Together with these, power dissipation limitations and packaging restrictions have been the historical limiters.

To meet the above challenges, innovative device designs such as the LIGBT, have resulted in reduced die sizes by a factor of 3 to 4 and therefore lower cost. Figure 2.11 shows a comparison of a Switch Mode Power Supply PIC [7], implemented with an LDMOS in one case (a) and an LIGBT in the other case (b). For switching frequencies below 30 KHz, the performances of the two implementations are comparable but clearly the die size and cost of the LIGBT version is considerably smaller. It should be noted that the design of LIGBTs for any application must account for peak current requirement and in many cases the design can be latch-up current limited. Furthermore, since an LIGBT has a higher forward drop compared to an LDMOS at low currents, power dissipation can be higher with LIGBTs. At high switching frequencies LIGBTs exhibit increased switching loss adding to the power dissipation.

In high current applications, trench based power devices are being developed for silicon die size reduction by more than a factor of 2. Figure 2.12 shows the

a b

Fig. 2.11 Switch Power Supply PIC **a** implemented with an LDMOS power device, and **b** with LIGBT power device

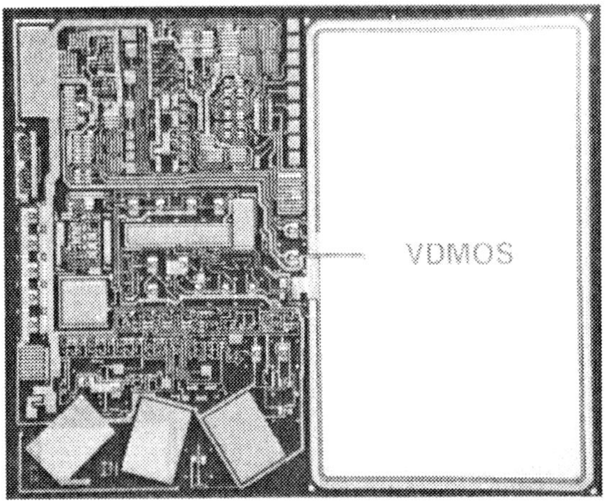

a

b

c

Fig. 2.12 Automotive high-side switch implemented in three generations of technology. **a** VDMOS based vertical PIC process. **b** Compact VDMOS and CMOS based PIC process. **c** Trench-DMOS and CMOS based PIC process

comparison of an automotive high-side switch implemented with three generations of power devices and vertical technologies. The first generation technology (a) utilized conventional VDMOS and associated control circuits, whereas the second generation (b) made use of denser CMOS controls and scaled VDMOS power device. The third generation (c) used a trench based DMOS. Again advances in devices and processing technology have aimed at cost reduction. Additionally, the second and third generation ICs have implemented sophisticated features like reverse battery protection and a variety of protection circuits (overtemperature, short circuit, etc.) [18]. These have been possible in part due to the higher density of control circuitry available in the process coupled with innovative circuit design techniques.

In certain applications where both high currents (above 5 A) and high voltages (above 500 volts) are involved it has been found that discrete power devices arranged in full or half bridge configurations are necessary. In these cases high voltage ICs have been developed with floating high voltage wells to provide drives for both the high and low side discrete power devices [19].

2.2.2 Dielectric Isolation (DI) and Silicon-on-Insulator (SOI) Technologies

Dielectric Isolation technology provides effective electrical isolation between different circuit components, low leakage even at high temperatures, reduced parasitic capacitance and can provide high packing density. Figure 2.13 shows the cross-section of a PIC process employing an early form of dielectric isolation [20]. Here, the individual devices are built in dielectrically-isolated tubs, thereby preventing conductive coupling between the components through the substrate. This results in improved performance of power control circuits (especially circuits employing double injection devices) compared to other isolation methods.

As an alternative to the DI process, Silicon on Sapphire (SOS) as shown in Figure 2.14 and more recently, Silicon on Insulator (SOI) has been employed for PICs. Of these, the SOI technology, employing relatively thin tubs or islands, promises a more cost effective implementation. With SOI, all the advantages of

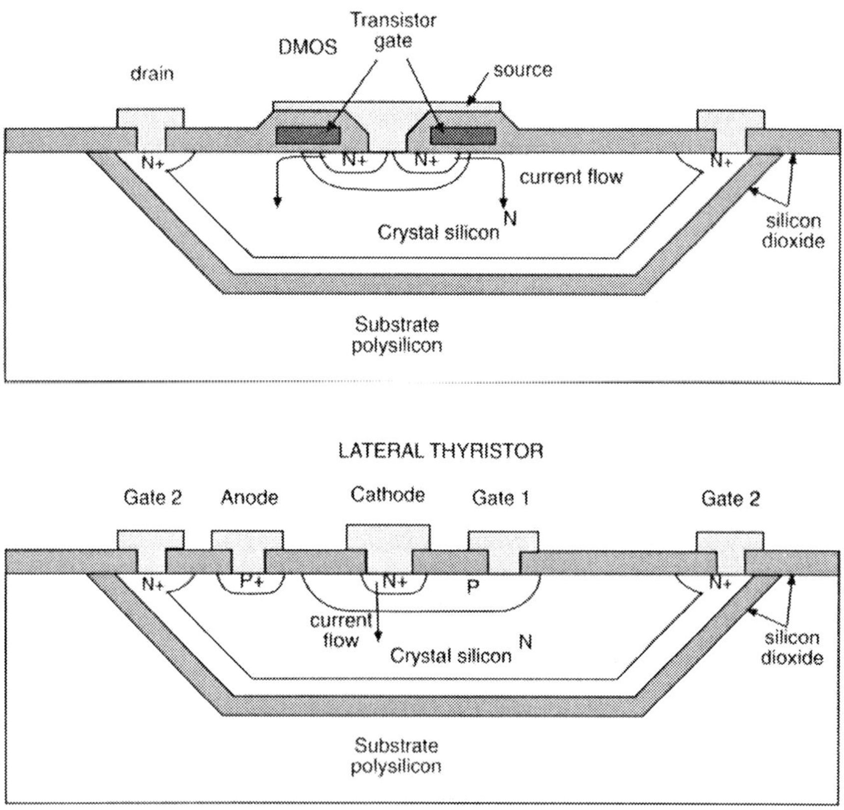

Fig. 2.13 Dielectric- Isolation (DI) based high voltage IC process

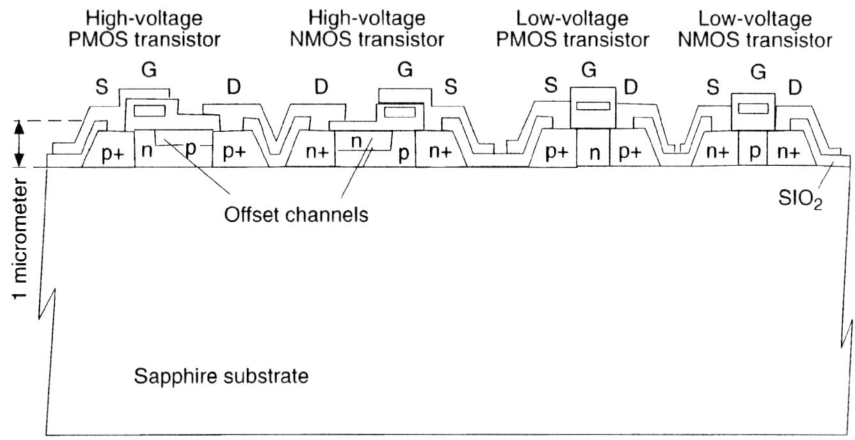

Fig. 2.14 A high voltage IC process based on SOS

the DI technology are maintained and circuits can be more compact due to the smaller isolation space required. Compared to the early DI methods, SOI requires different device structures for high voltage applications beyond 500 volts, and promises improved device performance in many cases. A number of interesting device structures have been proposed using thin SOI material [21, 22, 40].

Several methods of fabricating SOI starting materials have been explored [23, 24]. They include, silicon isolation by oxygen implantation (SIMOX), zone melting and recrystallization (ZMR), full isolation by oxidized porous silicon (FIPOS) and wafer bonding and etchback (BESOI) or thinning. Among these, wafer bonding and thinning or etchback appear to be promising options for producing cost effective starting material needed in Power IC applications. Wafer bonding, in comparison with the other approaches, allow more appropriate buried oxide properties (particularly breakdown field strength and thickness) for high voltage applications. Controlling the thickness of the SOI layer is the critical step in the wafer bonding approach. Recently a plasma etching technique called plasma assisted chemical etching (PACE) [25] has been used to thin the SOI layers down to a few hundred angstroms with accuracy of a few percent. This technique and other polishing or etchback techniques combined with wafer bonding can lead to low cost SOI starting material.

A comparison of bulk and SOI-based high voltage IC technologies is shown in Figure 2.15 . For maximum voltages of up to 300 volts, the choice of the SOI and buried oxide layer thicknesses is determined by thermal constraints arising from transient power dissipation in the power devices. Practical devices can be fabricated with around 2 microns of buried oxide and between one to two microns of silicon as the active layer. For process simplicity, thin SOI layer is preferred because lateral isolation of devices is easily accomplished by plasma etching and dielectric refilling. A relationship between the SOI layer thickness, buried oxide thickness and maximum voltage capability is shown in Figure 2.16

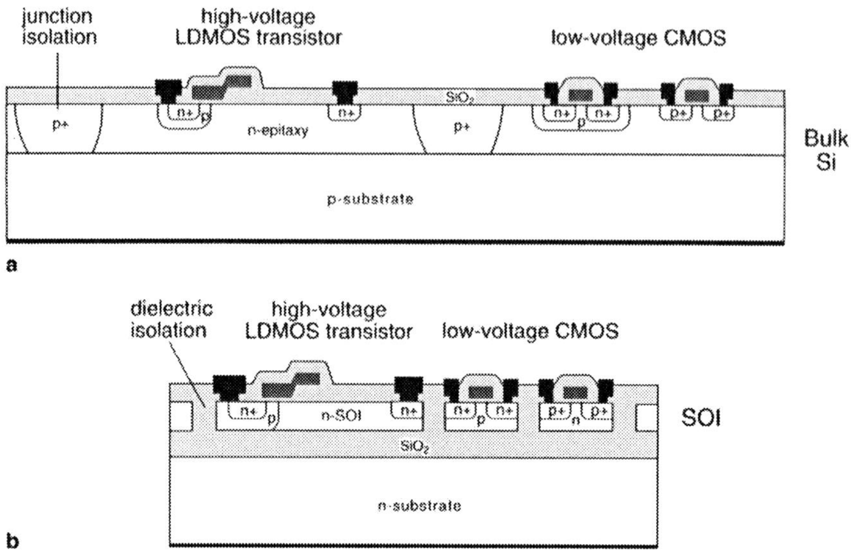

Fig. 2.15 Comparison of bulk silicon **a** and SOI **b** based high voltage IC processes

[36]. This is based on a one dimensional calculation of electric field across the buried oxide and the SOI layer while maintaining continuity of electric displacement across the interface. The breakdown voltage of this structure is reached when the field in the silicon layer reaches the critical breakdown field strength. From this figure it is seen that higher maximum voltage requires higher SOI layer thickness and/or the buried oxide thickness. However, for thin SOI layers (less than 0.5 micron), the critical breakdown field strength in silicon is effectively increased due to the limited ability of the charge carriers to create impact ionization in the thin layer, and this makes high voltage devices possible in thin SOI layers [21, 34]. Figure 2.17 shows a plot of breakdown voltage in thin layer SOI structures as a function of silicon layer thickness. Here the increase in breakdown voltage with decreasing silicon thickness is primarily due to the limited ability of the carriers to produce impact ionization and consequent avalanche multiplication as the silicon layer thickness decreases.

One of the key features of SOI or DI process is the ability to implement high voltage half or full bridge circuits where the high-side transistors can have source voltages up to the maximum allowable voltage in the technology. Unlike bulk high voltage ICs, it has been shown [26] that the on-resistance of SOI high side lateral devices is independent of source voltage above a certain source voltage value. This is due to inversion of the buried oxide silicon interface at a certain voltage that limits the spread of the depletion layer in the drift region. Figure 2.18 a shows the effect of source high condition on SOI based LDMOS in comparison with bulk, and Figure 2.18 b shows the comparison of on-resistance under different source bias conditions between SOI and bulk LDMOS.

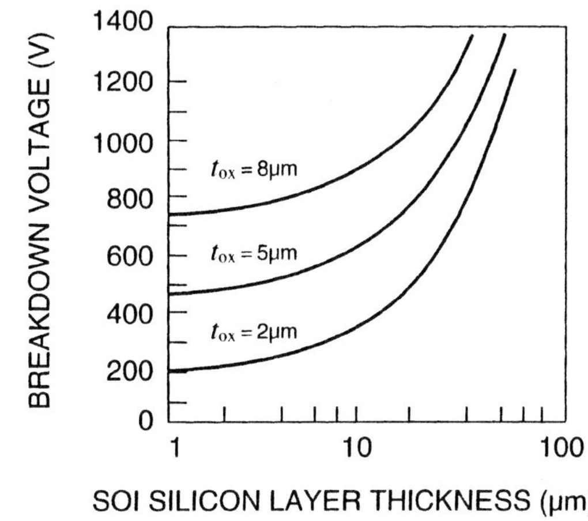

Fig. 2.16 Breakdown voltage vs SOI layer thickness for different buried oxide thickness based on constant critical breakdown field strength in silicon

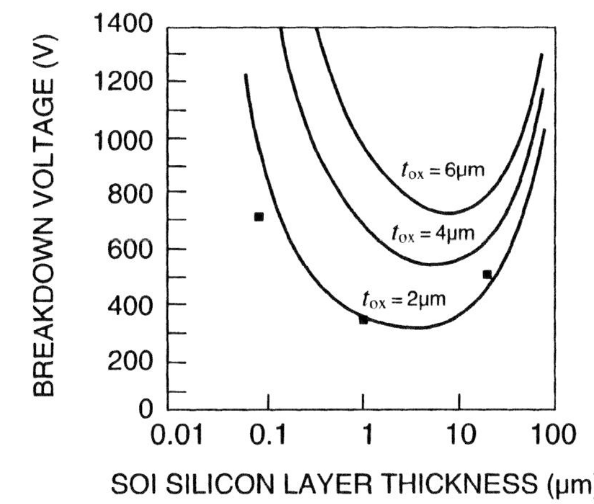

Fig. 2.17 Breakdown voltage vs SOI layer thickness for different buried oxide thicknesses based on silicon layer thickness dependent critical breakdown field strength. Experimental points for several devices are indicated [21]

In SOI or DI processes, it is also possible to use conductivity modulated devices such as LIGBTS in source high configuration without adverse parasitic effects that occur in bulk technologies. In thin layer SOI technologies, switching performances of LIGBTs are improved over their bulk counterpart due to the confinement of carriers in a thin layer [27].

High voltage (600 volts) IC processes with relatively thick SOI layers (20 microns) and thick buried oxide (3 m) have been reported in the literature [28, 29]. These technologies require deep trench etching and oxide refill and planarization to achieve dielectric isolation in the lateral direction. In these processes high voltage LDMOS as well as LIGBTs are implemented using the RESURF principle.

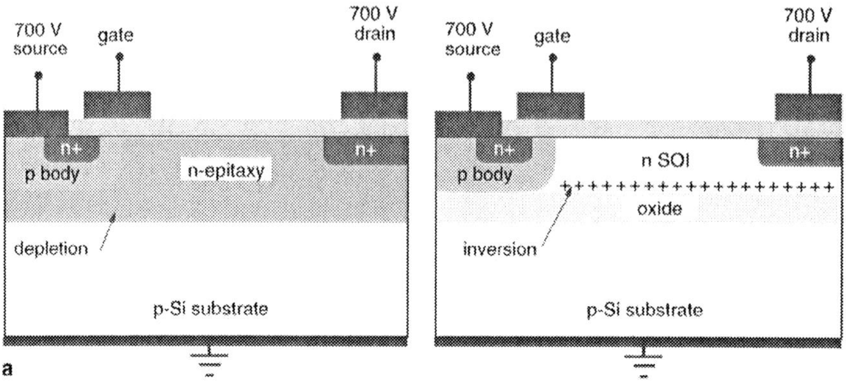

a

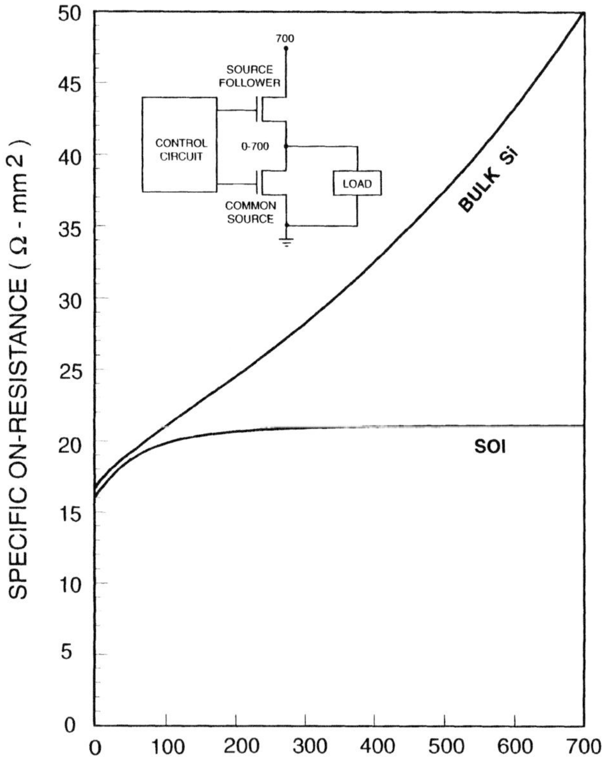

b SOURCE TO SUBSTRATE VOLTAGE

Fig. 2.18 **a** Effect of source high condition on bulk and SOI LDMOS, showing depletion into the drift region for bulk LDMOS and inversion layer formation in the SOI LDMOS. **b** Specific Ron vs source voltage for bulk and SOI LDMOS

2.2.2.1
Applications of DI Processes

From the circuit design perspective, dielectric isolation provides several advantages not available with bulk technologies. In PICs handling mixed signals of widely varying levels and slew rates, one of the major concerns is unwanted interactions between different sections of the circuit. DI or SOI processes can substantially minimize this interaction and thereby provide simpler and more robust designs. Additionally, dielectric isolation allows higher operating temperatures without circuit performance degradation due to junction leakage. Also conductivity modulated devices can contribute to reduction of die sizes without increasing interaction problems with control circuitry. Finally, with SOI it is possible to obtain higher operating speeds in the control circuits due to the reduction of parasitic capacitances. Some of these issues, although similar to those in VLSI, are more severe in the case of PICs.

The initial applications of DI technology appeared primarily in telecommunications ICs [30, 31] where integration with minimal interaction between devices was an overriding factor, and the relatively high cost of DI was not a limitation compared to the cost advantage gained in the overall systems. These applications ranged from 300 volts to as high as 600 volts and in many cases used integrated SCRs to gain area advantage.

More recently, DI technology has found applications in other areas, including television circuits [32]. The ability to integrate bipolar and MOS devices including thyristors, IGBTs and GTOs, and high voltage and high frequency operation made possible with dielectric isolation is exploited in video-amplifier ICs for colour television. Novel current sensing schemes stable over a wide temperature range is made possible in DI technologies for applications in DC brushless motor drives [33] and other high voltage controls.

The advantages in device isolation and eventually low cost promised by SOI technology is expected to open up new areas of applications not addressed so far. These include applications in electronic lighting controls, miniature motor controls, high precision power supplies and automotive electronics. Additionally, with SOI, telephone and telecommunications circuits are expected to reach broader application areas.

2.3
Conclusion

To address the wide variety of possible applications for Power Integrated Circuits, ranging from computers to home appliances to fluorescent lamps, a number of process technologies have been developed. Due to the maturity of bulk silicon processing techniques most of the development has been in this area. However, dielectric isolation and more recently Silicon-On-Insulator technologies are promising significant advantages both in terms of performance and cost. Several developments in new devices such as LIGBTs and Trench DMOS have also addressed the cost issue.

In bulk and DI technologies the complexity of PICs is growing but they are mostly confined to custom chips for special applications. To address this market, computer aided design methodologies are being utilized to increase flexibility and turn around time. With this trend, PICs are expected to grow significantly in the coming years.

Acknowledgement

I wish to thank the members of the Microelectronic Devices department of Philips Laboratories for direct and indirect contribution to this paper and Barry Singer for his continued support and foresight in this field. Special thanks are due to Ted Letavic, Steve Merchant and Emil Arnold for reviewing the manuscript.

References

1. Rumennik V (1985) Power Devices are in The Chips, IEEE Spectrum, pp 42–48
2. High Voltage Integrated Circuits, (ed) Baliga BJ, IEEE press 1988
3. Long EL and Frederiksen TM (1971) A High Gain 15-W Monolithic Power Amplifier With Internal Fault Protection, IEEE J Solid-State Circuits, vol SC-6, pp 35–44
4. Murari B, Cini C and Prestileo V (1974) A High-Power Hi-Fi Monolithic Amplifier, IEEE Trans. Broadcast Telev. Receivers, vol BTR20, pp 311–321
5. Andreini A, Contiero C and Galbiati P (1986) A New Integrated Silicon Gate Technology Combining Bipolar Linear, CMOS Logic, and DMOS Power Parts, IEEE Trans on Electron Devices, pp 2025–2030
6. Ludikhuize AW (1991) A Versatile 700–1200 IC Process for Analog and Switching Applications, IEEE Trans on Electron Devices, pp 1582–1589
7. Mukherjee S, Amato M, Wacyk I and Rumennik V (1987) LDMOS and LIGTs in CMOS Technology for Power Integrated Circuits, IEDM Tech Digest
8. Appels JA and Vaes HMJ (1979) High Voltage Thin Layer Devices (RESURF devices), IEDM Tech Digest, pp 238–241
9. Stupp EH, Colak S and Ni J (1981) Low Specific On-Resistance 400 V LDMOST, IEDM Tech Digest, p 426
10. Mukherjee S, Amato M and Rumennik V (1987) Influence of Device Structure on the Transient and Steady-State Characteristics of LIGT, Electrochemical Soc Spring Meeting, Abstract vol 87-1, pp 158–159
11. Robinson AL, Pattanayak DN, Adler MS, Baliga BJ and Wildi EJ (1985) Lateral Insulated Gate Transistor With Improved Latching Characteristics, IEDM Tech Digest, pp 744–747
12. Sin JKO, Salama CAT and Hou LZ (1986) The SINFET-A Schottky Injection MOS-Gated Power Transistor, IEEE Trans Electron Devices, vol ED-33, pp 1940–1947
13. McArthur D, Mullen R, High Voltage MOS, Transistor Having Shielded Cross-over Path For a High Voltage Connection Bus, US Patent # 5,040,045
14. Elmoznine A, et al. (1990) The Smart Power High-Side Switch: Description of a Specific Technology, its Basic Devices, and Monitoring Circuits, IEEE Trans Elec Dev, pp 1154–1161

15. Wong SL, Venkitasubrahmanian S, Kim MJ, and Young JC (1992) Design of a 60-V 10A Intelligent Power Switch Using Standard Cells, IEEE Journal of Solid-State Circuits, vol 27, No 3, pp 429–432
16. Ueda D, Takagi H and Kano G (1985) IEEE Trans. Electron Devices, ED-23,2
17. Mukherjee S, Kim M, Tsou L, Wong S and Young JC (1991) A High Current Power IC Technology Using Trench DMOS Power Device, IEDM Tech Digest, pp 145–148
18. Wong SL, Kim MJ, Young JC and Mukherjee S (1991) A Scaled CMOS-Compatible, SMART Power IC Technology, ISPSD, pp 51–55
19. Wildi EJ, Chou TP (1984) A 500 volts Junction-Isolated Bimos high voltage IC, Proc of Electro., session 25, paper #4
20. Bean KE and Runyan WR (1977) Dielectric Isolation: Comprehensive, Current and Future, J Electrochemical Soc, vol 124, 5c
21. Merchant S, Arnold E, Baumgart H, Mukherjee S, Pein H and Pinker R (1991) Realization of High Breakdown Voltage (> 700 V) in Thin SOI Devices, ISPSD, pp 31–35
22. Qin Lu, Ratnam P, Andre C, Salama T (1991) High Voltage Silicon-On-Insulator MOSFETS, ISPSD, pp 36–39
23. Partridge SL (1986) The Current Status of Silicon-On-Insulator Technologies, IEDM Tech Digest, pp 428–430
24. Ohashi H, Ohura J, Tsukakoshi T and Simbo M (1986) Improved Dielectrically-Isolated Device Integration by Silicon-wafer Direct Bonding(SBD) Technique, IEDM Tech Digest, pp 210–213
25. Mumola PB, Gardopee GJ, Feng T, Ledger AM, Clapis PJ and Miller PE (1993) Plasma-Thinned Silicon-On-Insulator Bonded Wafers. Proc of the Second International Symposium on Semiconductor Wafer Bonding: Science Technology and Applications. (ed) MA Schmidt, T Abe, CE Hunt, H Baumgart, pp 410–421
26. Arnold E, Merchant S, Amato M, Mukherjee S and Pein H (1992) Comparison of Junction-isolated and SOI High Voltage Devices Operating in Source Follower Mode, Proc ISPSD, pp 242–243
27. Pein H, Arnold E, Baumgart H, Egloff R, Letavic T, Merchant S and Mukherjee S (1992) SOI High Voltage LDMOS and LIGBT Transistors with a Buried Diode and Surface P-Layer, IEEE International SOI Conference Proceedings, pp 146–147
28. Nakagawa A, Yasuhara N and Baba Y (1990) New 500 V Output Device Structures for Thin Silicon Layer on Silicon Dioxide Film, Proc ISPSD, pp 97–101
29. Stoisiek M, Oppermann KG, Schwalke U and Takacs D (1995) A Dielectrically-Isolated High-Voltage IC-technology for Off-line Applications, ISPSD, pp 325-329
30. Beason JD (1977) High Voltage Dielectric Isolation SCR Integrated Circuit Process, IEDM Tech Digest, pp 178–180
31. Weston HT, Becke HW, Berthold JE, Gammel JC, Hartman AR, Kohl JE, Shibib MA, Smith RK and Wong YH (1982) Monolithic High Voltage Gated Diode Cross- Point Array IC, IEDM Tech Digest, pp 85–88
32. Koji Kawamoto, Tsuyosi Tanaka, Sigeru Sugayama, Hideaki Ohki, Shun Terada and Masanori Kamia (1991) High Voltage Video Amplifier IC for Colour television, ISPSD Proceedings, pp 231–236
33. Hideki Miyazaki, Naoki Sakurai, Kenichi Onda, Tomoyuki Tanaka, Mutshuhiro, Masayuki Wada and Hitoshi Matsuzaki (1991) A Novel High Voltage Three-Phase Monolithic Inverter IC with Two Current Levels Sensing. ISPSD Proceedings, pp 248–253
34. Merchant S, Arnold E, Baumgart H, Egloff R, Letavic T, Mukherjee S and Pein H (1993) Dependence of Breakdown Voltage on Drift Length and Buried Oxide Thickness in SOI RESUF LDMOS Transistors, Proc. ISPSD, pp 124–128
35. Amato M, Bruning G, Mukherjee S and Wacyk IT, Power Integrated Circuits, Philips Tech Rev 44, No 8/9/10, pp 310–320
36. Arnold E (1994) Silicon-on-Insulator Devices for High Voltage and Power IC Applications, J Electrochem Soc, vol 141, No. 7, pp 1983–1988
37. Rumennik V (1992) A 1200 volt BiCMOS Technology and its Applications, ISPSD, p 322

38. Ajit JS, Dan Kinzer and Niraj Ranja (1993) 1200 V High-Side Lateral MOSFET in Junction-Isolated Power IC Technology Using Two Field Reduction Layers, ISPSD, pp 230
39. Simpson MR, Gough PA, Hsieh FI, Rumennik V (1985) Analysis of the Lateral-Insulated Gate Transistor, IEDM Tech Digest, pp 740–743
40. Huang YS and Baliga BJ, Extension of RESURF Principle to Dielectrically-Isolated Power Devices, ISPSD 1991, pp 27

CHAPTER 3

Smart Discrete Technologies

J. Tihanyi

3.1 Introduction

"Smart Discretes" are the newest classification of semiconductor devices. They are primarily dedicated to application in automotive systems which are controlled by microprocessors. It is dificult to define essential differences between "Power ICs" and "Smart discretes", since both are integrated circuits whose function is power control. "Smart discretes" are defined in this chapter as microcontoller compatible power switches, which are able to protect themselves and their environment against disturbances such as shorted load, over-voltage, and over-heating. They are a modern form of relay, with built-in intelligence providing the required protection functions.

3.2 History

Electromechanical relays have proven to be adequate for the switching of electric power since the middle of the last century. They did not have any serious competition until the advent of semiconductor devices. This is not to say that they still do not enjoy a large market share which will extend into the future, only that they will come under increasing pressure from semiconductor switches.

At the inception of the semiconductor age, there were only discrete devices. The very first were bipolar transistors made of germanium. They were excellent, but one could not use a radio implemented with them on the beach since the temperatures there often exceeded those allowed for proper operation. The introduction of devices fabricated from silicon alleviated these problems and they operated reliably in environments with temperatures up to 150 degrees Celsius or higher.

A very important innovation was the Thyristor or "Silicon Controlled Rectifier". It was in contrast to the bipolar transistor, an AC switch, which was quickly developed for heavy machines such as those used in electric trains, automated factories, and power plants. They are extremely rugged. In fact they are so rugged that in the case of a short circuit in a power circuit, normal fuses can be used for protection.

The silicon bipolar transistors were very small at first and used primarily in HF applications. There were, however, also some power bipolar devices developed for the output stage of audio equipment. In the sixties, the Integrated Circuit began its evolution and MOS appeared. A result of the very fast developments within the semiconductor industry is that microelectronics has become a determining factor in the progress of human society. The major part of semiconductor production is made up of Integrated Circuits in whatever complexity. There is, however, a surprisingly large market for novel discrete devices which fulfill requirements not presently economically realisable in Integrated Circuit form.

In today's power applications, thyristors are used for switching high AC currents and voltages, while high DC voltages and moderate or small currents are best handled with discrete darlingtons, vertical DMOS-FETs or IGBTs. They are not, in contrast to thyristors or mechanical relays, inherently rugged. There is an old saying (one known interpretation of the Parkinsons-law) that a MOS or bipolar power transistor reliably protects the fuse which was intended to protect the transistor.

Thus, we come to the actual subject of "Smart Power". Smart power devices not only switch and control electric power but must be rugged enough to withstand fault conditions and continue functioning normally upon removal of the fault. The need for such power switches began to arise in the early eighties and has expanded continuously.

That these devices must be "Smart" also arises from the fact that almost all digital systems today require compatibility with a microprocessor or other dedicated control circuit. The fields of application are smart switches in automotive and automation electronics with DC supply voltages and control of heavy machines for AC networks. A growing need is also seen for power IC applications in a variety of switch mode power supplies and electrical ballasts for luminescent lamps.

This paper addresses the subject of relatively low voltage DC power switches. The topic of heavy power and AC motor drive using IGBTs belongs to the family of "Intelligent Modules" and will not be discussed here. In addition, the power ICs for Switched Mode Power Supplies and lamp ballasts fall outside of the scope of this chapter.

The field of "intelligent" or "smart" power devices is in a continuous state of development. The degree of intelligence covers a range, up to and including, output voltage regulation in switch mode power supplies or monolithically integrated microcontrollers. Others, dedicated to simpler tasks, are only moderately intelligent. They contain integrated functions for self-protection and some diagnostic capability, but no more. These are the "Smart Discretes" as defined earlier. The first generation of smart discrete devices has mainly found application in automotive and industrial systems with fourteen or twenty-four Volt DC power sources. They function as current switches without tightly defined analogue parameters. That is, they essentially fulfill the requirements placed on electromechanical relays, without the disadvantages of the associated limited life, arcing, and slow switching rates.

There is some confusion in terminology which arises due to assembly options. If a "Smart Discrete" chip is encapsulated in a housing similar to an electromechanical relay, it is designated as a semiconductor relay. When it is encapsulated in a TO220 five pin or SOT 223 plastic package, then it is sold as a Power IC or in some cases again as a semiconductor relay. Also, if the plastic encapsulated chip is assembled in a relay case, it is again designated a semiconductor relay.

Now, following this subjective historical swing by the author, let us examine the technical issues.

3.3 Requirements for Intelligent Power Devices

Let us begin by surveying the performance of "pedestrian" power switches which, until now, have widely been used for switching DC currents. The well-known mechanical relay has served the electronics industry for some time and even now finds numerous applications where there is not a better solution, particularly when price is the main issue. It has very low on-resistance, having metallic elements, and can be protected during short circuit conditions by a simple fuse. There are high-side and low-side configurations Fig. 3.1. The relays are available with various circuitry; e.g. they have single or multiple contacts. A high-side configuration is defined as a switch which connects the load to the "high" or positive terminal of the battery, while the load is electrically connected to the "low" terminal or chassis of an automobile. This type of switching has been prevalent in the automotive industry for decades because of the reduced corrosion in the standby mode: there is no voltage across the load if the switches are in the "off" state. The relays used in automobiles have a meaningful drawback: they require approximately 200 mA coil current when activated with a battery voltage of 14 volts; a power dissipation of roughly 3 watts in the activated control circuit alone. Other disadvantages are the slow switching rates and the limited life time due to degradation of the mechanical parts (e.g. the contacts).

The vertical power MOSFET is the only semiconductor alternative to the mechanical relay in the voltage range of 14 to 24 volts available today. A decade earlier, bipolar power transistors found acceptance as switching devices but have been swept from the scene by the advent of power MOSFETs. The remainder of the bipolar spectrum of ICs are still on the market, but new solutions favor MOSFETs. The primary advantage of a power MOSFET is that it can be held in the "on" state without input current and this means that there is essentially no power dissipation in the control circuit. The only meaningful heat loss results from the on-resistance of the switch. The bipolar transistor, in contrast, requires approximately 5% of the load current as a control current and has a built-in voltage drop. The essential features of a bipolar and a MOS power switching transistor for voltages below 50 Volts is contrasted in Fig. 3.2. The chip areas for the transistors are comparable. The built-in voltage drop of the bipolar device excludes a reduction in the power dissipation; even if the chip size is increased, and the power dissipation is roughly proportional to the load current. This is

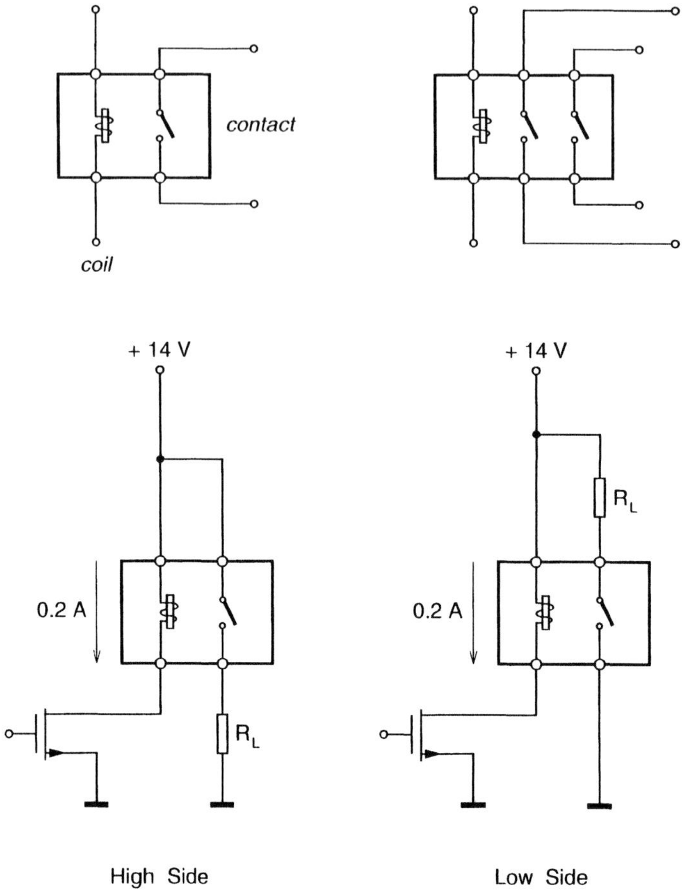

Fig. 3.1 High-side and low-side switches with relays

true even though the heat dissipation of the control circuit can be reduced by using a bigger chip for the same load current.

The power MOSFET does not need control power and its on-resistance is inversely proportional to the chip area. It has no built-in voltage drop and is truly ohmic. This means, if one is willing to use a larger silicon chip, he can freely reduce the overall load dissipation of a MOSFET switch. Since the cost of processed silicon is dropping continuously on the world market, this property of power MOSFETs alone makes them very attractive in the development of future economical solutions in the field of power microelectronics. In fact, practically all new applications being developed in the laboratories today are MOS based solutions. In addition, the potential for handling more current per square mm using the photolithographic techniques of sub-micron technologies is very

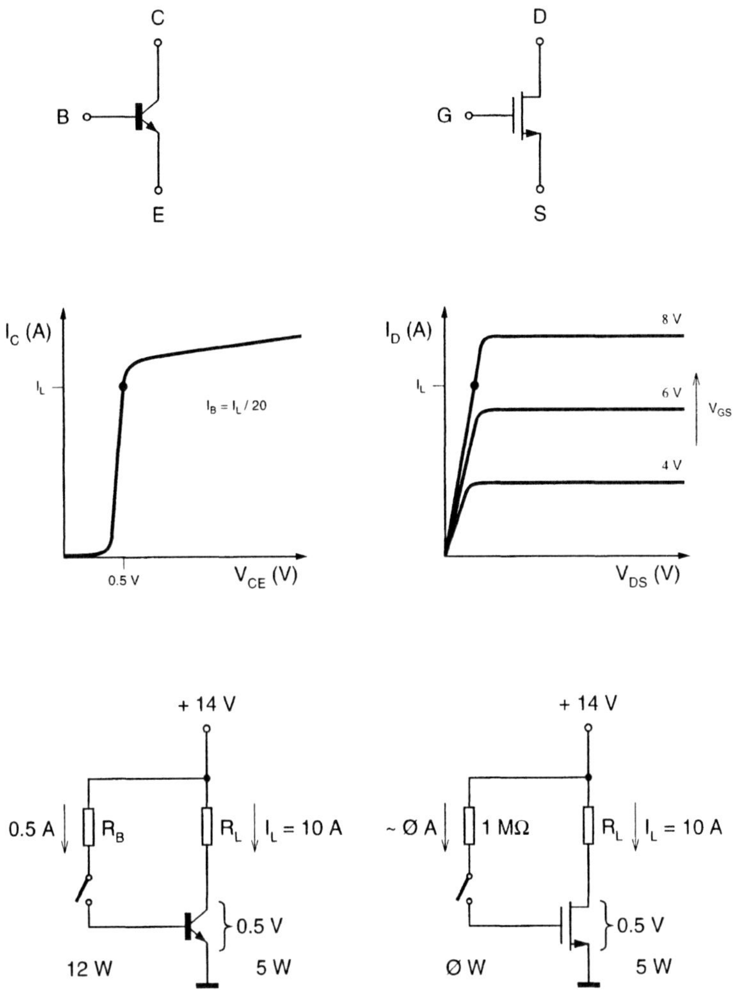

Fig. 3.2 Power MOSFETs require no power from controller

promising. New semiconductor power switches handling high currents, well in excess of 10 Amperes, and with prices lower than those for equivalent mechanical relays will appear on the market in the near future. Generally, it can be said: if a MOSFET based power switch dissipates less than 2 watts, it will win out in a design over a relay when it is competitively priced. Semiconductor power devices have an even larger edge if parameters other than price and power dissipation are considered important (e.g. high switching speeds as in ABS systems or engine controls).

The next question to be asked after the previous considerations is "what can be made with the power MOSFET?".

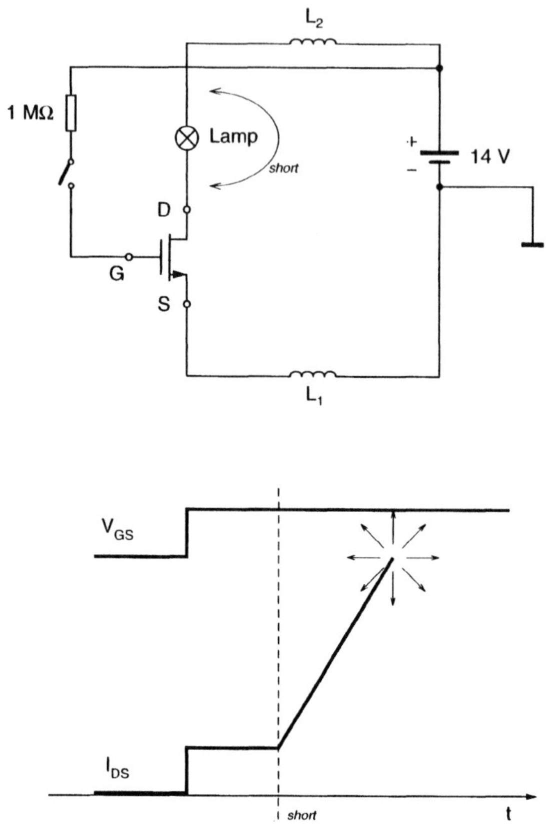

Fig. 3.3 Low-side switch with power MOSFET

Let us first discuss the low side circuit configuration using a MOSFET. To facilitate discussion, a circuit to control a lamp, is shown in Fig. 3.3. One need only apply a positive potential on the gate and a current flows in the load (designated as point A on the DC characteristics). No current flows when 0 volts is on the gate (point B). In order to portray a realistic situation, the stray inductances of the wiring, L_1 and L_2 are included. A shorting switch is included to allow simulation of a foreign object, such as a screwdriver, accidentally falling onto the circuit during repair work under the hood of an automobile.

During a shorted condition, the current begins to rise according to the expression $di/dt = Vb/(L_1 + L_2)$ and continues to rise until the power MOSFET is destroyed. It perhaps then continues to smoulder or burn (a bipolar device would, by the way, have done the same). To avoid such an unpleasant event, some intelligence has to be incorporated in the MOSFET, i. e. it must be "Smart". The device must detect that a short circuit condition has occurred and must shut itself off before the temperature at the hottest point exceeds the dangerous level of approximately 250 degrees Celsius. Consideration as to how to alleviate this problem in the simplest way, led to the birth of the Tempfet (R).

3.4 Low Side Smart Discretes

The Tempfet (R) (*Tem*perature *P*rotected *FET*) idea had its origin in a thyristor. For a number of years we have produced a very sensitive optically coupled triac which consists of two anti-paralleled optothyristor chips. They are specified up to 150 degrees Celsius and above that they are also on, even if no light is applied. The Tempfet (R) concept was to use an opto-thyristor chip with a defined turn-on temperature of 150 degrees Celsius as a sensor, glue it to the top of a power MOSFET, connect the anode to the gate electrode and the cathode to the source. If the thyristor temperature rises above 150 degrees Celsius, the MOSFET's gate is shorted to its source and the device is subsequently shut off. The Tempfet (R) was introduced in the middle 80s, was successful on the market, and has been

Fig. 3.4
TEMFET structure

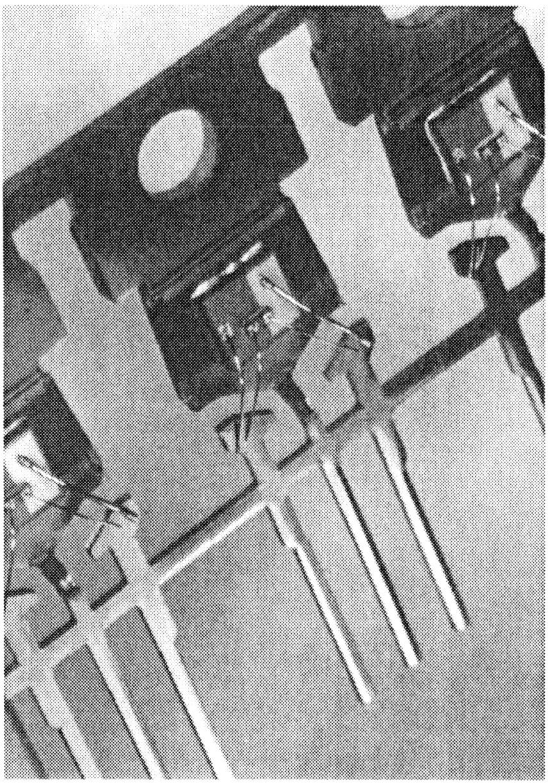

sold in substantial quantities up to this time. Fig. 3.4 is a photomicrograph of a Tempfet (R) which shows its construction. The small thyristor chip is electrically isolated but tightly thermally coupled to the MOSFET base chip. Should the surface temperature of the base chip rise (as in the case of high power dissipation due to a short circuit condition), the thyristor or "sensor chip" is also heated to this temperature after a time delay, and acts to switch off the MOSFET. It should be noted, however, that the temperature rise of the base chip must not be too fast, so that it remains in an area of safe operation (see Fig. 3.5). Examining the case when the gate voltage is greater than 10 Volts, the short circuit current is high, the heating is fast, and the sensor chip cannot follow the temperature rise

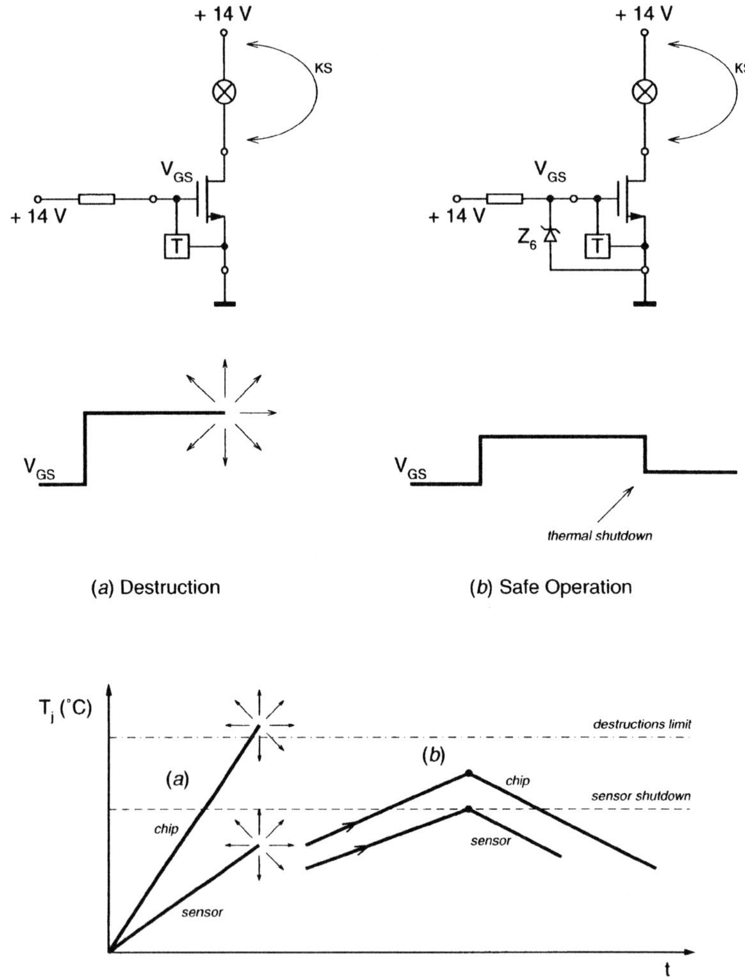

Fig. 3.5 Safe operation of a TEMFET under short circuit condition

of the base chip with a sufficiently small time delay, it is to be expected that the device will be destroyed. To achieve safe operation, the gate voltage and consequently the short circuit current must be lower and the rate of increasing temperature moderate. The sensor chip then reaches 150 degrees Celsius prior to the MOSFET temperature becoming too high and the device turns off and is safe.

This very simple Tempfet (R) can be termed a smart discrete device with the lowest degree of intelligence. It can only protect itself and the load circuit against short circuit conditions within limited conditions. In order to make it more intelligent, or "smarter", additional circuitry is required which is not implemented in the case of a Tempfet (R). For general use, Siemens suggests the circuit shown in Fig. 3.6.

The active zener diode between the drain and gate provides complete protection against over-voltage spikes regardless of wire length. It is of note that the ground reference of the driving circuit need not be connected to the source! The source pin sees an enormous voltage spike when the device is shorted, especially if the source to battery wire is very long. If the driver circuit is improperly connected to the negative battery terminal, it can be damaged or destroyed. As an illustration of what may take place, let us refer to Fig. 3.7, where the wave forms of source, drain, and gate voltages are shown under short circuit condition two (the short occurs when the device is in the "on" state). This is in contrast to short condition one which is defined as the event of switching the device to its "on" state when the short has already occurred. The short condition two is the more dangerous and obviously will occur most often in practice according to

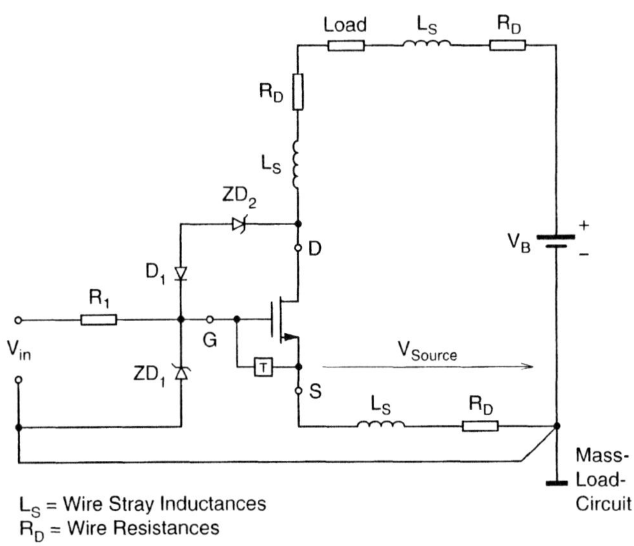

Fig. 3.6 Application diagram of a TEMFET under realistic operating conditions with long wires

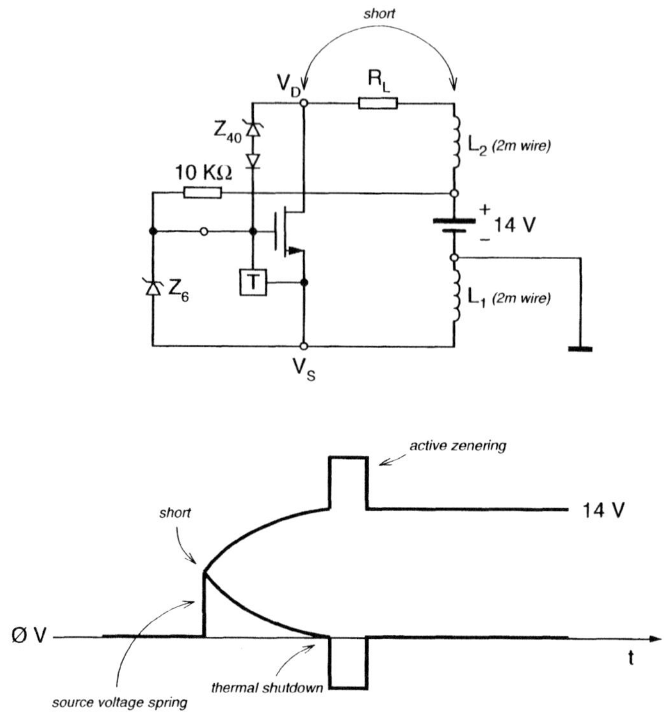

Fig. 3.7 Behavior of a TEMFET in the "on" state when a short circuit occurs

Parkinson's law. At the onset of a short circuit condition, the source voltage rises immediately to a positive value, depending on L_1 and L_2, and the current begins to rise. Following a transient period, the Tempfet (R) conducts a current which corresponds to the 6 volts gate voltage and it becomes hot. As it reaches the sensing temperature, it switches off and the voltage spikes on the inductances are clamped by the active zener diode. The voltage step at the source, which we spoke of above, is readily observed and has a destructive effect on the driving circuitry if it is not properly protected.

The application circuit of Fig. 3.6 allows a Tempfet (R) to be protected against system disturbance, even short circuit conditions and over-voltage spikes due to stray inductances. Tempfet (R) are flexible in their application and able to achieve the thermal limits of the device. The only cosmetic error of the device is that the gate voltage must be limited to a value lower than that for which a minimum on-resistance could be realized. The data sheet suggests a zener diode of 6 to 8 volts although optimum on-resistance would be achieved with 10 to 15 volts.

The development team began further improvement of the Tempfet (R) structure shortly after its introduction. It was not clear at its inception what level of acceptance these devices would find and consequently it was not known which direction the development should take. The resulting device is the Hitfet (R) *H*ighly *I*ntegrated *T*emperature Protected *FET*). The Hitfet (R) is essentially a Tempfet (R) with finer dimensions and with additional functions monolithically integrated into the base chip. It still has the old reliable thyristor chip bonded on top of and electrically isolated from the power MOSFET base chip. However, now the base chip is an integrated circuit. It has as a feature a finer cell pitch and an on-resistance of 0.25 $\Omega \cdot mm^2$. Additionally, it is constructed with a current limiting circuit which allows the current limit to be adjusted over a wide range by the user. Furthermore, there is a simple temperature sensor integrated on the chip which prohibits the chip temperature exceeding the danger level as illustrated in Fig. 3.8. The Hitfet (R), in contrast to the Tempfet (R), does not require additional external components to protect it against voltage spikes since the active zener function is also integrated. The only drawback is that the gate resistor also is integrated and thus determines the maximum switching speed of the device. The Tempfet (R) does not have this

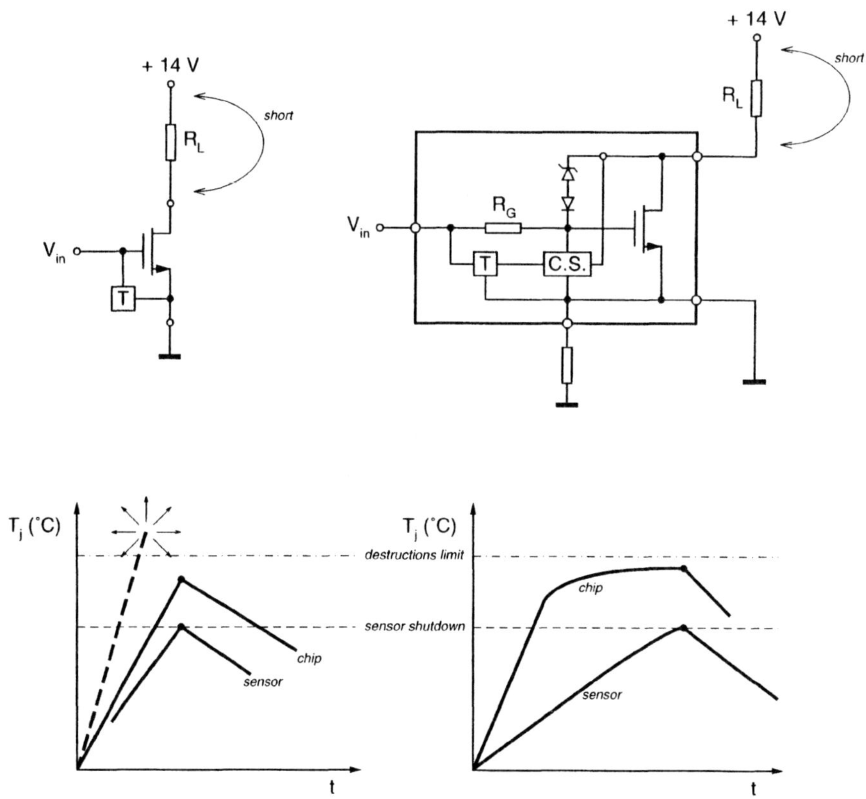

Figure 3.8 Comparison of TEMFET HITFET functions

restriction as its gate electrode is externally accessible and can be driven arbitrarily fast.

The questions could be posed: "Why is a two chip solution required at all?, What is the advantage of a chip-on-chip construction? and Why is the temperature sensing thyristor chip not integrated monolithically?". The reason is that any possible circuit having the function of the thyristor (latching at the sensing temperature of exactly 150 degrees Celsius and having an "on" impedance well below 100 Ohms) would require some voltage. (This voltage can neither be derived from the drain voltage since it is low in the "on" state nor from the input voltage since it would draw current and eliminate the advantage of the power MOSFET. The present, economical implementation of the Hitfet (R) requires essentially no input current for holding the device in the "on" state. The property of a normal MOSFET, that no input current is required, is fulfilled by both the Tempfet (R) and Hitfet (R) and can up to now only be achieved with a two chip solution and utilization of a thyristor to determine the exact temperature shutdown threshold.

Fig. 3.9 Photomicrograph of a HITFET

The structure of a Hitfet (R) is shown in Fig. 3.9. It is clear that the power MOSFET based chip is not as simple as that of the Tempfet (R). It is however simple enough to be classified as a "Smart Discrete".

Excellent device families from other companies, with a low side power switch configuration similar to Tempfet (R) and Hitfet (R) devices, have appeared on the market in the meantime. One such is the TOPFET (R) (*T*emperature and *O*ver voltage *P*rotected *FET*) family, manufactured by Phillips. They are very clever smart discretes with on-resistance down to 28 mOhms and similar performance to the Tempfet (R) and Hitfet (R) devices according to the published data sheets. They are, in contrast to the Tempfet (R) and Hitfet (R) monolithic devices. The TOPFET (R) family is offered in three and five pin packages. The three pin device derives its power from the input pin and consequently draws a current, which is not negligible in comparison to the Tempfet (R) or Hitfet, (R) and also has limitations on its switching speed.

The five pin TOPFET (R) has direct access to the gate electrode. It derives its operating voltage from an additional low voltage power supply and provides a status output in the case that a short circuit condition occurs. The primary difference with respect to the Hitfet (R) is that the switched current level is a built-in function, whereas in the Hitfet (R) the level can be adjusted by an external resistor.

Another family of low side switches for high currents and with similar performance to the Hitfet (R) and TOPFET (R) devices has recently been announced by International Rectifier.

All of these smart discretes switch relatively high currents; i.e. they are essentially large vertical MOSFETs which have some level of built-in protection in the case of short circuit conditions in order to avoid damage in the application or to the device itself. The technology for manufacturing them is a simple self-isolation technology. This is an outgrowth of the basic VMOS technology, where lateral devices have been added to the vertical power MOSFET. This self-isolation technology also finds application in numerous types of high side switches and will be discussed, in detail, later.

Junction-isolation technology is also used to fabricate low-side switches. There are companies with strong bipolar power IC tradition, such as ST, Siemens, and Motorola which sell low side switches with higher and lower complexity, mainly in multiple channel configurations assembled in packages with larger pin counts. These are either devoted to lower current levels (generally less than 2 Amperes) and are highly complex, having not only protective functions but also diagnostic capabilities. They are then not designated as smart discretes as we have defined them, but are power ICs and will be discussed under that heading. We will now progress to the high side smart discrete devices.

3.5 High Side Smart Discretes

Another option, generally used in automotive electronics, is the high side switch. The first implementation was a mechanical relay (see Fig. 3.10 a). A replacement

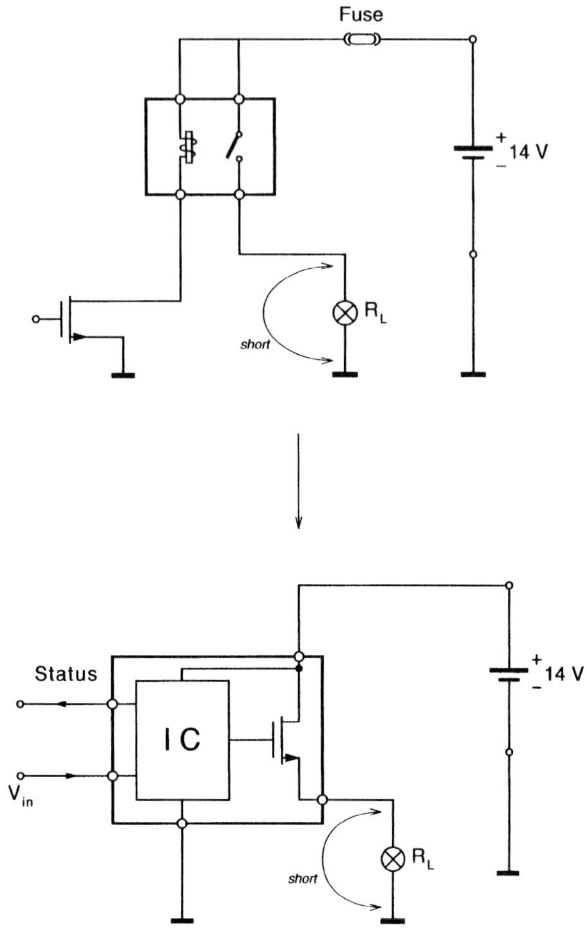

Fig. 3.10 Smart high-side switch as an electromechanical relay replacement

is the microcontroller compatible smart discrete (see Fig. 3.10b) which consists of a main switch and some intelligence, preferably monolithically integrated. A desired high side switch obviously has full protection against fault conditions, a built-in fuse function, can survive all system disturbances, and has no negative influence on the controller even with a short across the load circuit. Furthermore, it is advantageous that the high side switch provide some means of diagnostic feedback indicating the type of failure in the load circuit.

The simplest way to achieve this would be to take a p-channel Tempfet (R) which is essentially a high side switch. However, a p-channel power MOSFET has about twice the on-resistance of a n-channel device with the same area and is therefore prohibitively expensive for the intended applications. Due to the cost factor alone, n-channel devices are also employed as high side switches.

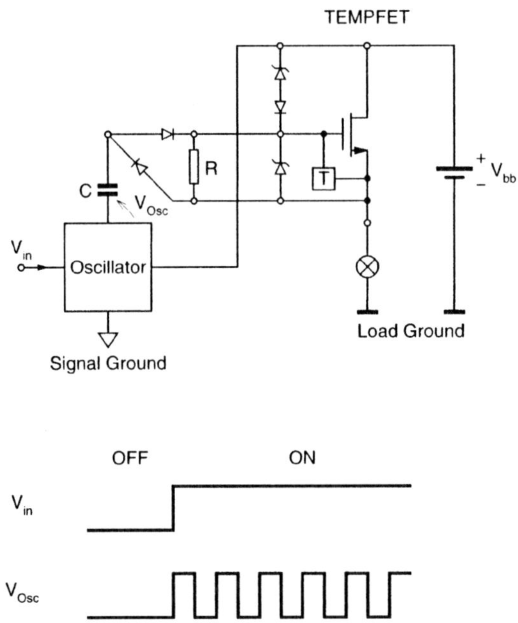

Fig. 3.11 Simplest circuit for high-side switching with a power MOSFET (no short circuit protection)

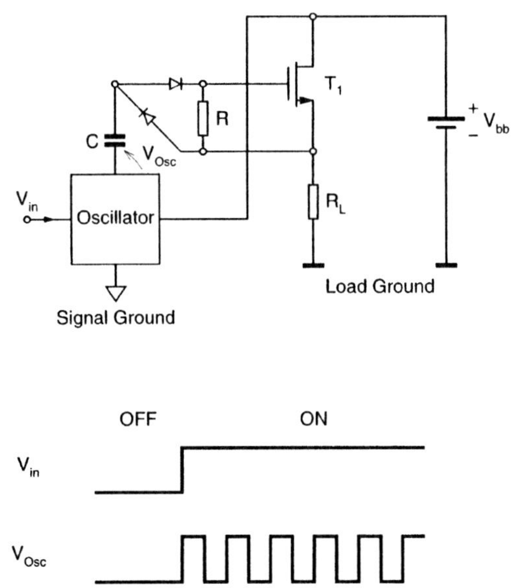

Fig. 3.12 Short circuit protected high-side switch using a TEMFET

To realize a high side switch with a n-channel power MOSFET, some basic functions must be added. A minimum configuration of a high side switch is shown in Fig. 3.11 and, as can be seen, includes a n-channel MOSFET and a charge pump. While a n-channel MOSFET requires a positive gate-to-source voltage to be in the "on" state, it must be accompanied by a charge pump circuit. This charge pump circuit consists of a controlled oscillator, at least one capacitor, and at least two diodes as it is shown in Fig. 3.11. Turn-off of the MOSFET is accomplished by discharging the gate capacitance through the resistor R and inhibiting the oscillator function. In this primitive circuit, no protective functions are implemented. Realization of a short circuit or overload protection is possible with the application of the circuit shown in Fig. 3.12. There, the earlier simple MOSFET is replaced with a Tempfet (R) or Hitfet, (R) which already has the protective functions. In the "on" state the charge pump circuit

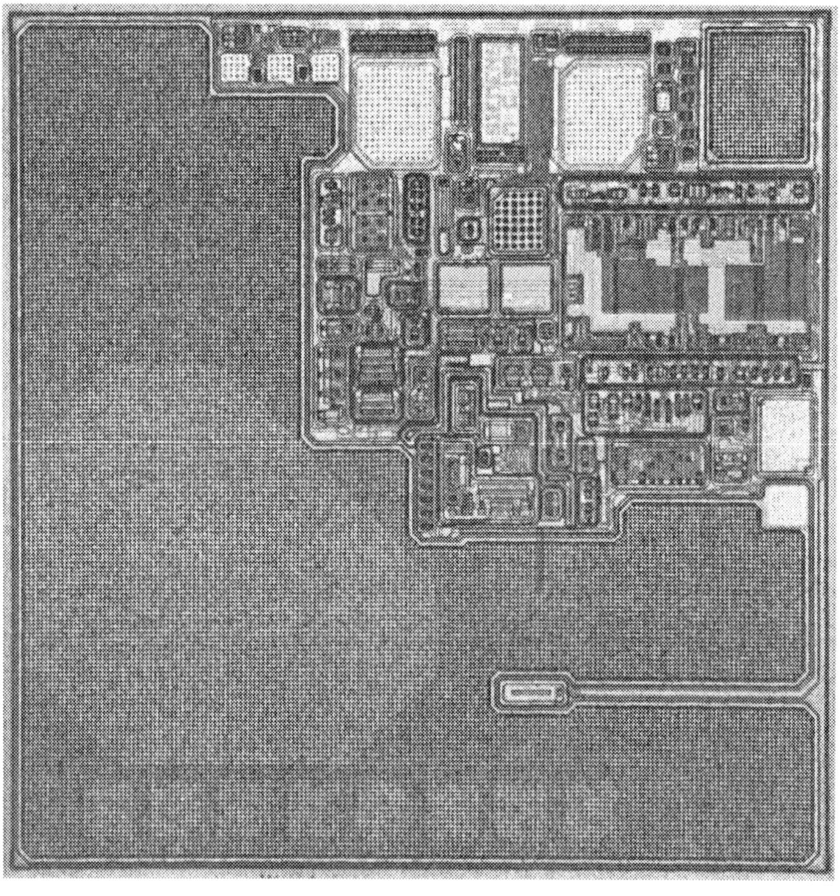

Fig. 3.13 BTS 426, a Smart SIPMOS high-side switch

provides the positive gate voltage, in the "off" state the gate capacitance is discharged through R, and in a short circuit condition the thyristor shorts the gate electrode to the source, thus switching the device off. The zener diode Z_1 limits the power in the device during a short circuit condition. The active zener function provides protection from over-voltage spikes and transients arising from switching inductive loads. The switching speed is dependent on the capability of the charge pump and the relationship of R to the gate and to source capacitance. Obviously further circuit blocks must be added if one wants to sense an open load in the "on" state or to detect if the device is in thermal shut-down due to a short circuit condition. When all of the functions are monolithically integrated and the circuit is not too complex, we have a smart discrete or, if the complexity increases, a smart IC.

Fig. 3.13 illustrates a novel intelligent high side switch which, although a smart discrete, borders on being a smart IC. It is a photomicrograph of the BTS426. The power MOSFET occupies the major part of the chip. The chip area is approximately 7 mm^2 and the on-resistance is less than 60 mΩ. The block diagram is shown in Fig. 3.14. In comparison to the simple high side switch of Fig. 3.12, there are additional function blocks such as over-voltage protection and open circuit and short to Vbb detection which increase its utility. The essential function blocks are common to both: the oscillator and capacitor C correspond to the charge pump of Fig. 3.14), the levelshifter/rectifier in Fig. 3.14 implements the function of the two diodes and R, and Z_1, Z_2, and D_3 correspond to the block "gate protection – current limiting".

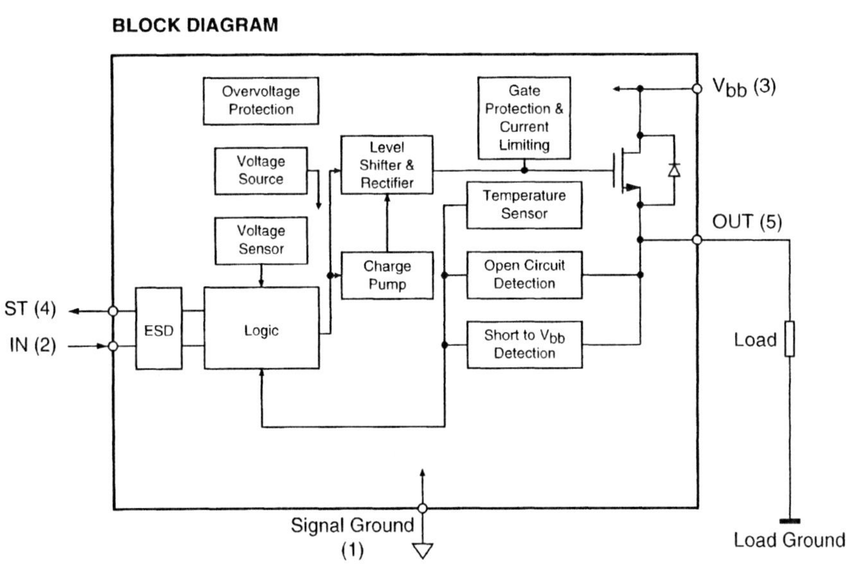

Fig. 3.14 Block diagram of a BTS426

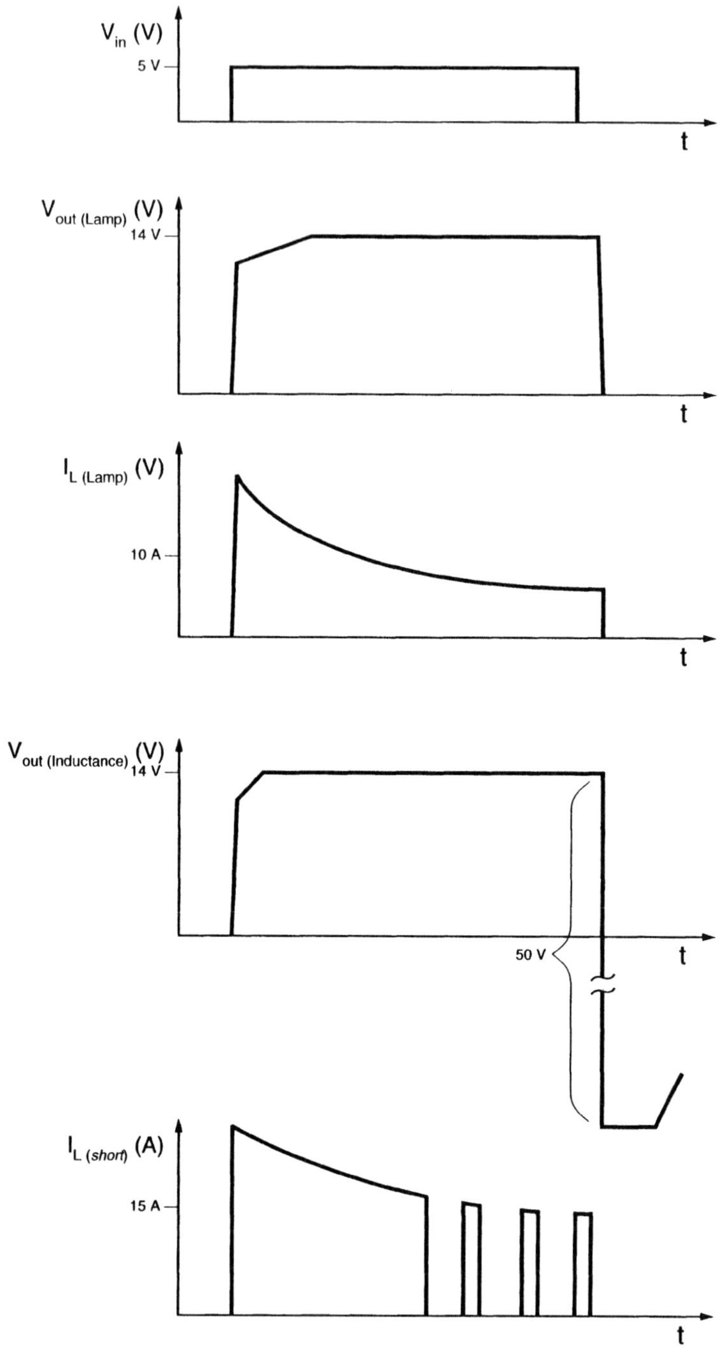

Fig. 3.15 Switching wave forms of a BTS426

Fig. 3.15 illustrates the behaviour of the BTS426 by showing the wave forms with a lamp, an inductance, and a short as the load conditions. The negative inductive clamp voltage is impressively high, the short circuit current is limited to approximately 15 amperes, and there is a thermal oscillation resulting, giving rise to an amount of heat which is dissipated due to the cooling of the device. The chip temperature of the device under short circuit conditions is approximately 160 degrees Celsius. Upon removal of the short, the device resumes normal operation. An open load detection is also built into the device.

Obviously, there are other manufacturers of high side switches which exhibit similar performance. The smart power high side switches from ST are well known. Motorola, Hitachi, International Rectifier, and others offer a range of power related products in the marketplace. The differences between the various devices does not essentially lie in their functionality but in the technology used to produce them.

Let us now briefly survey which processes are employed in the manufacture of smart discretes as well as power ICs.

3.6 Processing of Smart Power ICs and Smart Discretes

There are basically two main approaches today to constructing smart power for automotive and industrial applications. The one is the technology based on that used for old bipolar power ICs. The companies, which traditionally were in the field of bipolar devices, added during the 80s a power MOSFET to their earlier bipolar process. That this addition extended the available device palate to lateral CMOS at no extra cost, was even welcomed. To those companies and development teams belong such companies as ST, Motorola, and also within Siemens, the team working in the area of "Smart Power Technology" or SPT. The result is a mixed bipolar-CMOS-power MOS process with a large variety of devices. The evolution from bipolar IC to MOS-bipolar IC was retarded by the fact that bipolar oriented people first tried to prove that the bipolar power device is superior to the power MOSFET instead of acknowledging the advantages of the MOSFET. The transition to MOS was more or less complete by the end of the 80s. Today, there is no doubt that the future belongs to the power MOS based smart IC. In fact one type, the junction isolated device, has been dropped from the older bipolar IC technology.

The second approach for building power ICs is the self-isolating technology. Those companies who started with power MOSFETs have added some logic functions in order to give them a degree of intelligence and increase their utility. This includes such companies as Siemens, Phillips, and more recently, International Rectifier. They have extended their power MOSFET processes by adding additional steps and utilizing the epitaxial wafers, inherent to these devices, without the need for isolated islands. They are able to build CMOS logic devices and high voltage lateral devices simultaneously with the power MOSFET.

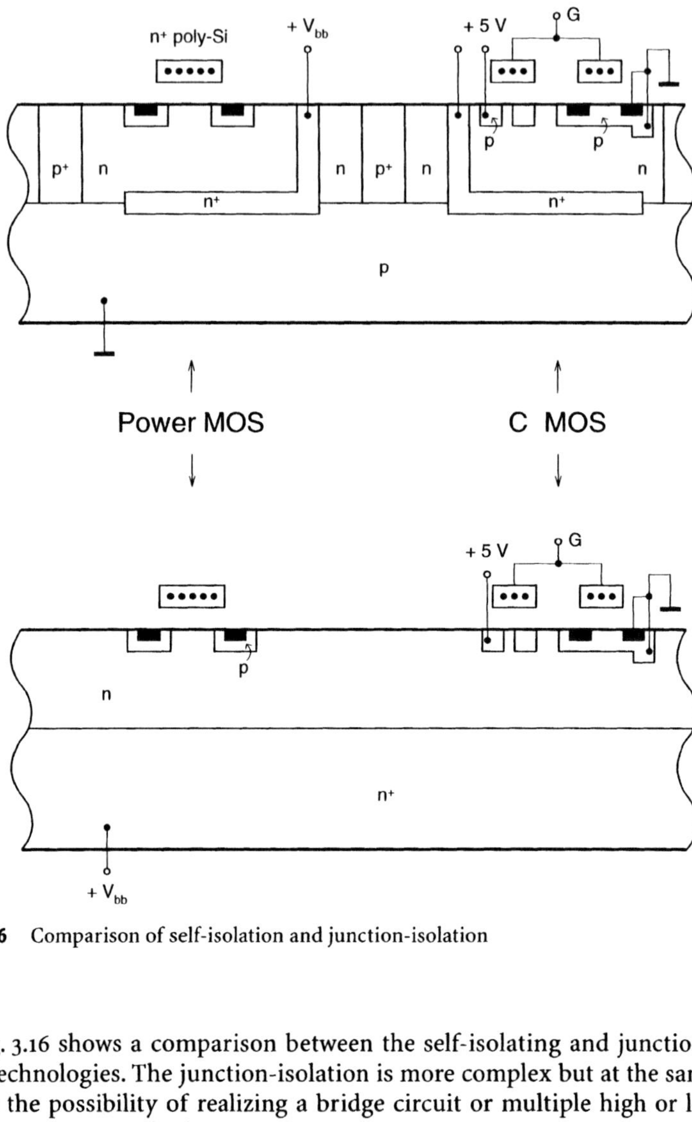

Fig. 3.16 Comparison of self-isolation and junction-isolation

Fig. 3.16 shows a comparison between the self-isolating and junction-isolation technologies. The junction-isolation is more complex but at the same time offers the possibility of realizing a bridge circuit or multiple high or low side switches. Due to its higher complexity, the on-resistance per unit area is not as low as that achieved using self-isolating technology. On the other hand, the self-isolating structures are simple, achieve the full power handling capability of the power MOSFET and are the preferred choice; particularly in the area of high-current, high-side switches. The low-side switches, Hitfets (R) and TOPFETs (R), are also fabricated using a self-isolating technology, thus demonstrating the flexibility of this approach.

A very interesting modification to the junction-isolation technology is the combination of it with a vertical power MOSFET instead of the power switches in isolated islands as seen in Fig. (3.16). A novel version of such a solution is designated "VIP" (*V*ertical *I*ntelligent *P*ower) and was introduced by ST.

The main issue here is that the power MOSFET has its drain on the back side of the chip, like a self-isolated structure, and thus achieves the conductance per unit area of the MOSFET while retaining the device palette of the junction-isolation technology.

All of these various technologies are now competing for application in automobile and industrial electronics. The smart discretes will definitely continue to build their market share in the foreseeable future, particularly by replacing more and more mechanical relays. They will become ever more cost effective and utile for the user. A very interesting trend in the field is represented by the saying "use silicon instead of a heat sink". This idea presents very attractive options and has found enthusiastic acceptance amongst the development team at Siemens.

3.7
Using Silicon Instead of a Heat Sink

To begin the discussion of this issue, let us begin by again noting the important characteristics of power MOSFETs. The drain current as a function of drain voltage is truly ohmic and for a given technology, the on-resistance is inversely proportional to the area of the device. This means that one can realize as low an on-resistance as desired if one is willing to pay the chip price. Furthermore, the size of a chip, for a given on-resistance, continues to decrease as a function of time due to application of VLSI innovation (e.g. trench technology). Today, devices which exhibit 20 $m\Omega \cdot mm^2$ with operating voltages to 60 volts are available. Devices with 10 $m\Omega \cdot mm^2$ have been reported. This has the significance that silicon chips will continue to become less expensive.

Simultaneous to the trend toward lower prices for silicon chips is the general use of surface mounting technology for system applications. The automatic insertion of components into printed circuit boards is fast, flexible and economical. There is however one drawback: the surface mounted power devices sitting on a printed circuit board are difficult to cool. A power dissipation level of roughly one Watt can be tolerated as this amount of heat is dissipated into the surrounding medium without an additional heat sink and subsequent costs. This is the point at which the trend toward cheaper power MOSFET silicon and the wish "do not heat the printed circuit board" come together. The solution is obvious: use a larger, lower on-resistance MOSFET to handle power when used in such applications as a more cost effective alternative to cooling a smaller, less expensive device. The rule of the thumb is very simply stated: for a given task, choose a MOSFET which does not dissipate more than one Watt under normal worst case conditions. One then asks "what if the possible conditions are not well defined or normal?". Then, a simple MOSFET is not a solution and one must select a smart device which can deal with conditions outside of the normal operating range.

The smart power SMD must be a smart discrete with somewhat more capability than seen in simple configurations. The most essential feature of an SMD smart power device is that it be capable of limiting the stress on printed circuit board traces to a non-destructive level. A Tempfet (R) or TOPFET (R) would not do the job because their reaction time is too long and they have too high a short-circuit current level. The SMD smart discretes must react quickly to a short-circuit condition. The short-circuit current must be tightly controlled to a level which does not exceed twice the current specified under normal operating conditions. The remaining functions (i.e. charge pump, over-voltage protection and diagnostic feedback) may be identical to those found in the usual smart discretes.

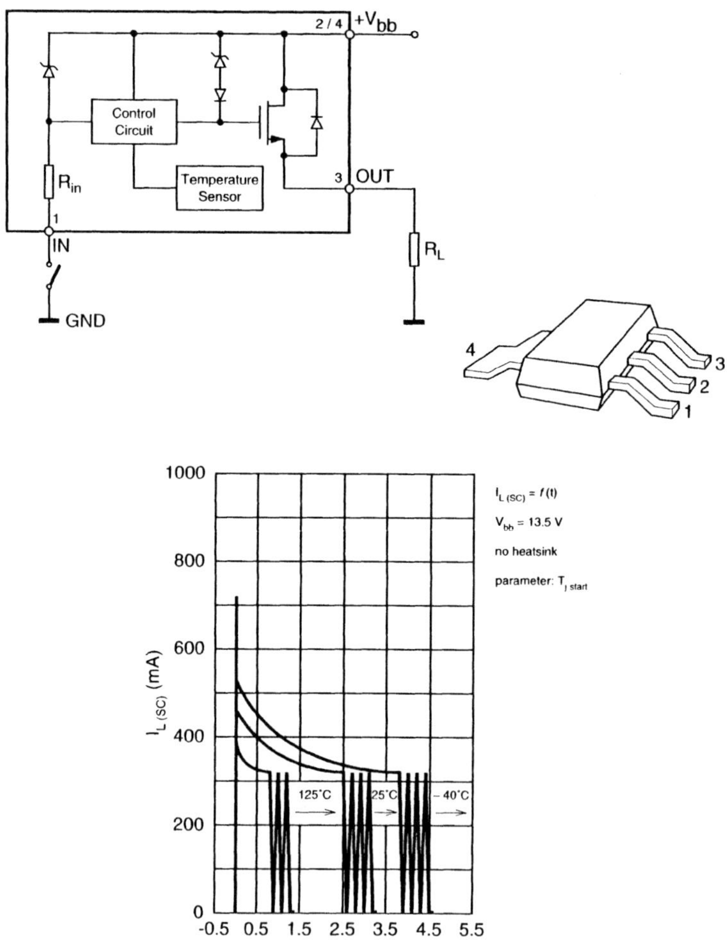

Fig. 3.17 Surface mounted smart discrete BSP350

PRODUCT SUMMARY

V_{bb} - V_{OUT} Avalanche Clamp		47	V
V_{bb} (operation)		5.0 ... 34	V
R_{ON}	2 x 65	or 1 x 33	mΩ
$I_{L(SCr)}$	2 x 12	1 x 12	A
$I_{L(NOM)}$	2 x 3.9	1 x 6.0	A

BLOCK DIAGRAM

Two Channels; Open Load Detection in ON State

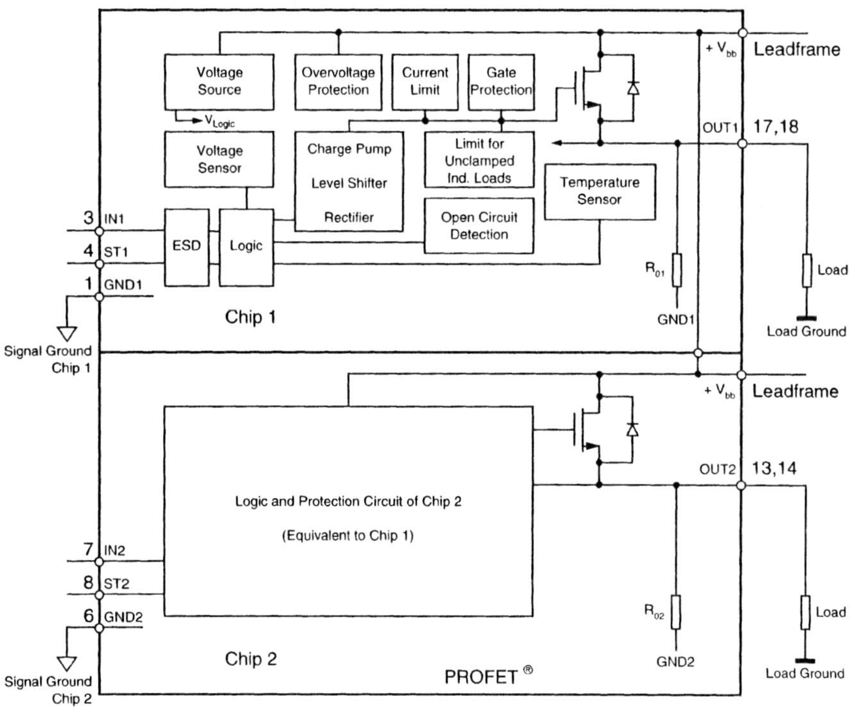

Fig. 3.18 Surface mounted BTS726L1 encapsulated in a 20 pin PDSO package

The first such device to appear on the market was the BSP350 from Siemens. It was introduced in 1992. Recently, other, larger area devices have also been introduced. It is not known to the author, at this time, if other manufacturers have similar products but it can be prophesied that they will offer such products in the near future.

Some specifics of the BSP350 data sheet are shown in Fig. 3.17. The device has an area of approximately 3 mm^2, an on-resistance of 5 ohms, and has an operating voltage range of 5 to 45 volts. It is protected against over-temperature. In the case of a short, the current is limited to approximately 0.5 ampere, the device heats up and at approximately 150 degrees Celsius the current level begins to oscillate so that the effective power dissipation is regulated to a level where the temperature remains constant (150 degrees Celsius). Upon removing the short, the device resumes normal operation. The BSP350 is presently widely used as a signal former for automotive sensors or as a protective device for long wires which may be short circuited in the event of a crash.

Information regarding the newly introduced BTS726L1 is presented in Fig. 3.18. It is a dual channel high-side switch in a normal 20 pin PDSO package. The dual channels are made up of two identical high-side switch chips, each with an on-resistance of 60 mΩ. When the channels are paralleled, the switch has an on-resistance of 30 mΩ. It can be used in the switching of lamps or any other type of ohmic, capacitive, or inductive load. It is protected against overvoltage, reverse battery, loss of ground, and loss of the positive battery voltage. It also can detect open load in the "on" state. The short circuit current is limited to approximately 10 amperes and the power dissipation is 1 Watt at the specified operating current of 5 amperes using one active channel. The two channels in parallel can switch 6 amperes under normal operating conditions. The thermal properties are similar to those described in connection to the device BSP350.

3.8 Future Perspectives

The "know how" for power ICs has now reached a level high enough to address nearly any application in the automotive or industrial control sectors, as long as the operating voltage remains under 60 Volts. Smart power ICs are gradually replacing electromechanical relays and thus open possibilities for new subsystems in automobiles which could not have been conceived earlier. The interaction between microcontrollers and power ICs promotes creativity in the development of better automobiles from an ecological and cost point of view. Certainly, the complexity of power ICs will continue to increase for some applications. The smart discretes, those smart ICs with self protecting features and limited logic functions, will continue to prosper in simpler applications. This is particularly true in the case of replacing electromechanical relays, for eliminating fuses, and for reducing the amount of wire in the wiring harnesses of automobiles.

The competition on the market will not diminish, but will lead to better and more cost effective solutions such as in the case of SMD devices. Summarizing, the author strongly believes that the family of smart discretes has a rosy future before it and will continue to produce surprises with innovative and clever new products in the near term.

References

1. Goldbacher A (1994) Totgesagte leben länger, Elektronik 25, pp 88–110
2. Throngnum-chai K (1992) An intelligent Discrete Power MOSFET with Shorted Load Protection Using Thin Film Bipolar Transistor. Proc ISPSD, pp 144–149
3. Morikawa M, Sakamoto K, Uchida K, Koda T, Satonaka K, Yoshida I (1992) A 30 V 75 m Ohm mm^2 Power MOSFET For Intelligent Driver LSIs. Proc ISPSD, pp 150–154
4. Sakamoto K, Yoshida I, Otaka S, Tsunada H (1992) Power MOSFET With Hold Type Thermal Shutdown Function. Proc ISPSD, pp 238–239
5. Rummenik V (1992) A 1200 V BiCMOS Technology and Its Applications. Proc ISPSD, pp 322–327
6. Seki Y, Hanada Y, Iwamun N, Kugamai N (1994) A New IGBT With a Monolithic Over-Current Protection Circuitry. Proc Ispsd 94, pp 31–35
7. Shimizu Y, Nakana Y, Kono Y, Sakurai N, Sugawara Y, Otaka S (1994) A High Performance Intelligent IGBT With Over-Current Protection Proc ISPSD, pp 37–41
8. Reinmuth K, Stut H, Lorenz L (1994) Intelligent Power Modules For Driving Systems. Proc ISPSD, pp 3–7
9. IR Preliminary Data Sheet 6,027. IRSF 3010. 7/93
10. Philips Power MOS Transistors including TOPFETs and IGBTs. Data Handbook SC 13. 1993
11. ST. VNP35N07, 1993, OMNIFET advanced data sheet
12. ST. Seminar. (1994) Chapter 4

Note: The above references were not quoted in the text. As they were referred to in general through-out the entire chapter.

Dielectric Isolation Technologies and Power ICs

Y. Sugawara

4.1
Introduction

In order to get high performance, low cost, small size and high reliability of electronic equipment, the development of monolithic power ICs has been promoted. Recent social trends regarding down-sized and environmentally friendly products are emphasizing the evolution and practical use of power ICs.

The remarkable improvements of power ICs recently include the realization of higher blocking voltage (≅1200 V), larger current capability (≅30 A) and higher intelligence, which have been achieved by the progress in semiconductor device and process technologies. As isolation technologies for power ICs, Dielectric Isolation (DI) technologies are lately receiving much attention as compared to junction-isolation (JI) technologies, because of the former's ideal electric isolation.

In this chapter, recent developments of both DI technologies and dielectrically isolated power ICs (DI-PICs) are dealt with.

In Section 4.2, the developed DI-ICs, are described comprehensively, while the features of DI technologies are discussed in Section 4.3. In Section 4.4, warpage reduction techniques of DI wafers are explained; warpage is an important problem of conventional DI technologies. In Section 4.5, high voltage techniques specific to DI-PICs are explained. In order to reduce the device size without deteriorating their performances, many devices have been developed for power ICs and their structures have been optimized. In Section 4.6, improved devices specific to DI-PICs are presented. Furthermore, increment techniques for electric power capability are discussed in Section 4.7; this capability is limited to a low value in conventional DI technologies. Finally, several typical DI-PICs are more exhaustively considered in Section 4.8.

4.2
Dielectrically-Isolated Power ICs (DI-PICs)

Figure 4.1 shows the main DI power ICs (DI-PICs) developed since 1981 and their applications [1, 2]. From the viewpoint of both voltage and current capabilities, they are classified into the following three kinds.

1) Low voltage, large current ICs: high-side switch ICs for automobiles, step motor driver ICs, printer driver ICs, etc.
2) High voltage, small current ICs: driver ICs for flat panel displays such as EL, PDP and VFD, subscriber line interface circuit ICs for exchanges, video output ICs, etc.
3) High voltage, large current ICs: power supply ICs, motor driver ICs for appliances, ignition ICs for automobiles, etc.

Figure 4.2 shows a typical block circuit diagram of these DI-PICs [1, 2].

In addition to output power circuits and their driver circuits, accident detection circuits of both the loads and the DI-PIC, as well as protection circuits are integrated. An accident here means an overvoltage, an overcurrent, an overtemperature of both short and open of loads. The trend shows that interface circuits, which send and receive diagnostic signals between the DI-PIC and microcomputers, are integrated and complex sequence operations are realized.

Figure 4.3 shows conventional DI technologies which were presented before 1985. The EPIC [3] is the only DI technology for power ICs in mass-production, in spite of a small complexity process. Most of the power ICs shown in Figure 4.1 have been fabricated with EPIC technology. Neither Spinel [4] nor SOS [5] are established mass production processes due to their poor yield in the hetero-epitaxy step. SIMOX [6] is an attractive and promising technology since its process is smart, but it is still in evolution because of poor through-put in fabrication of the buried isolation SiO_2 film.

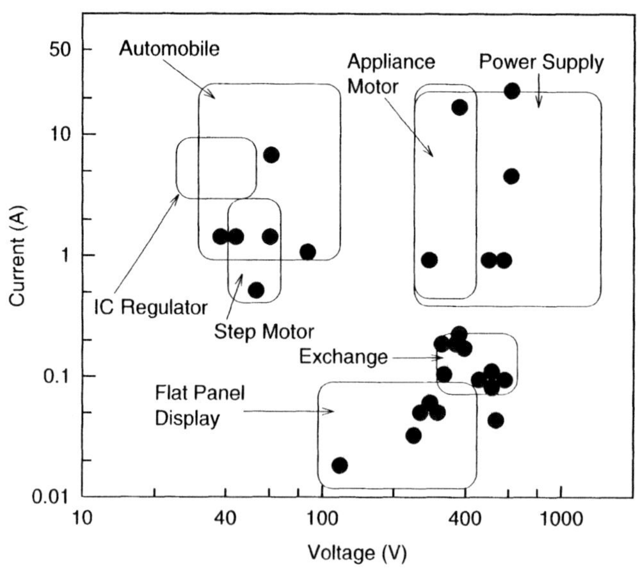

Fig. 4.1 Power DI-ICs developed since 1981 and their applications

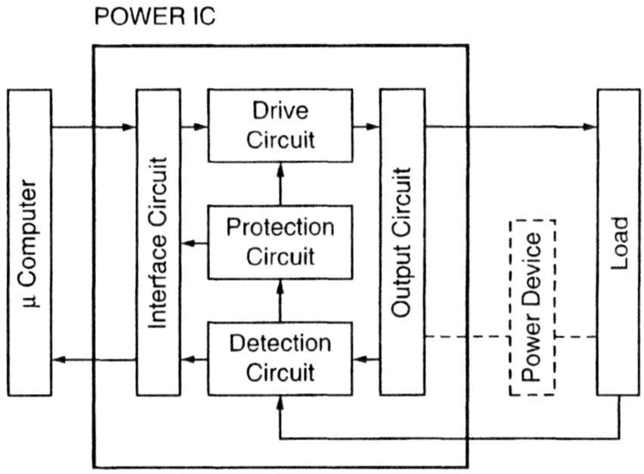

Fig. 4.2 Typical block circuit diagram of DI-PICs

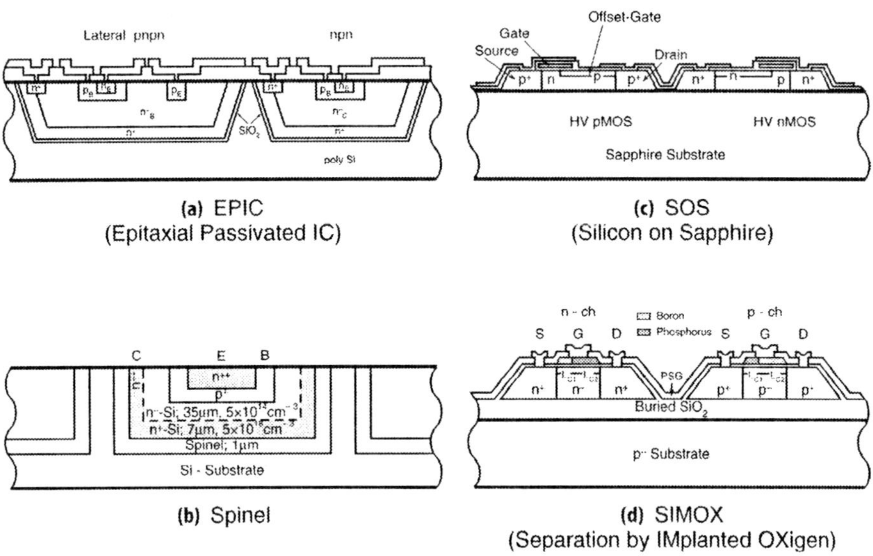

4.3 Features of DI Technologies

Figure 4.4 shows cross-sectional IC structures which allow the comparison of DI and JI technologies. DI has some merits as compared with JI [7].

The first merit is complete isolation. Since devices in the power IC are surrounded by a dielectric film, electric interference among them is completely eliminated. Therefore, high operation speed can be easily realized. And, both logic and control circuits which treat accurate small signals can be easily integrated together with power devices which treat high voltages and large currents and generate large surge currents and large surge voltages. Complete isolation allows simple integration of multi-output circuits of the totem-pole type and integration of latch-up devices such as thyristors, which are ideal switching devices for power ICs because of their very low on-state voltage at high current density. This results in low power dissipation and small chip size. In the case of the JI-ICs, it is difficult to integrate thyristors because of the latch-up action through the pn junction used for the isolation.

The second merit is a large temperature resistance [7]. Output power devices generate a lot of heat, which locally increases the IC temperature. This results in a large leakage current of the isolation junction in the case of the JI. The leakage current hinders accurate operations of linear circuits.Furthermore, it generates an increase in static power dissipation and the latch-up actions of the logic circuits. Since the leakage current of the SiO_2 film for the DI is negligible, DI has a large temperature resistance.

The third merit is compact chip size. For the JI, the width of the isolation area W must be larger than the total width of the p+ diffusion layer for isolation and the depletion layer around the p+n junction. Therefore, W becomes larger as the applied voltage increases. However, the isolation area of the dielectric isolation is independent of the depeletion layer width, and its width W depends on only the machining accuracy of the lapping and polishing steps in the fabrication process of the EPIC type DI substrate and W can easily be reduced to less than 5 μm. Since the isolation layer surrounds each device in the IC, its total area in the chip becomes very large with increased IC breakdown voltage in the case of

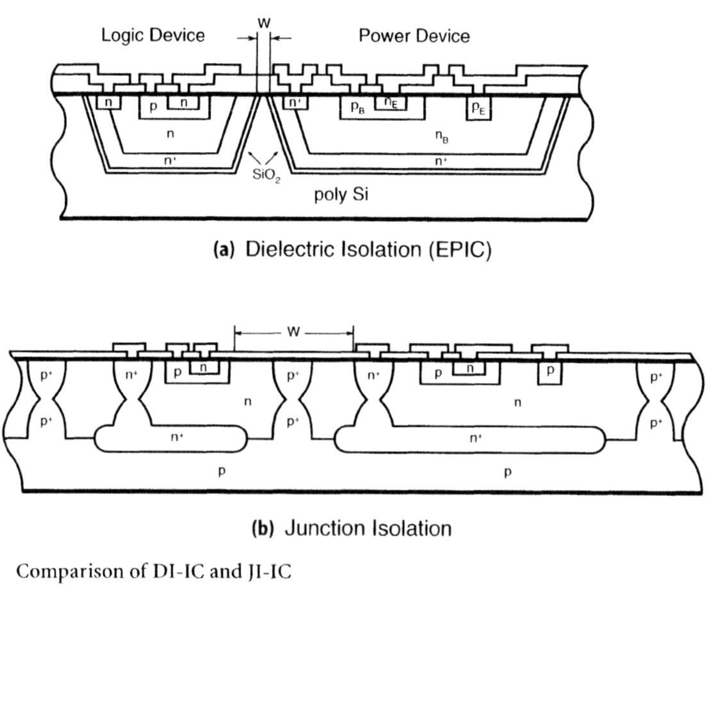

Fig. 4.4 Comparison of DI-IC and JI-IC

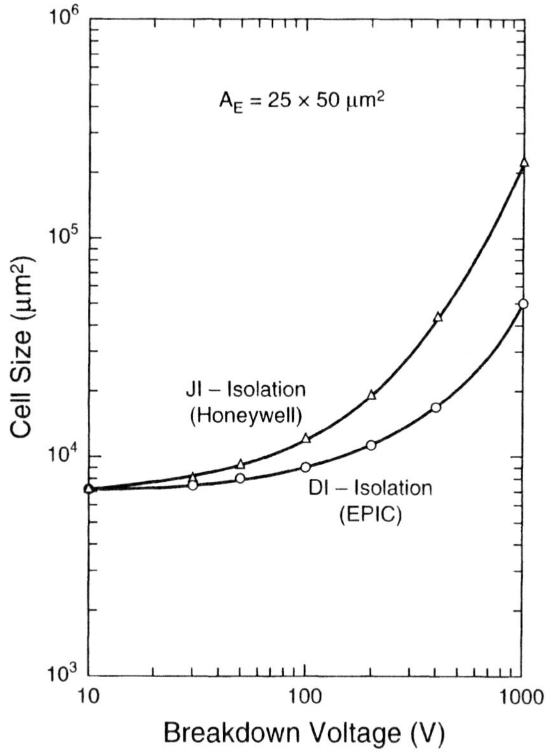

Fig. 4.5 Comparison of transistor cell size between DI and JI

the JI. This results in a larger chip size for JI, which balances the higher DI process cost.

Figure 4.5 shows a comparison of the typical transistor cell sizes between DI and JI [7]. With increased breakdown voltage, the DI cell size becomes larger due to an increase in the field region, but the JI cell size becomes even larger due to increases in both the isolation layer and the field region. The DI process cost seems to be 1.3–1.5 times higher than that of JI. Since the JI cell size is more than 1.3 times larger than that of DI at the breakdown voltage of more than 100 V, the higher DI process cost is balanced by the larger JI cell size. This seems to result in lower cost of DI-PICs than that of JI-PICs in the region of high breakdown voltage.

These merits of DI have recently led to posittive developments of its technologies. Much attention has been paid to overcoming the weak points of DI, such as large wafer warpage and large electric and thermal resistances. The warpage hinders attempts not only to lower the minimum device patterning size, but also to increase DI wafer size, which results in increased DI-PIC cost. The large resistances encumber attempts to increase current capability of DI-PICs.

4.4 Wafer Warpage Reduction Techniques

4.4.1 Wafer Warpage and Reduction Techniques of EPIC Type DI Wafers

Figure 4.6 shows the fabrication process flow of an EPIC type DI wafer [8]. After polishing, the DI wafer is subjected to conventional IC processing, i. e., oxidation and diffusion, etc.

Figure 4.7 shows the effects of the process steps on wafer warpage [8]. The wafer warps concavely toward the poly-Si layer after the poly-Si deposition and warps convexly toward the poly-Si layer during thermal oxidation. The sign of the warpage direction is defined as "plus" for the former and as "minus" for the latter. The degree of warpage after the poly-Si deposition increases when the thickness of the poly-Si deposition layer increases, and it increases as the deposition temperature decreases. What leads the warpage towards the "plus" direction is considered to be the tensile growth stress generated through the contraction of the growing poly-Si layer at high temperatures.

The degree of warpage after thermal oxidation is remarkable, and it decreases as the thickness of the poly-Si deposition layer increases, and grows as the deposition temperature rises. The warpage in the "minus" direction is attributed to the expansion stress generated through diffusion of a high concentration of oxygen along the grain boundaries of poly-Si into the surface region of the poly-Si deposition layer.

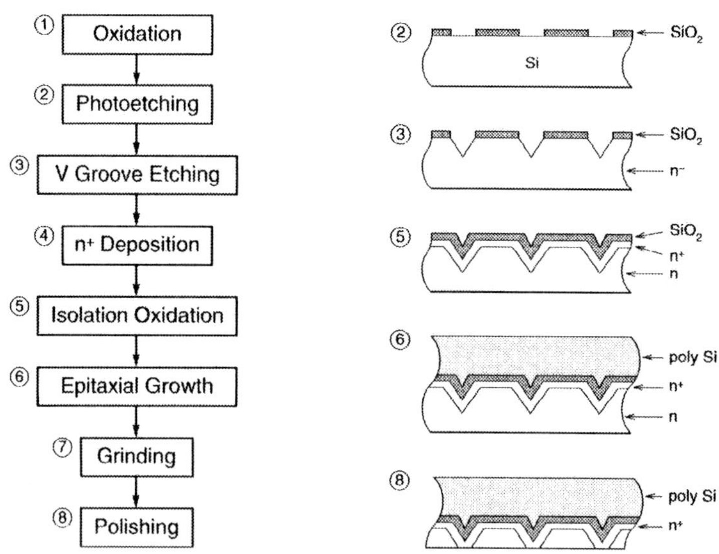

Fig. 4.6 Fabrication process flow of EPIC type DI wafer

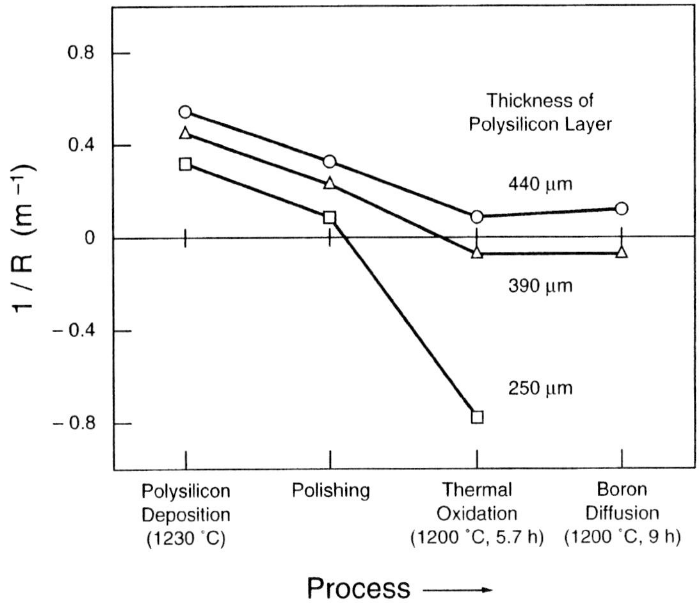

Fig. 4.7 Effects of process steps on wafer warpage. (Reprinted by permission of the Electrochemical Society)

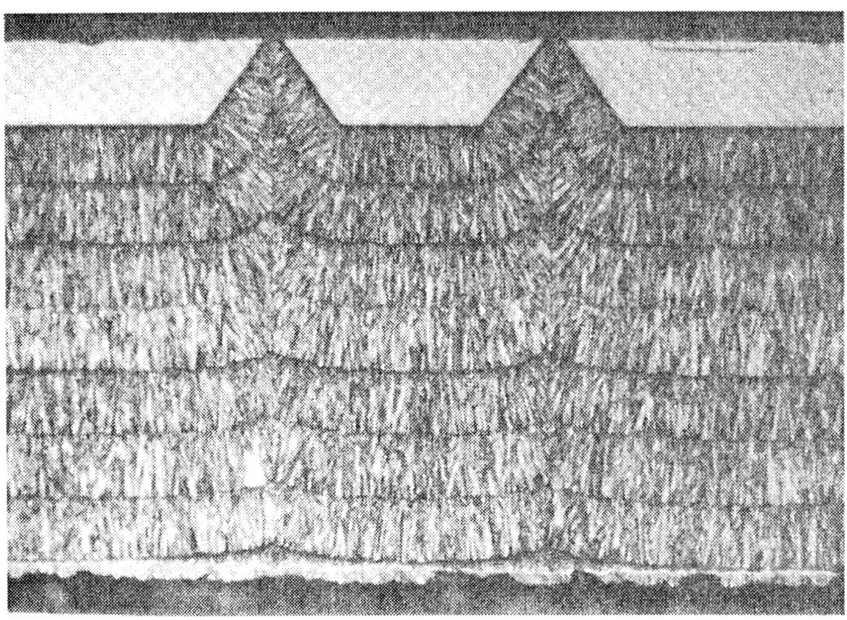

Fig. 4.8 Multilayer poly-Si support structure. (Reprinted by permission of the Electrochemical Society)

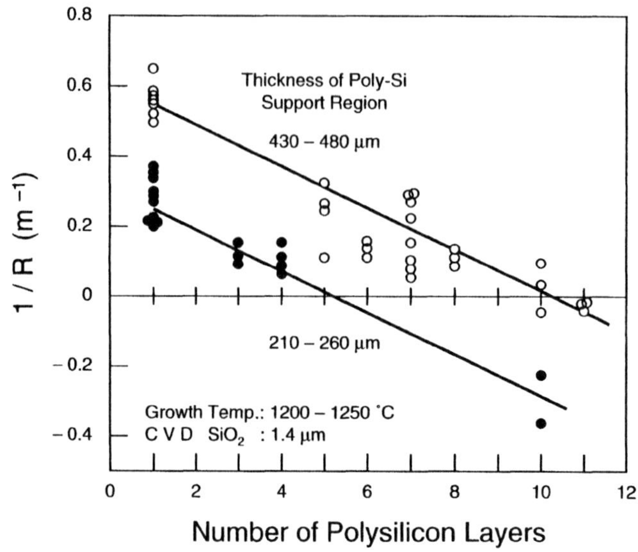

Fig. 4.9 Relation between the warpage and number of poly-Si layers in multilayer poly-Si support structure. (Reprinted by permission of the Electrochemical Society)

Warpage of the EPIC type DI wafer can be reduced by using the multi-layer poly-Si support structure composed of alternate layers as shown in Fig. 4.8, in which the contraction stress caused by poly-Si deposition is compensated by the expansion stress caused by the SiO_2 deposition and this compensation mechanism is repeated alternately [8].

Figure 4.9 shows the relation between the warpage and the number of poly-Si layers in a multi-layer poly-Si support structure. The warpage decreases with greater numbers of poly-Si layers. Although the warpage is smaller when the thickness of the poly-Si support region is smaller, the thickness per layer at which the warpage becomes zero is almost the same, i. e., 40 – 50 μm. Regarding this thickness per layer, the contraction stress caused by poly-Si is balanced with the expansion stress caused by the CVD SiO_2 layer of 1.4 μm thickness.

4.4.2
Wafer Warpage Reduction Technologies with a Single Si Substrate

Using the structures other than EPIC, the wafer warpage reduction for DI has been investigated. Fig. 4.10 shows four typical warpage reduction techniques, LEGO [9, 10], SODIC [11], SDB [12] and SPSDB [13]. A common point between them is that the single Si support layer is used in place of the poly-Si support layer which causes the wafer warpage. Therefore, SIMOX described in Section 4.2 has the effect of reducing the wafer warpage as well.

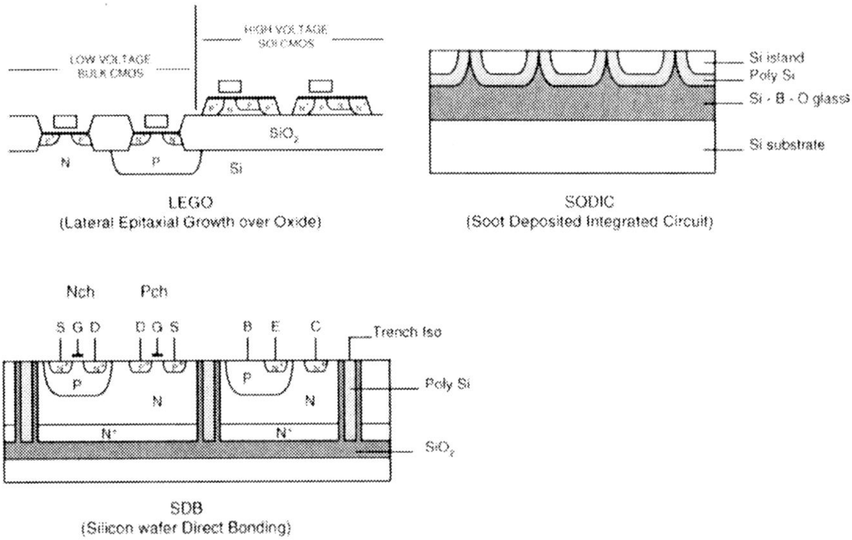

Fig. 4.10 Wafer warpage reduction techniques

	SDB (V – groove)	SDB (Trench)	SPSDB
Cross-section	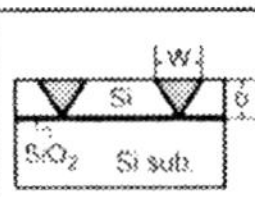	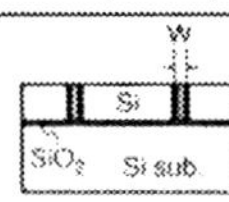	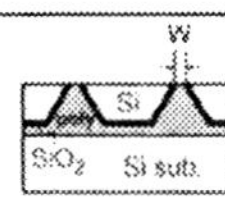
Isolation Width (W)	× (~ 57 μm)	○ (~ 3 μm)	○ (~ 5 μm)
Chip Size	×	○	○
Formation of D > 10 μm	○	×	○
Mass Productivity	○	△	○
Bonded Interface	Si / SiO$_2$	Si / SiO$_2$	Si / poly-Si

( ) : W for 400 V Power ICs, which need D > 40 μm

Fig. 4.11 Comparison of wafer direct bonding techniques

LEGO and SIMOX have some points to be investigated regarding both crystal quality for high breakdown voltage and through-put for mass production. SODIC has a problem of contamination in the DI-PIC fabrication process from boron which is out-diffused from the Si – B – O glass binder. Wafer direct bonding techniques such as SDB and SPSDB had a problem obtaining good direct bonding, but it has been solved by improvements in wafer polishing. Therefore, wafer direct bonding techniques seem to be the most atractive from the viewpoints of crystal quality for high breakdown voltage and process similarity to EPIC, which is the only technique with actual mass production experience.

Figure 4.11 shows a comparison of wafer direct bonding techniques [13]. The symbols ○, △, × means good, not bad, bad, respectively. The SDB techniques have some problems due to their isolation regions. In the case of the SDB with the V-groove isolation region, its isolation width, W, is very wide and becomes wider as the breakdown voltage increases, since the single Si thickness D becomes large. For example, since power ICs of a 400 V class need D of more than 40 μm, W becomes more than 57 μm. This results in a very big IC chip size. In the case of SDB with the trench isolation region, the isolation width, W, can be narrow, about 3 μm, since the trench is formed by unisotropic dry etching. But, it is not easy to form a deep trench of more than 10 μm with the present unisotropic dry etching technique. Therefore, mass productivity of this SDB is poor at present.

4.4.3 Single Si – Poly Si Direct Bonding Technique (SPSDB Technique)

The SPSDB DI [13] has an inverse V-groove isolation region and the sharp top edge of the inverse V-groove is used for isolation, so W is very narrow. W is independent of breakdown voltage; it depends on accuracy of the lapping and polishing machines. At present, it is easy to make W very narrow about 5 μm. Since the inverse V-groove is formed by the anisotropic wet etching technique, it is easy to make D deeper than 10 μm. And, since the formation process and formation machinery of the inverse V-groove are the same as those of EPIC, its large-scale production is good, as it is evinced by actual experience with the EPIC. Furthermore, SPSDB has another relevant merit: that the same design rule can be used as for EPIC under mass production.

Figure 4.12 shows the fabrication process flow of the SPSDB DI wafer. After inverse V-groove formation, the buried layer is formed and the SiO_2 film for isolation is grown by thermal oxidation. Then, poly-Si deposition is performed to fill the groove and lapping of poly Si is performed. Next, wafer bonding is performed and, with lapping , the SPSDB DI wafer is completed.

SPSDB has another feature that the deposited poly-silicon thickness is less than 20% that of EPIC. This means less raw material is needed and the high temperature deposition time is shorter. These result in not only high throughput of the poly silicon deposition process, but also high device packing density due to less out-diffusion of the n^+ buried layer, which is shown in Fig. 4.3a and is formed at the process step of Fig. 4.6.

Figure 4.13 shows 5-inch wafers imaged by ultrasound. The imagescope can distinguish voids a hundred microns in diameter. For the unimproved SPSDB DI wafer, some voids form at the bonded interface. But, for the improved SPSDB DI wafer, no voids can be found, which is also confirmed by the TEM observations. TEM can distinguish voids a few tens of angstroms indiameter but none are observed [13].

Figure 4.14 plots the relationships between bonding strength and wafer bonding temperature. For the measurements, the samples were attached using an adhesive to the apparatus for a tensile strength test. White and black symbols indicate samples which split at the bonded interface and at the adhesive,

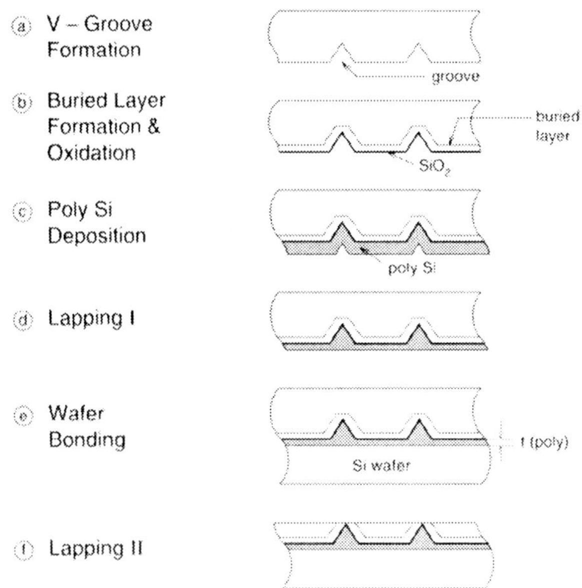

Fig. 4.12 Fabrication process flow of SPSDB DI wafer

respectively. Below 900°C, SPSDB wafers have lower bonding strength than SDB wafers with Si – SiO_2 bonded interface. Since more voids are formed at the interface of SPSDB wafers as compared with the SDB wafers, this seems to be the reason for their splitting. But, SPSDB wafers have the same bonding strength as that of SDB wafers with Si – Si bonded interface, and above 1000 °C, SPSDB wafers have a higher bonding strength than the adhesive. Thus the bonding strength of the SPSDB wafers is confirmed to be enough for the fabrication of power ICs.

Figure 4.15 shows the change of wafer warpage height during SPSDB DI wafer processes. After poly-silicon deposition of about 80 μm thickness, wafers warp concavely toward the poly-silicon layer side due to the contraction stress of the poly-silicon. After lapping I, wafers warp convexly toward the poly silicon layer due to decrease of contraction stress of the poly-silicon layer. After wafer bonding, wafers warp slightly and concavely toward the bonded silicon substrate side. Although warpage heights are very different between two kinds of wafers with different poly silicon thicknesses after lapping I, their deviation becomes small after wafer bonding. This is one of the most important effects of SPSDB. The reason seems to be that the stress remaining after the lapping I process is compensated for by the stress of both the poly-silicon re-contraction and the thick single silicon substrate. After the lapping II process, isolated single silicon islands are formed. At this time, wafers warp further concavely toward the substrate side. After the thick field oxidation for high breakdown voltage, the warpage height is unchanged. At this process step, warpage height is only 40 – 60 μm.

Fig. 4.13 5 inch SPSDB DI wafers imaged by ultrasound

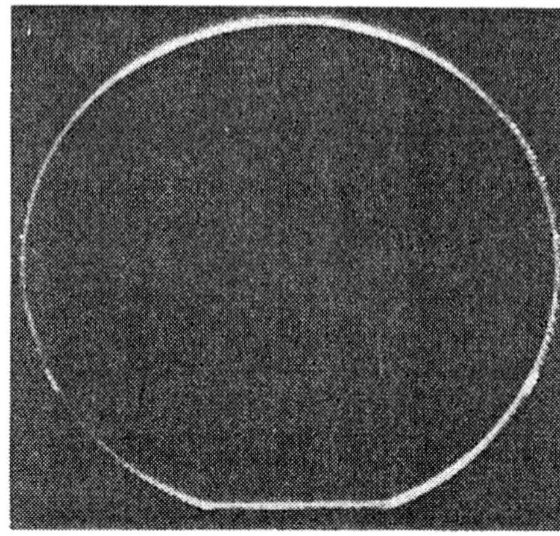

a Improved SPSDB DI wafer

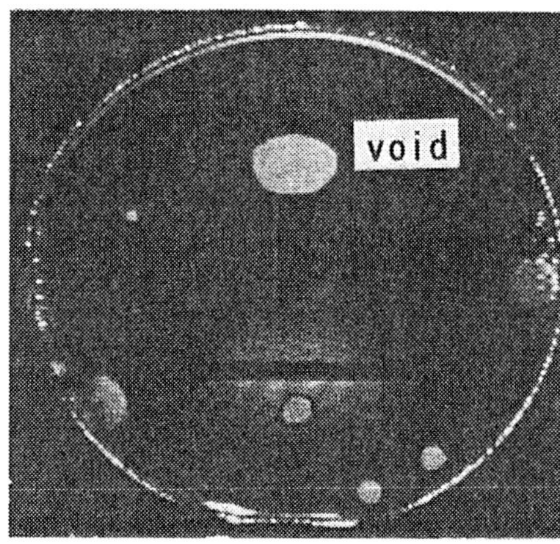

b Unimproved SPSDB DI wafer

Figure 4.16 shows changes of wafer warpage height, Δh, during thermal heat treatment in an oxidation atmosphere, which corresponds to the fabrication process of devices in power ICs. The poly-silicon thickness, t(poly), of the SPSDB wafers is 20 μm, and the temperature of thermal heat treatment is 1100 °C. The Δh is very small and is −5 to +15 μm even at long heat treatment times.

By using the SPSDB technique, a very low warpage height of about 40 μm and warpage change of less than 15 μm during IC fabrication can be obtained.

This is sufficient to realize accurate device patterning of less than 3 μm on DI wafers larger than 5 inches in diameter.

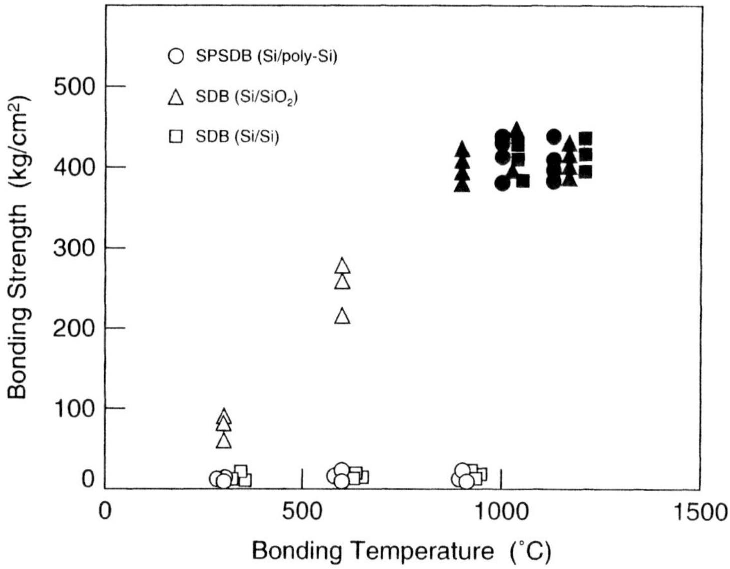

Fig. 4.14 Relationship between bonding strength and wafer bonding temperature

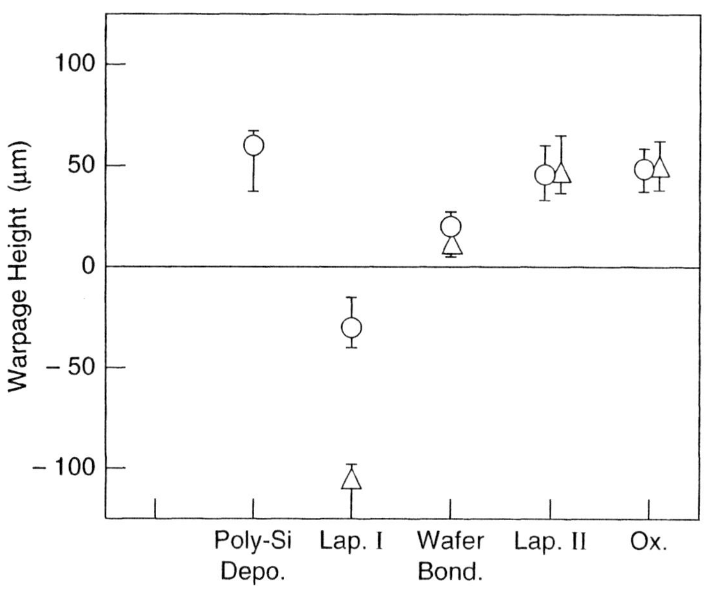

Fig. 4.15 Change of wafer warpage height during SPSDB DI wafer processes

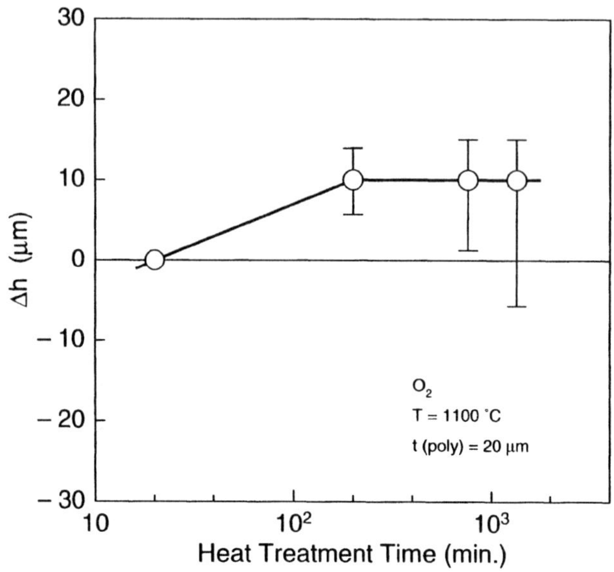

Fig. 4.16 Change of wafer warpage height during thermal heat treatment

4.5 High Voltage Techniques for DI-PICs

4.5.1 Overview

For the realization of high voltage power ICs, it is necessary both to achieve high breakdown voltages of the main pn junctions and to reduce deterioration of breakdown voltage by the field effects of interconnection layers. For the former, high voltage techniques such as extention junction termination [14], field plate [15], field limiting ring [16] and SIPOS [17] are used. These have been developed for discrete power devices. The RESURF technique [18], developed for power ICs, is also used. The latter is a specific problem for power ICs, and the high voltage techniques such as channel stopper, equi-potential ring [19] and SIPOS are utilized. In designing power ICs, combinations of these techniques may be applied to satisfy requests for chip area reduction, process simplification, and high reliability. In this Section, the high voltage techniques which are specific to DI-PICs are explained.

4.5.2 Field Plate and Inclined Buried Channel Stopper

The channel stopper is indispensable for DI-PICs in order to prevent the breakdown voltage from being affected by the potential of the poly-Si subtrate [20].

Fig. 4.17 shows the blocking voltage of a DI device as a function of the substrate potential. As the negative potential of the substrate increases, the blocking voltage decreases remarkably. This is because the channel generated through the bottom isolation SiO_2 film by the field effect of the substrate potential connects with the channel generated through the surface passivation SiO_2 film by the field effect of the Al layer potential, and a large leakage current flows from the p anode to the n cathode. When the positive potential of the substrate increases, the blocking voltage decreases. This is because the channel generated by the field effect of the Al layer potential contacts with an accumulated n+ layer generated by the field effect of the substrate potential at the edge of the DI island under the Al layer, and intense concentration of electric field occurs. By forming the channel stopper shown in Fig. 4.18b, the channel along the Al layer is interrupted and the deterioration of the blocking voltage is eliminated.

The combination of the field plate and the channel stopper is a suitable high voltage technique for PICs because of their small area. However, for this combination, the breakdown voltage is limited by the electric field concentration around the surface of the channel stopper when the thickness of the SiO_2 passivation film on the channel stopper is not very thick. This results in a low breakdown voltage, complex fabrication process and poor reliability.

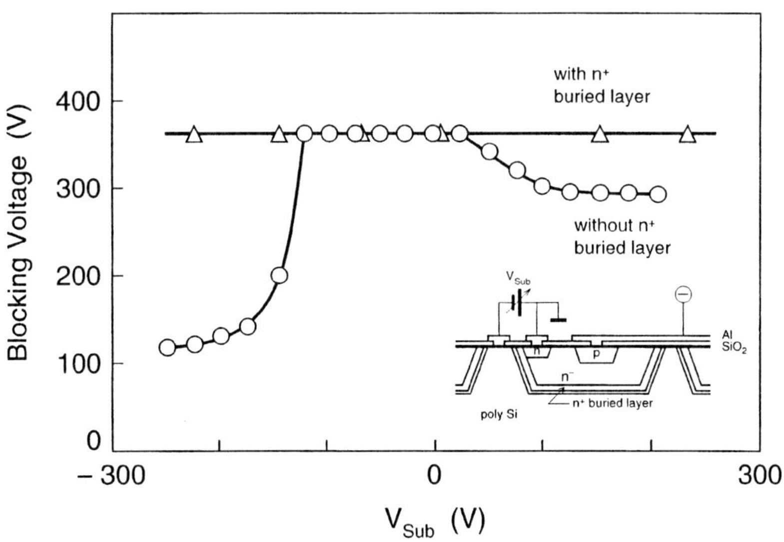

Fig. 4.17 Blocking voltage of a DI device as a function of substrate potential

Fig. 4.18 Comparison of channel stopper structures. The symbols ○, ● indicate sites a and b respectively

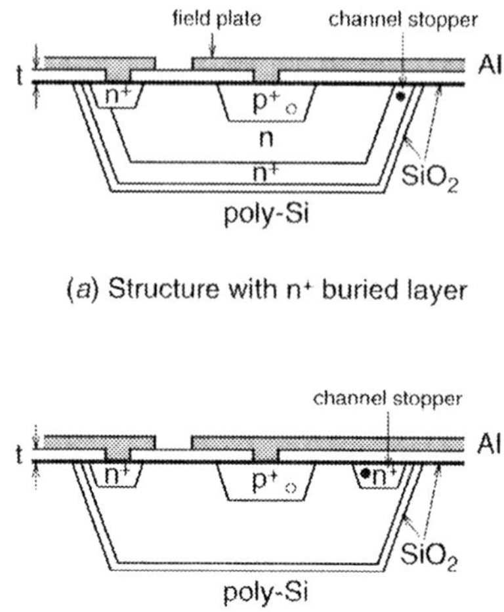

(*a*) Structure with n⁺ buried layer

(*b*) Structure without n⁺ buried layer

a) Inclined Buried Channel Stopper (IB-channel Stopper)

Figure 4.18 b shows a diode with the conventional channel stopper structure, which is fabricated by diffusion from the surface of a single Si island after the poly-Si deposition process in the DI substrate fabrication process. Figure 4.18 a shows a diode with the inclined buried chanel stopper (IB-channel stopper), which is the specfic channel stopper developed for DI-PICs [20]. This channel stopper is built by diffusion from the bottom surface of a single Si island before the poly-Si deposition process. The electric field at the channel stopper is limited by the impurity concentration gradient of its surface. As the side of the bottom face inclines against the surface and the impurity is diffused from this inclined side, the impurity gradient parallel to the surface of the IB-channel stopper is smaller than that of the conventional structure for the same diffused impurity profile. Therefore, the latter structure allows the reduction of the electric field around the channel stopper.

Figure 4.19 shows the calculated maximum electric field Emax at applied voltage of 400 V related to the surface oxide thickness for diodes which have the same diffused impurity profile. The p+ junction depth is 4 μm and thickness of the n^+ channel stopper is 3 μm. The Emax in the structure with the n^+ IB-channel stopper (Fig. 4.18 a) is lower than that of the conventional channel stopper (Fig. 4.18 b), therefore, a higher breakdown voltage can be achieved for the same oxide thickness. It is evident from Fig. 4.19 that Emax for site b decreases, but Emax of site a increases as the oxide thickness increases. Therefore, to achieve a high breakdown voltage, it is very effective to form the oxide layer with different thicknesses; a thin oxide layer beneath the field plate, and a thick oxide layer on the n^+ channel stopper.

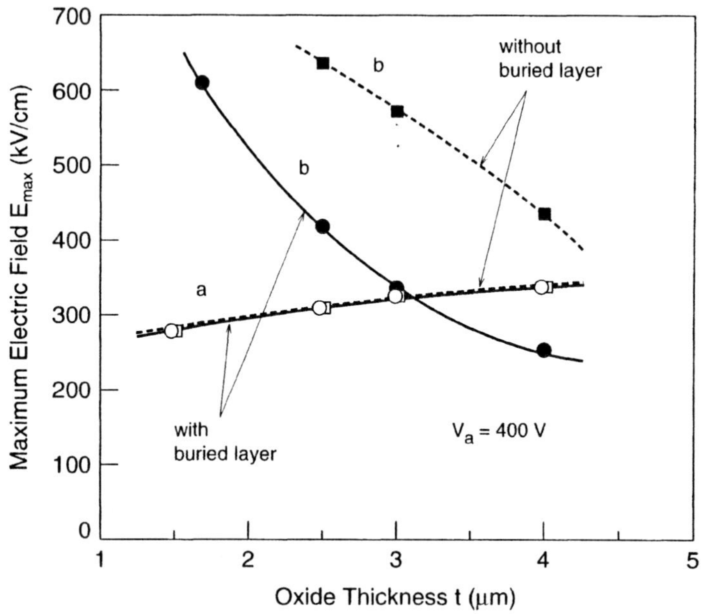

Fig. 4.19 Maximum electric field vs surface oxide thickness of 4 μm junction depth diodes (applied voltage: 400 V)

b) Charge Control in Oxide Film

Charge density in the oxide Q strongly influences the breakdown voltage of the diode shown in Fig. 4.18 a. Figure 4.20 shows the relationships between the calculated breakdown voltage and the total charge density in the oxide Q/q of the diode with IB-channel stopper and with the non-uniform thickness oxide and the n+ buried layer. When Q/q is lower than each peak point of the curves in Fig. 4.20, the breakdown voltage is limited by the electric field of site b shown in Fig. 4.18 a because the depletion layer extends along the interconnection more easily. When Q/q is higher than each peak point, the breakdown voltage is limited by the field at the corner of the junction because the electric field reduction by the field plate becomes less effective. Therefore, Q/q must be controlled within a limited range so as to obtain the specified breakdown voltage. This range can be expanded by increasing t_1 and reducing the electric field of site b, but this range becomes smaller as the junction depth becomes shallower.

Figure 4.21 shows experimental relationships between breakdown voltage of diodes and N_{FB} in the passivation oxide. N_{FB} was examined by controlling the temperature and ambient of the final heat treatment in the IC fabrication process; the fixed charge control technique by Deal et al. [21] was applied and oxide charge density Q/q was controlled. The breakdown voltage of the diode with an 18 μm junction depth increases when N_{FB} increases. The breakdown voltage of the diode with a 3.5 μm junction depth decreases as the N_{FB} increases.

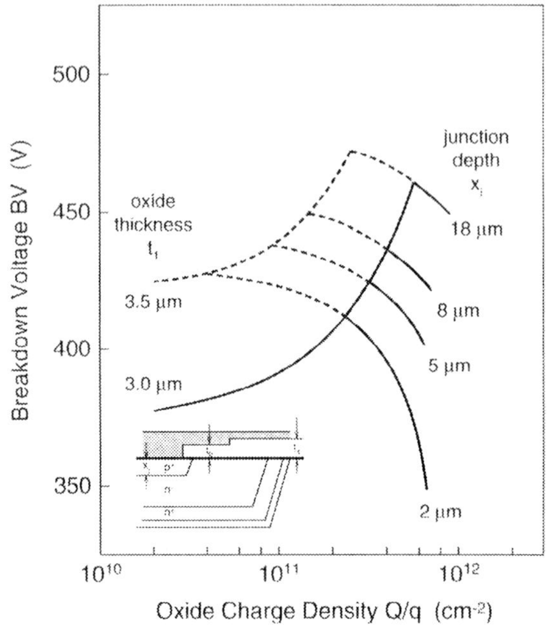

Fig. 4.20 Calculated breakdown voltage vs oxide charge density for diodes with IB-channel stopper and with nonuniform oxide thickness

The calculated relationships (Fig. 4.20) for the experimental structure agree with the experimental results within 10 %, confirming the validity of the former.

4.5.3 Field Reduction Region (FRR)

To reduce the electric field at the IB-channel stopper further and to realize higher voltage, the Field Reduction Region (FRR) technique is developed [22].

Figure 4.22 shows a cross-sectional view of the DI-PIC with FRR. The symbols ●, ■, ▲ indicate sites a, b, c, respectively. The depth of the pn junction X_{jn} is about 5.0 µm. To implement a high breakdown voltage by employing this shallow junction, the structure of the field plate is optimized, namely both the fixed charge density in the oxide and the oxide thickness under the field plate are decreased by using the experimental data in 4.5.2. Thus, the depletion layer near the junction is spread out by the field plate and the electric field at the corner of the junction is reduced. On the other hand, the electric field at the n^+ IB-channel stopper becomes higher, especially under the interconnections, because the depletion layer spreads along the Si surface easily and is halted at the n^+ IB-channel stopper. Thus, when the higher breakdown voltage is needed, it is limited by the concentration of the electric field at the channel stopper.

To reduce the electric field at the channel stopper, a wider field region is necessary. This results in larger device areas. Therefore, FRR is introduced to reduce the width of the field regions, W_F. This layer must be designed so that it has a higher

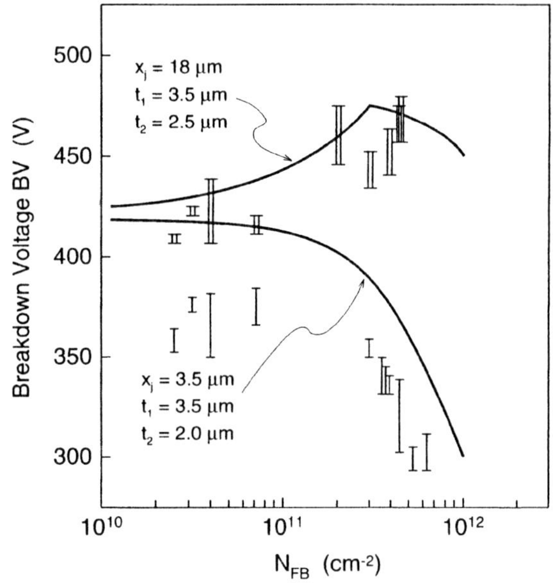

Fig. 4.21 Experimental relationships between breakdown voltage of diodes and N_{FB}

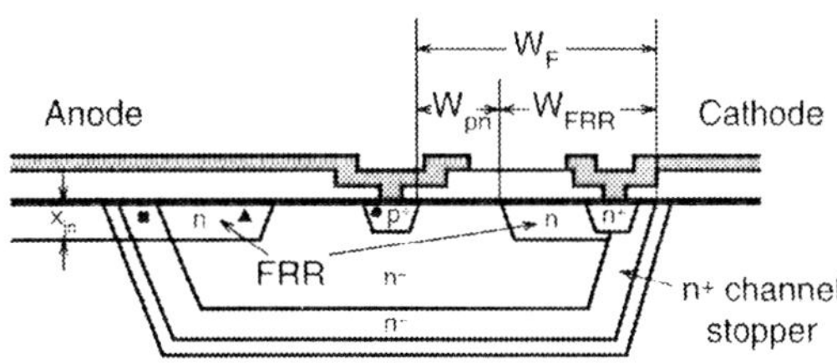

Fig. 4.22 Cross-sectional view of the DI-PIC with FRR. (The symbols WF, WFRR, Wpn and Xjn indicate the field region length, the FRR length, the p-n region distance and the FRR junction depth, respectively)

impurity concentration than the n^- layer and it is depleted completely when a high voltage is applied to the device. Thus, the applied voltage shared in by the depleted FRR becomes higher than that shared in by the depleted n^- layer with same width, because impurity concentration of FRR is higher. This results in that FRR can reduce the electric field at the n^+ IB-channel stopper. When the electric field at the channel stopper is kept at a constant value, the width of the field regions can be reduced. But, since breakdown occurs at the FRR when its impurity concentration is too high, it is necessary to optimize the impurity concentration.

Figure 4.23 shows the experimental relationship between the breakdown voltage and the arsenic dose of FRR. When the arsenic dose is about 5×10^{12}cm^{-2}, breakdown voltage is at maximum. When arsenic dose is smaller than that value, the breakdown is determined by the reach-through mechanism. Namely, the electric field has its maximum at the Si surface of the IB channel stopper and breakdown voltage is limited by the electric field at site b. Then, since FRR is completely depleted and its electric field is much lower than that at site b, it is

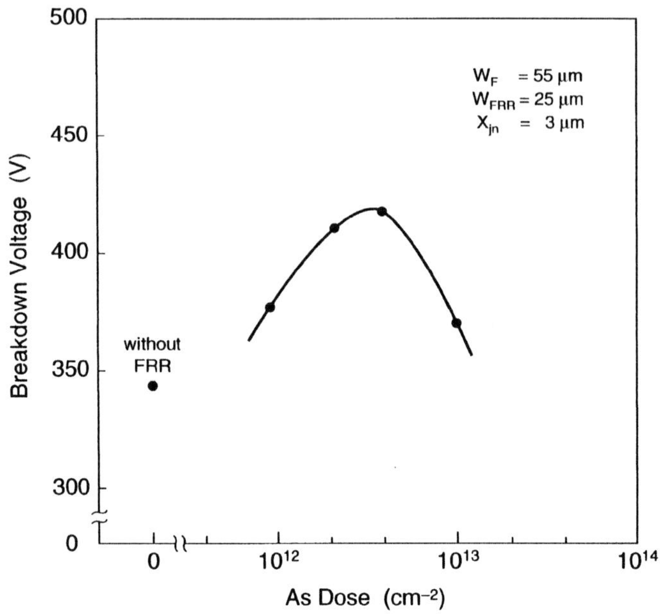

Fig. 4.23 Experimental relationship between breakdown voltage and arsenic dose of FRR

possible to increase the electric field at FRR by increasing the arsenic dose without increasing the electric field at site b. This means the breakdown voltage can be increased by increasing the arsenic dose. Near the arsenic dose corresponding to the maxmum breakdown voltage, the electric field at the junction corner becomes larger than that at the surface of the channel stopper; therefore, the breakdown is determined by the effective radius-of-curvature of the field plated p^+ region, specially site a. When the arsenic dose is more than 5×10^{12} cm^{-2}, breakdown voltage is limited by the electric field at site c and becomes a lower value.

Figure 4.24 shows the relationships between breakdown voltage and the field region width W_F when the oxide thickness t is 3 µm and junction depth is 5 µm. Breakdown voltage is larger than that of the diode without the FRR by about 80 V, when WF is 55 µm. The smallest W_F, which is necessary to achieve a high breakdown voltage of more than 350 V, is reduced by the FRR.

4.5.4 SOI-RESURF

In the case of SOI substrates built with SIMOX [6], LEGO [9] and SDB [12] techniques, the trench is used to isolate the integrated devices laterally. A deep trench is needed for the realization of high breakdown voltage since a thick Si layer is needed in the SOI substrate for this purpose. But, fabrication of the deep trench is difficult with today's dry etching technique, as described in Section 4.4.2. There-

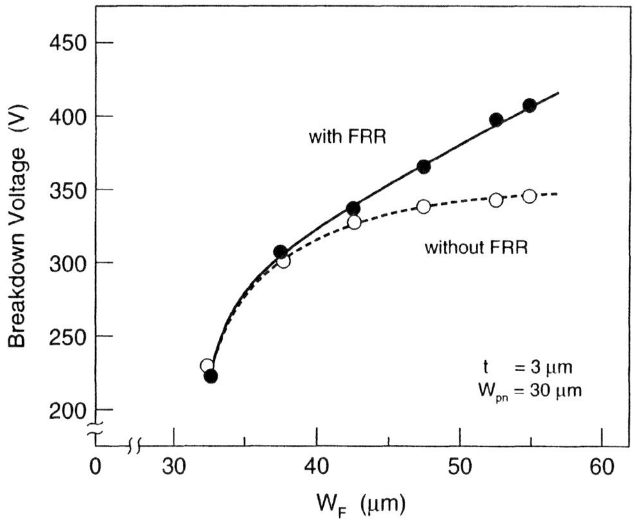

Fig. 4.24 Relationship between breakdown voltage and field region width W_F

fore, other ways have been actively studied to realize a high breakdown voltage with a thin Si layer, which can be isolated by a shallow trench easily formed by dry etching. Since the RESURF technique [18] is essential to obtain a high breakdown voltage for thin Si layer devices, these are named SOI RESURF devices.

Figure 4.25 shows a typical SOI RESURF device and its relationships between breakdown voltage and Si thickness [23]. The impurity concentration of the thin n type Si has a linear distribution toward the direction from p^+ to n^+. Breakdown voltage increases the buried SiO_2 thickness increases, because the applied voltage is shared in not only the depeletion layer in the thin Si but also in the buried SiO_2. Breakdown voltages become a minimum at Si thicknesses between 5 μm and 10 μm, and increase as the Si thickness increases above about 10 μm and as the Si thickness decreases below about 1 μm. A 700 V diode has been realized at the Si thickness of 0.1 μm and the buried SiO_2 layer thickness of 2 μm, and an 860 V LDMOSFET has been realized at the Si thickness of 0.2 μm and buried SiO_2 layer thickness of 4.4 μm [24].

Since it is difficult to form the linearly distributed impurity concentration, the technique for the realization of high breakdown voltage is developed by using the thin SOI-Si layer with uniform impurity concentration [25]. Figure 4.26 shows its structure. The SIPOS film is formed on SiO_2 passivation film to realize a uniform horizontal electric field in the thin Si layer with uniform impurity concentration. Figure 4.27 shows the experimental relationships between breakdown voltage and Si thickness. Thicknesses of SIPOS film and bottom oxide layer are 1 μm and 3 μm, respectively. A linear relationship is observed. The breakdown voltage increases by about 80 V with every 1 μm increase in buried oxide thickness, and a breakdown voltage of 680 V is achieved at the buried oxide thickness of 3 μm.

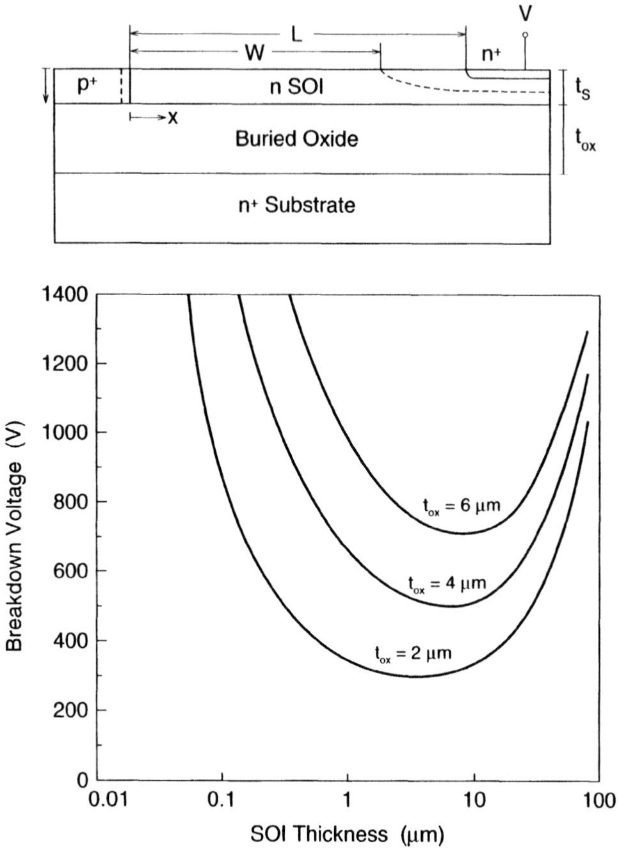

Fig. 4.25 SOI RESURF device with linearly graded impurity Si layer and dependence of breakdown voltage on Si thickness. (Reprinted by permission of IEEE)

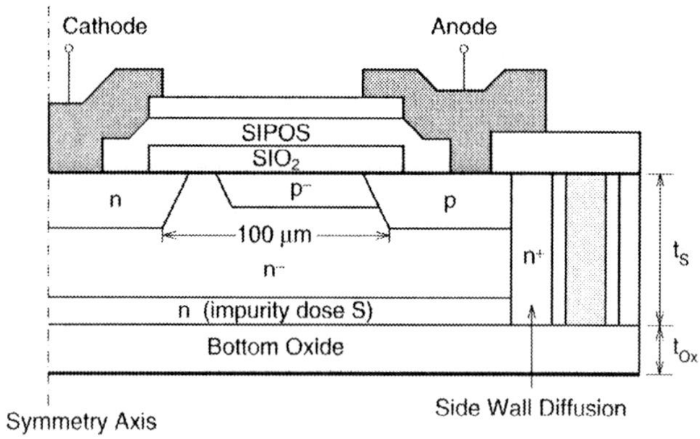

Fig. 4.26 SOI RESURF device with uniform impurity Si layer. (Reprinted by permission of IEEE)

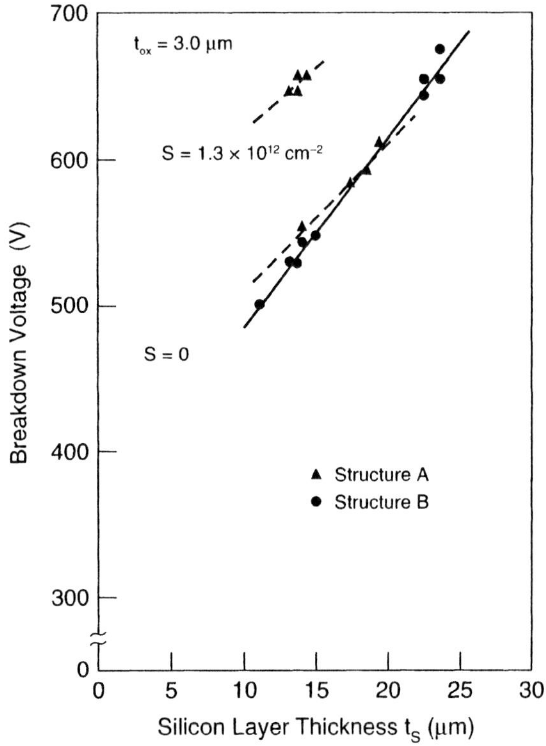

Fig. 4.27 Experimental relationships between breakdown voltage and Si thickness (S: impurity dose of bottom n layer). (Reprinted by permission of IEEE)

4.6 High Voltage Devices for DI-PICS

Many kinds of vertical and lateral high voltage devices, such as diodes, transistors, thyristors, MOSFETs, IBGTs, MOS thyristors and light activated devices are intgrated into DI-PICs. Compared to vertical devices, lateral devices have very simple fabrication processes, but they occupy large device areas and their electrical performance is poor, specially when used as high voltage devices. In order to reduce their sizes without deterioration of their electrical characteristics, many improvements have been made. In this section, the improved devices specific to DI-PICs are introduced.

4.6.1 Diodes

(a) Lateral DI Diodes

Figure 4.28 shows a lateral DI diode with breakdown voltages up to 600 V [26]. Thicknesses of Si layer and buried oxide layer are 5–32 μm and 1–3 μm,

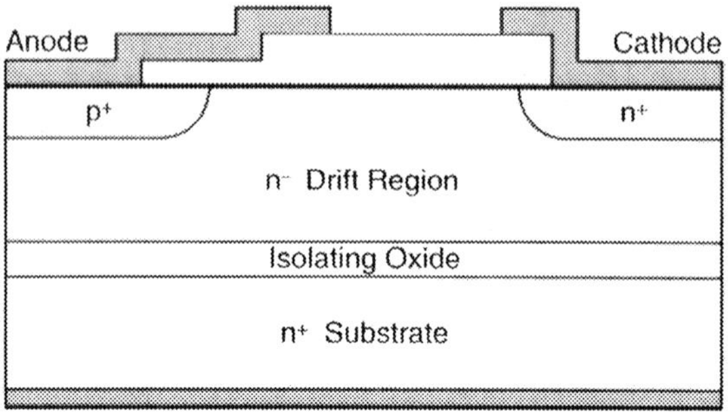

Fig. 4.28 Lateral DI diode of 600 V class. (Reprinted by permission of IEEE)

respectively. The drift region length is 10–100 μm and the resistivity of the Si layer is 45–65 Ωcm. The resistivity of the n^+ substrate under the buried oxide layer is 0.01 Ωcm.

Figure 4.29 shows a comparison of forward characteristics of DI and JI diodes with the same structures except those related to the isolation [26]. To achieve a high breakdown voltage for JI, the RESURF technique is applied in addition to the field plate. The forward voltage drop of the DI diode is larger than that of the JI diode, although it decreases as the thickness of the Si layer increases. And the reverse recovery time of the DI diode (200–400 μm) is much smaller than that of the JI diode (1400–1700 μm). These are caused by the elimination of carrier injection into the DI substrate. For the JI diode, there is a large amount of carrier injection deep (about 100 μm at 200 A/cm^2) into the substrate during the on-state. This results in a larger cross-sectional area for lateral current flow leading to a lower forward voltage drop and a longer time for carrier recombination during turn-off leading to a slower switching speed. Furthermore, although the JI diode has a negative temperature coefficient for the forward voltage drop, the DI diode has a positive temperature coefficient.

(b) Soft and Fast Recovery Diodes (SFDs)

In order to improve the reverse recovery characteristics, an SFD of 550 V · 1 A class with the structure shown in Fig. 4.30 is developed [27]. The SFD has a hybrid anode which is composed of conventional p+ anodes and extremely thin p anodes. The latter is formed by Al diffusion, from an Al-Si electrode at about 500 °C. Figure 4.31 shows excess hole distributions [28]. Impurity concentration and depth of the thin p layer are 1×10^{16}/cm^3 and 20 nm, respectively. Minority carrier lifetime is 1.2 μs. By using the thin p anode, the hole density in the n^- layer close to the pn junction is reduced. Therefore, as compared with the conventional p^+n diode, both reverse recovery time and reverse recovery peak current can be reduced, although forward voltage drop increases slightly. Figure 4.32 shows the recovery waveform of the 550V · 1A SFD [27]. Reverse recovery time and reverse recovery peak current are 200 ns and 1.7 A respectively, and these are reduced to 1/1.5 and 1/2.5 of the pn diode values, respectively, which means power switching loss is reduced drastically (about 1/4). Its recovery

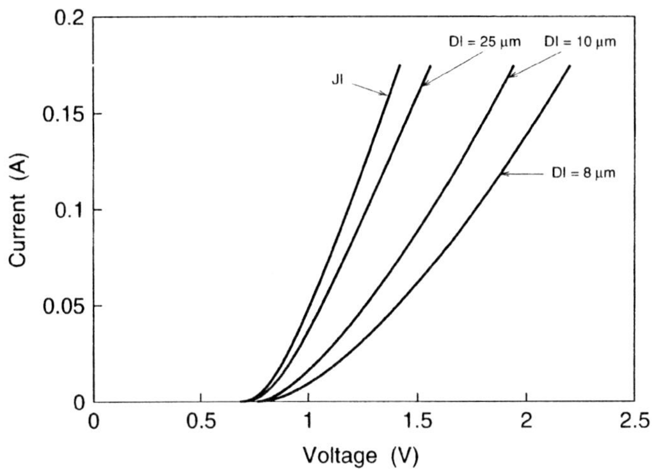

Fig. 4.29 Comparison of forward characteristics of DI and JI diodes. (Reprinted by permission of IEEE)

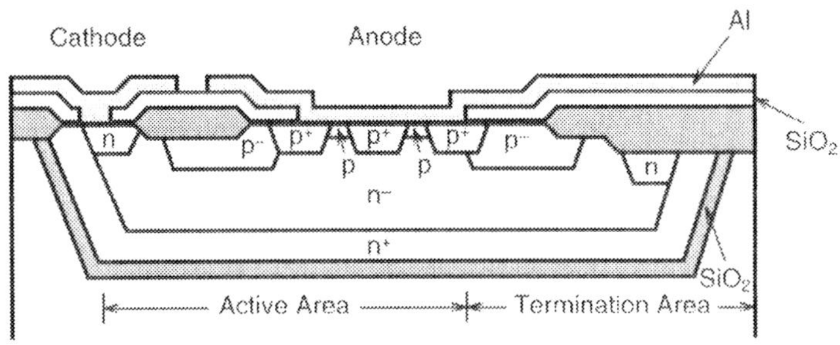

Fig. 4.30 Soft and fast recovery diode (SFD) of 550 V · 1 A class

Fig. 4.31 Excess hole distributions

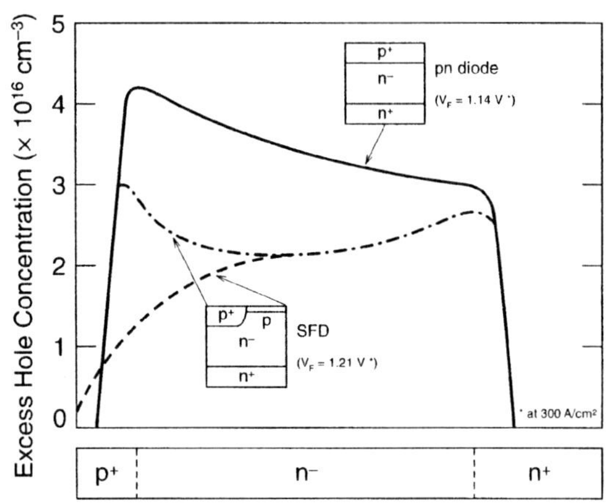

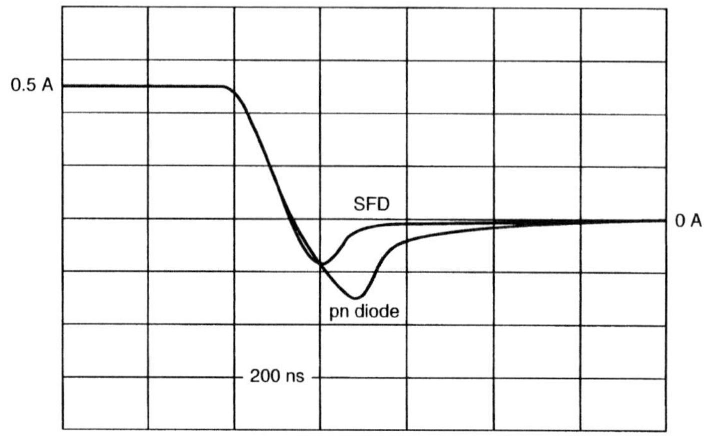

Fig. 4.32 Recovery waveform of SFD

current change is also reduced to 1/3, which means electrical noise due to an inductive voltage is reduced to 1/3. The inductive voltage is generated by parasitic inductance L_p and has a value of $L_p \times dI_r/dt$.

4.6.2 Lateral Transistors

(a) Lateral DI pnp Transistor

Figure 4.33 shows a lateral DI pnp transistor of 380 V class. The resistivity and the depth of the Si layer are 25 Ωcm and 40 μm, respectively. The width of the base ($n_{\bar{B}}$) between p emitter (p_E) and p collector (p_C) is 60 μm. The current gain h_{FE} of the lateral pnp transistor becomes larger when its emitter approaches more closely the side wall of the island integrating the transistor and when the island becomes smaller [29]. By utilizing these findings, device size can be reduced and h_{FE} can be increased.

Figure 4.34 shows a comparison of pnp transistor h_{FE}. Transistor (a) is a conventional type and its length L from the side wall oxide to the diffusion pattern of the p emitter is 80 μm. Transistor (b) is an improved type which has a shortened L of 22.5 μm. The h_{FE} of the latter is about 30 which is 3 times that of the former. This is because the injected holes are effectively reflected by the potential of n-n^+ junctions not only along the bottom of the island, but also along the inclined side wall of the island [29].

Transistor (c) is a diagonally disposed type with two p emitters and L of 22.5 μm. The emitters of transistor (c) are surrounded by the n^+ buried layer, except in the direction toward the collector, and its island area is reduced to 72% that of transistor (b). Its maximum h_{FE} is about 50 and becomes larger. It seems that, in addition to the effect from shortening L, there is an effect from reducing

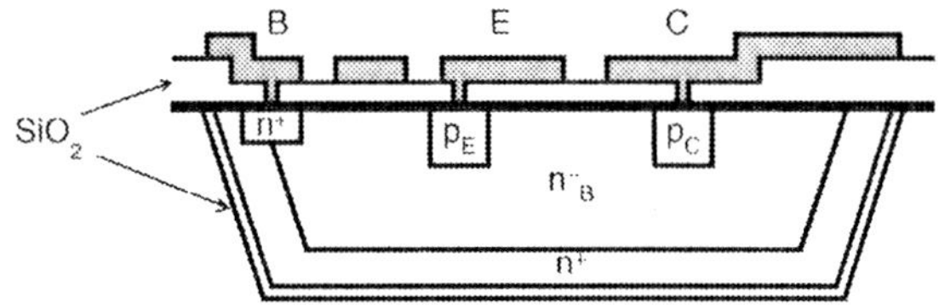

Fig. 4.33 Lateral DI pnp transistor of 380 V class

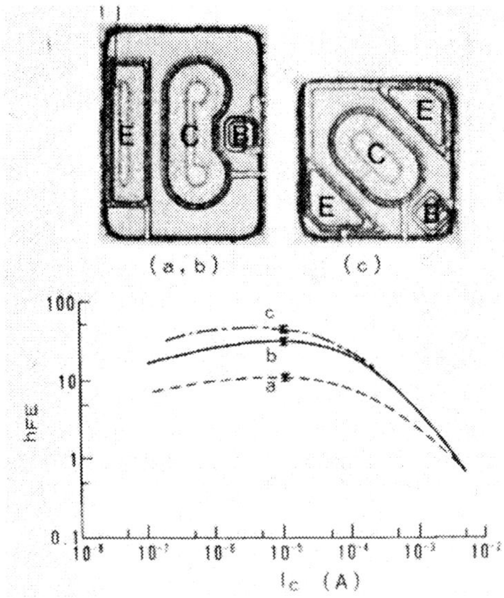

Fig. 4.34 Comparison of pnp transistor's h_{FE}. (The symbols E, B and C indicate emitter, base and collector, respectively)

the island size, which is caused by reducing the recombination of the injected holes at the surface and in the bulk of the island.

(b) SADDL pnp Transistors

To realize high voltage analogue ICs, complementary npn and pnp high voltage transistors are necessary. In such applications, a combination of vertical npn transistors and lateral pnp transistors are used because of the simple fabrication process. But, as compared with the vertical npn transistor, the general lateral pnp transistors including the transistor described in Section 4.6.2 (a) have poor performance in the high voltage region. Since the n base of the lateral transistor is doped lightly, a wide n base is required to avoid punch-through between the emitter and the collector and to achieve a high breakdown voltage. This results in low current gain h_{FE}, low cut-off frequency f_T and low Early voltage.

To achieve the excellent complementarity with the vertical npn transistor, a self-aligned double diffused lateral (SADDL) pnp transistor is developed [30–32].

(b-1) Device Structure and Operation Mechanism

Figure 4.35 shows a cross-sectional view of the SADDL pnp transistor [30–32]. To achieve high h_{FE} and f_T, a narrow n base is necessary like that in the vertical transistor. Therefore, the narrow and heavily doped n base is formed by using the self-aligned technique, namely, the p emitter (p_E) of the SADDL transistor is diffused from the same photo-etched SiO_2 hole that its n base (n_B) is diffused from. To implement the simple fabrication process, it is desirable to diffuse p_E and p collector (p_C) simultaneously by the same boron diffusion process as for the p base of the vertical npn transistor.

To achieve high BVceo, the p^- layer ($p_{\bar{C}}$) is introduced. Its surface impurity concentration (NA) is designed to be lower than n_B and higher than the n^- substrate. Since the $p_{\bar{C}}$ must act as the collector in the range of the operating voltage, $p_{\bar{C}}$ should not be completely depleted. But, when a high voltage which is equal to the breakdown voltage is applied to the SADDL transistor, the depletion layer is fully extended into the $p_{\bar{C}}$ layer, but only slightly into the n_B layer. Thus, punch through between the emitter and the collector is avoided in spite of the narrow n_B, and, at the same time, the electric field at the corner of the $n_B p_{\bar{C}}$ junction is also reduced and high BVceo is achieved.

(b-2) Calculated Breakdown Voltage and Current Distribution

Figure 4.36 shows the calculation model structure and the calculated relationships between the breakdown voltage and NA for the SADDL transistors with three $p_{\bar{C}}$ depths. The symbols ○, □, △ indicate site a, b and c, respectively. When NA is much lower than each peak point of the curves, the breakdown voltage is limited to a small value by the electric field at site b shown in the calculation model structure, because the depleted layer extends into $p_{\bar{C}}$ more easily and is abruptly stopped at p_C. When NA closely approaches the value of the peak points, the breakdown voltage becomes limited by the electric field at site a and becomes a large value. When NA is higher than that of the peak points, the breakdown voltage is limited by the electric field at site c and abruptly becomes small. By using Fig. 4.36, an optimum NA can be obtained. By increasing the distance between n_B and p_C, breakdown voltage can be increased further because the electric field at site a is reduced.

Figure 4.37 shows the calculated current distributions for four kinds of lateral transistors, which have the surface impurity concentrations and the distances from p_E to p_C as shown in Table 4.1. The current distributions of the SADDL transistors concentrate near the Si surface as compared with the conventional transistor D.

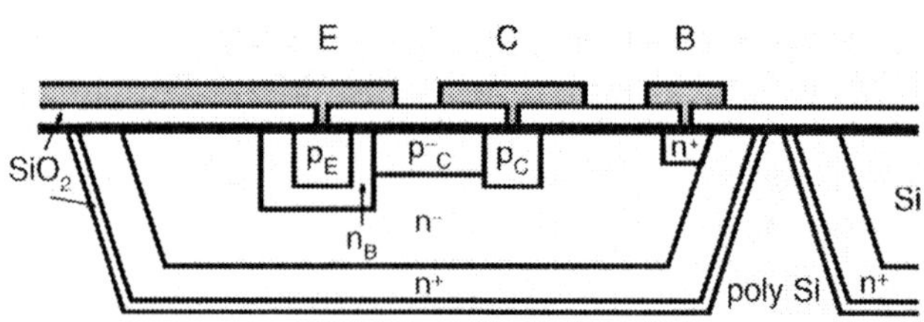

Fig. 4.35 SADDL pnp transistor

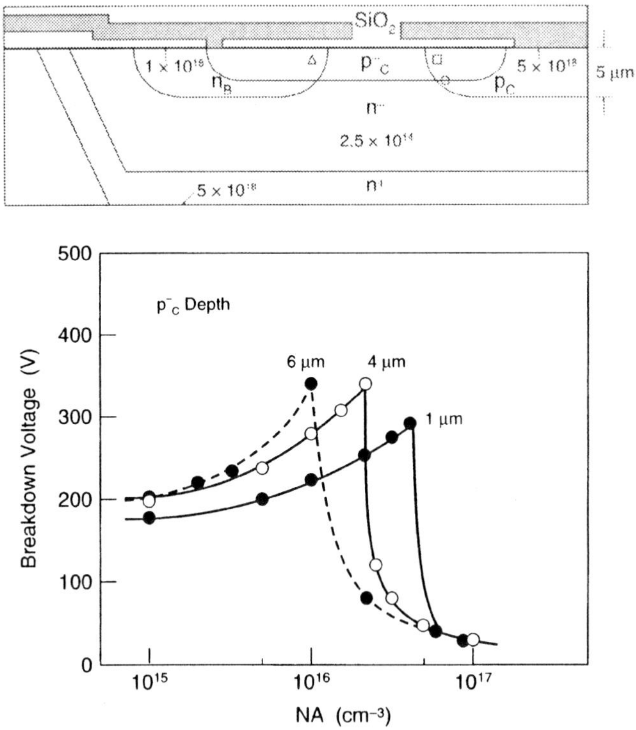

Fig. 4.36 Calculated relationships between breakdown voltage and impurity concentration of p^- layer

From the figure, it is clear that the $p_{\bar{C}}$ which has a low impurity concentration to achieve a high breakdown voltage acts as the collector and the SADDL transistor operates as an ideal lateral transistor. As the impurity concentrations of p_E, n_B and $p_{\bar{C}}$ become low, the current distribution of the SADDL transistors expands into the bulk and approaches that of the conventional transistor.

(b-3) Dependence of Electrical Characteristics on Ic and Vce

Figure 4.38 shows typical experimental relationships between the electrical characteristics and Ic of the lateral pnp transistors. BVceos of the transistors are 250 V. The cutoff frequency f_T of the conventional transistor (type D of Table 4.1) is 0.25 MHz. The SADDL transistor area is about 0.045 mm^2 and its emitter area is about 1200 µm. To reduce the transistor area, the field reduction technique FRR in Section 4.5.3 is utilized. The h_{FE} and f_T of SADDL transistor are very high as compared with those of the conventional transistor. Current capacity of the former is also increased.

Figure 4.39 shows the dependence of Ic on V_{CE} of the lateral pnp and vertical npn transistors [31]. In the SADDL pnp transistor, the dependence of Ic on V_{CE} has been remarkably improved. The Early voltage of the conventional transistor

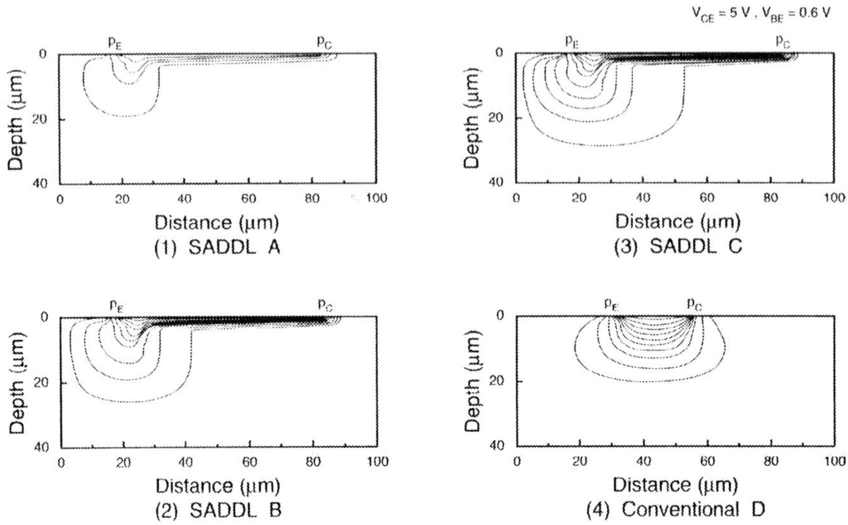

Fig. 4.37 Calculated current distributions

Table 4.1 Surface impurity concentrations and distances for calculated SADDL transistor

	$p_E \cdot p_C$	n_B	$p_{\bar{C}}$	Distance from p_E to p_C
SADDL A	1×10^{19} cm^{-3}	1×10^{18} cm^{-3}	1×10^{16} cm^{-3}	50 μm
SADDL B	5×10^{18} cm^{-3}	1×10^{18} cm^{-3}	8×10^{15} cm^{-3}	50 μm
SADDL C	5×10^{18} cm^{-3}	5×10^{17} cm^{-3}	1×10^{16} cm^{-3}	50 μm
Conventional D	5×10^{18} cm^{-3}	5×10^{14} cm^{-3}	-----	20 μm

is about 200 V, but that of the SADDL transistor is larger than 1000 V. This output characteristic is suitable for obtaining accurate high voltage analog ICs.

(b-4) Trade-Off Between f_T and BVceo

Generally, the lateral transistor shows the tendency that h_{FE} and f_T become high, but BVceo becomes low, according to decreases in the n base width. Figure 4.40 shows the trade-off between f_T and BVceo for three kinds of lateral transistors. The solid line and the dotted line show the trade-off for the SADDL transistor and the conventional transistor, respectively. The broken line shows the trade-off for the modified transistor which has a structure with n_B removed from the SADDL transistor, and having $p_{\bar{C}}$ only. To achieve high breakdown voltage, the n base width of the modified transistor must be wide. It is designed with a value from 17 to 47 μm. This modified transistor has a trade-off of the same level as that for the other modified lateral transistors which have been reported [33–35]. The trade-off for the SADDL transistor has been improved about ten times over that for the modified transistor and about a hundred times over that for the conventional transistor.

Fig. 4.38 Relationships between electrical characteristics and Ic

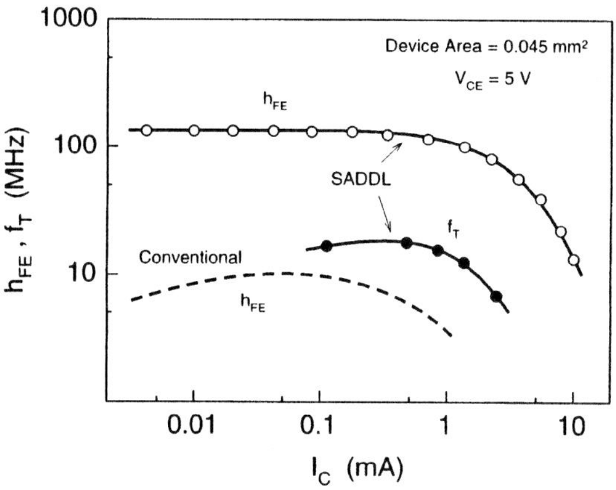

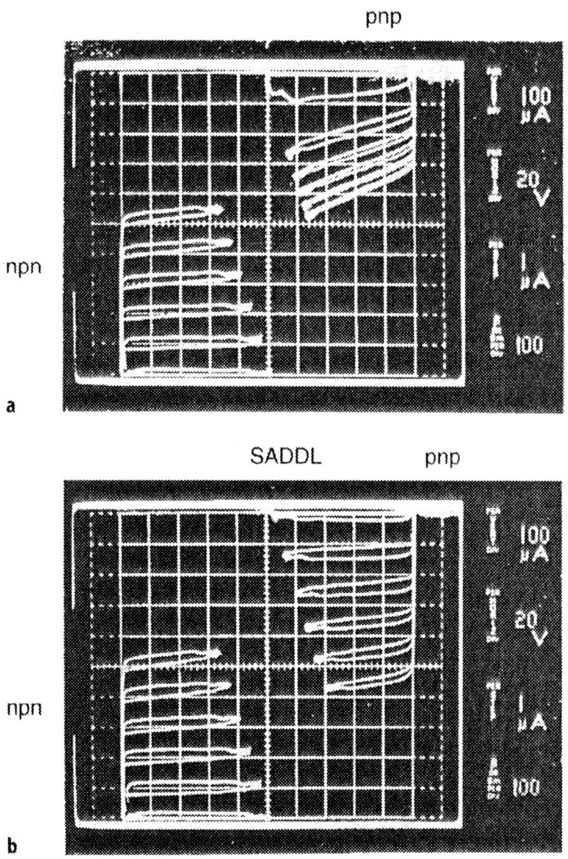

Fig. 4.39 Comparison of output characteristics of npn and pnp transistors

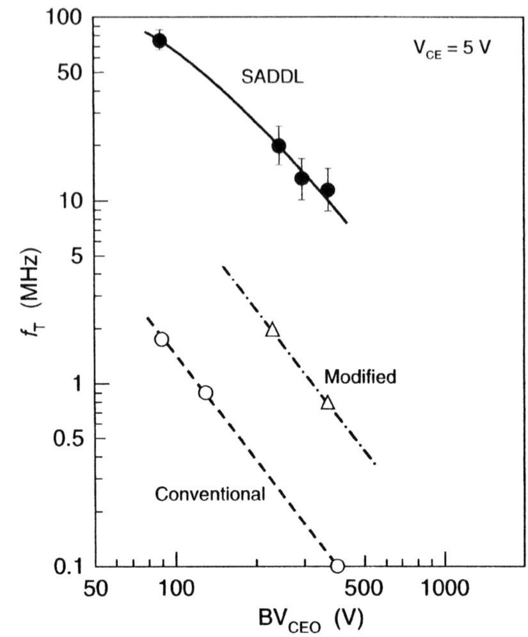

Fig. 4.40 Trade-off between f_T and BV_{CEO} for lateral transistors

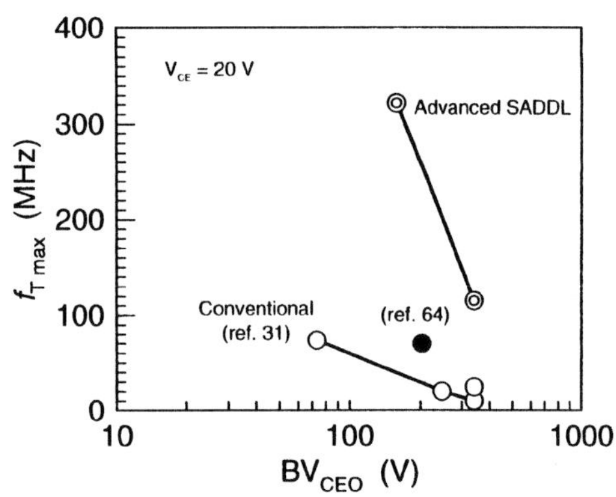

Fig. 4.41 Trade-off between f_T and BV_{CEO} for the advanced SADDL transistor

By reduction of junction capacitances; by making the junction depths shallower, and by changing the base electrode position from the n^+ buried layer to the n^+ base contact layer, the trade-off has been developed further [36]. Figure 4.41 shows the trade-off which has been improved about 5 times compared to the that of the SADDL transistor shown in Fig. 4.40. For the SADDL transistor with BV_{CEO} of 150 V, f_T and h_{FE} are 320 MHz and 100, respectively. And, for the SADDL transistor with BVceo of 340 V, f_T and h_{FE} are 120 MHz and 50, respectively.

4.6.3
Lateral Thyristors and GTOs

To improve electrical characteristics of lateral thyristors, their dependences on device structures are investigated.

Figure 4.42 shows the structure of a lateral DI gate turn-off thyristor (GTO) and Fig. 4.43 shows the relationships between the characteristics and base width [29]. Since on-resistance R_{ON} (at $I = 30$ mA) depends on the surface length not only between the p emitter and p base, but also between n^+ emitter and n-base, the length W_B between the n^+ emitter and p emitter is adopted as the horizontal axis of Fig. 4.43. On the other hand, breakdown voltage BV and interruption current I_{OFF} depend only on the length between the p emitter and p base, so the

Fig. 4.42 Lateral DI gate turn-off thyristor (GTO)

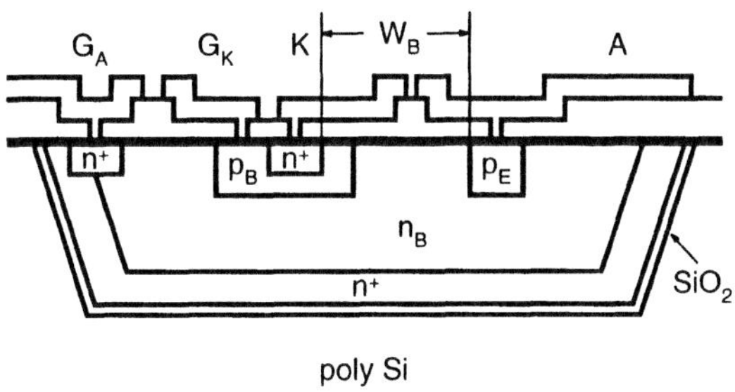

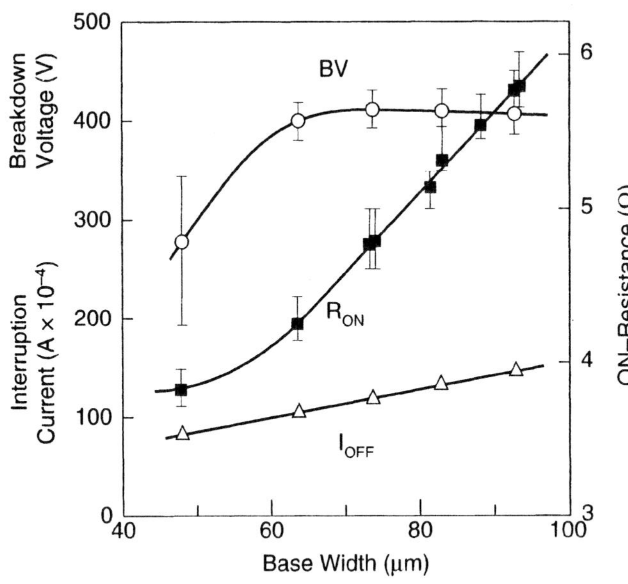

Fig. 4.43 Relationships between GTO characteristics and base width

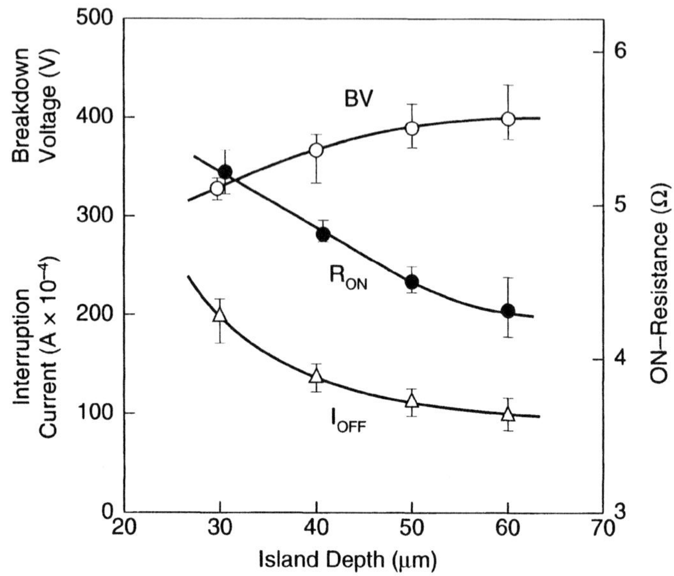

Fig. 4.44 Relationships between characteristics and island depth

surface length between the n^+ emitter and n- base is fixed at a constant value for the samples that are measured. R_{ON} decreases as the base width decreases, and it is much smaller than 10 Ω which is the practical limit for widespread applications. Therefore, size limits of W_B are decided by the breakdown voltage. The length of the p emitter is 150 μ, but R_{ON} is so small as compared with the limits that the length can be shortened.

Figure 4.44 shows the relationships between the characteristics and island depth [29]. By decreasing the island depth, both I_{OFF} and R_{ON} increase, but BV decreases. According to these data, the island depth should be decided by BV. The thyristor which is in the on-state seems to behave like a pin diode [37], and the n-base seems to correspond to the i layer of the diode. Since the resistance of the n-layer increases by increasing W_B and decreasing D, R_{ON} of the thyristor increases. By decreasing the thyristor length, BV does not change, but I_{OFF} and R_{ON} increase. Therefore, the thyristor length is decided by R_{ON}.

From these data, practical size limits of a lateral DI thyristors can be decided.

4.6.4
Lateral IGBTs

Figure 4.45 shows a DI lateral IGBT (LIGBT) fabricated in a SOI-substrate with a Si layer of 5–32 μm thick and a buried oxide layer of 1–3 μm [38]. The resistivity of the Si layer is 45–65 Ωcm and that of the n^+ substrate is 0.01 Ωcm. Figure 4.46 shows comparison of the on-state voltage between JI and DI LIGBTs

Fig. 4.45 Lateral DI IGBT (LIGBT). (Reprinted by permission of IEEE)

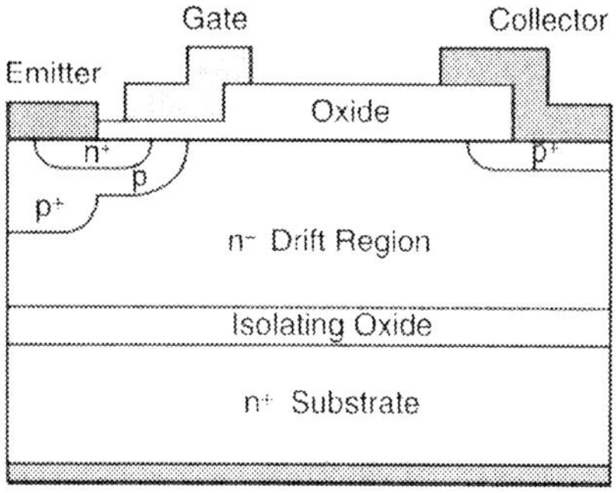

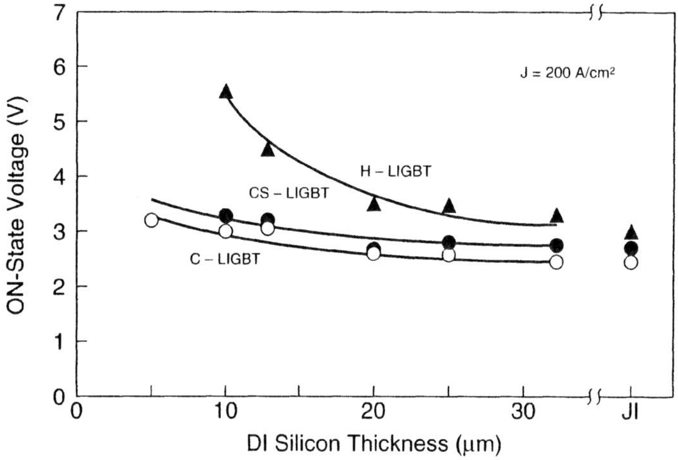

Fig. 4.46 Comparison of on-state voltages between JI and DI LIGBTs with various Si thickness. (Reprinted by permission of IEEE)

with various Si thicknesses. Data for three kinds of LIGBTs are shown in the figure, which are the conventional LIGBT (C-LIGBT), the collector shorted LIGBT (CS-LIGBT) and the hybrid Shottky LIGBT (CH-LIGBT). Although the on-state voltages decrease as the Si thickness increases, these are slightly larger than those of the JI LIGBTs. The reason is because a parasitic vertical pnp transistor with high h_{FE} exists for the JI LIGBTs and a large fraction of the collector current flows into the substrate and reduces the on-state voltage. The latching current of the DI IGBTs is slightly lower than that of JI LIGBTs due to the elimination of the current flow into the substrate. The turn-off time of the DI LIGBTs (about 1 μs) is shorter by one order of magnitude as compared with that of JI LIGBTs. Because the current flows into the substrate, a large number of carriers is stored in the substrate and this makes the switching speed very slow for the JI IGBTs.

Figure 4.47 shows the on-state voltage and short circuit withstand capability Tp as a function of the collector emitter distance L_{CE} [39]. Tp is measured as the time until the LIGBT is destroyed when a load of the circuit is shorted at the gate

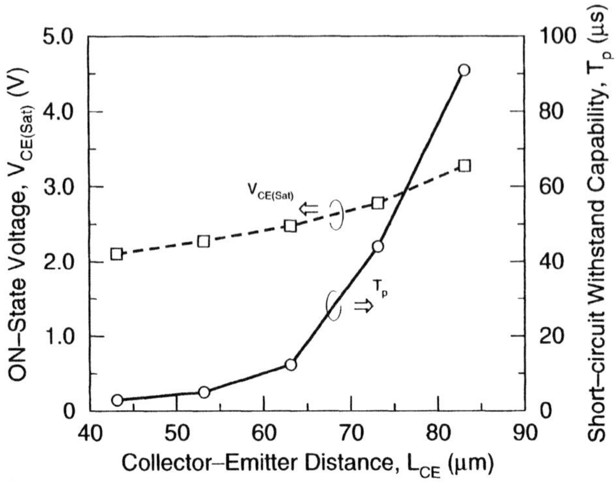

Fig. 4.47 On-state voltage and short circuit withstanding capability as a function of collector-emitter distance

voltage V_G of 15 V and the applied voltage V_{CE} of 280 V. The on-state voltage increases when the L_{CE} increases because the resistivity of the n drift layer becomes large. Tp increases abruptly with L_{CE}. In the short-circuit condition, a large saturated current of the LIGBT flows and junction temperature is raised by the IGBT's own power loss. This results in the parasitic thyristor that the LIGBT is operated and the LIGBT is destroyed by latch-up. Therefore, Tp depends strongly on the saturated current and latch-up generation. The saturated current decreases as L_{CE} decreases because the h_{FE} of the pnp transistor formed by the p^+ collector layer, the n drift layer and the p base layer becomes lower. It becomes harder for latch-up to occur with long L_{CE} because h_{FE} is lowered and hole current flowing under the n+ emitter layer is reduced. Therefore, the LIGBT is harder to destroy with long L_{CE}. The Tp has to be longer than the time in which the protection circuit for the short-circuit operates.

4.6.5 MOS Gate Thyristors (COGMOS Thyristors)

Since high voltage power ICs are used under high surge voltage and the large surge current as compared with conventional ICs, electrical isolation between input circuits and output power circuits is very important to reduce electric interferences between them. Additionally, floating operations are also important. For these purposes, relays, transformers and photo-coupled semiconductor devices and capacitor coupled devices [40, 41] are commonly used, but only the latter devices can be integrated into an IC chip. Capacitor coupled devices have two problems: they cannot be driven by DC drive signals and they dissipate large drive powers. To overcome these problems, the COGMOS (Complementary Gated MOS) thyristor has been developed [42, 43].

(a) Device Structure and Operation Mechanism

Figure 4.48 shows an equivalent circuit of a COGMOS thyristor which is composed of six elements, pnp and npn transistors, p and n channel MOSFETs and two diodes [42]. The COGMOS thyristor and its drive circuit are isolated by the gate oxides of the p and n channel MOSFETs. When an input signal is given to the control terminal C, the supply voltage Vs is applied to the gate electrodes of the MOSFETs Q3 and Q4. Under a forward bias condition, the thyristor, which is composed of pnp and npn transistors, is triggered by the p channel MOSFET Q3 when the potential V_K of its cathode electrode K is higher than that of the drive circuit V_S, and it is triggered by the n channel MOSFET Q4 when the potential V_A of its anode electrode A is lower than that of the drive circuit. Thus, the COGMOS thyristor can be driven under the floating bias condition by the drive circuit with a constant low supply voltage Vs. Figure 4.49 shows the structure of the COGMOS thyristor. The cell size is 325 μm × 280 μm and cell size reduction technique FRR [22], explained in Section 4.5.3, is utilized. The six elements shown in Fig. 4.48 are merged in one device. The lateral p channel MOSFET and the vertical n channel MOSFET are fabricated in parallel with the lateral pnp transistor and the vertical npn transistor by making MOS gate electrodes on the $n_{\bar{B}}$ base and the p_B base, respectively. By connecting the n contact of the $n_{\bar{B}}$ base to the gate electrode Ga, p_E emitter and p_B base junctions are utilized as diodes D1 and D2, respectively. The n_E and p_B of the vertical n channel MOSFET under the MOS gate electrode Gk are fabricated with a self-aligned technique to obtain high transconductance and a low threshold voltage. To achieve high blocking voltage, the junction termination extension technique [14] is used under the gate electrodes so that the lateral p channel MOSFET has double off-set gates not only on the side of the p_E source but also on the side of the p_B drain. The combination technique of the field plate and IB channel stopper in Section 4.5.2 [20] is used except under the gate electrodes.

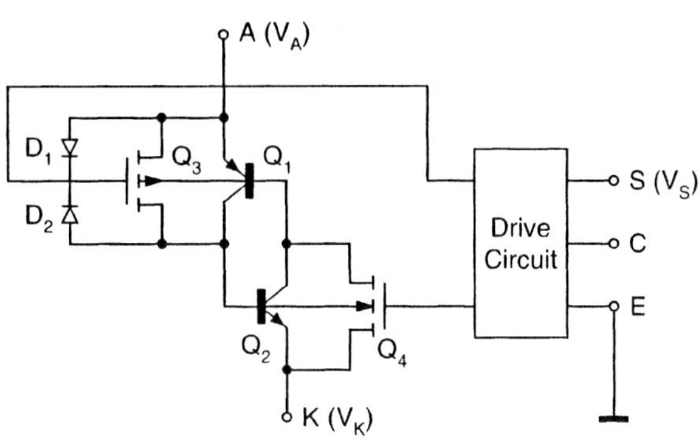

Fig. 4.48 Equivalent circuit of a COGMOS thyristor

Fig. 4.49 Structure of COGMOS thyristor

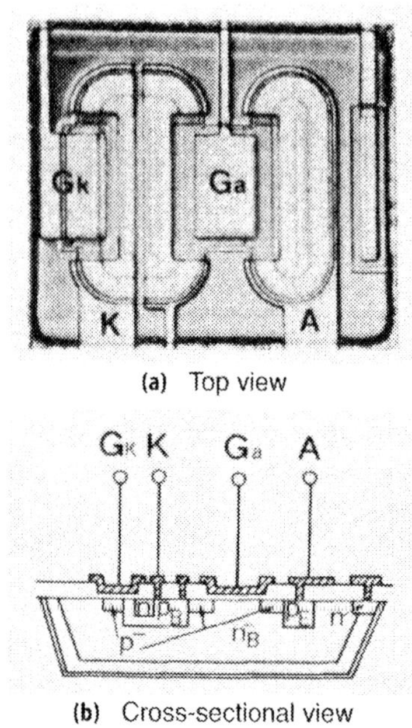

(b) Blocking Voltage and Triggering MOS Gate Voltage

In conventional lateral power MOS devices, high forward blocking voltages are realized by sacrificing the reverse blocking voltages, namely by connecting between the n^+ source and p body. In the COGMOS thyristor, both good forward and reverse blocking voltages must be realized. As shown in Fig. 4.48, a specific method using short diodes D1 and D2 is developed to protect the gate electrode from floating and to reduce generation of channels under the gate electrode. By connecting the contact of the $n_{\bar{B}}$ to the gate electrode, p_E emitter and p_B base junctions of the thyristor act as short diodes D1 and D2, and the potential of the gate electrode is automatically fixed to the highest potential by these diodes and generation of a channel under the gate is reduced. Thus, high bidirectional blocking voltages of about 350 V are achieved. The blocking voltage increases by increasing the length of the p^- layer because the electric field reduction effect due to the p^- layer is increased.

Figure 4.50 shows experimental relationships of isolation voltage or triggering MOS gate voltage related to the thickness of the gate oxide film. The isolation voltage and the triggering MOS gate voltages become larger in proportion to the increased thickness. When the thickness of the oxide film is 0.7 μm, p and n channel triggering MOS gate voltages of about 5 V and isolation voltages of more than 500 V can be realized at the same time. Forward voltage drop is 1.8 V at an average current of 200 mA and on resistance is 7 Ω. The $^{dv}/_{dt}$ capability is 1000 V/μs.

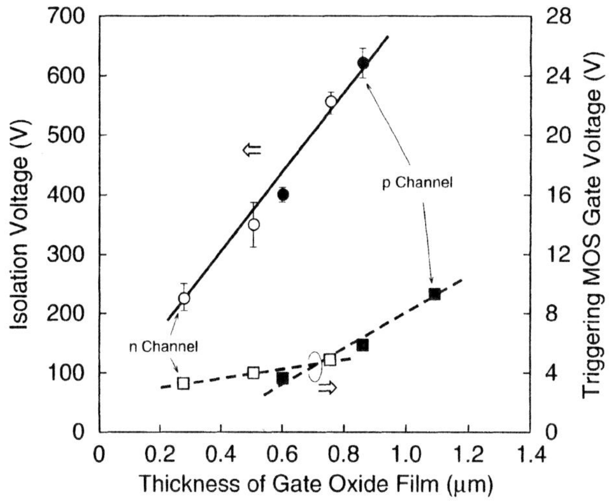

Fig. 4.50 Experimental relationships of isolation voltage or triggering MOS gate voltage vs. thickness of gate oxide film

Fig. 4.51 Floating bias operation of COGMOS thyristor

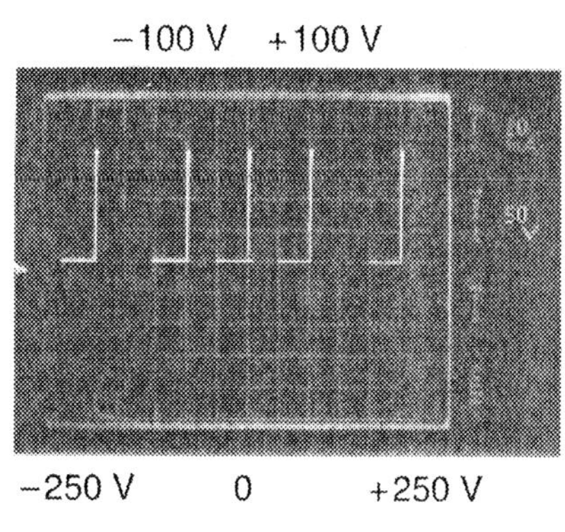

(c) Floating Bias Operation

Figure 4.51 shows the floating bias operation of the COGMOS thyristor. The drive circuit is a flip-flop type [43] and is composed of the high voltage p and n channel MOSFETs which have the same structure as the MOSFETs in the COGMOS thyristor. Although the voltage of the cathode is changed from −250 V to +250 V, the COGMOS thyristor can be turned on, and the floating bias operation is obtained. The main electrodes A or K (Fig. 4.48) and the electrodes of the drive circuit S, C or E (Fig. 4.48) are isolated electrically by the gate oxides of both p and n channel MOSFETs Q3 and Q4, the diodes, and the MOSFETs of the drive circuit. An isolation voltage of more than 350 V is achieved between the COGMOS thyristor and the terminals of the drive circuit.

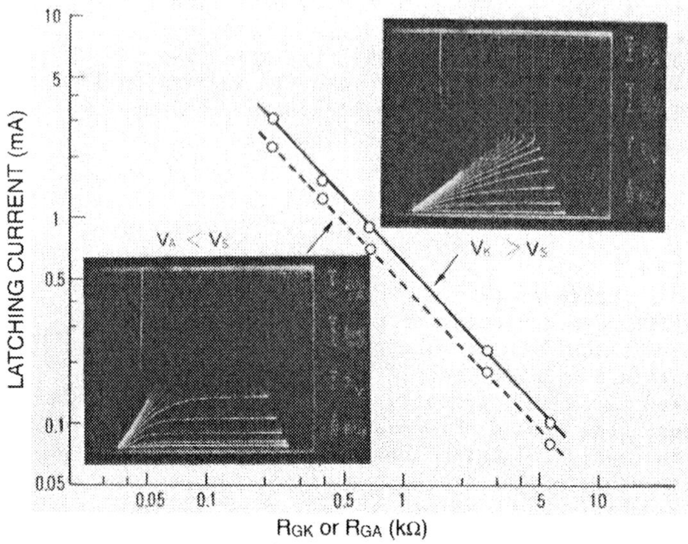

Fig. 4.52 Output characteristics and latching current vs. R_{GK} and R_{GA}

For telecommunications uses, linearity of output characteristics before latching and a large latching current are needed in addition to the electrical characteristics described above. As shown in Fig. 4.52, the necessary good linearity is realized and the latching currents are controlled by the resistance between the p gate and the cathode, R_{GK} and that between the n gate and the anode, R_{GA}. These electrical properties of the COGMOS thyristor make many applications possible in place of photo and capacitor coupled devices, relays and transformers.

The concept of this COGMOS thyristor can be applied to other COGMOS devices, such as COGMOS transistors.

4.6.6
SOI-RESURF Devices

High voltage devices are developed using the SOI-RESURF high voltage techniques described in Section 4.5.4. In this section, the devices which are fabricated in very thin SOI-layers are introduced.

(a) SOI-RESURF MOSFETs

Figure 4.53 shows the SOI-RESURF lateral DMOSFET, and Fig. 4.54 shows its experimental relationships between breakdown voltage and drift length with different buried oxide thickness (b) [44]. Thickness of the SOI-Si layer (t) is 0.2 μm. The SIPOS film is deposited on the thin Si layer to realize a uniform horizontal electric field in the Si layer. Breakdown voltage approaches an ideal

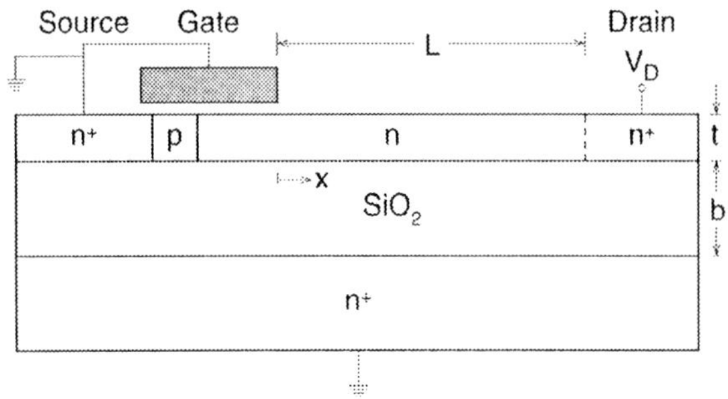

Fig. 4.53 SOI-RESURF lateral DMOSFET (b: buried oxide thickness). (Reprinted by permission of IEEE)

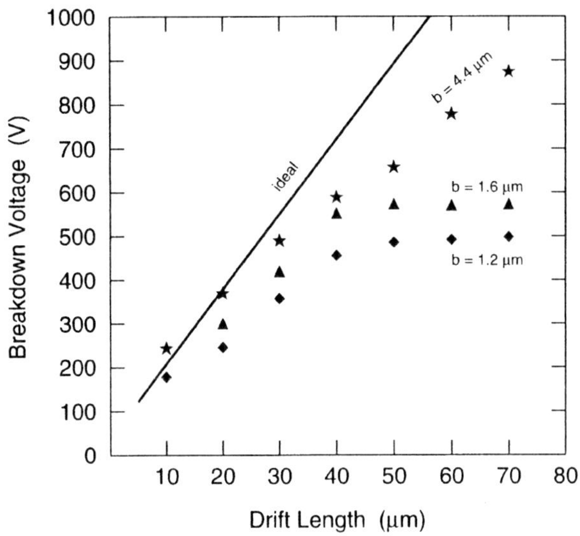

Fig. 4.54 Experimental relationships between breakdown voltage and drift length. (Reprinted by permission of IEEE)

curve when the buried oxide becomes thick and the SOI-Si layer becomes thin. The deviation from the ideal curve in the long drift length region is caused by the charge being less than the optimal value in that region. When thickness of the buried oxide layer is 4.4 μm and the drift length is 70 μm, a breakdown voltage of 860 V is realized. On-resistance is 140 Ω. Simulations show that on-resistance becomes large and saturates to a 20% larger value when substrate potential becomes negative [45, 46]. The reason is that the n drift region near the buried oxide is depleted. To supress the influence of the substrate potential, the Si layer has to be made thicker than the effective inversion layer thickness. In this case, the p channel generated on the buried oxide can shield the subtrate bias. When substrate potential becomes positive, breakdown voltage of SOI-MOSFET deteriorates. But, when SIPOS or Si_xN_y films are used as passiva-tion films in place of SiO_2 film, deterioration of the breakdown voltage is suppressed [47].

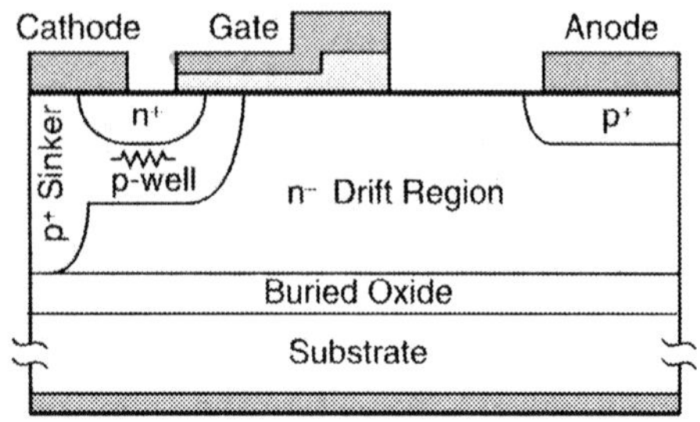

Fig. 4.55 SOI-RESURF LIGBT

(b) SOI-RESURF IGBTs

Figure 4.55 shows an SOI-RESURF LIGBT [48]. Thickness of the Si layer and the the buried oxide layer are 2 μm and 1 μm, respectively, and the drift region length is 25 μm. It has a one-sixth shorter turn-off time (520 ns), and has a small 10 % increase in on-resistance (2.94 Ωmm^2), as compared with those values of an equivalent LIGBT formed in bulk Si substrate [48, 49]. The shorter turn-off time is due to the buried oxide layer which prevents accumulation of charge in the substrate and reduces the volume of stored charge and which can support the increasing anode voltage immediately when the device turns off. The small increase in on-resistance is due to the restricted current flow in the thin drift region due to the presence of the buried oxide; in the bulk LIGBT, the current flow expands in the epi-substrate. The latch-up current of the SOI-RESURF LIGBT is increased to 1335 A/cm^2 which is six times that of the bulk LIGBT, when the pinch resistance of its p-well is reduced by a dual implantation. After the the standard implantation, a second implantation is performed with higher dose (1×10^{15}/cm^2) and higher energy (100 keV) in the region only under the n$^+$ source, using a self-alignment technique. Therefore, deterioratation of its threshold voltage can be prevented [50].

(c) SOI-RESURF ESTs

An SOI-lateral EST of the 500 V class is also developed in a SOI-substrate with Si layer of 5–32 μm thickness and buried oxide layer of 1–3 μm [51, 52]. But, its forward voltage drops and turn-off times are similar and not much superior to those of LIGBTs. One reason is because the cell pitch of the SOI-lateral EST is 40 % larger than that of LIGBT due to incorporation of the floating n$^+$ region. A second reason is because a larger fraction of the anode current flows through the MOSFET channel in the former and the voltage drop in the channel becomes large.

4.6.7 Photo-Coupled Devices

Power ICs are generally used under the conditions of high voltage, large current and large surge voltage and surge current. Photo-coupled devices are suitable for utilization under such conditions as compared with current drive devices, since they have a high isolation voltage between input and output circuits and they can operate accurately under floating bias conditions. Junction isolation is unsuitable for photo-coupled devices, because photo currents are generated at isolation junctions and this results in electrical interferences between integrated devices. Therefore, photo-coupled ICs are developed using only dielectric isolation.

(a) Photo-Coupled Cross-Point Switch IC for Exchanges

Figure 4.56 shows a photo-coupled cross-point switch for exchanges [53]. The IC is mainly composed of a thyristor, resistor R_{GK} and transistors Q1 and Q2. The

Fig. 4.56 Photo-coupled crosspoint switch for exchanges. (Reprinted by permission of IEEE)

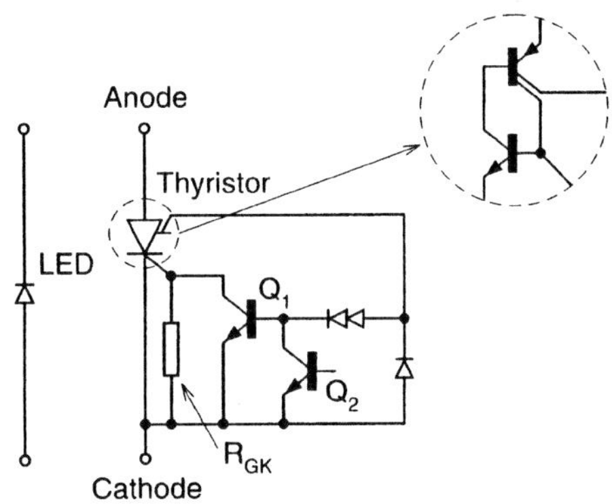

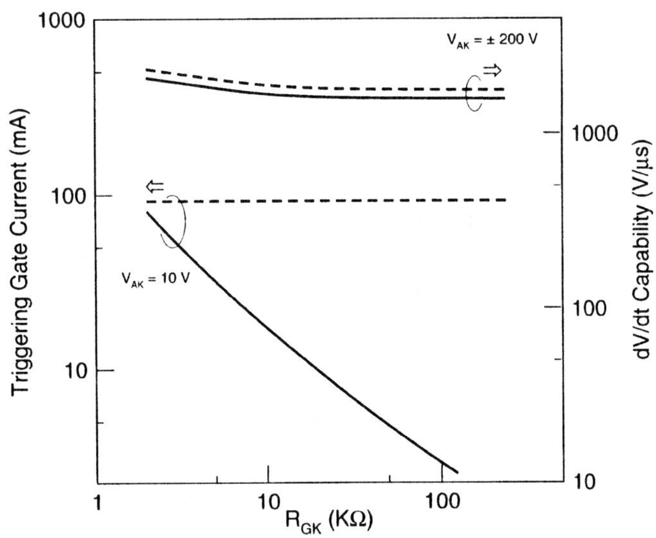

Fig. 4.57 R_{GK} dependence of triggering gate current and allowable dv/dt capability. (Solid lines: circuit in Fig. 4.56, Dotted lines: circuit without Q2)

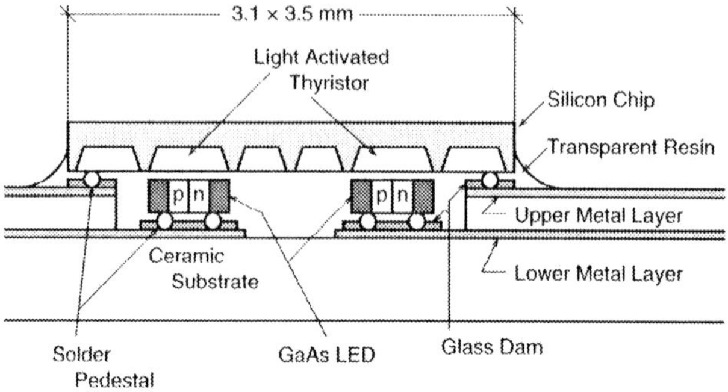

Fig. 4.58 Photo-coupled crosspoint DI switch

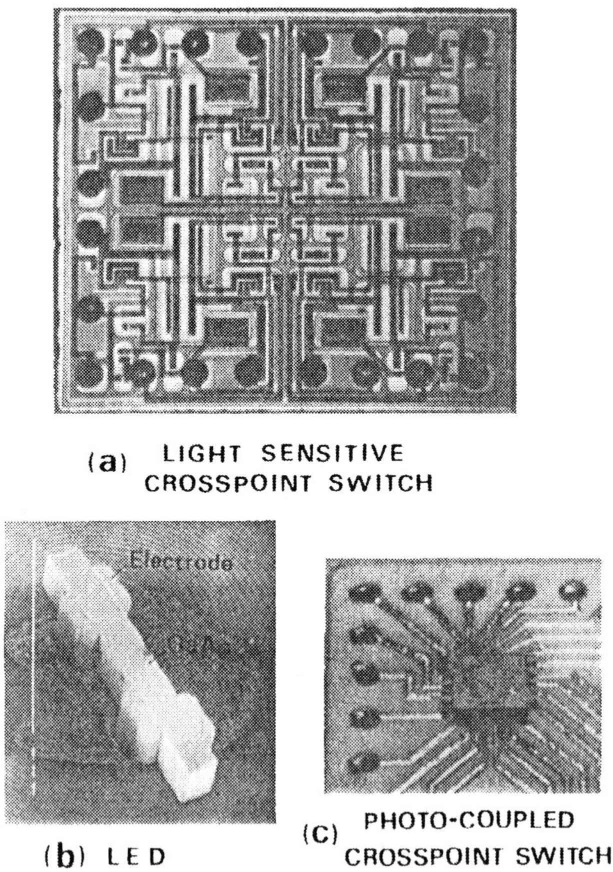

Fig. 4.59 Devices in the photo-coupled crosspoint DI switch

transistor Q1 is driven by a displacement current induced through a change in the anode voltage of the thyristor and functions for increase of allowable dv/dt capability. The resistor R_{GK} is designed very high, 40 Ω, to increase triggering sensitivity of the thyristor. When the thyristor is triggered by the light of the LED, Q1 is also driven by stray light. This results in a decrease of triggering sensitivity. The photo-transistor Q2 acts to keep high triggering sensitivity of the thyristor. When the thyristor is triggered by the LED, Q2 is also driven and it shorts the emitter junction of Q1. Thus, Q1 cannot act and the triggering sensitivity can be kept. Figure 4.57 shows R_{GK} dependence of the triggering gate current and allowable dv/dt capability. Solid lines refer to values obtained with the circuit in Fig. 4.56 and dotted lines refer to the circuit from which Q2 is removed. Low triggering gate current and high dv/dt capability are realized at the same time.

Figure 4.58 shows a cross-sectional view of the photo-coupled cross-point switch and Fig. 4.59 shows devices which compose the switch. Four units of a bi-directional switch which is composed of the two circuits shown in Fig. 4.56 are integrated into a light- sensitive cross-point switch IC. Two bi-directional switches, namely 4 thyristors, are simultaneously driven by only one LED. Each bidirectional switch can treat AC signals, such as a ringing signal and a howler signal. To realize efficient photo-coupling, an edge emitting LED and an accurate alignment technique utilizing solder tension are used. The LED length is about 1.9 μm. Blocking voltage is ± 250 V. Average current is 200 mA and on-resistance is 8 Ω. Triggering gate current for 4 thyristors is 15 mA. Turn-on and turn-off times are 15 μs and 8 μs, respectively. The dv/dt capability is more than 1000 V/μs and isolation voltage is more than 1000 V.

(b) Photo-Coupled Switching IC for Cordless Telephones

Figure 4.60 shows a circuit diagram of a high voltage photo-coupled switching IC which controls the switch hook in cordless telephones [54]. The IC is mainly composed of two pairs of MOSFETs, thyristors and photo diode arrays. MOSFETs are connected bi-directionally to realize bi-directional linear output characteristics. The thyristors function as surge protection devices. Photo-diode arrays are connected between the source and the drain of the MOSFETs.

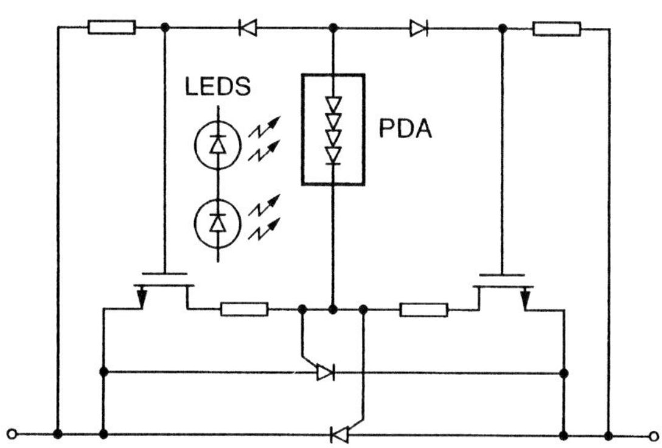

Fig. 4.60 High voltage photo-coupled switching IC for cordless telephones

Two series connected LEDs which are mounted on the IC activate the photodiode arrays and the MOSFETs are driven. Blocking voltage is 150 V. Control current is 2 mA and on-resistance is 20 Ω. Surge capability is 10 A for a 1 ms pulse and isolation voltage is 2500 V.

4.7 Large Electric Power Capability Techniques for DI-PICs

4.7.1 Incrementation of Electric Power Capability

Allowable power dissipation of power ICs is limited by heat radiation, to about 25 W in the case of natural cooling. When power dissipation is over the limit, discrete power devices for output circuits have to be taken out of the IC chip and a module has to be composed as shown in Fig. 4.2 But, an increase of the allowable power dissipation by a monolithic structure has recently been experimented by reduction of both electric and thermal resistances of the power ICs.

Figure 4.61 shows the rising trend in electric power capability from 1982 to 1992 [2]. The electric power capability is defined as the product of the rating voltage and the rating current. By using IGBTs as output devices, the capability

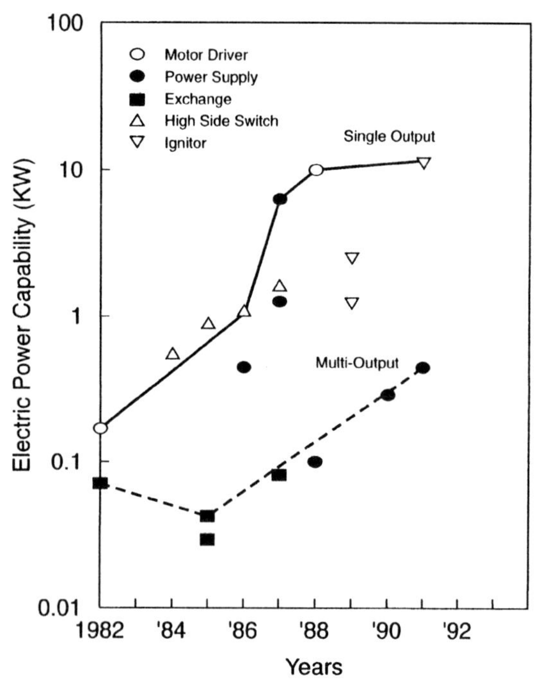

Fig. 4.61 Increment trend of electric power capability. Applications of each ICs are distinguished with symbols shown in the figure

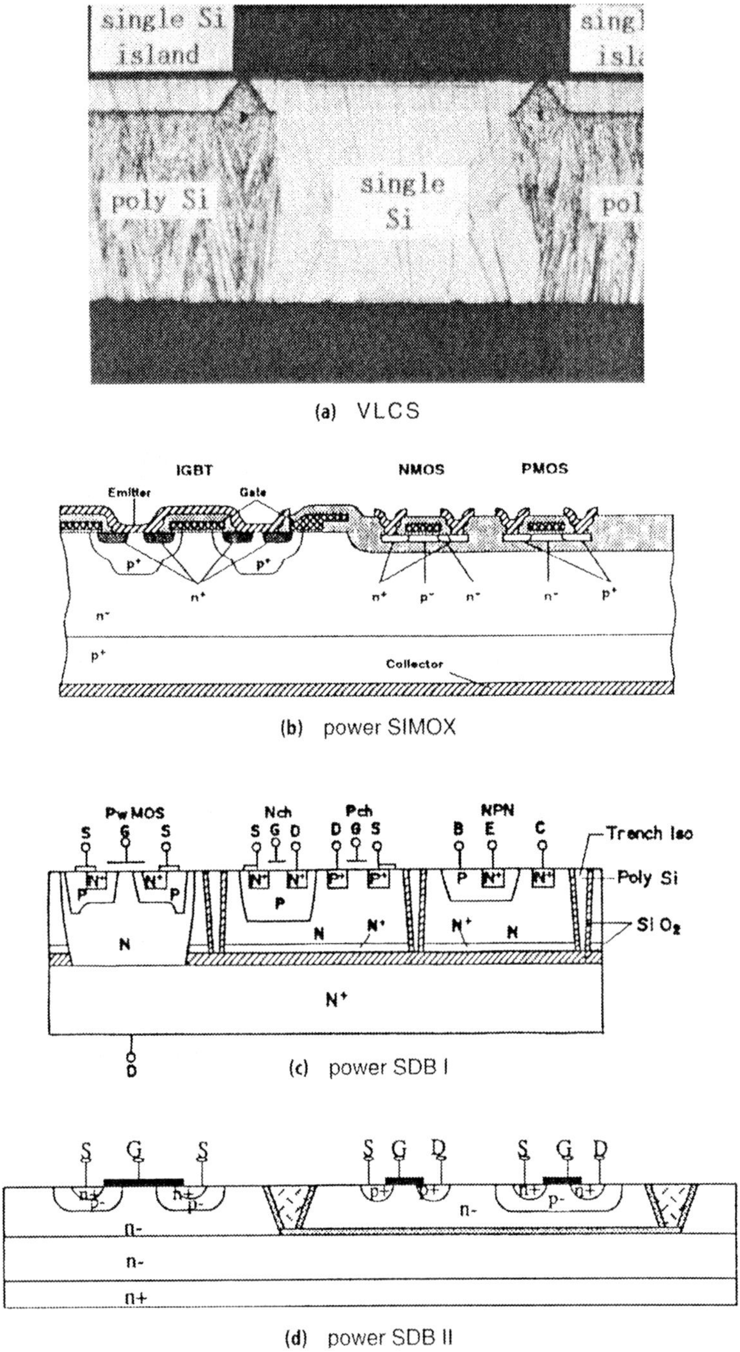

Fig. 4.62 Increment techniques for electric power capability

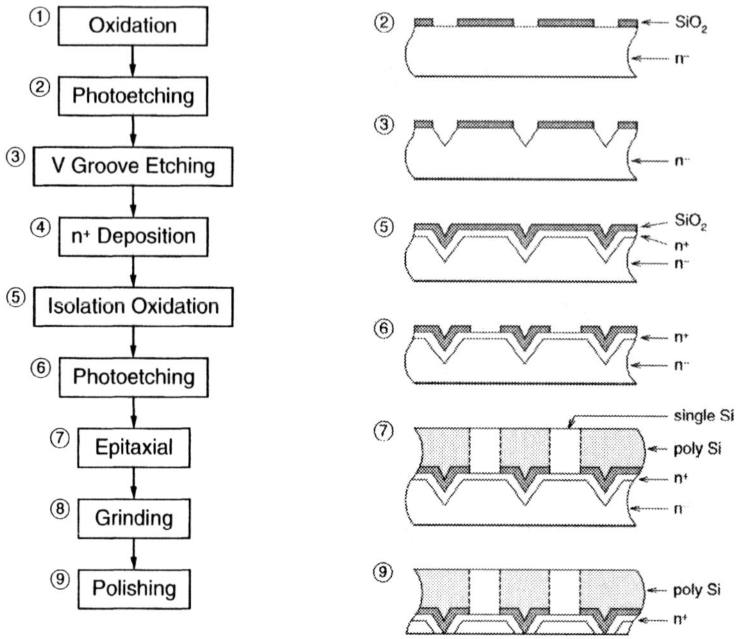

Fig. 4.63 Process flow for VLCS

Fig. 4.64 Bowing height of VLCS DI wafer vs. vertical single Si area rate

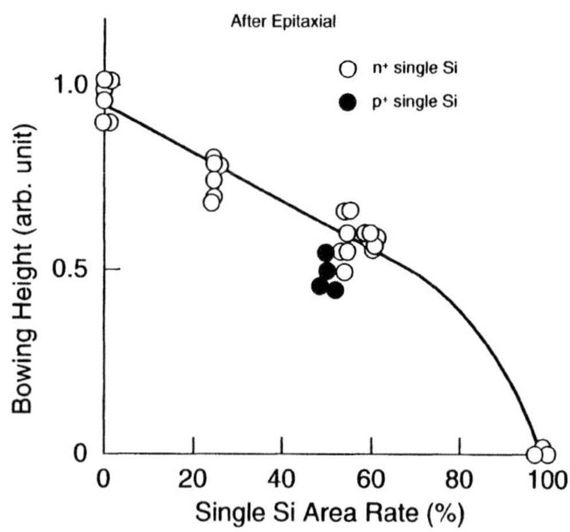

Fig. 4.65 Bowing height of VLCS DI wafer vs. epitaxial temperature

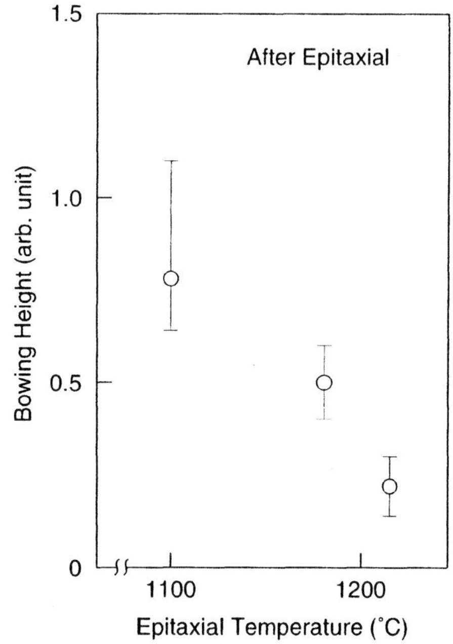

Fig. 4.66 Comparison of EPIC and VLCS

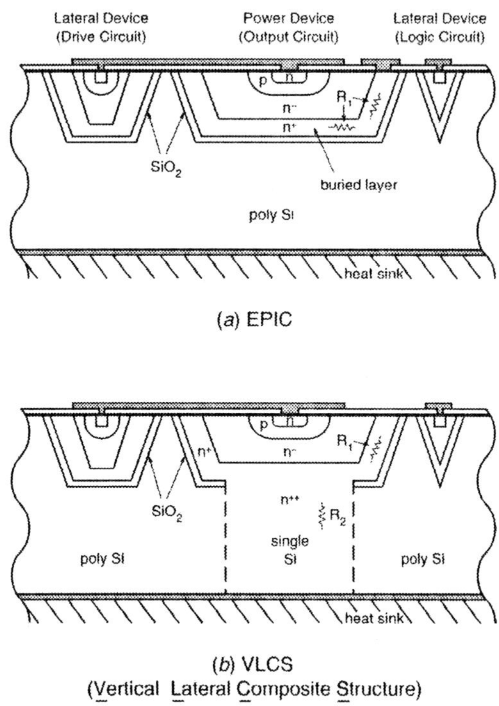

is increased in the case of both multi- and single output ICs. In the case of the single output IC, the capability is close to that of the modules. The capability increment techniques for the DI-PICs are shown in Fig. 4.62. They are VLCS (Vertical Lateral Composite Structure) [55, 56], power SIMOX (Silicon IMplanted Oxygen) [57, 58] and power SDBs (Silicon Direct Bondings) [59, 60]. In all of these techniques, vertical output devices are integrated in place of lateral output devices to increase their capability. The power SIMOX and power SDB I [59] have problems to be considered in relation to mass productivity in the fabrication process for the buried SiO_2 isolation film and for the vertical single Si region, respectively. SDB II [60] is attractive because of its simple process, but it is still under development. In this section, these techniques are introduced.

4.7.2 Vertical Lateral Composite Structure (VLCS)

The VLCS shown in Fig. 4.62a [55, 56] can be formed by addition of one photo-etching process before the poly-Si deposition to the EPIC process as shown in Fig. 4.63. By removing the bottom SiO_2 film of the power device island, single Si can be automatically formed on this removed region in the epitaxial process. The VLCS is effective for the reduction of DI wafer bowing. Fig. 4.64 shows the bowing height of the VLCS DI wafer. By increasing the area of the single Si region, the bowing height is reduced and then more accurate photo-etching becomes easier. The bowing height is almost the same between the VLCS DI wafer with the n^+ epitaxial single Si and with the p^+ one. Fig. 4.65 shows the bowing height dependence on the epitaxial temperature of the vertical single Si region. The bowing height can be reduced by increasing the epitaxial temperature.

Figure 4.66 shows a comparison of EPIC and VLCS [55]. Spreading resistance of the n^+ buried layer R1 becomes very large on increasing the size of the power device for EPIC, but resistance of n^{++} single Si region R2 becomes small on increasing the size of the power device for VLCS. Furthermore, thermal resistance of EPIC is large because both the SiO_2 film and poly-Si layer with large thermal resistances exist between the power device and the heatsink, but the thermal resistance of VLCS is small because only the n^{++} single Si region exists between them. Figure 4.67 and 4.68 show the comparison of output characteristics and thermal resistances for 600 V npn transistors, respectively [56]. Both resistances are reduced drastically, and it is clear that the npn transistor with a very large electric power capability of the 600 V · 20A class can be integrated by using VLCS.

4.7.3 Power SIMOX

The power SIMOX shown in Fig. 4.62b can be formed by implanting oxygen through the SiO_2 mask in the conventional SIMOX process [57, 58]. The oxygen is

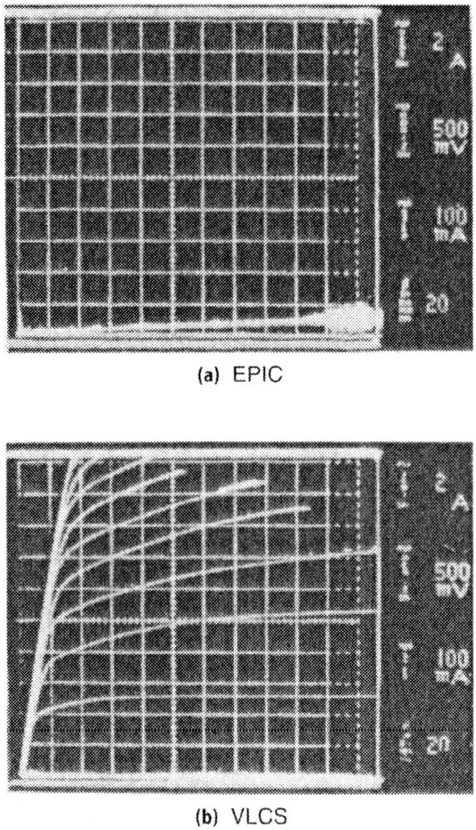

Fig. 4.67 Comparison of output characteristics

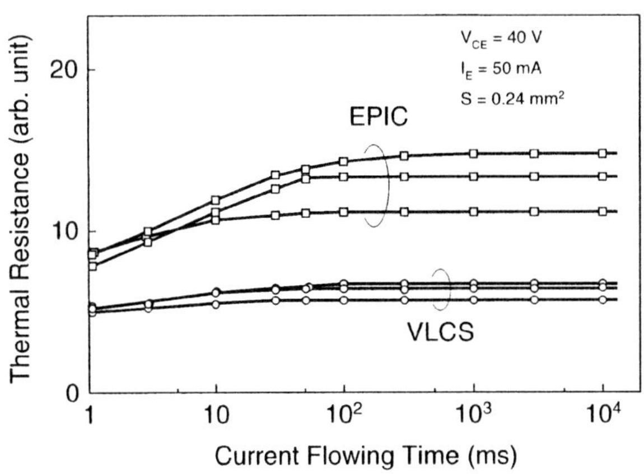

Fig. 4.68 Comparison of thermal resistances

implanted with an energy of 200 keV at a wafer temperature of about 620 °C and the implantation dose is about $2 \times 18^{18}/cm^3$. After reaching 1300 – 1350 °C for recrystallization of the oxygen-doped silicon, devices are fabricated. Typically, the buried oxide thickness is 400 nm and that of the recrystalised single silicon is 250 nm. By using the power SIMOX structure, a power IC of the 500 V · 10 A class for a motor drive is developed [57]. Its output device is an n-DMOSFET with on-resistance of about 0.5 Ω and its turn-on delay time and turn-off delay time are 0.4 μs and 0.2 μs, respectively.

4.7.4 Power SDBs

(a) Power SDB I

Figure 4.69 shows the fabrication process flow of the power SDB I [59]. After oxidation of the n^+/n wafer and n^+ wafer, the two wafers are bonded directly. Next, the silicon and oxide film where the power device is formed, are removed by wet etching. After the etched place is buried with Si epitaxial growth, the lapping and the polishing are performed. Next, lateral isolation regions are formed by the conventional steps of groove etching, thermal oxidation for isolation film formation and refilling of the groove with poly-Si.

By using this power SDB I technique, a high-side switching IC of the 30 V · 8 A class is developed for automotive applications [59]. Its output device is an n-DMOSFET with on-resistance of about 0.1 Ω and its on-switching time and off-switching time are 80 μs and 2 μs, respectively. The on-switching

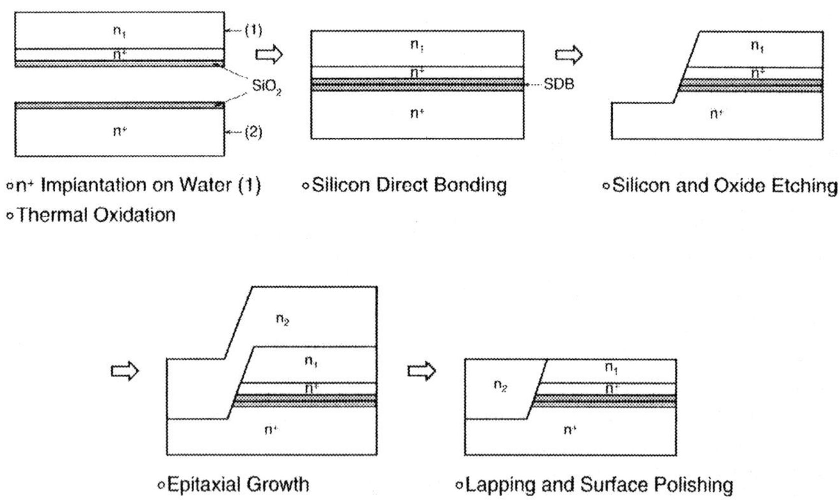

Fig. 4.69 Process flow of power SDB I. (Reprinted by permission of IEEE)

time is long because of the charge-pump circuit for the gate drive of the MOSFET.

(b) Power SDB II

Figure 4.70 shows the fabrication process flow of the power SDB II [60]. After a pan 3 μm deep is formed, an implantation is performed for the n^+ buried region. Next, thermal oxidation is carried out to form an isolation film of more than 2 μm and the wafer is polished until the Si-SiO_2 coexistent optically flat surface is exposed. For a voidless SDB II wafer, a flatness of less than 20 nm has to be ensured. Then, wafer bonding is performed at room temperature. It is important to remove the naturally oxidized film and to bond the wafers in a deoxidized atmosphere to minimize the amount of oxygen at the Si-Si bonding interface. Next, the bonded wafer is subjected to a heat treatment at 1100 °C in a nitrogen atmosphere. After that the side of the bonded wafer on which devices are fabricated is ground and polished to a thickness of several microns. Then, lateral isolation regions are formed by the conventional steps of groove etching, thermal oxidation for isolation film formation and refilling of the groove with poly-Si.

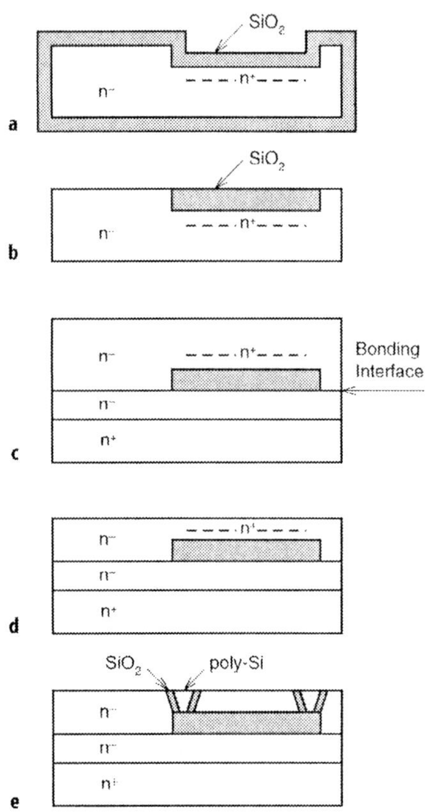

Fig. 4.70 Process flow of power SDB II. (Reprinted by permission of IEEE)

By using this power SDB II technique, a PWM switching regulator IC of the 600 V · 5 A class can be developed [60]. Its output device is the n-DMOSFET with on-resistance of about 0.6 Ω and its switching frequency is 650 kHz.

4.8 Dielectrically-Isolated Power IC

In this section, typical DI-PICs are introduced, which have been developed and in some cases, put into practical use in recent years. Packages used are the SP-23 type, except for the EL driver ICs with FP-48 type packages.

4.8.1 High-Side Switch ICs for Automotive Devices

For down-sizing and high reliability of electric components of automobiles, a number of low-side switches and high-side switch ICs have been developed and put into practical use.

Figure 4.71, 4.72 and Table 4.2 show a circuit block diagram, an IC chip photograph and the electrical characteristics of a 4-channel high-side switch of the 50 V · 1.5 A class, respectively [61]. Owing to the exclusion of charge pump circuits which occupy large chip areas, p channel power MOSFETs are

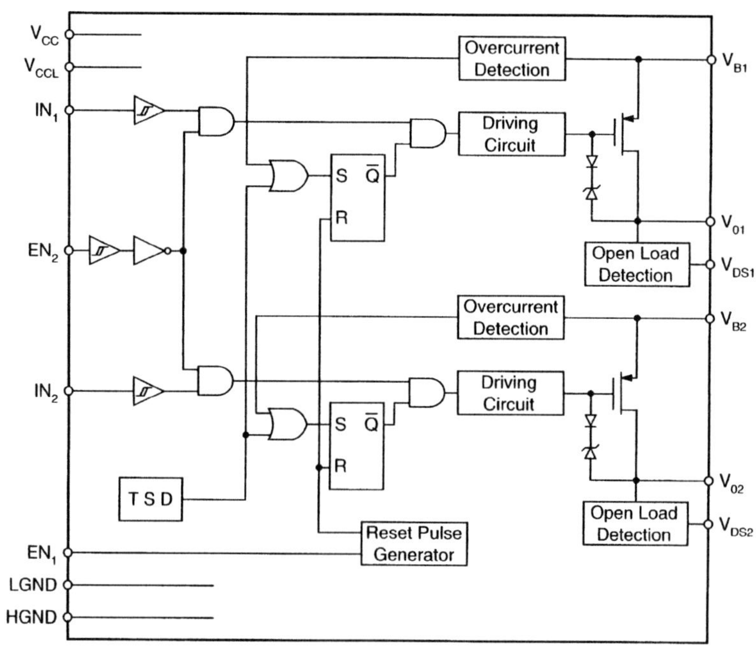

Fig. 4.71 Circuit block diagram of 4-channel high side switch (50 V · 1.5 A)

Fig. 4.72 IC chip photograph

Table 4.2 Electrical characteristics of 4 channel high side switch (50 V · 1.5 A)

No.	Parameter		Typical Value	Unit
1	Supply Voltage		35 -16	V
2	Input Voltage		-0.3 ~ 7	V
3	Output Current		1.5	A
4	Standby power Supply current		10	mA
5	On resistance (IN = 1.5 A, V_B = 12 V)		1	Ω
6	Delay time	t_d on t_d off	15 15	μ sec μ sec
7	Over Voltage Protection		35	V
8	Over current Protection		4	A
9	Open load detection		100	KΩ
10	Over temperature protection		170	°C
11	Input Voltage	V_{IH} V_{IL}	≥33 ≤1.5	V
12	Storage Temperature		-55 ~ 150	°C
13	Junction Temperature		150	°C

utilized as output devices. In addition to the driver circuit of the MOSFETs, some protection and diagnostic circuits for overcurrent, overvoltage and overtemperature have been integrated into the chip. These protection circuits operate at 4 A, 35 V and 170 °C, respectively. Since the EPIC type DI is utilized, electric interference between the channels is removed and operation under a high temperature of 150 °C is possible. Furthermore, chip size is small, being 3.9 mm × 4.8 mm.

4.8.2
Driver ICs for EL Flat Panel Displays

For down-sizing and low cost of EL flat panel displays, multi-channel row and column driver ICs are put into practical use.

Figure 4.73, Fig. 4.74 and Table 4.3 show a circuit block diagram, an IC chip photograph and the electrical characteristics of a 34-channel row driver IC of the

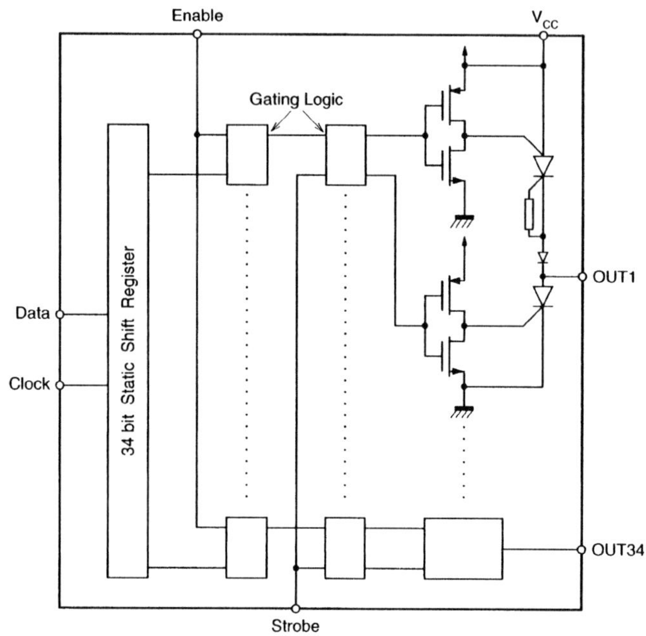

Fig. 4.73 Circuit block diagram of 34 channel row driver IC (280 V · 0.2 A)

Fig. 4.74 IC chip photograph

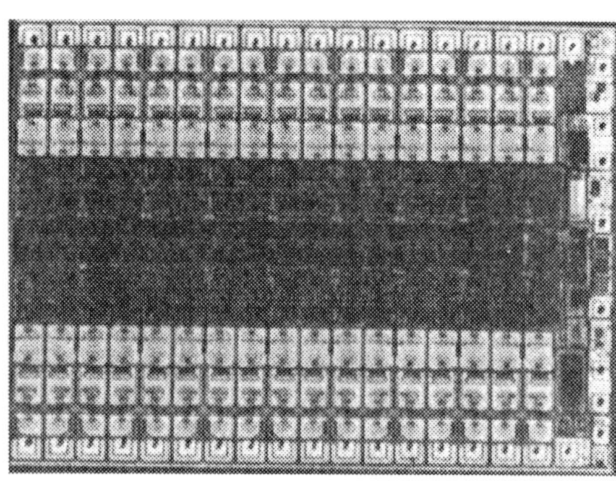

Table 4.3 Electrical characteristics of 34 channel row driver IC (280 V · 0.2 A)

Items	Raw driver	Column driver
Supply Voltage	280 V	70 V
Output Source Current	-200 mA	-30 mA
Output Sink Current	200 mA	10 mA
Clock Frequency	250 KHz	6 MHz
Clock Pulse Width	80 ns	80 ns
Output Delay Time	< 3 μs	< 2 μs
dV/dt Immunity	> 200 V/μs	> 50 V/μs

280 V · 0.2 A class [13]. The output circuit of each channel is the push-pull type and is composed of high voltage lateral thyristors explained in Section 4.6.3, because thyristors have low on-state voltage and a latch-up function which are suitable for reductions of chip size and power dissipation. Turn-on and turn-off circuits are composed of a push-pull type high voltage CMOSFET. The pull-up thyristor is operated by the anode drive mode and the pull-down thyristor is operated by the cathode drive mode. In addition to the turn-on and turn-off circuits of the thyristors, CMOS gating logics and a 34-bit static serial shift resistor are integrated. In spite of the integration of many high voltage channels, chip size is small because the EPIC type DI is used. The chip is 3.74 μm × 2.75 μm. Since thyristors are used for both pull-up and pull-down, not only output source current, but also output sink currents are large, and are 200 mA. So, the driver IC can be applied to large-scale EL panels.

4.8.3 Subscriber Line Interface Circuit for Exchanges

For down-sizing and high reliability of exchange, the subscriber line interface circuit of exchanges is integrated into two chips, a high voltage analog IC and a low voltage analog/digital CMOS LSI.

Figure 4.75 shows a circuit block diagram of the high voltage analog IC for the subscriber line interface circuit [63]. It has functions, such as sending bell signals, test signals and facsimile calling signals, reversing line current polarity for charge payment, battery feeds for telephone receivers, loop monitoring for ring trip, etc. Figure 4.76 shows an IC chip photograph. In Table 4.4, the electrical characteristics of the high voltage analog IC are summarized. It is an IC of the 350 V · 0.2 A class. Output devices of the signal sending circuits are composed of high voltage lateral thyristors of the 400 V· 0.2 A class (cf. Section 4.6.3) and those of analog circuits are composed of high voltage high f_T pnp and npn complementary transistors of the 360 V · 25 MHz class [32]. As the pnp transistors, SADDL pnp transistors (cf. Section 4.6.2) are used. By using the EPIC type DI technique, a small chip size of 4.9 μm × 5.0 μm and highly accurate analog characteristic are realized without electrical interference between devices.

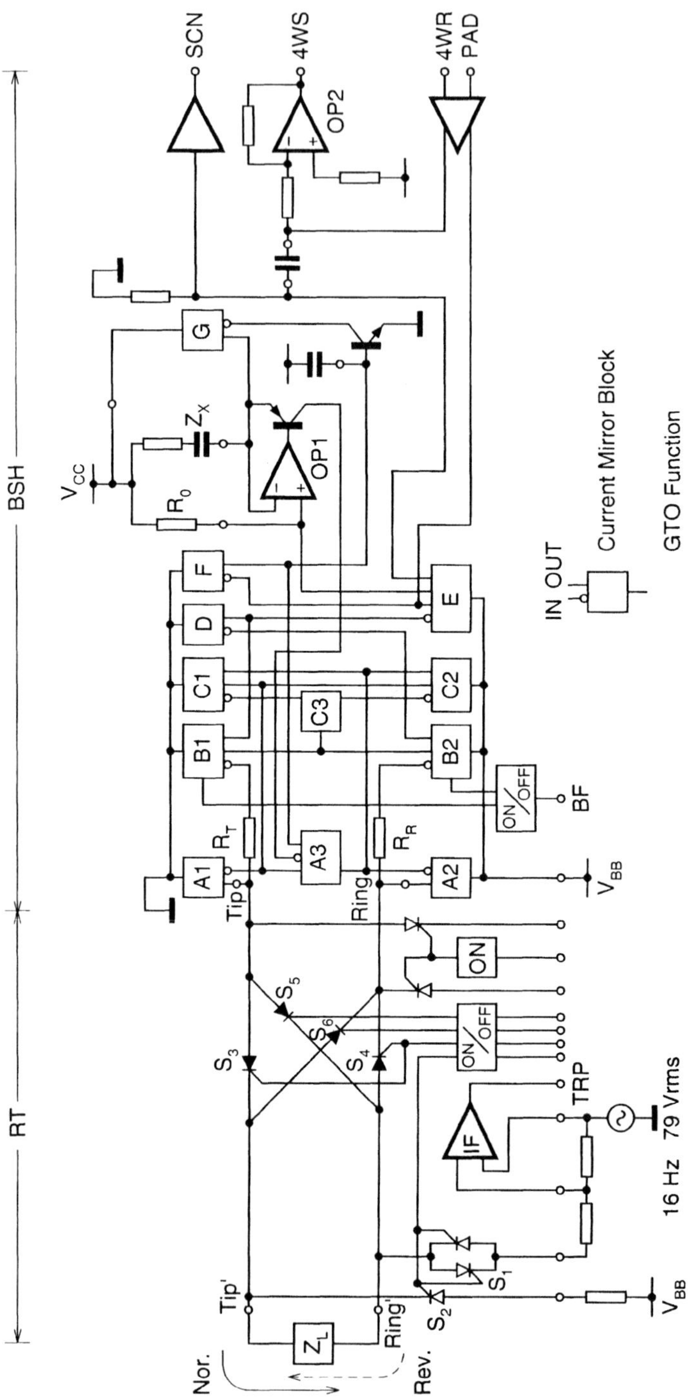

Fig. 4.75 Circuit block diagram of high voltage analog IC (350 V · 0.2 A). (Reprinted by permission of IEEE)

Fig. 4.76 IC chip photograph

Table 4.4 Electrical characteristics of high voltage analog IC (350 V · 0.2 A)

Blocking Voltage	± 350 V
DC Current Capability	250 mA
On Resistance	7 W
dv/dt Capability	3000 V/μs
Longitudinal Balance (2w/4w)*	50 dB/51 dB
Idle Channel Noise (2w/4w)*	–79 dB/–86 dB
S/N (4w-2w/2w-4w)*	60 dB/60 dB
Return Loss*	42 dB
Transhybrid Loss*	28 dB
PSRR*	37 dB
Level Linearity*	± 0.5 dB

* Values at loop current 50 mA

Furthermore, in spite of the high IC voltage, highly accurate analog characteristics such as high longitudinal balance and low ideal channel noise are realized because of high h_{FE}, high f_T and low noise of the complementary transistors. The ICs were put into use for digital exchanges about seven years ago.

4.8.4 Motor Driver ICs for Appliances

Variable speed control by inverters is needed for motors to realizeboth high performance and high energy savings in small-sized motor applications. But, conventional inverters composed of a printed board with discrete devices, are the same size as the motor, so down-sizing is limited. However, 3-phase inverter ICs of the 250 V · 1 A and 550 V · 1 A classes, which can be directly connected to rectified AC 100 V and AC 200 V respectively, have been developed [27]. By using the ICs, not only the installment of the inverters in motors, but also the high frequency operation of 20 kHz above the audible frequency region of humans

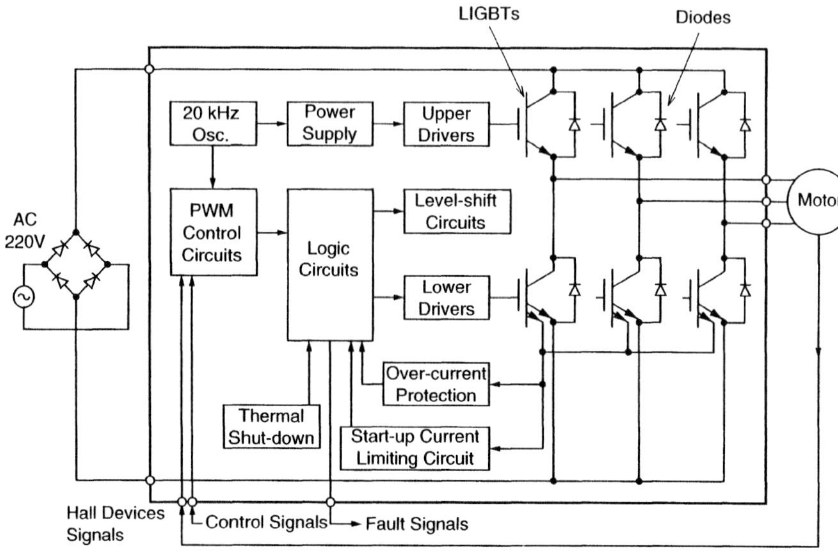

Fig. 4.77 Circuit block diagram of 3-phase inverter IC (550 V · 1 A)

Fig. 4.78 IC chip photograph

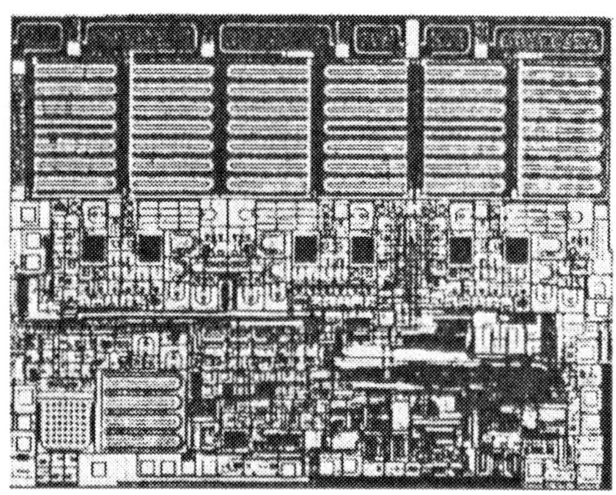

become possible, and both down-sizing and noise reduction of motor application products are advanced.

Figure 4.77, Fig. 4.78 and Table 4.5 show a circuit block diagram, an IC chip photograph and the electrical characteristics of the 3-phase inverter IC of the 550 V · 1 A class, respectively. Its output circuit is composed of 6 pairs of lateral IGBTs (Section 4.6.4) and new soft recovery high speed diodes SFD (Section 4.6.1). The on-state voltage of the lateral IGBT is reduced to 2.8 V at the rating output current of 0.5 A, in spite of its large short-circuit withstand capability of more than 30 μs. The FVD of the diode SFD is also reduced to 1.3 V at 0.5 A.

Table 4.5 Electrical characteristics of 3 phase inverter IC (550 V · 1 A)

Blocking voltage	550 V
Maximum peak current	1.0 A
Rating output current	0.5 A
On-state voltage of IGBTs	2.8 V at 0.5 A
Foward voltage drop of diodes	1.3 V at 0.5 A
Turn-on delay	0.5 μs
Turn-off delay	0.3 μs
PWM frequency	20 KHz
Power supply	12 V
Standard voltage supply	7 V
TSD operation temperature	125°C
Start-up current limit	0.5 A

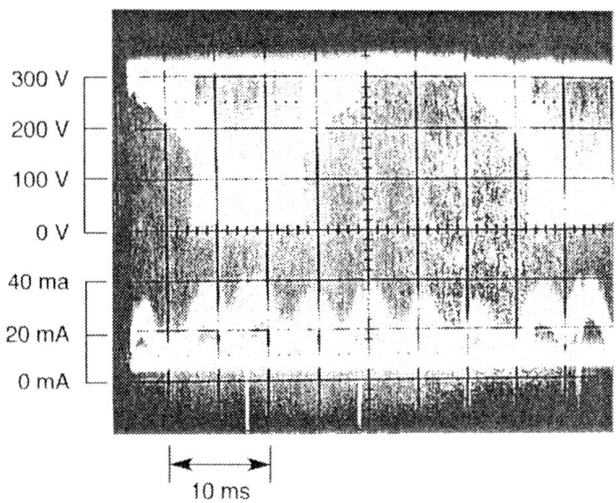

Fig. 4.79 Waveforms of output voltage and current

Furthermore, since the reverse recovery current and reverse recovery time of the diode SFD can be reduced to about 40% and 50% of those values for the conventional diode, power dissipation of the inverter is reduced to about 25%. Some kinds of protection circuits such as for overcurrent and overtemperature, 20 kHz oscillator, PWM control circuit, power supply and start-up current limit circuits are integrated.

The IC fabricated is directly supplied by the rectified AC 220 V, and it can correctly control a motor of 50 W class. From Fig. 4.79, it is clear that PWM frequency is 20 kHz. The turn-on and turn-off delay times are very fast, below 0.5 μs, although the signals are transferred from the logic circuit to the upper drivers through the level shift circuits. At start-up of the motor, a large current flows because no electromotive force occurs for the motor. But, the current is limited to 0.5 A by the start-up current limit circuit. In spite of high

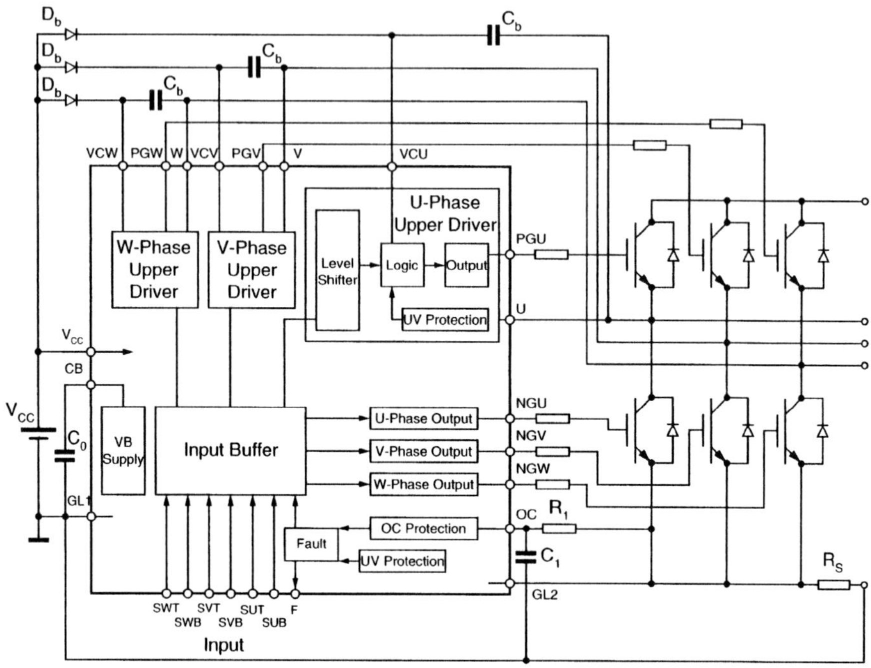

Fig. 4.80 Circuit block diagram of pre-driver IC (620 V · 0.5 A)

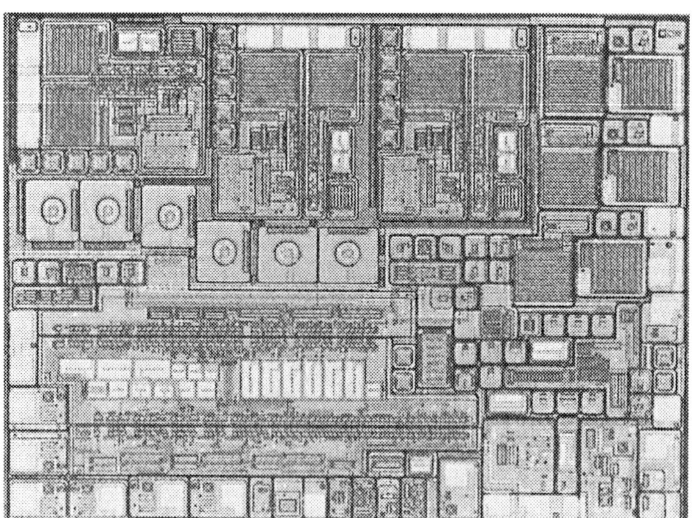

Fig. 4.81 IC chip photograph

Table 4.6 Electrical characteristics of pre-driver IC (620 V · 0.5 A)

Items	Unit	Ratings
HV-Device Breakdown Voltage	V	620
Source Output Current	A	0.25
Sink Output Current	A	0.50
Operating Frequency	KHz	20

electric power capability and high intelligence, chip size is very small, 4.4 mm × 5.8 mm, due to the utilization of the EPIC type DI. Since this inverter IC can be mounted in a motor, motor system size and weight can be reduced drastically.

The pre-driver IC for 3-phase IGBT inverters is also developed. Fig. 4.80 and 4.81 and Table 4.6 show a circuit block diagram, an IC chip photograph and the electrical characteristics of the pre-driver IC. To reduce the warpage of the DI wafers and to improve the IC design rule, the SPSDB technique (Section 4.4.3) is utilized. Devices for output driver circuits are the 15 V CMOSFETs, which are composed of the n and p lateral MOSFETs. And, devices for the level shifter are the 620 V n DMOSFETs. This pre-driver IC can drive the 3-phase IGBT inverter of the 30 A class, and integrates the protection circuits for overcurrent and undervoltage.

4.8.5 Ignitor ICs for Automobiles

Figure 4.82, Fig. 4.83 and Table 4.7 show a circuit block diagram, an IC chip photograph and the electrical characteristics of the ignitor IC of the 600 V · 25 A class for automobiles, respectively [62]. As an output device, a vertical IGBT is

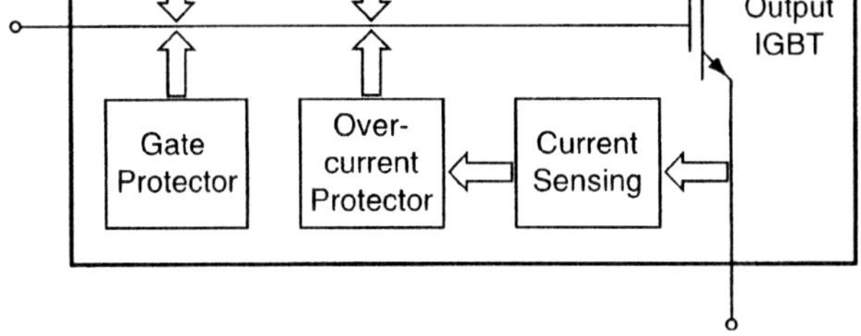

Fig. 4.82 Circuit block diagram of ignitor IC (600 V · 25 A)

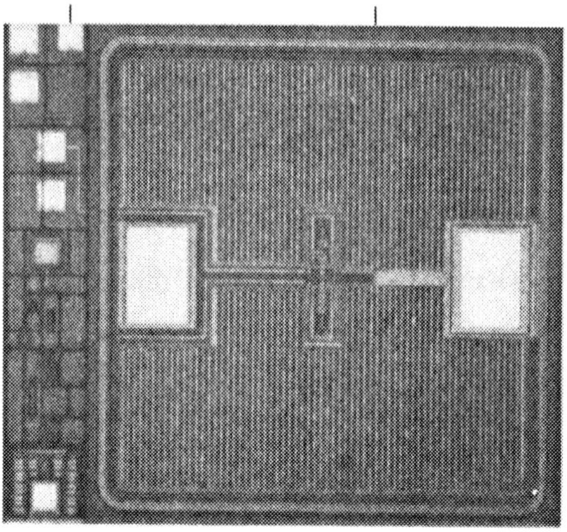

Fig. 4.83 IC chip photograph

Table 4.7 Electrical characteristics of ignitor IC (600 V · 25 A)

	Temperature	Low V_{th} device	High V_{th} device
V_{th} (V)	25°C	2.4	3.7
BV_{CE} (V)	25°C	600	600
FVD (V) at $I_C = 100\ A/cm^2$ $V_{GS} = 12\ V$	25°C	1.2	1.5
J_L (A/cm^2)	150°C	480	900
Fall time (μs) at $I_C = 100\ A/cm^2$	150°	7	3

integrated. Protection circuits for both overcurrent and overvoltage are integrated in addition to a gate driver circuit and a gate protection circuit. Latch-up current of the IGBT at 150 °C is 27 A and the IGBT can turn-off below 25 A without latch-up. To reduce power dissipation, on-state voltage of the IGBT is designed low, i. e. 1.5 V at 100 $^{A}/_{cm^2}$. This makes it easy for a large current to flow during load shortages and for the IC to be broken by latch-up. But, the gate voltage of the IC is reduced from 15 V to 7 V by the over-current protection circuit in the case of the loads shortage and output current is limited below 20 A. Thus, the IC is protected from damage. Its chip size is very small, 3.5 mm × 4.0 mm, in spite of its high electric power capability and high intelligence, due to utilization of the VLCS type DI (Section 4.7.2).

4.9 Conclusion

DI technologies and DI-PICs have received much attention because of their ideal electric isolation, and their development has been active. To overcome the weak points of EPIC which is today's only DI technique in mass production, new techniques such as wafer bonding techniques for the reduction of wafer warpage and vertical power device integration techniques for the increase of electric power capability have been developed. Futhermore, many high voltage techniques and high voltage improved-devices specific to DI-PICs have been developed and put into practical use.

Consequently, increases in both power capability and intelligence will be accessed under the trends of down-sizing, energy saving, etc. For these aspects, functions for protection and diagnosis will need to be more accurate, and communication functions, memories and processors should be integrated. In addition, both power supplies and inverters of more than 10 A and 400 V can be expected to be integrated into one DI-PIC within the near future.

References

1. Sugawara Y (1988) J IEE Japan, Vol 108, No 2, p 152
2. Sugawara Y (1993) Trans IEE of Japan, Vol 113-C, No1, p 5
3. McWilliams D, Fa C, Larchian G and Maxwell O (1964) J Electrochem Soc, Vol 111, p 153
4. Ihara M, Arimoto Y, Jifuku M, Yamaoka T and Kurokawa K (1981) ISSCC Digest of Technical Paper, p 210
5. Ronen RS, Splinter MR and Treamain RE (1976) IEEE J Solid-State Circuits, Vol SC-11, p 431
6. Izumi K (1978) Electro Lett, Vol 14, p 593
7. Sugawara Y and Shimura T (1987) Proceedings of Symposium on High Voltage and Smart Power Devices by Electrochem Soc, Philadelphia (USA), p 96
8. Suzuki T et al. (1980) J Electrochem Soc, Vol 127, No 7, p 1537
9. Celler GK and Trimble LE (1987) Proceedings of Symposium on High Voltage and Smart Power Devices by Electrochem Soc, Philadelphia (USA), p 241
10. Kawamura S et al. (1987) IEDM Technical Digest, p 758
11. Sawada R and Watanabe J (1988) Proceedings of ISPSD, Tokyo, p 141
12. Ohashi H, Ohura J, Tsukakoshi T and Simbo M (1986) IEDM Technical Digest, p 210
13. Sugawara Y, Inoue Y, Ogawa S and Kurita S (1992) Proceedings of ISPSD, Tokyo, p 316
14. Temple VAK (1977) IEDM Technical Digest, p 423
15. Grove AS, Leistiko O and Hooper WW (1967) IEEE Trans Electron Devices, Vol ED-14, p 157
16. Kao YC and Walley ED, Proceedings of IEEE, Vol 55, p 1409, 1967
17. Matsusita T, Aoki, Ootsu, Yagi H, Hayashi H, Okayama and Kawana (1979) IEEE Trans Electron Devices, Vol ED-23, p 826
18. Appels J et al. (1979) IEDM Technical Digest, p 238
19. Conti E and Conti M (1972) Solid-State Electronics, Vol 15, p 93
20. Sugawara Y et al. (1983) Proceedings of Conference on Solid-State Devices, Tokyo, p 77, 1982; Japanese Journal of Applied Physics, Vol 22, Supplement 22-1, p 77

21. Deal BE, Sklar M, Grove AS and Snow EH (1967) J Electrochem Soc, Vol 114, p 266
22. Sugawara Y and Kamei T (1987) IEEE Trans Electron Devices, Vol ED-34, p 1816
23. Merchant S et al. (1991) Proceedings of ISPSD, Baltimore, p 31
24. Merchant S et al. (1993) Proceedings of ISPSD, Monterey, p 124
25. Yasuhara N, Nakagawa A and Furukawa K (1991) IEDM Technical Digest, p 141 [26] Sridhar S, Baliga BJ, Tandon S and Reisman A, Proceedings of ISPSD, Tokyo, p 245, 1992
26. Sridbar S, Huang YS and Baliga BJ (1992) IEDM Technical Digest, p 245
27. Sakurai N, Nemoto M, Arakawa H and Sugawara Y (1993) Proceedings of ISPSD, Monterey, p 310
28. Mori M, Yasuda Y, Sakurai N and Sugawara Y (1991) Proceedings of ISPSD, Baltimore, p 113
29. Sugawara Y, Kamei T, Hosokawa Y and Okamura M (1983) IEDM Technical Digest, p 412
30. Sugawara Y and Kamei T (1984) Proceedings of Conference on Solid State Devices, Kobe, p 317
31. Sugawara Y and Kamei T (1986) IEEE Trans Electron Devices, Vol ED-33, p 23
32. Sugawara Y, Miyata M and Okamura M (1986) IEEE Trans. Electron Devices, Vol ED-33, p 2002
33. Imaizumi I, Kimura M, Yoshimura M and Inoue F (1980) IEEE Trans Consumer Electron, Vol CE-26, p 367
34. Beason J (1983) IEDM Technical Digest, p 424
35. Krishna S and Ramde A (1982) IEEE J Solid-State Circuits, Vol SC-17,2 p 781
36. Sugawara Y et al. (1994) Proceedings of ISPSD, Davos, p 419
37. Nakagawa T (1963) J IEE Japan, Vol 83, p 1765
38. Huang YS, Baliga BJ, Tandon S and Reisman A (1992) Proceedings of ISPSD, Tokyo, p 245
39. Sakura N, Nemoto M, Arakawa H and Sugawara Y (1993) Proceedings of ISPSD, Monterey, p 310
40. Sugawara Y and Shimura T (1987) Proceedings of Symposium on High Voltage and Smart Power Devices by Electrochem Soc, Philadelphia (USA), p 96, May
41. Mattheus WHA (1981) ISSCC Digest of Technical Paper, p 238
42. Sugawara Y (1986) IEDM Technical Digest, p 226
43. Murakami S, Sugawara Y and Kawahata S (1989) Proceedings of Symposium on High Voltage and Smart Power Devices by Electrochem Soc, Los Angeles, p 288
44. Merchant S, Arnold E, Baunmgart H, Egloff R, Letavic T, Mukherjee S and Pein H (1993) Proceedings of ISPSD, Monterey, p 124
45. Matsudai T and Nakagawa A (1992) Proceedings of ISPSD, Tokyo, p 272
46. Arnold E, Merchant S, Amato M, Mukherjee S and Pein H (1992) Proceedings of ISPSD, Tokyo, p 242
47. Held R, Serafin J, Fullmann M, Constapel R and Korec J (1993) Proceedings of ISPSD, Monterey, p 130
48. Disney D and Plummer J (1992) Proceedings of ISPSD, Tokyo, p 48
49. Huang YS, Baliga BJ, Tandon S and Reisman A (1992) Proceedings of ISPSD, Tokyo, p 40
50. Disney D and Plummer J (1993) Proceedings of ISPSD, Monterey, p 254
51. Huang YS, Srifhar S and Baliga BJ (1993) Proceedings of ISPSD, Monterey, p 259
52. Neubrand H, Serafin J, Fullmann M and Korec J (1993) Proceedings of ISPSD, Monterey, p 264
53. Sugawara Y et al. (1979) Proceedings of Conference on Solid State Devices, Tokyo, p 405, 1978; Japanese Journal of Applied Physics, Vol 18, Supplment 18-1, p 405
54. Gammel JC (1985) Electro '85, 24/1
55. Sugawara Y and Shirasawa T (1987) IEDM Technical Digest, p 762
56. Sugawara Y and Shirasawa T (1988) Proceedings of ISPSD, Tokyo, p 121
57. Boguszewicz R et al. (1990) Mikroelektronik, Vol 6, No 4, p 256

58. Weyer J and Vogt H (1992) IEDM Technical Digest, p 225
59. Ohata T et al. (1986) Tech Digest of IEEE Costume IC Conf, p 443
60. Ohoka T, Yoshitake T, Kikuchi H and Okonogi K (1993) Proceedings of ISPSD, Montery, p 119
61. Suda K, Ozeki S, Matsuzaki H and Kawamoto K (1990) Proceedings of ISPSD, Tokyo, p 49
62. Mizoguchi T, Shirasawa T, Mori M and Sugawara Y (1991) Proceedings of ISPSD, Baltimore, p 40
63. Ohno T, Sakurai T, Inabe Y and Koinuma T (1984) ISSCC Digest of Technical Paper, p 230
64. Masquelier M and Okada D (1991) Proceedings of ISPSD, Baltimore, p 56

Power Mosfets Driving Circuits and Protection Techniques

D. Rossi

5.1 Introduction

Microelectronic technologies combining CMOS, Power DMOS and often Bipolar transistors have in the recent past represented the major technology breakthrough in the Smart Power technical arena allowing very often lower power dissipation and a by far higher circuit complexity if compared to standard, pure Bipolar technologies.

These two features have considerably changed the way to approach applications that, for a long time, did not achieve great evolution and allowed the integration of complete systems or subsystems in true single chips solutions.

This chapter, omitting voluntarily both the physics and the technology of these new structures wants to represent firstly an attempt to emphasise the pros and cons of Power Mosfet transistors; secondly it wants to introduce circuit solutions suitable to efficient driving and protections; and finally, even if not exhaustively, it spreads over problems related to the interference between power and signal parts, and induced by undesired but unavoidable injection of parasitic currents typical of any inductive load driving.

5.2 Power DMOS Input Characteristics

For a better understanding of how a power DMOS behaves and its best utilisation, it is worth referring to Fig. 5.1 where the device equivalent circuit is shown together with its cross section.

Here all the most important parasitic elements are taken into consideration. These are:

- the inductances of each one of the pin connections;
- the capacitances between gate and source metalizations (C_{gsm}) and between gate and source diffusion (C_{gso});
- the capacitances between gate and body region and between gate and epi-layer;
- the capacitances of all the p-n junctions;
- the npn parasitic transistor whose emitter base and collector are respectively source, body and drain.

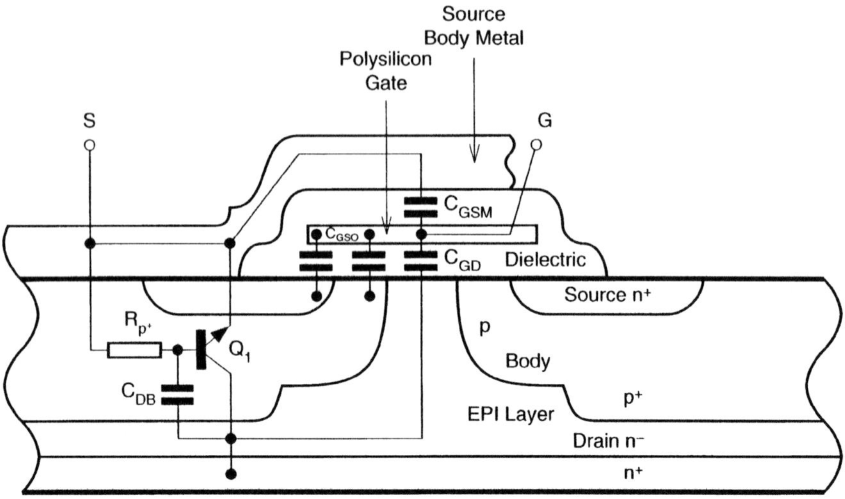

a

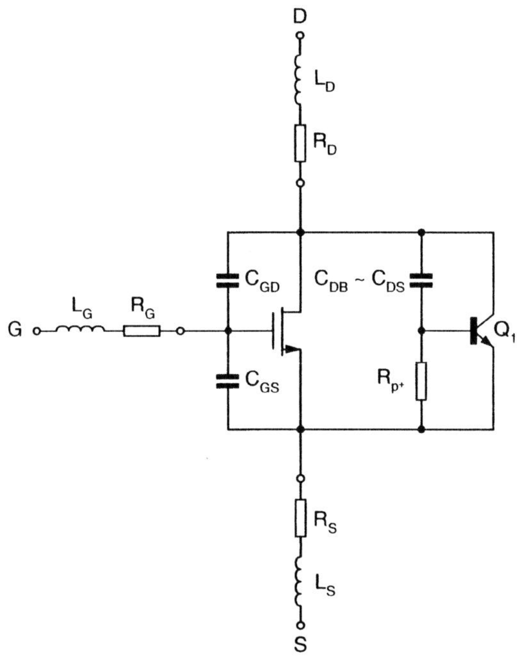

b

Fig. 5.1 Power Mosfet cross section **a** and Equivalent Circuit **b**

All the mentioned capacitances contribute to and are often implicitly expressed by the capacitances C_{iss}, C_{oss}, C_{rss} as in the following:

- C_{iss}: is known as input capacitance and is the parallel sum of C_{gd} and C_{gs}. Thus.

$$C_{iss} = C_{gd} + C_{gs} \tag{5.1}$$

- C_{oss}: is known as output capacitance and represents the total capacitance between drain and source with gate and source shorted. Thus

$$C_{oss} = C_{ds} + C_{gd} \tag{5.2}$$

- C_{rss}: is known as reverse transfer or Miller capacitance and simply represents the gate-to-drain capacitance with $V_{gs} = 0$. Thus

$$C_{rss} = C_{gd} \tag{5.3}$$

This capacitance, C_{gd}, largely depends on the condition of the semiconductor surface directly faced to the gate metalization that, depending on V_{gd}, can be essentially in:

- accumulation
- depletion.

Referring to the cross section in Fig. 5.1 [1], an accumulation layer is formed when V_{gd} is > 0. A positive voltage on the gate attracts electrons toward the silicon surface. The more the surface is accumulated, the thinner the accumulated layer will be; hence, the capacitance will approach asymptotically the value associated with the pure oxide. Thus, the capacitance in accumulation, Cacc., approaches

$$C_{acc} = A\, C_{ox} = \frac{\varepsilon_{sio2}\, \varepsilon_0}{t_{ox}} A \tag{5.4}$$

Where:

A = gate area
ε_{sio2} = relative permittivity of SiO_2
tox = oxide thickness

When the gate voltage becomes more positive, electrons are repelled from the silicon surface and the system is in the depletion-bias condition. Under this condition the overall capacitance corresponds to the capacitance that is obtained by a series connection of the oxide capacitance and the capacitance across the surface depletion region.

$$C_{dep} = \frac{1}{1/C_{ox} + X_d/\varepsilon_s} A \tag{5.5}$$

where Xd is the depth of the surface depletion layer, which, in its turn, depends upon gate to epi bias as well as upon the doping and oxide properties.

Since the capacitance of the system decreases as the depletion layer widens, a straightforward analysis shows that C_{dep} decreases increasing the gate to source voltage. This, theoretically, continues until the gate bias is high enough to invert the surface attracting holes toward the silicon surface. Actually, V_{gd} never reaches the inversion threshold voltage over the entire voltage range of the component to avoid a direct current path from source to drain.

- C_{gs} (gate-to-source capacitance) is physically due to the overlap capacitance gate to channel and to source diffusion and is basically voltage independent.
- C_{ds} (drain-to-source capacitance) is basically given by the p (body) – n (epi-layer) junction capacitance and varies inversely with the drain to source voltage.

Fig. 5.2 Typical Mosfets Capacitances V_s Drain-to-Source Voltage

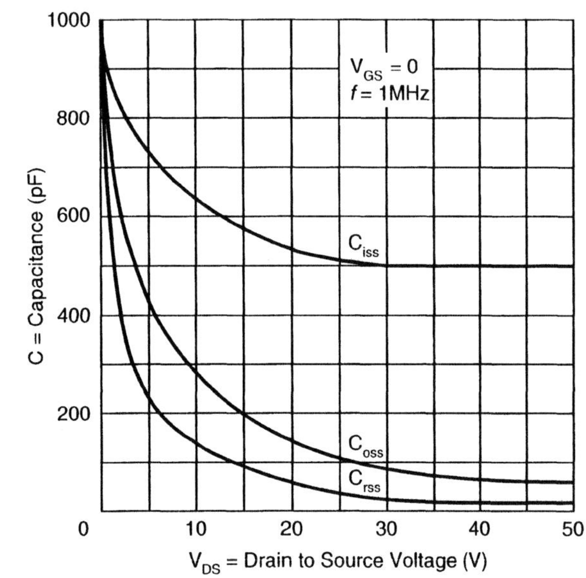

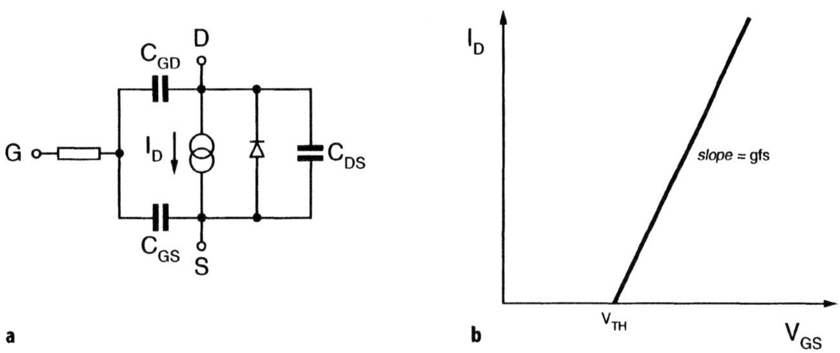

Fig. 5.3 **a** Power Mosfet Equivalent Circuit; **b** Simplified Transfer Characteristics

Because of the heavy voltage dependence (see Fig. 5.2) knowing C_{iss}, C_{oss} and C_{rss} is often not enough to allow a straightforward design of any driving circuit even when the load characteristics, voltage ratings and commutation times requirements are exactly known (to get around this problem refer to par. 5.3 where the concept of total amount of charge is introduced and explained).

Apart from this voltage dependence, the power MOS operation can be simply represented by the simple circuit of Fig. 5.3.

Like a standard MOS component a power Mosfet is characterised by zero DC input bias current while the drain current is, at a first assumption, just controlled by the gate to source voltage and given by:

1) $I_d = 0$ when $V_{gs} < V_{th}$
2) $I_d = V_{ds}/R_{ds\text{-}on}$ when $V_{gs} > V_{th}$ and the Power MOS operates in triode region ($gfs(V_{gs}\text{-}V_{th}) > V_{ds\text{-}on}/R_{ds\text{-}on}$)
3) $I_d = gfs\ (V_{gs}\text{-}V_{th})$ and the Power MOS operates in saturation region ($V_{ds} > I_d\ R_{ds\text{-}on}$)

Even if simple, this equivalent circuit is often used for a rough analysis of the transient behaviour and useful to understand advantages and draw-backs of Mosfet based power stages.

5.3 MOSFET Vs Bipolar Power Transistors

If, from the design standpoint, knowing the pros and cons of the Power Mosfets in terms of power dissipation, simplicity of associated driving and protection circuitry, component ruggedness is usually enough; a better comprehension of all these features passes through an analysis of two different mechanisms governing the current conduction.

In Power Mosfets, only majority carriers are responsible for the current flow, while in Bipolar transistors the current is controlled and regulated by minority carrier injection.

In the case of n-channel Power Mosfets, conduction starts when the input capacitance has received a charge amount high enough to create an electron channel.

Once switched on, the Mosfet does not need (see also the equivalent circuit of Fig. 5.3) any extra power to keep the channel existing and the current flowing.

On the contrary, in npn Bipolar transistors, conduction happens when the injection of the holes from the base creates a positive charge which, in respect to the charge neutrality principle, has to be compensated by an inflow of electrons. Basically, in bipolar transistors, conduction starts thanks to the recombination that, taking place in the base region, requires a base current flowing as long as the transistor has to be driven on.

This base current is usually not negligible at all, especially if bipolar transistors are driven in heavy saturation in the attempt to lower the voltage drop between collector and emitter. The higher the base current, the lower the

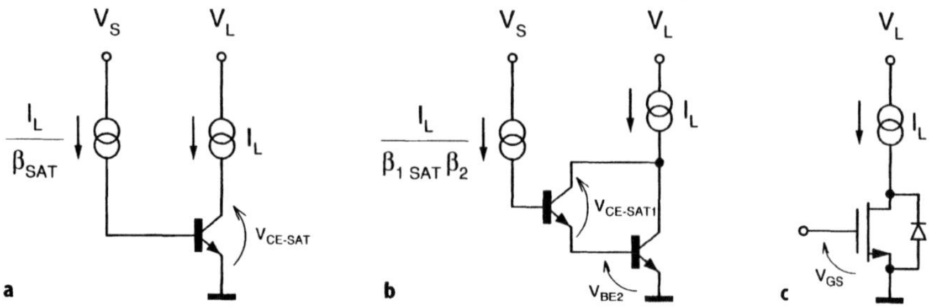

Fig. 5.4 Bipolar and Power Low Side Drivers

collector to emitter voltage ($V_{ce\text{-}sat}$) is with great disadvantage of the transistor current gain.

As a matter of fact, this lower voltage drop across the bipolar power switch causes higher power dissipation in the associated driving circuitry.

Additional disadvantage associated to heavy saturation is a longer turn-off time because of the delay time spent to extract the excess of minority carriers stored into the base region.

As a major consequence, Power Mosfets beyond featuring low static power dissipation are also capable of higher operating frequency.

For a better understanding of the impact of all the previous considerations, three different low-side power switches [2], as shown in Fig. 5.4 are compared.

For judgement objectivity all of them handle the current level I_L over the same output voltage swing V_s.

Figure 5.4a refers to a single transistor output stage, Fig. 5.4b to a composite Darlington stage, Fig. 5.4c finally to a N-channel DMOS [1].

For the stage of Fig. 5.4a the total power dissipation is given by:

$$P_{d\text{-}npn} = (V_{ce\text{-}sat})\, I_L + (I_L/\beta_{sat})\, V_s \tag{5.6}$$

where

$V_{ce\text{-}sat}$ = voltage drop across the power element at $I_c = I_L$

I_L/β_{sat} = base current

V_s = supply voltage of the driving part

For the stage of Fig. 5.4b

$$P_{d\text{-}darl.} = (V_{ce\text{-}sat\,1} + V_{be2})\, I_L + (I_L/(\beta_{1\text{-}sat}\,\beta_2))\, V_s \tag{5.7}$$

where

$V_{ce\text{-}sat\,1} + V_{be2}$ = voltage drop across the power element

$IL/(\beta_{1sat}\,\beta_2)$ = current required in the driving stage

as well known, a lower input current is here needed to keep the power element conductive; bigger is, however, the voltage drop across the element.

For the stage in Fig. 5.4c

$$P_{d\,mos} = R_{ds}\, I_L^2 \tag{5.8}$$

If R_{ds} I_L is kept by design lower than $V_{ce\text{-}sat}$, being on the other hand negligible the power lost in the driving stage, Mosfet power stages offer evident advantages.

The consideration, often used in the past of bipolar components showing smaller area for a given ON state voltage drop is, in most cases, not true any more.

Since [3] its ON resistance strongly depends on the L/W (length to width ratio), a power Mosfet could benefit in act of finer and finer microlitography in contrast with power bipolar transistors whose emitter efficiency is mainly determined by the emitter area, which remains unchanged.

As support to this assertion, in Fig 5.5 curves reporting current densities at 1 V voltage drop across the power elements (both Bipolar and Mosfets) V_s the maximum operative voltage are shown.

Additional advantage of ICs technologies centred around DMOS transistors, even not directly influencing the power components, is related to density of the signal CMOS and Bipolar parts.

In the case of integrated npn transistors (see Fig. 5.6), the collector to emitter breakdown voltage with base shorted (BV_{ces} is usually made equal to BV_{cbo}) mainly depends on the breakdown voltage of diodes D_1 and D_2.

The net epitaxial thickness W_1, its resistivity and the reach-through mechanism define the breakdown voltage of D_1 while the breakdown of D_2 mainly depends on the radius of curvature of the base diffusion [4].

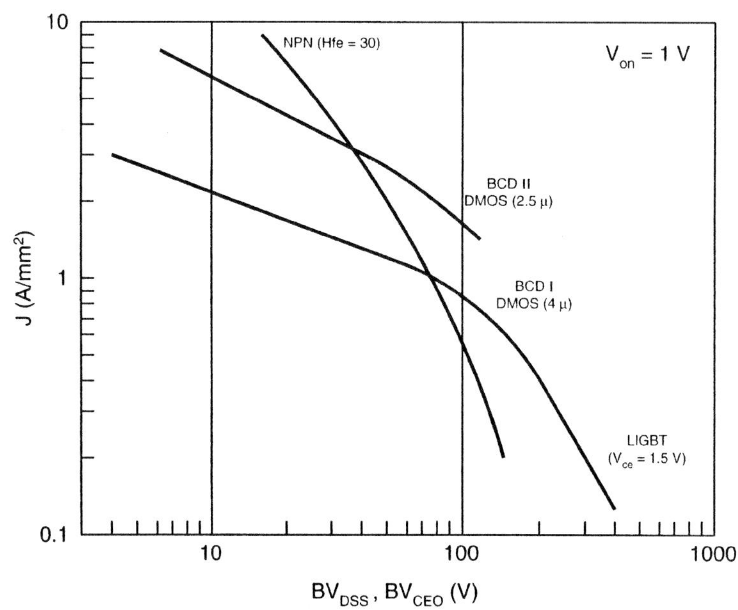

Fig. 5.5 Current Densities versus Breakdown Voltage for Bipolar and Power Mosfet, with different microlitography, Transistors

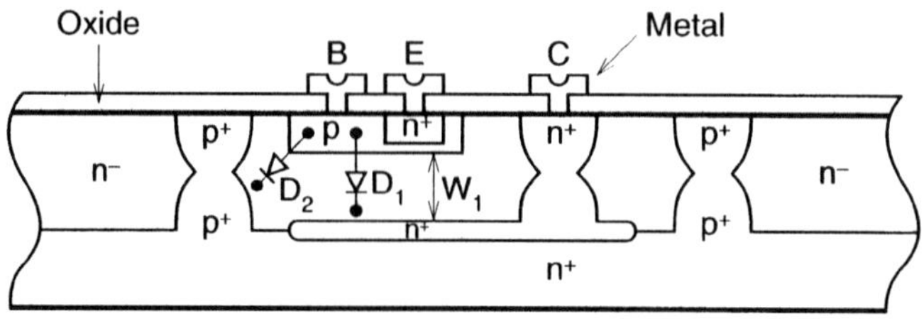

Fig. 5.6 Voltage Limitations in Standard Bipolar Transistors

In standard Bipolar processes an increase in the process voltage capability is achieved by increasing the thickness and the resistivity of the epitaxial layer.

The higher these two values, the bigger the lateral diffusion of the isolation with evident drawbacks in the size of all the junction isolated components (all bipolar) no matter if even used for signal processing purposes.

Once fixed BV_{cbo}, the voltage capability of a bipolar technology is defined by the breakdown voltage BV_{ceo} (emitter to collector breakdown voltage with open base).

Since BV_{ceo} is lower than BV_{cbo} and given by

$$BV_{cbo} = BV_{ceo} \sqrt[n]{\beta} \tag{5.9}$$

a bipolar transistor can be regarded as incapable to fully exploit the maximum voltage the technology permits.

On the contrary, a DMOS component (see again Fig. 5.1b) is capable of working at a breakdown voltage BV_{dss} equal to the BV_{cbo} of the parasitic npn component provided that its base to emitter short circuit is good enough to avoid any spurious turn-on.

To some extent, DMOS based mixed technologies, compared to pure bipolar technologies featuring the same breakdown voltage in the power part, offer by far smaller bipolar elementary components.

As an additional advantage, DMOS structures can be made compatible with standard CMOS transistors able to guarantee high density levels in all the logic parts.

If lower D_C power dissipation is, by its own, a very important feature, even more evident is the benefit of using power Mosfet in the case of switching applications.

To evaluate the power losses typical of any switching inductive load driving it is worth referring to the circuit in Fig. 5.7 [2].

The load here represented can be an electrical machine, a motor, a transformer.

From an initial condition characterised by $I_L = 0$ and the switch OFF, once the power element is switched on the system evolves in such a way that the load current grows to a rate controlled by V_s/L (V_s is the supply voltage; L the load inductance).

Fig. 5.7 Circuit Driving Clamped Inductive Loads

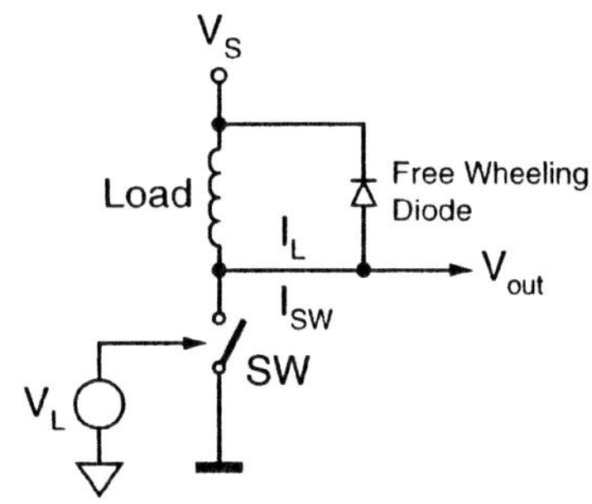

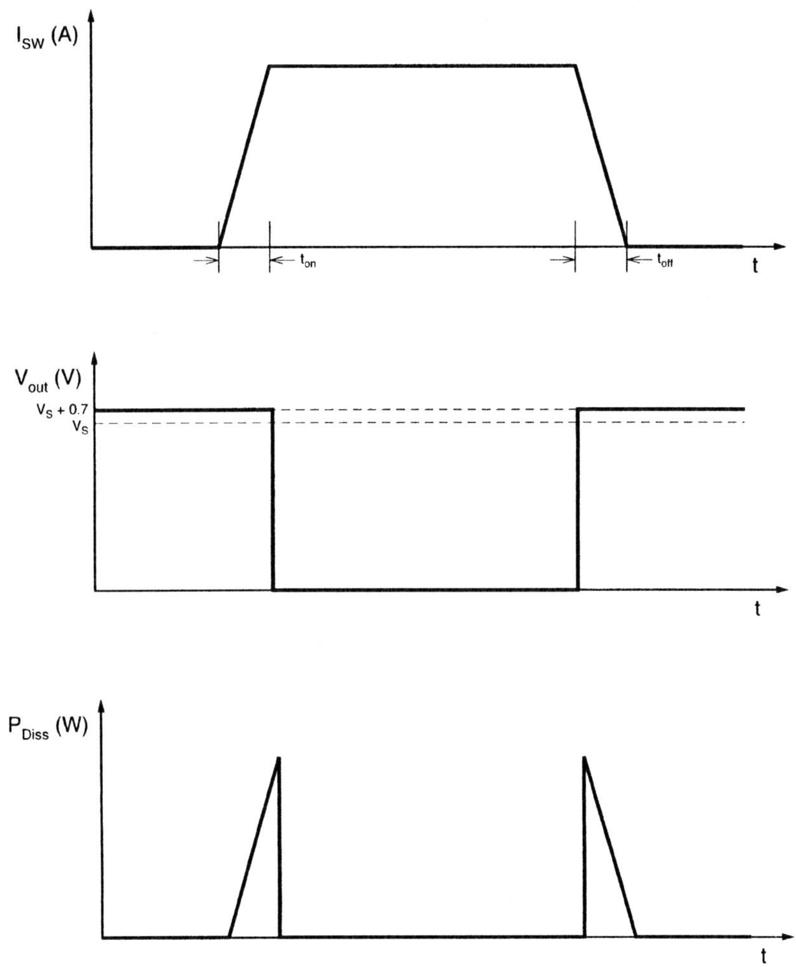

Fig. 5.8 Turn-on and Turn-off Current and Voltage Waveforms for the Circuit of Fig. 5.7

During the turn off transition, the inductive nature of the load in the attempt to keep its current constant, forces a fast output transition until the output voltage reaches $V_s + 0.7$ V, while the current decays with the transistor typical turn-off switching time.

Assuming that during all this phase the current decay is negligible, at a first assumption, the same current level still flows when the power switch is turned on again.

During all this transition, the voltage across the switch is clamped to $V_s + 0.7$ V until the switch is able to handle all the load current. Only at this point the diode turns off allowing the output voltage to fail.

The voltage and current waveforms during all these phases are reported in Fig. 5.8 together with the power instantaneously dissipated by the power switch.

A simple computation, once known the repetition rate and the t-on and t-off commutation times, gives the average dissipated power as:

$$P_{diss} = 1/2\, V_s\, I_L\, fsw\, (t_{on} + t_{off}) \tag{5.10}$$

where:
V_s = supply voltage
fsw = switching frequency
t_{on}, t_{off} = turn-on and turn-off switching times

(the only basic assumption here made is a linear output voltage transition).

As shown, once fixed I_L, V_s and fsw, P_{diss} depends only on t-on and t-off. Since Power Mosfets, whose t_{on} and t_{off} can be as low as few nanoseconds, are faster than bipolar transistors they can be considered either capable of higher switching frequency or less dissipative for a same switching frequency.

Another advantage the power DMOS offers is the possibility of using the intrinsic Body-to-drain diode as free wheeling components in switching inductive load driving.

Figure 5.9 shows an application circuit where the inductive load is alternatively switched between ground and the main supply voltage V_s.

If the switching frequency is high enough to keep basically the load current constant, disregarding the commutation losses and the reverse recovery losses of the diode, the average power dissipation is:

$$P_{d1} = R_{ds\text{-}on}\, I_L^2\, (t_{on}/T) + V_d\, I_L\, (t_{off}/T) \tag{5.11}$$

where
I_L = average output current,
V_d = diode forward voltage drop,
t_{on} = power Mosfet ON time,
t_{off} = power Mosfet OFF time (during this time the load current recirculates through the diode),
T = switching period.

An enormous amount of power can be saved by replacing the diode with a properly driven Power Mosfet as shown in Fig. 5.10.

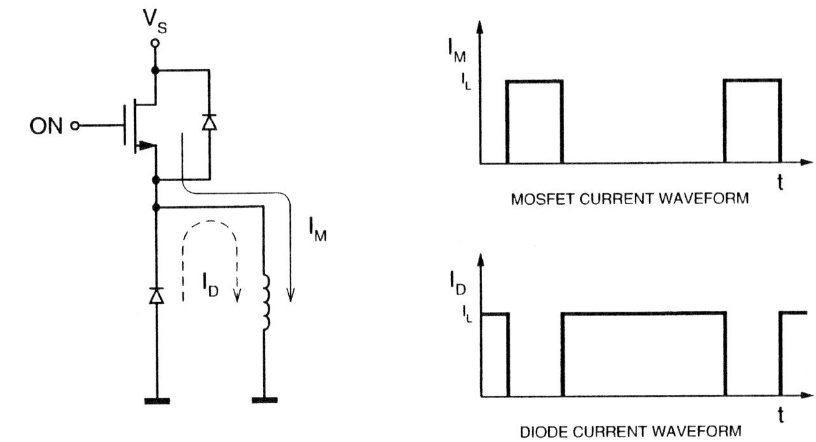

Fig. 5.9 PWM Current Control

Fig. 5.10 Power Mosfet Push-Pull Current Control

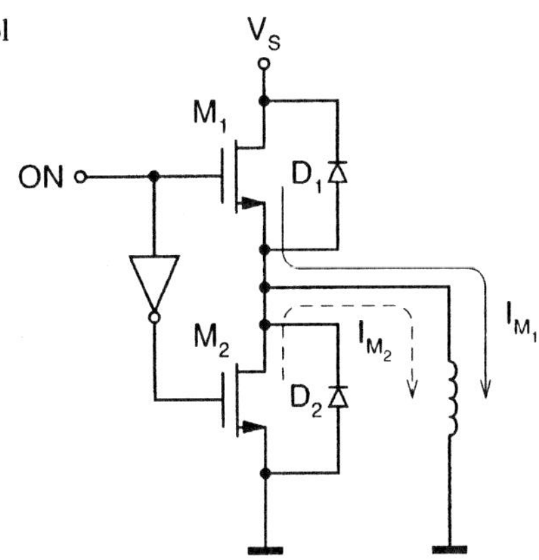

Power Mosfets driven with a positive gate-to-source voltage can, in fact conduct negative currents (negative stands for a current flowing from source to drain) only through the Mosfet channel until the voltage drop ($R_{ds\text{-}on}\ I_L$) is insufficient to forward bias the Body-to Drain diode (in this last case current flows partially through the channel and partially through the diode).

In this case the dissipated power becomes:

$$P_{d2} = R_{ds\text{-}on}\, I_L^2 \tag{5.12}$$

assuming both the low and the high-side drivers having the same ON resistance. The power saved if the diode is replaced by a synchronous rectifier is then given by:

$$P_{saved} = V_d\, I_L\, (t_{off}/T) - R_{ds}\, I_L\, (1 - (t_{on}/T)) \tag{5.13}$$

(in case of 25% duty cycle; $R_{ds\text{-}on} = 0.3$ Ohm; $I_L = 3$ A; $V_d = 1$ V $P_{saved} = 4.725$ Watt).

The only assumption here made is that the dead time (during this time both the upper and the lower power transistors are kept OFF to avoid rail to rail cross conduction) is short enough not to cause an increase in the dissipated power.

Solutions like this have been widely used especially in Smart Power ICs in the attempt to decrease the overall system power consumption and/or use low dissipation packages.

The only disadvantage is that power Mosfets can, sometimes, fail when the intrinsic Body-Drain diode is used (for more detailed information refer to Par. 5.3.1).

Now most Mosfets manufacturers provide this information in the discrete components data sheet and Body-Drain diodes are often used as free wheeling components.

The additional Schotky of Fig. 5.11a or the antiparallel diodes of Fig. 5.11b are no longer needed, unless very high switching frequency and the diode reverse recovery current lead to an excessive power dissipation.

If in many PWM power controllers properly driven Body-Drain diodes bring many benefits, in other fields like the automotive one, this diode represents an undesired component potentially causing permanent damages or spurious load activation in the case of reverse battery connection [5].

The most severe situation occurs when a 24 V battery, used to jump start a car, is connected with the wrong polarity.

In this condition, Power Mosfets, even if switched off (see Fig. 5.12a), usually allow the current flowing from the battery, through the diode to ground causing unpredictable load behaviour and, most likely, device failure.

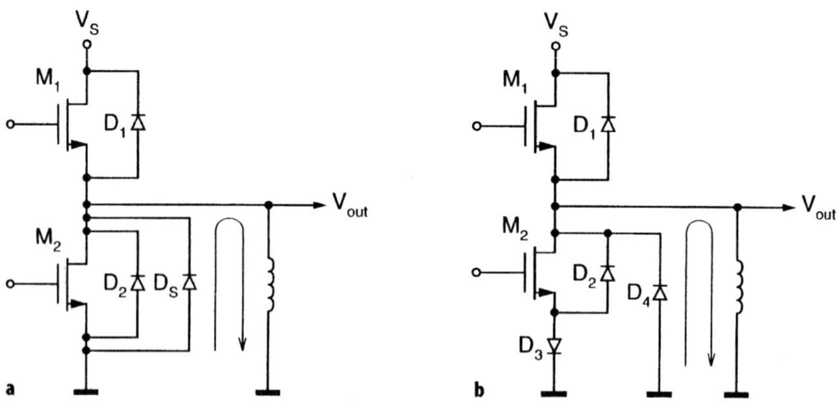

D_S = Schottky Diode

Fig. 5.11 Reduction of High Commutation Losses by Using Low Reverse Recovery Current Diodes

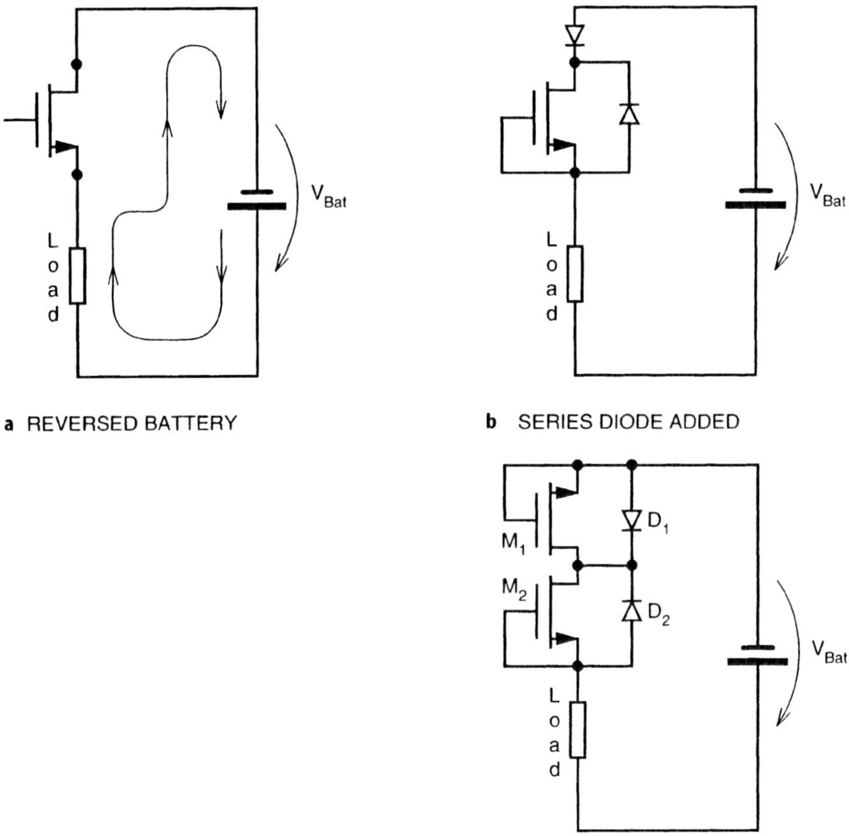

Fig. 5.12 Mosfet High-Side Drivers in Reverse Battery Condition

c SECOND MOSFET ADDED

There are two possible solutions to this problem: the first (Fig. 5.12b) is the addition of a diode blocking the reverse current flow (evident disadvantage is an increased power dissipation); the second, an additional Power Mosfet connected in the configuration in Fig. 5.12c.

In reverse battery conditions, both the two components being switched off and the upper Mosfet M_2 showing a breakdown voltage higher than 24 V are enough to block the unwanted current flow.

Vice versa, in normal bias conditions, the composite component can be switched on assuring to both the two gates a positive gate-to-source voltage.

In this bias condition the device, see Fig. 5.13a, shows a piece-wise linear triode characteristic having two different $R_{ds\text{-}on}$ values.

At low levels, (Fig. 5.13b, zone 1) all the current flows through the Mosfets channel and the total $R_{ds\text{-}on}$ is given by:

$$R_{ds\text{-}on}\ \text{total} = R_{ds\text{-}on}\ M_1 + R_{ds\text{-}on}\ M_2 \tag{5.14}$$

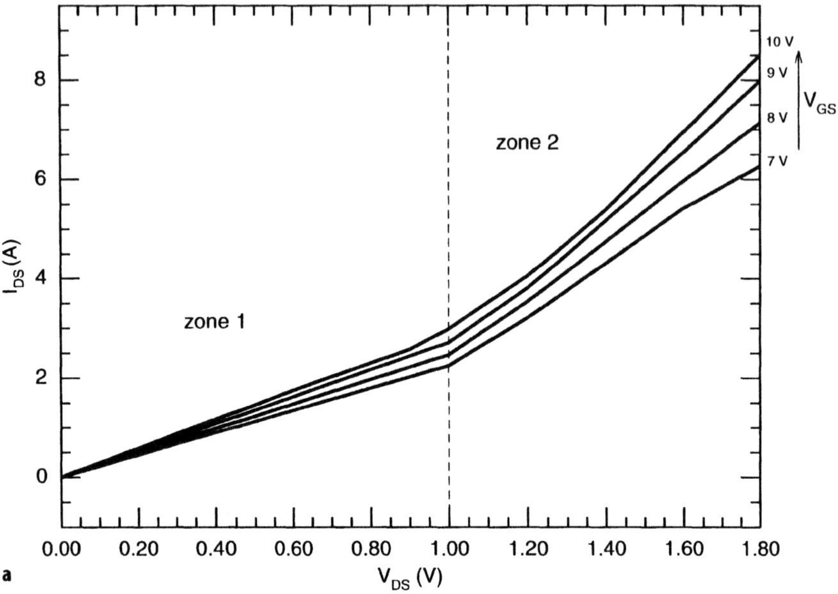

a

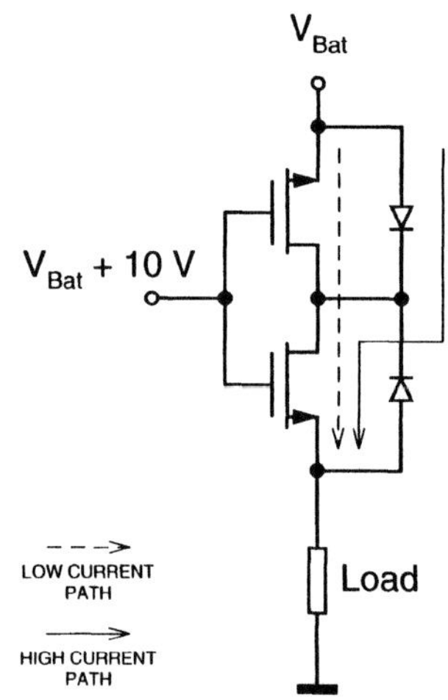

b

Fig. 5.13 **a** Dual Gate Power Mosfet (DGDMOS) Characteristics in Triode Region, **b** DGDMOS High-Side Driver

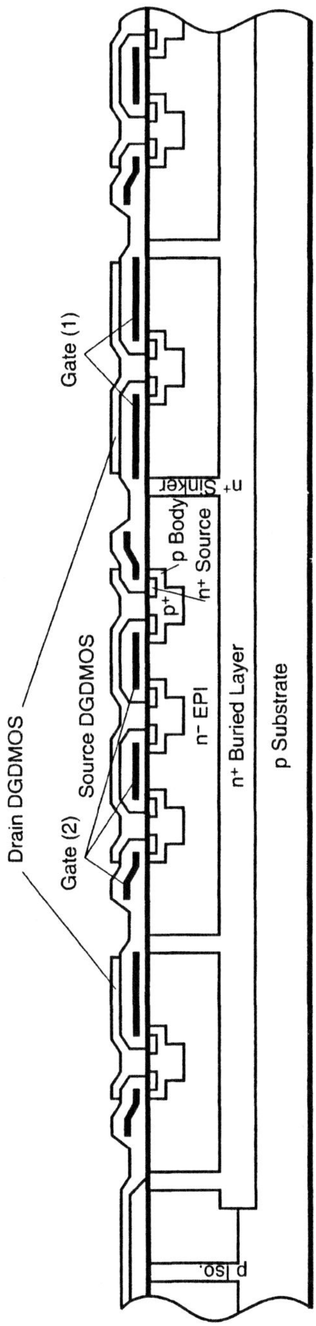

Fig. 5.14 Latero-Vertical Dual Gate DMOS

At higher levels (Fig. 5.13b, zone 2) when the intrinsic diode clamps the Drain-to-Source voltage of M_2 the $R_{ds\text{-}on}$ is lower and given by:

$$R_{ds\text{-}on}\ \text{total} = R_{ds\text{-}on}\ M_2 + R_s\ \text{diode} \quad (5.15)$$

where
R_s diode = series resistance of the intrinsic diode.

The main disadvantages of this technique are by far a bigger silicon area (4 times bigger) and an increased circuit complexity.

Partial solution to this increased silicon area can be found under the assumption that the device has to withstand two different breakdown voltages (60 to 80 V in normal and 24 V in reverse connection condition).

This situation allows in fact the combined usage of vertical (Higher $R_{ds\text{-}on}$, higher BV_{dss}) and lateral (lower $R_{ds\text{-}on}$, lower BV_{dss}) components to optimise the $R_{ds\text{-}on}$ by Area factor.

In Fig. 5.14 the cross section of this component is reported.

5.4
MOSFETS Power Stages and Associated Driving Circuits

Even if switching from non conducting to conducting state seems to be just a question of raising the gate-to-source voltage, the non linearity of many of the intrinsic capacitances makes the gate driving circuit design quite difficult to be managed by conventional methods or just taking into account the input capacitance and the equivalent gate series resistance.

Especially the gate-to-drain capacitance that, beyond being considerably non linear, is sometimes amplified by the Miller effect (this effect well known by all electronic designers gives a capacitive multiplication because of the feedback path between gate and drain provided by the gate-to-drain capacitance) leads to an input capacitance by far bigger than the sum of the static capacitances.

To account for both gate-to-source and gate-to-drain capacitances, a parameter known as "GATE-CHARGE" has been introduced [6].

5.4.1
GATE-CHARGE Factor

This parameter, basically insensitive to the drain current and quite independent of the temperature, allows an easy design of the driving circuitry and enables an easy estimation of the charge, hence the energy required to turn the Power Mosfet on and off.

A schematic representation of the test circuit, used to measure the "GATE-CHARGE" is shown in Fig. 5.15.

In this circuit a constant current source (I_g) is charging the gate of the device under test while an other current (I_L) is applied to its drain.

Measuring V_{gs}, once noted that the horizontal scale is proportional to the supplied charge, the plot of Fig. 5.16 offers a direct indication of the energy

Fig. 5.15 Basic Gate Charge Test Circuit

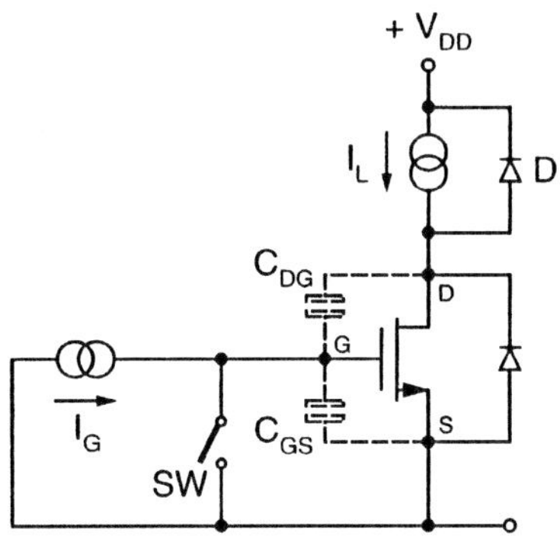

Fig. 5.16 Gate Voltage Vs Gate Charge Characteristics

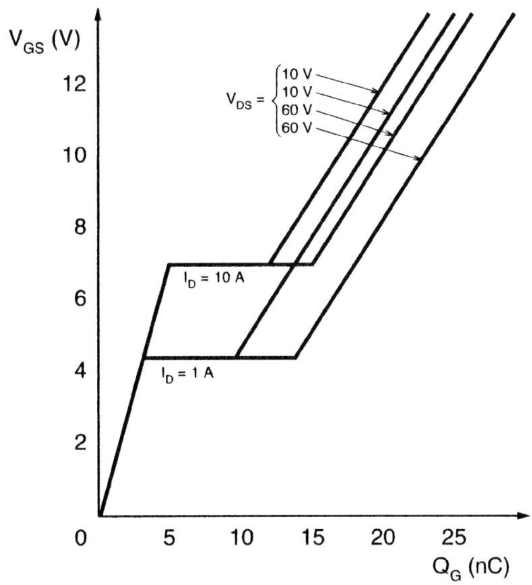

spent to switch I_L over the supply voltage V_{dd} (Fig. 5.17 is then a plot of V_{gs} Vs Charge).

The typical gate-to source voltage swing is shown in Fig. 5.17a together with drain voltage and current transients (Fig. 5.17b).

Before time t_o, switch SW is closed, the DUT withstand all the supply voltage V_{dd}, the gate-to-source voltage and the drain current are both zero.

SW_1 is opened at time t_o, the gate-to-source and the gate-to-drain (C_{gd} is minimum because charged to V_{dd}) capacitances start to charge and the gate-to-source voltage increases.

No drain current flows until the V_{gs} exceeds the Power Mosfet threshold voltage; this happens at time t_1.

Fig. 5.17 Gate Voltage and Output Voltage and Current Waveforms

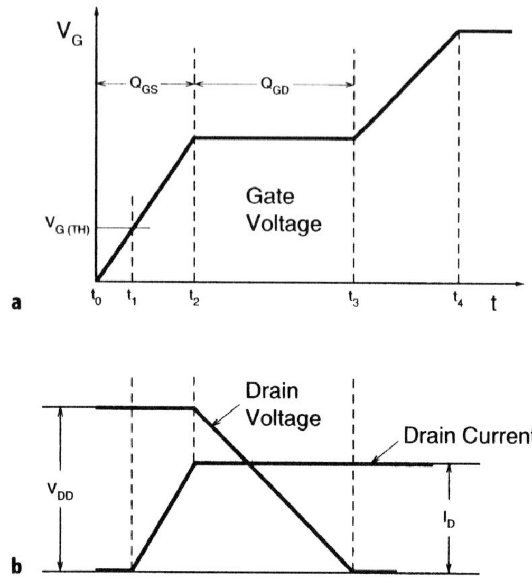

During the time interval t_1-t_2, the input capacitance continues to charge while the voltage across the power element remains equal to V_{dd} since the power element is not yet able to conduct I_L.

During all this period, the current taken by the gate-to-drain capacitance is usually small, often negligible since C_{gd} at $V_{ds} = V_{dd}$ is by far smaller than C_{gs}. The slope of the gate voltage is then dominated, during all this time period, by C_{gs}.

At time t_2, the drain current reaches I_l; the drain no longer tied to V_{dd} starts to fall while the drain current remains constant.

At the beginning, as long as the device works in saturation no current flows into the gate-to-source but only into the gate-to-drain capacitance on the other hand amplified by the Miller effect.

At time t_3, the Power Mosfet enters into triode region and the drain-to-source voltage becomes equal to $R_{ds\text{-}on} \times I_L$.

The gate-to-source voltage is now free to raise with a slope controlled by the charging current and by an input capacitance that, in the meanwhile, has reached its maximum value (C_{gs} is maximum because $V_{ds} = 0$).

This situation continues, until time t_4, when the gate-to-source voltage becomes equal to the maximum allowable value. Since, as already pointed out, the horizontal axis (time) is directly proportional to the delivered charge, time interval t_0-t_2 represents the charge Q_{gs} delivered to the source to gate capacitance; time interval t_2-t_3 represents the charge transferred to the gate-drain or Miller capacitance, while the charge transferred from time t_3 on is simply an excess of charge that, for practical reasons, represents a gate-to-source voltage overdrive needed to guarantee the minimum drain-to-source ON-resistance.

The great advantage of the gate-charge parameter is that, as shown in Fig. 5.16, it is quite independent of the drain current, independent of the supply voltage and relatively independent of the temperature.

The gate-charge parameter offers then a straightforward method to determine the level of current needed to achieve a desired switching time or, vice versa it is useful to define the right trade-off between current required then, dissipated power, and achievable switching time.

Referring to Fig. 5.18 and taking 20 nanocoulomb as charge required to switch drain current of 1 A over a voltage swing of 80 V, an easy computation leads to 2 A peak current in the case of 10 ns switching time; 20 mA in the case of 1 μs and so on.

The gate-charge parameter allows also a quick estimation of the average drive power.

$$P\ drive = Q_c\, V_{gs}\, fsw \tag{5.16}$$

where

Q_c = gate-charge factor
V_{gs} = gate-to-source voltage
fsw = switching frequency

Taking fsw = 100 KHz, Q_c = 20 nC, V_{gs} = 10V then P drive = 20 mW

As outlined by Eq. 5.16, the power dissipated in driving the Mosfet is independent of the charging current (the higher the current, the shorter the time spent in charging the input cap.) but only proportional to the switching frequency.

Even if in fact 1 A drive current may appear quite considerable, it flows for such a short time that the average power is negligible (still 200 mW at 1 MHz in the above example outlining once again the advantage Power Mosfets offer in high frequency switching applications).

5.4.2
Driving Power Mosfet Low-Side Drivers

As seen in the previous paragraph, short switching times require driving circuit able to source and sink peak current as high as 1 A, while a gate-to-source voltage high enough is needed to guarantee the minimum ON state resistance.

Many different circuits suitable for direct driving of Mosfet low side drivers are shown in Fig. 5.18; all of them: pure CMOS, Totem-pole Bipolar and BiCMOS are here fed by the auxiliary supply voltage V_{aux}.

This voltage reference, often obtained from the main supply rail, is just a simple series regulator that has to fulfil two basic requirements: the first related to the peak current it has to deliver, the second to its voltage level.

About current the most important parameter, more than the average current usually negligible, is the peak current that can be very high especially if short turn-on times are needed.

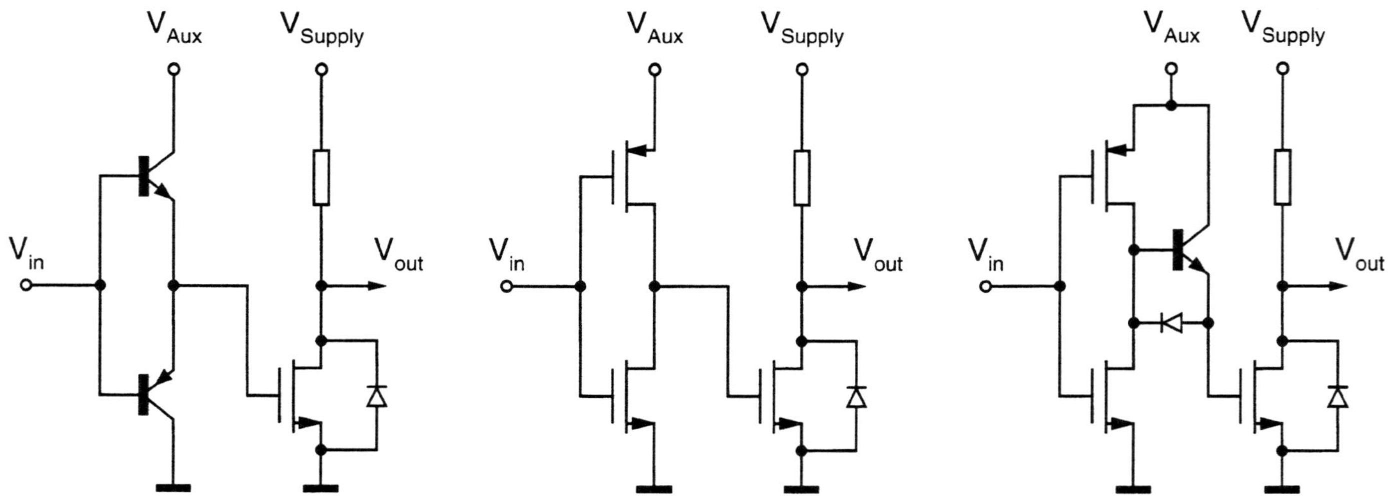

Fig. 5.18 Typical Low-Side Mosfet Switches Driving Circuits

For this reason an auxiliary capacitor C_{aux}, intended to deliver the required charge as well as to filter V_{aux}, is usually connected between V_{aux} output and ground.

In this situation V_{aux} has just to restore the adequate voltage level in the minimum OFF time.

About voltage, V_{aux} has to be high enough to guarantee an ON state voltage drop across the power switch as low as possible avoiding anyway heavy overdriving.

Overdriving the gate, in the attempt to minimise conduction losses, since the energy lost in driving is directly proportional the gate-to-source voltage (the same as V_{aux} in all the circuits of Fig. 5.18), leads to an increased power dissipation.

5.4.3 Driving Power Mosfet High-Side Drivers

Since N-channel components are smaller and less expensive in terms of process complexity, an efficient driving (efficient stands for able to guarantee minimum ON-resistance) has to deliver a gate voltage adequately higher than the positive power supply rail.

This voltage, usually not available at board or system level, has to be "ad hoc" created for every different application.

All most popular solutions can be classified in two different categories: the first that, using the voltage bootstrapping technique is able to deliver just temporarily this voltage level; the second, vice versa, that is able to keep indefinitely the high-side Mosfet ON.

- Bootstrap techniques

A simplified circuit diagram, applying bootstrap technique to drive the high side switch of a Power Mosfet push-pull stage is shown in Fig. 5.19 where just the fundamental components, for the sake of simplicity, are analysed.

These are:

- C_{bs} (bootstrap capacitor)
- D_{bs} (bootstrap diode)
- V_{aux} (auxiliary voltage).

The bootstrap capacitor is here charged or better refreshed to V_{aux} every time the output, forced by the recirculation of an inductive load or by the lower Mosfet switching on, approaches ground.

In Fig 5.19, the current flowing during C_{bs} voltage refresh is shown.

When the upper Mosfets starts switching on, the output pin rises, the bootstrap diode switches off, while the C_{bs}, still feeding the upper switch driving circuitry, allows the upper Mosfet gate voltage to rise up to $V_{supply} + V_{aux}$.

As long as C_{boot} remains charged to a voltage high enough, the upper switch is properly driven and the output voltage is $V_{supply} - R_{ds}\, I_L$ (I_L = load current).

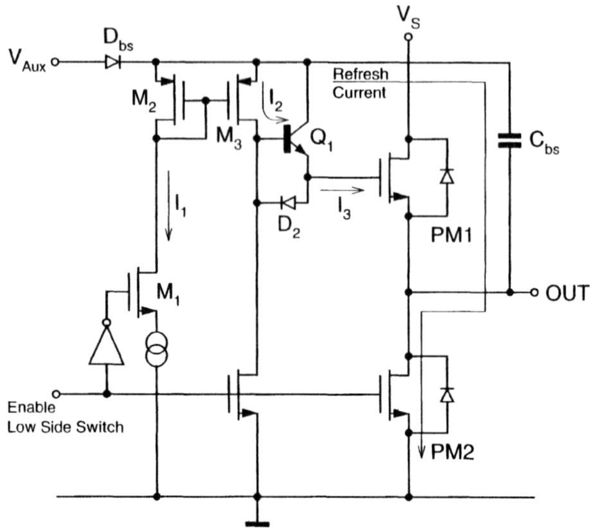

Fig. 5.19 Bootstrap Technique Applied to High-Side Switch Driving

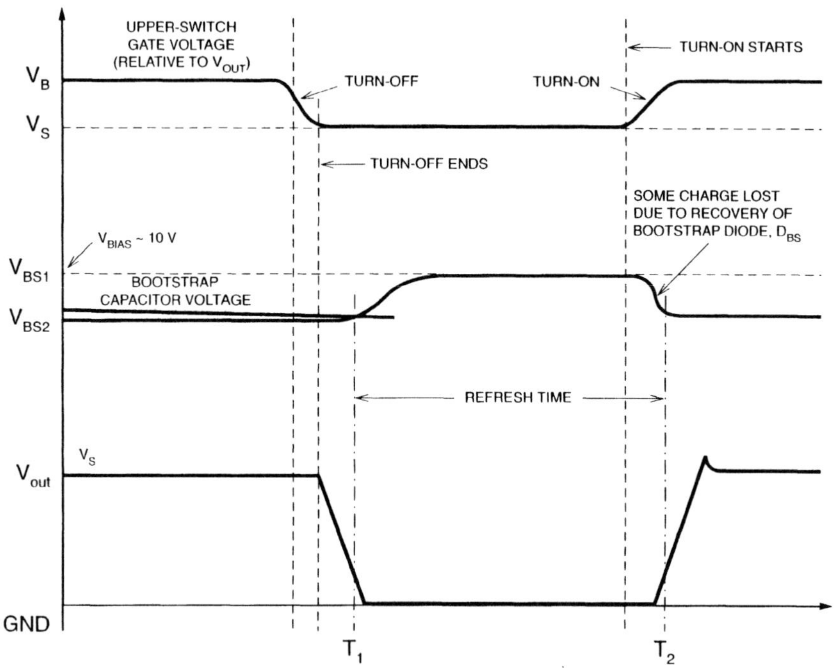

Fig. 5.20 Typical Voltage Waveforms for the Circuit of Fig. 5.19 when used to drive inductive Loads (refresh starts when D_{bs} is forward biased at t_1 and ends when D_{bs} is reverse biased at t_2)

To understand better how the circuit works, all the most significant voltage waveforms of the circuit of Fig. 5.19 are shown in Fig. 5.20.

The major responsible for the sharp decrease in the capacitor voltage occurring, when the upper transistor is switched on, is the charge transfer from C_{bs} to the power Mosfet input capacitance (the reverse recovery current of the bootstrap diode is just responsible for a by far smaller voltage drop).

Additional factors responsible for a slow, but constant (this is the reason why such a technique is not suitable for DC driving) voltage decrease are the quiescent current of the driving stage and the reverse leakage current of the bootstrap diode.

Minimising the bootstrap capacitor value, while keeping a gate to source voltage high enough for all the time between two successive turn on transitions, is basically finding a solution to the following charge balance equation [7].

$$\Delta V_{cbs} C_{bs} = Q_c + Q_{rr} + (I_{dr} + I_{qu}) \Delta t \tag{5.17}$$

Where

ΔV_{cbs} = allowable voltage drop across the bootstrap capacitor
C_{bs} = Bootstrap capacitor value
Q_c = Gate charge of power device
Q_{rr} = Bootstrap diode reverse recovery current
I_{dr} = Bootstrap diode leakage current
I_{qu} = upper driver quiescent current
Δt = upper switch maximum ON time

Assuming for the sake of simplicity, a duty cycle close to 100%, this equation solved for C_{bs} delivers :

$$C_{bs} = \frac{Q_c + Q_{rr} + (I_{dr} + I_q)/fsw}{\Delta V_{cbs}} \tag{5.18}$$

Solving this equation allows an easy estimation of the bootstrap capacitor size.

As far as the value of V_{aux} is concerned, voltage and current capabilities are the same as outlined in the low-side driver circuitry requirements.

Once again care must be taken in targeting the right peak current level able to restore the right voltage level across C_{bs} in the minimum time slot (minimum time occurs when, with heavily unbalanced duty cycles, the lower transistor is ON for a short time).

As an additional example of bootstrap technique used to drive Mosfet high side switch, Fig. 5.21 shows the output stage of a step down SMPS (switched mode power supply) [8].

All the basic considerations, done around the push-pull output stage are still valid and applicable.

- Charge pump techniques.

To understand how a charge pump circuit behaves, it is worth referring to the circuit in Fig. 5.22.

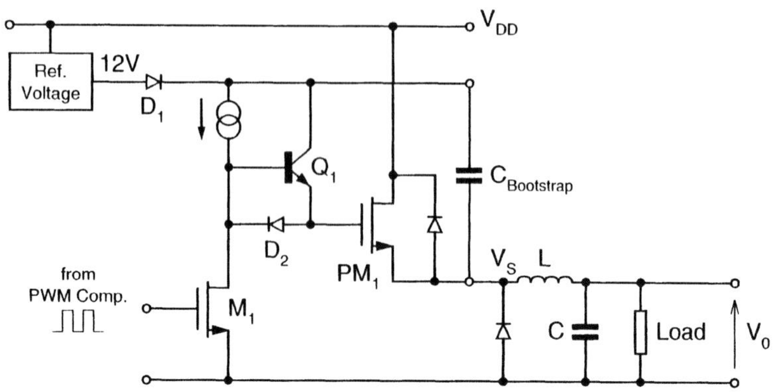

Fig. 5.21 Bootstrap technique applied to SMPS step down Mosfet Output Stage

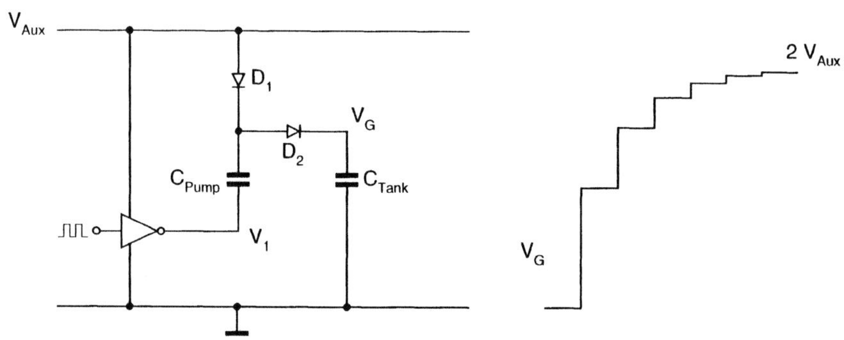

Fig. 5.22 Basic Charge Pump (voltage doubler) Circuit

All these charge pumps, named capacitive, use two different capacitors (C_{pump} injecting charge into C_{tank}), unidirectional switches (here represented by D_1 and D_2) and an auxiliary oscillator.

When V_1 is tied to ground C_{pump}, through D_1, whose voltage forward drop is here neglected, it is charged to V_{bat}.

C_{pump}, when V_1 is then brought to V_{bat}, transfers, through D_2, charge to C_{tank} that, in steady state condition, reaches a voltage level equal to $2\,V_{bat}$ (for this reason the circuit is called voltage doubler).

The output voltage transient is basically controlled by the clock frequency and by C_1/C_2 ratio.

Since Power Mosfets show purely capacitive input characteristics, C_2 in Fig. 5.22 has been practically replaced by the power element itself as shown in Fig. 5.23 [9].

When ENABLE is LOW and M_1 ON, C_{pump} is charged to V_{aux} when M_1 is switched off, the current generator I_2 brings point A to V_s allowing C_{pump} to

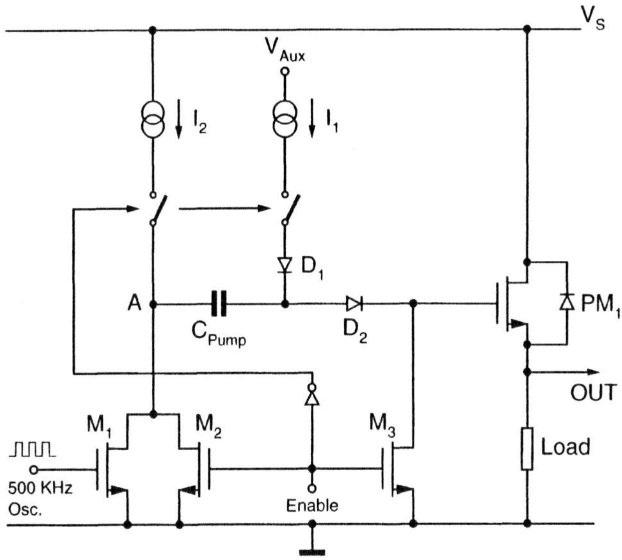

Fig. 5.23 Charge Pump Techniques Applied to High-Side Switch Driving

Fig. 5.24 Typical Capacitive Voltage Tripler

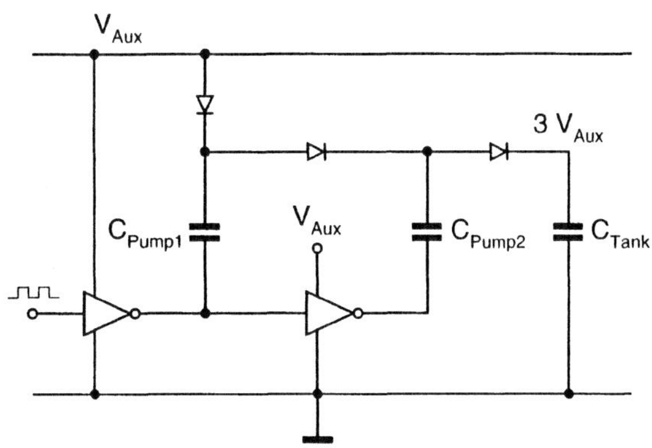

transfer charge to the gate of the power Mosfet able then to reach, at the voltage transition end, a level equal to $V_{supply} + V_{aux}$.

From this condition on, the oscillator, running as long as the power switch is ON, periodically refreshes the gate voltage compensating all the present losses.

In the case of circuits intended to work at lower supply voltage, the doubler can be modified into a voltage tripler roughly represented in Fig. 5.24.

Always disregarding all diodes forward voltage drop, C_2 is charged to $2\ V_{aux}$, C_3 to $3\ V_{aux}$.

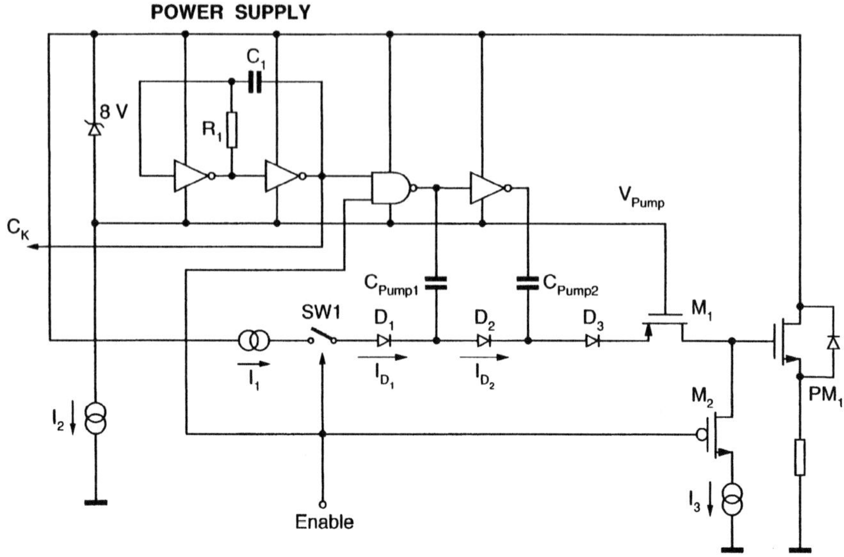

Fig 5.25 Voltage Tripler used to properly drive High-Side Switch in case of Low Supply Voltage

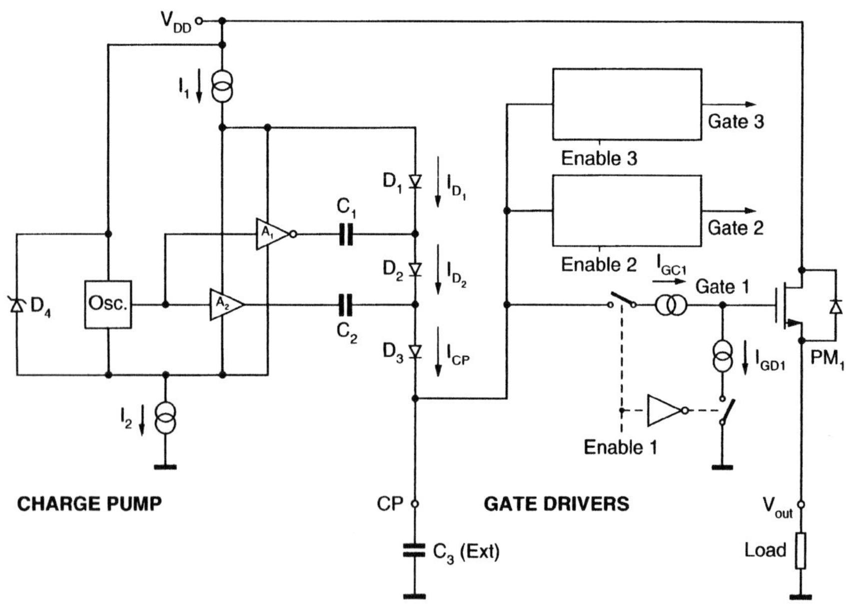

Fig. 5.26 Single Charge Pump for Multiple High-Side Switches Driving

The practical realisation of a voltage tripler, used to properly drive Mosfet high side driver is shown in Fig. 5.25. As a major difference in comparison to the circuit in Fig. 5.25, most circuits are floating and tracking the main supply voltage (the two supply rails are V_{supply} and V_{supply} – 8 V).

C_2 and C_3, on the order of 100 pF are integrated on chip; the current generator I_1 is introduced to limit the current charging the gate of the Power Mosfet in the attempt to limit the output voltage slew rate.

Controlling this transient, as well as avoiding current spikes from the mains strongly reduces in fact all the radio frequency radiation sometimes causing troubles at system level.

Always referring to the circuit in Fig. 5.25, P-channel M_1 is used to control the access to the power switch allowing transfer of charge when SW_1 turns on and M_2 off, and basically isolating the gate that is now free to go down to ground when ENABLE goes LOW (M_1, in this condition, practically avoids any stress by overvoltage to C_{pump1} and C_{pump2}).

An alternative solution directly, derived from the circuit of Fig. 5.25 and needing just one charge pump to drive multiple high side switches, is shown in Fig. 5.26.

Referring to this figure, C_3 = 0.1 uF is external; C_1 and C_2 external.

In every state it acts as a reservoir of charge that, through the enable block is used to independently drive all power Mosfets.

The size of C_3, as well as the oscillator frequency (usually ranging from 200 KHz to 1 MHz) control the rise time of the charge pump output voltage.

- Combined Bootstrap and Charge Pump Techniques.

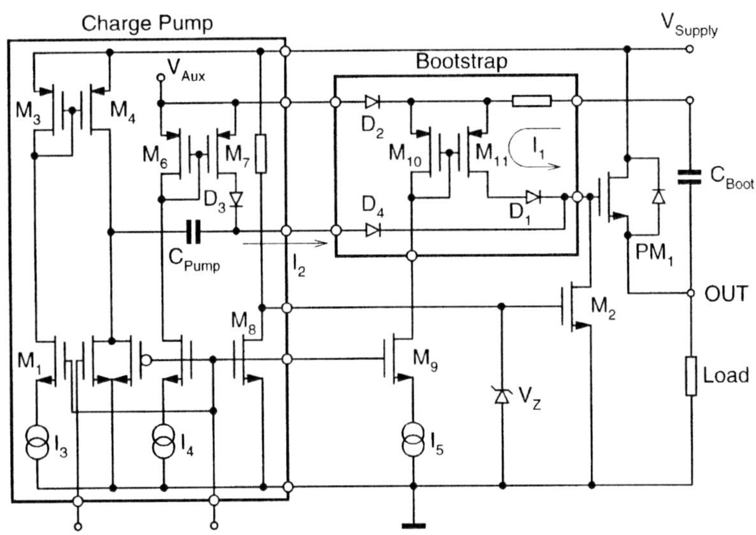

Fig. 5.27 Combined Bootsrap and Charge Pump Techniques applied in driving Power Mosfets High-Side Drivers

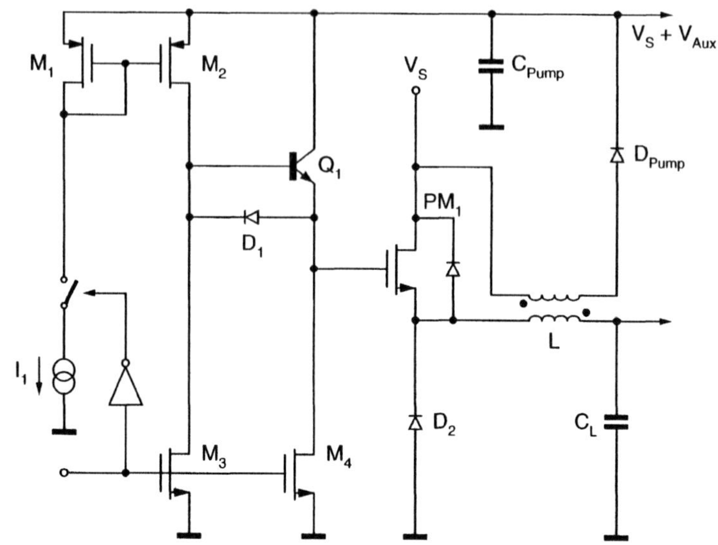

Fig. 5.28 Inductive Coupled Charge Pump Circuit

In the following, techniques able to guarantee fast as well as time indefinite turn on are described.

Figure 5.27 shows, as an example, a circuit solution applying bootstrap (able to guarantee fast turn-on time) and charge pump techniques (able to keep the Power Transistor "forever" On) previously introduced.

An alternative solution, easily applicable to step down SMPS is shown in Fig. 5.28 [12].

5.4.4
Biasing Class AB Power Mosfet Output Stages

Even if most likely used in PWM switching applications, Power Mosfets can properly act as power elements is linear AC and DC amplifiers. Fig 5.29 presents a push-pull output stage realised only using N-channel components.

The lack of complementary devices (N and P channel) and the essential constraint of low output impedance when the output transistors work either in common source (the upper) or in common drain (the lower) configuration resulted into a close loop configuration applied around the lower Power Mosfet transistor [13].

The operational amplifier A, providing a local feedback and setting a unity gain transfer function between input and output, considerably lowers the output impedance and allows a precise control of the quiescent current.

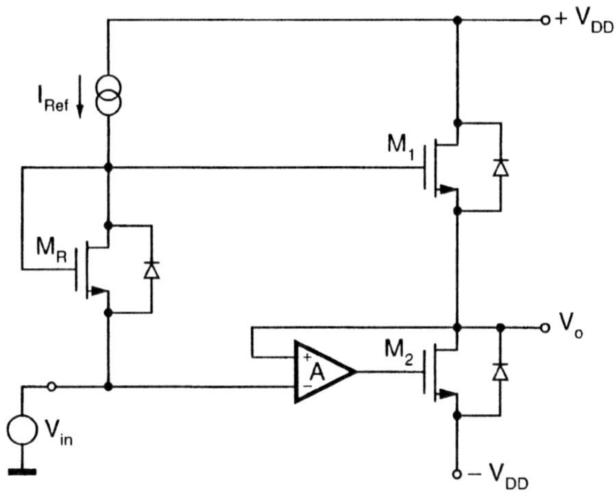

Fig. 5.29 Principle Schematic of Power DMOS Unity Gain Output Buffer

Since $V_{out} = V_{in}$ fixed by the local loop, V_{gs} of M_{ref} and M_1 are the same. Thus:

$$I_q = I_{ref} \frac{(W/L)\ M_1}{(W/L)\ M_{ref}} \tag{5.19}$$

given by the product of the reference current by the mirror ratio between M_1 and M_{ref}.

Once fixed M_1 by the desired $R_{ds\text{-}on}$ requirements and the above adimensional mirror ratio, a proper setting of I_{ref} allows an easy control of the quiescent current that can be then made independent of any technology process spread and insensitive to any temperature variation. As far as V_{aux} is concerned, it can be obtained by using bootstrap techniques in all AC applications or an additional continuous voltage in all DC applications.

5.5 Power Mosfet Ruggedness

Before introducing circuit solutions able to keep Power Mosfets within safe limits, their limitations on maximum power dissipation, voltage and current ratings are here described.

In addition, before introducing the SOA (Safe Operative Area) of Power Mosfets, emphasis is dedicated to a comparison between Bipolar and Power Mosfets in terms of ruggedness related features.

Bipolar transistors are well known among power electronic designers to be affected by the second breakdown failure mechanism.

This phenomenon, typical of any Bipolar transistor, occurs whenever the absorbed energy exceeds critical values creating hot spots in the transistor area [14].

The term "Second Breakdown" has ever since been used to outline the collapse of the collector-to-emitter voltage to a level by far lower than "first breakdown" voltage (first breakdown is the self-clamping voltage occurring across the base-collector p-n junction and caused by avalanche or carrier multiplication) and, depending on the transistor bias condition, is classified as Forward or Reverse Bias Second Breakdown.

5.5.1 Forward Biased Safe Operative Area

For forward biased Bipolar transistors (Forward Bias Second Breakdown), the practical mechanism responsible for this failure mechanism is the thermal run-away and the relative current crowding.

Since Power Mosfets show negative feedback forcing current density to decrease when a hot spot occurs (a decrease in the threshold voltage is in fact over compensated by a higher decrease in the carrier mobility), they can be though free from forward second breakdown failures.

For Forward Biased Power Mosfets, the basic limitations concurring to define the SOA (SOA defines the region of all allowed points in the plane log I_c, log V_{ce}) boundaries are (see Fig. 5.30):

- Maximum current ($I_d < I_{dmax}$) tolerable by the bonding wires (stretch B to C of Fig. 5.30)

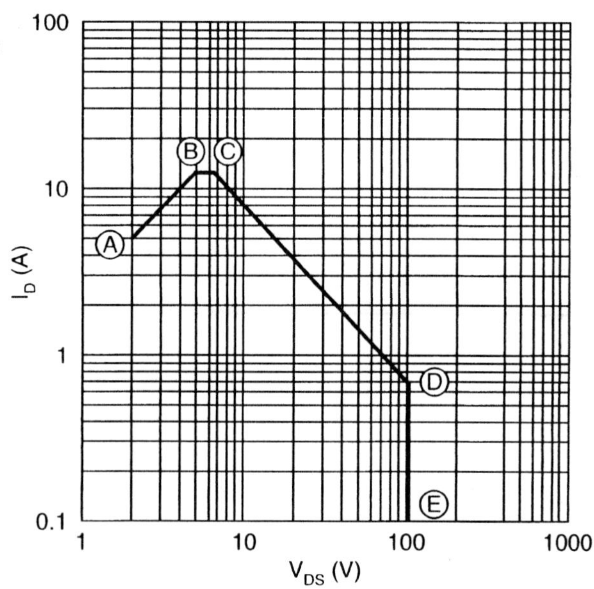

Fig. 5.30 Typical Power Mosfet Safe Operating Area

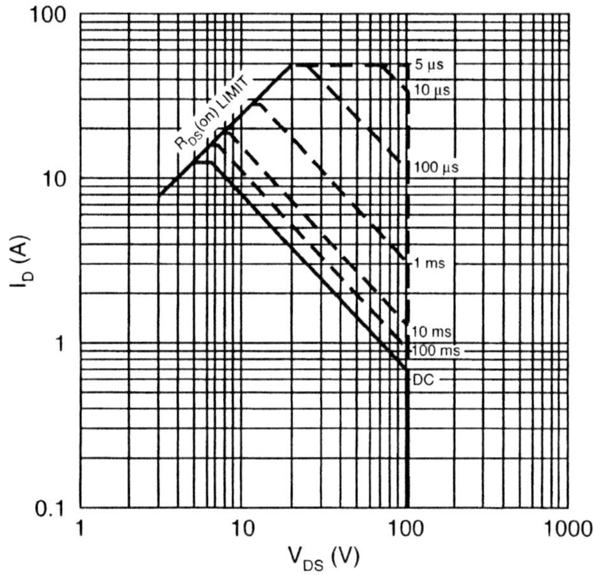

Fig. 5.31 Typical Power Mosfet Safe Operating Area in DC and Pulsed Conditions

- Maximum drain-to-source breakdown voltage ($V_{ds} < BV_{dss}$) (stretch D to E)
- Hyperbola of maximum affordable power dissipation ($P_{diss} = V_{ds} \times I_d <$ const. with const. mainly dependent on the thermal impedance of the system Silicon-Package and on the maximum junction temperature (150 °C usually)) (stretch C to D)
- Maximum conveyable current ($I_d < V_{ds}/R_{ds\text{-}on}$) (stretch A to B).

As shown in Fig. 5.30, stretch C to D is characterised by only one slope because of the mentioned absence of any Forward Second Breakdown. Figure 5.30 refers to DC condition and to a given ambient temperature; in the case of single power pulses, SOA extends beyond the limits of Fig. 5.30 as shown in Fig. 5.31.

In the case of single, short enough, power pulses, bonding wires can, in fact, tolerate higher level of current and the thermal stress is lower (the thermal resistance or better the thermal impedance can be by far lower).

On the contrary, the diagram in Fig. 5.32 considers how the SOA limits can eventually change at different ambient temperatures. This is basically obtained once observed that the basic constraint to be met is:

$$P_{diss} < \frac{T_{max} - T_{amb}}{RTh_{j\text{-}amb}} \qquad (5.20)$$

(once T_{max} is fixed the lower T_{amb}, the higher the level of power is)

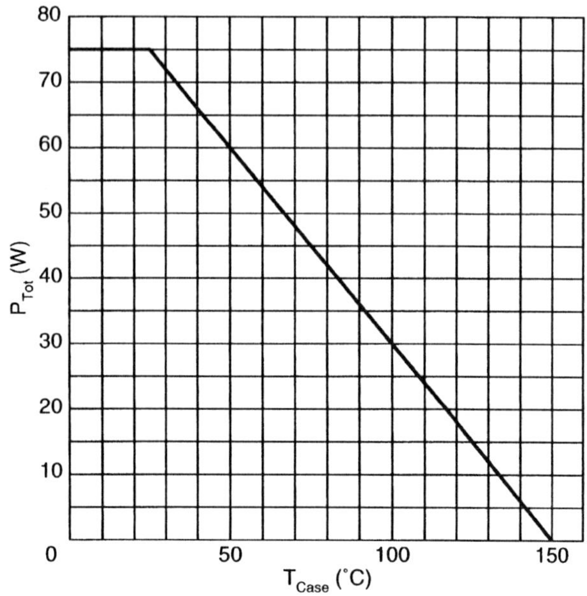

Fig. 5.32 SOA Derating Curve giving Max Allowed Power Dissipation Vs. Case temperature

5.5.2 Static dV/dt Failure

Forward Bias Safe Operating Area is sometimes not enough to avoid any failure especially when Power Mosfets are used in fast switching circuits.

Subjected in fact to high voltage slew rate between drain and source as, for instance during load driving turn-off, Power Mosfets can experience failure mechanism close to Bipolar transistor Reverse Bias Second Breakdown.

This phenomenon, called "Static dV/dt failure" (static because the Power Mosfets experiencing failure is virtually inactive and the drain-to-source voltage variation induced by external elements), can be studied recalling the Mosfet equivalent circuit in Fig. 5.1.

The Power Mosfet, subjected to high V_{ds} voltage slew rate, can spuriously switch on because of:

- Power Mosfet itself switching on
- Parasitic Bipolar transistor switching on.

Figure 5.33 a shows the equivalent circuit used to explain the spurious turn on of the Power Mosfet.

If the drain-to source voltage transient is fast enough to induce a currents through C_{gd} as expressed by the equation (Fig. 5.33 b):

$$I_{Cgd} = C_{gd} \times (dV_{ds}/dt) \tag{5.21}$$

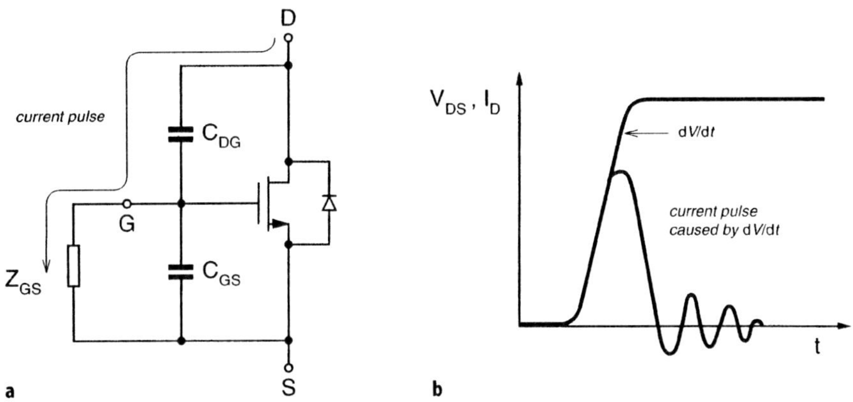

Fig 5.33 a Equivalent circuit explaining conduction by spurious turn-on of Power Mosfets, b Typical V_{ds} and I_{ds} waveforms induced by fast V_{ds} voltage transients

high enough to create across R_{gen}, a voltage drop greater than the Mosfet threshold the device can, in its turn, switch ON.

According to the eq. 5.21 and the above considerations, this phenomenon beyond on dV_{ds}/dt also depends on:

- R_{gen} (generator resistance); the higher R_{gen}, the higher the possibility of undesired conduction;
- C_{gd}; spurious turn on more likely occurs in devices with high C_{gd} (for instance the phenomenon is more evident in large devices with low BV_{dss});
- V_{th} the lower the threshold voltage, the higher the possibility.

Stress by dv/dt is not destructive in this situation but only responsible for higher power dissipation and unwanted source of interferences.

For these reasons, always in the case the parasitic Bipolar component does not switch on, Power Mosfet can often be used in unclamped inductive load switching (Fig. 5.34).

For this to happen all the energy stored in the inductor when the Power transistor is ON has to be entirely dissipated in breakdown condition without any failure to occur.

Figure 5.35 shows, on the contrary, the circuit that evidences how the parasitic Bipolar transistor can switch on.

In the same condition as before a capacitive current can result into a voltage drop across the base emitter junction according to the equation:

$$V_{be} = Rbb' \, C_{gd} \, (dV_{ds}/dt) \tag{5.22}$$

If V_{be} is high enough (0.65 V about) to switch the Bipolar transistor on, this can fail when the drain-to-source voltage "snaps back" to its sustaining voltage and second breakdown occurs.

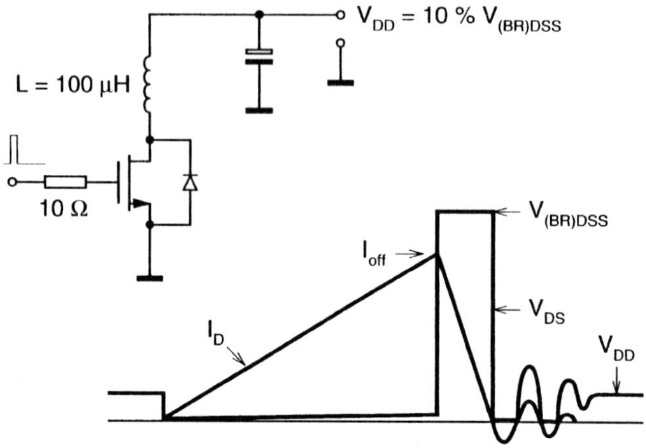

Fig. 5.34 V/I waveforms in the case of unclamped inductive loads driving

Fig. 5.35 Equivalent circuit explaining how conduction occurs by spurious turn-on of the parasitic bipolar transistor

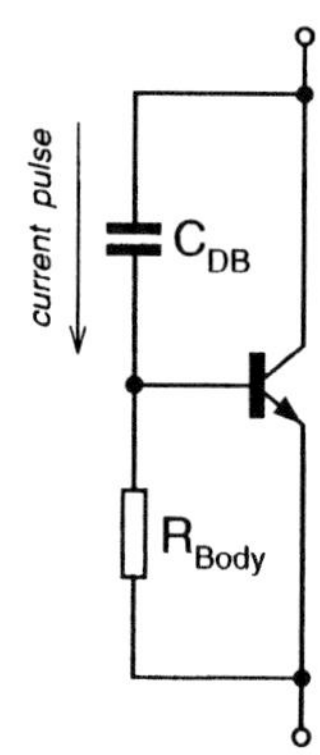

Mainly dependent on the breakdown characteristics of the BJT parasitic transistor, this phenomenon can be minimised by reducing and equalising Rbb′ among all distributed npn transistors (for instance increasing the Body area).

Referring to the unclamped inductive load driving, in case dv/dt is fast enough to trigger the Bipolar transistor different protection techniques can be used. These are:

- snubber networks lowering dv/dt slew rate to the disadvantage of increased switching losses (Fig. 5.36 a)
- free wheeling diodes across the inductive load or clamping zener diode between gate and drain (Fig. 5.36 b) automatically switching on the Power Mosfet before the BJT component avalanche is enabled.

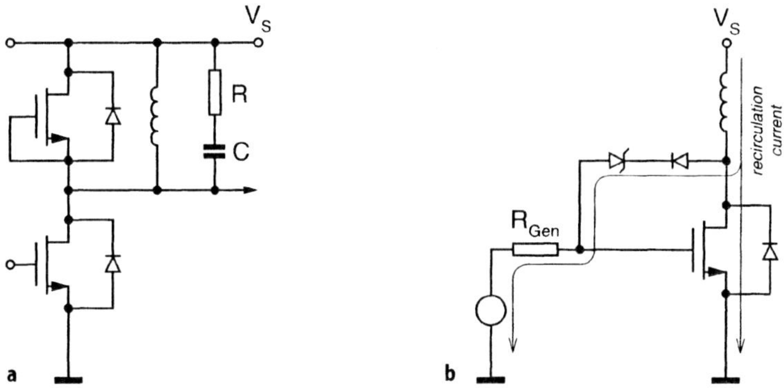

Fig. 5.36 Power Mosfets Over voltage Protections

Fig. 5.37 Low-Side Switch using Body-Drain Diodes acting as free wheeling components in Clamped Inductive Load Driving

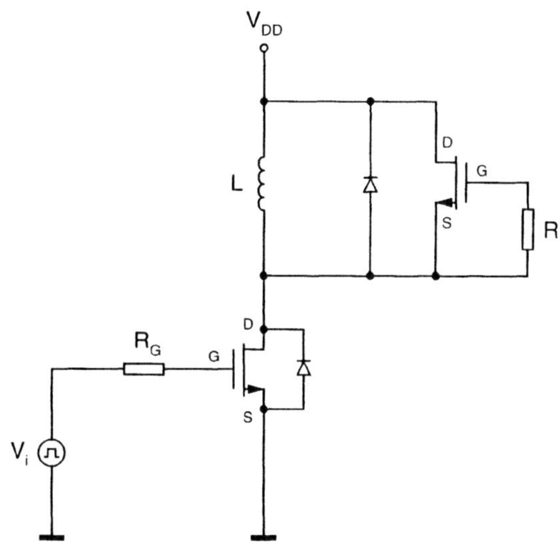

More influential than all the two previous failure mechanisms is the situation occurring when the intrinsic body-drain is used as a recirculation diode.

As all forward biased pn-junctions, this diode stores an excess of charge to be then fully removed in order to switch back to OFF.

This charge, besides increasing the power dissipation (the diode can be assumed behaving as an undesired short circuit for most of reverse recovery time), is also responsible for an extra current adding to the normal capacitive current induced by C_{db}.

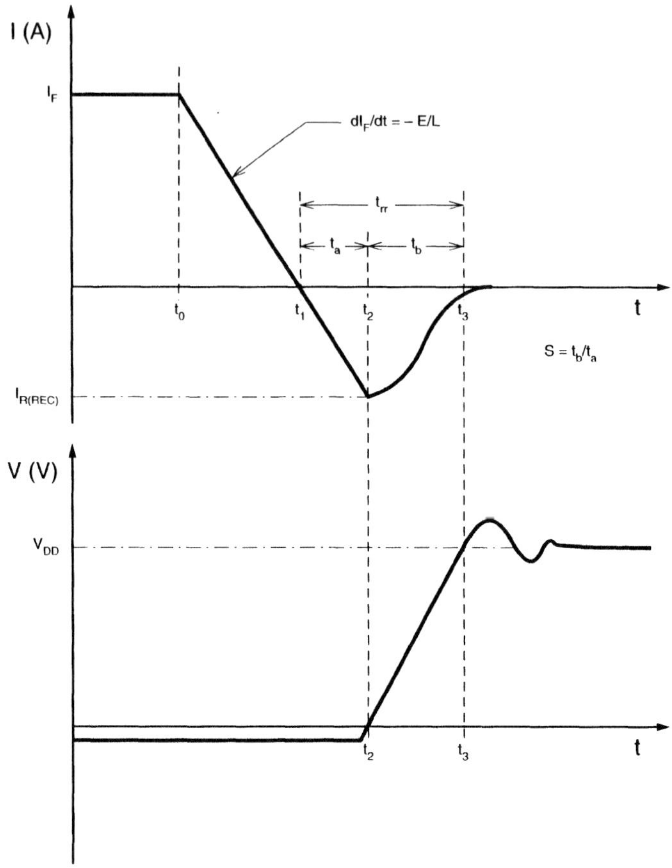

Fig. 5.38 Expanded Views of Current and Voltage waveforms for the free wheeling diode of Fig. 5.37

To better understand how this phenomenon is enabled, it is worth referring to Fig. 5.37 and 5.38. Fig. 5.37 shows a clamped inductive load driving with the free wheeling diode basically replaced by the body-drain diode, whereas Fig. 5.38 shows all the most important transients for the circuit of Fig. 5.37: voltage across body-drain junctions and commutation current. Starting at time t_0, the diode is switched off at a current rate dI_f/dt that depends, in the application of Fig. 5.37, mainly on the turn on time of the lower Power Mosfet.

During time t_1-t_2, the stored charge is able to supply more current than the circuit requires (the rectifier behaves like a short circuit) then, during time t_2-t_3, the output voltage starts increasing until reaching V_{supply}.

During time t_2-t_3, the recovery current is still high and the BJT, subjected at high dv/dt, can fail.

This failure is still dV/dt sensitive because the higher the voltage slew is the higher the capacitive current and because the diode reverse recovery current considerably sensitive to the current slope at zero crossing (the lower di/dt; the shorter trr).

As a consequence, the peak of reverse recovery current is strongly influenced by the switching time of the lower transistor (the faster the switching time, the higher the peak of reverse recovery current).

Once optimised C_{db} and Rbb′, in specific applications the switching times must be then properly controlled to avoid any failure.

All the basic over-voltage protections are the same as dealt with in Fig. 5.36a and b.

5.6 Over-Current Protections

The maximum current allowed to flow in the power element is usually sensed by measuring the voltage drop across a sense resistor R_s crossed by all or part of the output current.

This voltage can be then processed by a differential amplifier or by a comparator that, once exceeded a pre-set threshold, enables the over-current protection.

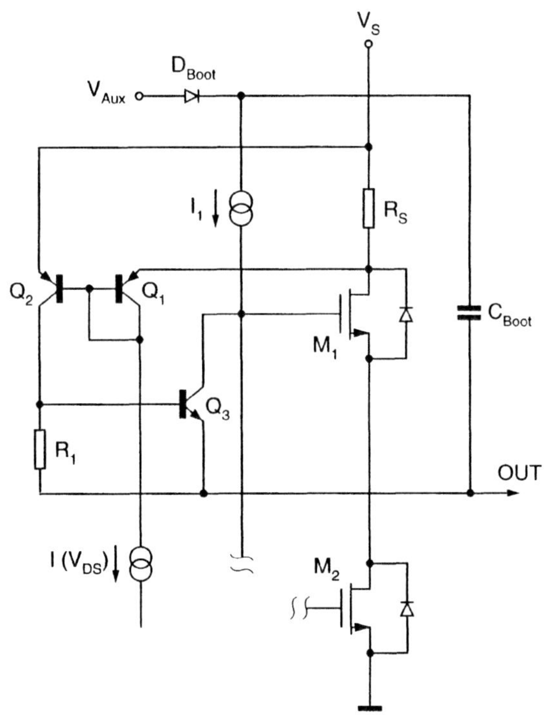

Fig. 5.39 Circuitry limiting Maximum current flowing in the Mosfet Based Output Stage of Linear Amplifiers

Here, in the following, over-current protections suitable to properly drive linear as well as switching amplifiers are introduced and explained together with different techniques used in sensing the output current.

5.6.1 Over-Current Protections in Linear AC and DC Amplifiers

As explained above, maximum current can be limited, after measuring the output current flowing in the power transistor, by using the circuit in Fig. 5.39 where M is the power transistor to be protected.

Transistors Q_1 and Q_2 form a simple differential amplifier with built-in, controlled offset voltage.

When the voltage across R_S exceeds this offset, Q_3 is driven on limiting then the Power Mosfet gate overdrive as well as the output current.

The value of I_{max} depends on the value of R_S, Q_1 to Q_2 emitter area ratio and on the value of the current source $I_{c1} = I_{c1}(V_{ds})$.

The feedback loops that results is such that

$$I_{omax} < I_{lim}\ (V_{ds}) \tag{5.23}$$

with I_{max} that can be found according to the following equation:

$$V_{be2} = V_{be1} + R_s\ I_{omax} \tag{5.24}$$

Since

$$V_{be} = (KT/q)\ \ln\ (I_c/A_e\ I_s) \tag{5.25}$$

with
A_e = transistor emitter area
I_s = transistor saturation current

equation 5.23 gives

$$I_{omax} = (1/R_s)\ (KT/q)\ \ln \frac{I_{C2}\ A_1}{I_{C1}\ A_2} \tag{5.26}$$

Furthermore if

$$I_{c2} = \frac{V_{be3}}{R_1} \tag{5.27}$$

$$I_{c1} = I\ (V_{ds}) \tag{5.28}$$

it is enough that $I_{ds} = I_{ds}\ (V_{ds})$ being properly dependent on supply voltage and on temperature, to have

$$I_{omax} = \frac{1}{R_s}\ \frac{KT}{q}\ \ln \left(\frac{A_1}{A_2}\ \frac{V_{be3}}{R_1}\ \frac{1}{I\ (V_{ds})}\right) \tag{5.29}$$

meeting all the constraints fixed by the SOA limits.

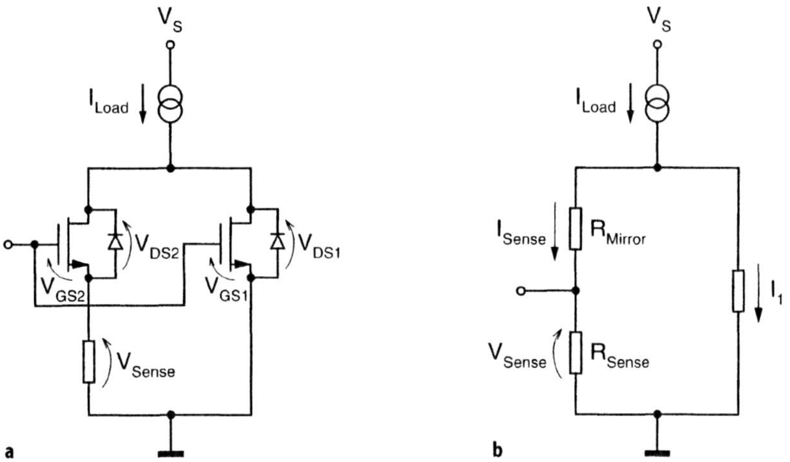

Fig. 5.40 Current-mirror Power Mosfets and associated equivalent circuit. Sense resistor R_{sense} causes the real current ratio to differ the theoretical value

Low ohmic values for R_s (R_s value lies in the 10 to 50 mOHM range not to reduce the output voltage swing as well as not to increase the overall dissipated power) are usually obtained by using a metal strip behaving as a resistor connected between the supply rail pad and the power element.

As results from eq. 5.29, the input stage of the comparator here introduced is also able partially to cancel the temperature variation of this metal resistor by the almost identical $\Delta V_{be}/\Delta T$ variation.

A fully different methodology of sensing the output current, applicable just in the case of Mosfets and not possible at all in the case of Bipolar components, because of their by far higher transconductance, is shown in Fig. 5.40 a.

Once observed that Power Mosfets are basically an array of thousands of individual, identical, elementary transistors all connected in parallel, the circuit in Fig. 5.40 b can be basically modelled with two transistors: a low $R_{ds\text{-}on}$ power transistor and an high $R_{ds\text{-}on}$ sense transistor. The overall drain current is distributed between the two parts according, at first approximation, to the cell ratio "n".

Since n can be about 1/1500, this current mirroring allows an easy and power dissipation free sensing of the output current.

As a matter of fact, the current mirroring depends not only on the cells ratio, but also on the sense voltage.

To understand exactly, how the composite element behaves it is worth referring to Fig. 5.40 b.

In case both the two transistors work in triode mode, their drain currents can be expressed as in the following [15]:

$$I_d = K\left((V_{gs} - V_{th})\, V_{ds} - 1/2\, V_{ds}^2\right) \tag{5.30}$$

with

$$V_{ds2} = V_{ds1} - V_{sense} \tag{5.31}$$

$$V_{gs1} = V_{gs2} - V_{sense} \tag{5.32}$$

Substituting the expressions of V_{ds2} and V_{gs2} in the square-law equation 5.30 and performing some factoring, I_{sense} can be found as:

$$I_{sense} = K_{sense}\left((V_{gs1} - V_{th} - V_{sense})(V_{ds1} - V_{sense}) - 1/2\,(V_{ds1} - V_{sense})^2\right) \tag{5.33}$$

that, as noted before, is function of the sense voltage and not only dependent on the cells number ratio.

It is worth then considering the mirroring factor "n′" as expressed by:

$$n' = \frac{I_{sense}}{I_1} \tag{5.34}$$

to exactly define the ratio of the currents flowing into the power and into the sensing cells. Substituting I1 and Isense square law equations in (5.34) n′ results in:

$$n' = \frac{K_{power}\left((V_{gs1} - V_{th})\,V_{ds1} - 1/2\,V_{ds1}^2\right)}{K_{sense}\left((V_{gs1} - V_{th} - V_{sense})(V_{ds1} - V_{sense}) - 1/2\,(V_{ds1} - V_{sense})^2\right)} \tag{5.35}$$

where K_{power}/K_{sense} reflects both the source periphery ratio as well as the impedance ratio R_{power}/R_{sense}.

Since K_{power}/K_{sense} is the cell ratio "n", the above expression can be reduced to the simpler version:

$$n' = \frac{n}{1 - \dfrac{V_{sense}\,(V_{gs1} - V_{th} - 1/2\,V_{sense})}{V_{ds1}\,\,(V_{gs1} - V_{th} - 1/2\,V_{ds1})}} \tag{5.36}$$

From eq. 5.36, if V_{sense} and V_{ds1} are smaller than V_{gs1} and V_{th}, the current mirror ratio can be approximated by

$$n' = \frac{n}{1 - (V_{sense}/V_{ds1})} \tag{5.37}$$

Equation 5.37 shows that, as the sense voltage increases in respect to $V_{ds\,(on)}$ of the power part, the denominator decreases and the current mirror ratio increases.

From the physics standpoint, increasing the sense resistor's value in the current mirror causes the sense voltage also to increase.

This increase subtracts from the mirror cell's drain-to-source voltage, thereby increasing the equivalent resistance of this cell.

This increase in the resistance causes a decrease in the mirror current, hence an increase in the current mirror ratio.

Figure 5.41 clearly shows the flattering of the sense voltage occurring when, increasing the value of the sense resistor, V_{sense} becomes just the Power device $V_{ds\text{-}on}$.

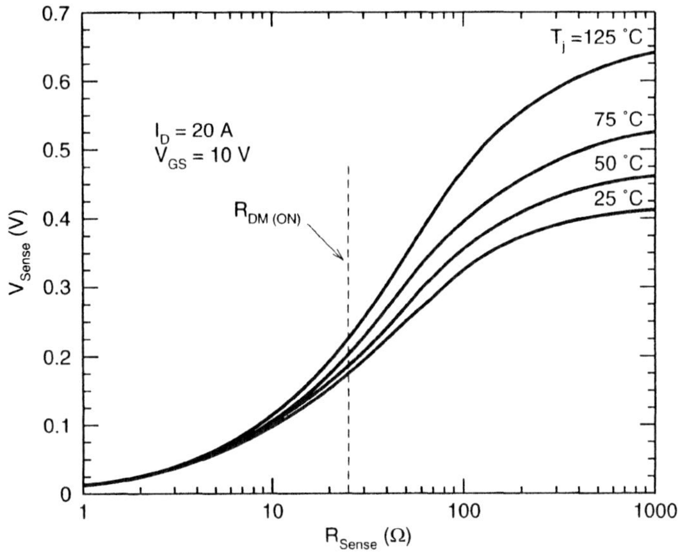

Fig. 5.41 Sense Voltage as function of the sense resistor: mirroring ratio and temperature dependence

As from Fig. 5.41, it is common practice to have R_{sense} not exceeding the $R_{ds\text{-}on}$ of the power part at 25 °C.

To accurately predict the sense voltage temperature variation, it is worth re-arranging eq. 5.37 as in the following:

$$n' = n + \frac{R_{sense}}{R_{ds\text{-}on}} \tag{5.38}$$

(note that $V_{sense} = R_{sense} \times I_{sense}$; $V_{ds1} = R_{ds\text{-}on} \times n' \, I_{sense}$).

Since the on-resistance of a Power Mosfet roughly doubles between 25 and 150 °C, this can be expressed by the equation:

$$R_{ds\text{-}on}(T) = R_{ds\text{-}on}(25\,°C) \exp\left(K(T - 25)\right) \tag{5.39}$$

where

$$K = (1/125) \ln\left(\frac{R_{ds\text{-}on}(150\,°C)}{R_{ds\text{-}on}(25\,°C)}\right) \tag{5.40}$$

Substituting 5.41 and 5.38 in 5.34, V_{sense} can result in:

$$V_{sense}(T) = \frac{I_{ds} \times R_{sense}}{n + \left(\frac{R_{sense}}{R_{ds\text{-}on}(25\,°C) \exp\left(K(T - 25\,°C)\right)}\right)} \tag{5.41}$$

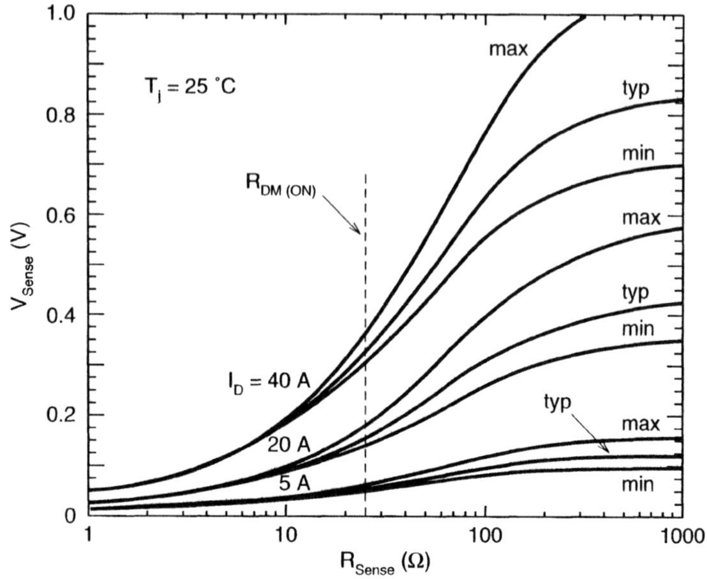

Fig. 5.42 Sense Voltage as function of the sense resistor: dependence on $R_{ds\text{-}on}$ process distribution

Plotting this expression, in the case of R_{sense} external and not tracking the junction temperature, demonstrates how the sense ratio becomes more and more temperature dependent as the value of R_{sense} becomes comparable to or greater than $R_{ds\text{-}on}$ (25°C) (see again Fig. 5.41).

As well as with the temperature, device-to-device $R_{ds\text{-}on}$ spreads affect the mirror ratio precision as shown in Fig. 5.42 where the sense voltage as a function of R_{sense} for typical and limit devices is considered.

Even in this case, sense resistors lower than $R_{ds\text{-}on}$ give sense voltages less sensitive to $R_{ds\text{-}on}$ spreads.

If in most applications this constraint is not a problem, in others, when the current is low and higher resistor values have to be used, the virtual-ground current sensing method, (see Fig. 5.43), can bring benefits.

If its major advantage is that the sense ratio becomes independent of the sense resistor and junction temperature, the major drawback is the usage of an extra op-amp often limiting the bandwidth of the current sensing circuitry (this is a heavy limitation, for instance, in high frequency switching applications).

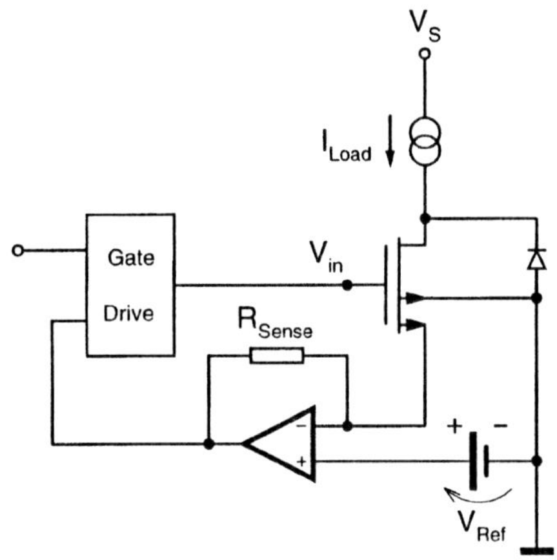

Fig. 5.43 Virtual-ground sensing, $V_{sense} = I_{sense} \times R_{sense}$ (V_{sense} must be referred to the Kelvin-source)

5.6.2 Over-Current Protections in Switching Amplifiers

Figure 5.44 shows a switching power amplifier together with the associated current limiting circuit.

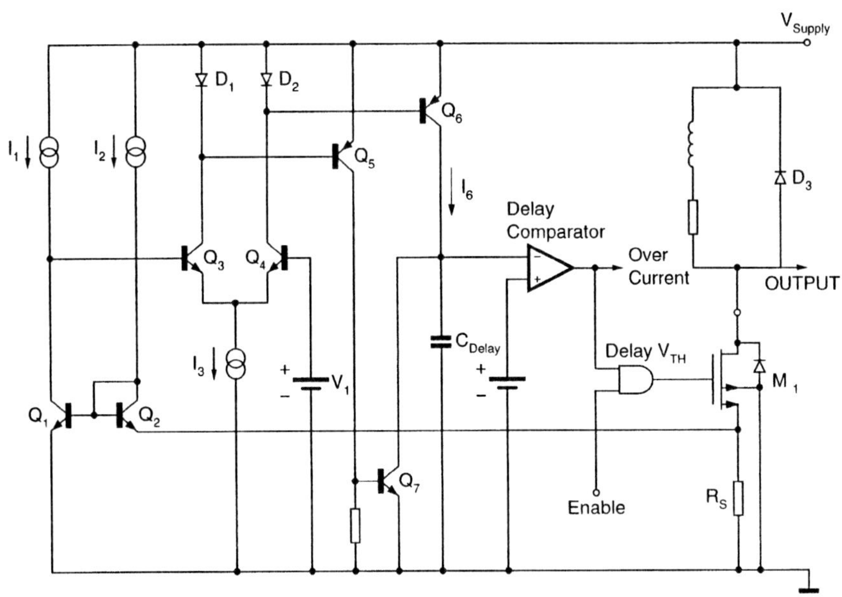

Fig. 5.44 Switching Amplifier with Over-current protection circuit

I_{max} is measured by using as sense resistor Rs and then limited to the level given by the following equation.

$$I_{max} = \frac{KTq}{Rs} \ln \frac{A_2 I_1}{A_1 I_2} \tag{5.42}$$

When I_{out} reaches the current threshold corresponding to I_{max}, Q_3 turns OFF and consequently Q_6 begins charging the delay capacitor C_{delay} because, in the meanwhile, Q_7 has switched OFF (T_7 is necessary to discharge C_{delay}).

Only if I_{out} exceeds I_{max} for a time longer than the time the delay capacitor takes to exceed the delay threshold voltage, the delay comparator changes state ($1 \rightarrow 0$) turning OFF the power transistor M_1.

This delay time is often necessary to mask D_3 excess recovery current flowing in M_1, when this turns ON.

Since the value sensed by R_S during all the reverse recovery time

$$I_{sense} = I_{rr} + I_{load} \tag{5.43}$$

is usually higher than or comparable to I_{Load}, this delay (depending on I_{Load}, this time is normally in the range of 0.5 usec) is used to avoid any spurious triggering of the over current protection.

5.7 Over-Temperature Protections

As seen in par. 5.4, the SOA of Power MOS transistors is fixed by the maximum dissipation curve in the I_d-V_{ds} plane, whose points meet the constraint $V_{ds}.I_{ds} < P_{dmax}$.

Since $P_{d\text{-}max}$ is, at its turn, dependent on the maximum reachable junction temperature, ambient temperature and thermal impedance of the systems, it is evident that the maximum allowed dissipation has to be function of the junction temperature of the dissipating element.

Here in the following, temperature methods and over-current protection techniques are discussed both for switching and linear power amplifier.

5.7.1 Over-Temperature Protection with Hysteresis

In switching amplifiers, it is common practice switching the Power Element Off as soon as the junction temperature exceeds t_{jmax} and, once the temperature has fallen below a lower temperature level (for this reason this protection is said with Hysteresis), to switch it On again.

Figure 5.45 shows a circuit implementing this function.

Transistors arrays Q_1,Q_2 and Q_3 to Q_5 work as thermal sensors, both of them delivering, in fact, two voltage levels V_1 and V_2 that, compared by Comp., give an over temperature output signal LOW when the junction temperature exceeds t_{max}.

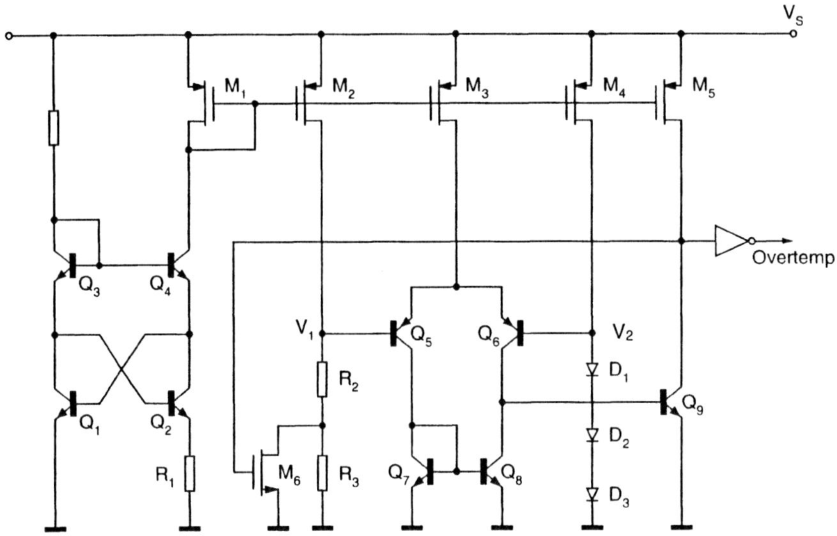

Fig. 5.45 Over-temperature protection with Hysteresis

In detail when $T_j < t_{max.}$ V_1 and V_2 as expressed in the following equations:

$$V_1 = (\Delta V_{be1\text{-}2})\ \frac{R_2 + R_3}{R_1} \tag{5.44}$$

$$V_2 = 3\ V_{be} \tag{5.45}$$

have to satisfy the inequality

$$V_2 > V_1 \tag{5.46}$$

Since V_1 and V_2 have opposite thermal variation coefficients (V_1 positive, V_2 negative), properly setting V_1 and V_2 at 25 °C allows the comparator to change its output from Low to High when the junction temperature exceeds $t_{max.}$ (for instance $t_{max.} = 150$ °C) and the Power element to switch Off.

In the meanwhile this transition, switching M_1 On, changes V_1 in this case given by

$$V_1 = (\Delta V_{be1\text{-}2})\ \frac{R_2}{R_1} \tag{5.47}$$

providing, in such a way, a thermal hysteresis.

Once exceeded t_{max}, until the junction temperature lies in between the two thresholds levels, the power element is kept OFF allowing the system to cool down.

Only after the T_j has reached the lower threshold level the power part can restart.

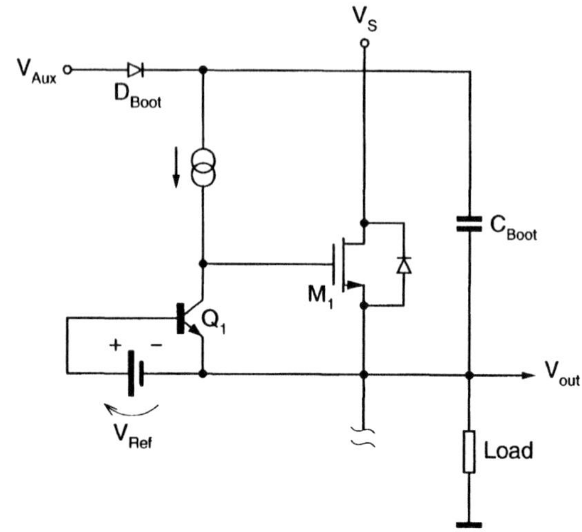

Fig. 5.46 Temperature Sensing Element Used to avoid Over-temperature in Linear Power Amplifiers

5.7.2 Over-Temperature Limiting at T_j Constant

A different way of implementing over- temperature controls, often used in linear power amplifies, just relies on the possibility of decreasing the output power when the temperature exceeds T_{max}.

The simplest and most efficient solution is represented in Fig. 5.46; V_{ref} is a fixed voltage, whose value is such as to keep Q_1 below its conduction threshold at ambient temperature.

As the local temperature of the power element increases, the collector current of Q_1 will eventually add to Q_3 of Fig. 5.39 introducing a thermal feedback loop able to keep $T_j < T_{max}$.

Always referring to the circuit in Fig. 5.46, this circuit is such to keep

$$I_o < I_{omax}(V_{ds}, T)$$

with I_{omax} decreasing function of T as commonly required.

5.8 Controlling Parasitic Current in Junction-Isolated ICs

Parasitic currents enabled by inductive loads are among all the unrequired, parasitic effects, the most peculiar, by far the least predictable and the most difficult to manage.

The inductive nature of many power loads can, in fact, drive the output pins beyond the power rails (below the substrate and above the positive supply voltage) resulting in both hole and electron injections [16].

Power applications using Mosfet Power transistors or, in more detail, the intrinsic Body-Drain as free wheeling diode can activate parasitic components that, if not properly controlled, can cause difficulties ranging from improper operations to catastrophic failures.

In order to better study how these phenomena can be triggered, a power Mosfet H-bridge, used to drive inductive loads, is shown (for which in the following H-bridges are the most difficult to managed) (Fig. 5.47).

In free wheeling condition, with all Power Mosfets OFF, the load drives OUT_1 above V_{dd} and OUT_2 below ground letting the current find through D_1 and D_2 a proper recirculation path.

In such a situation, both Q_1 and Q_2, whose real structure is shown in cross section in Fig. 5.48, are enabled and parasitic current is injected.

Q_2 emitter, going below ground, forward bias the n-epi-tub/p-substrate junction forcing electrons either to diffuse to the backside contact or laterally to be collected by other n-tubs that are reverse biased.

Fig. 5.47 Power Mosfets H-bridge with parasitic npn and pnp transistors shown

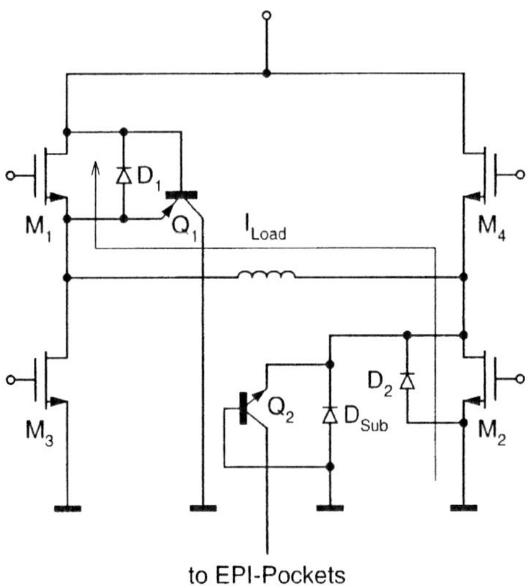

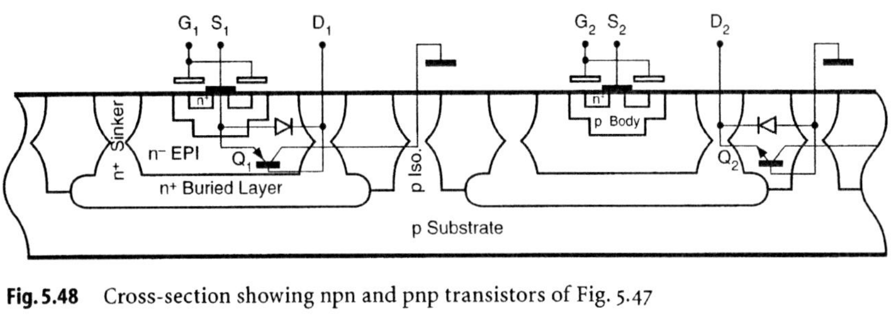

Fig. 5.48 Cross-section showing npn and pnp transistors of Fig. 5.47

Minimising the supply of holes (avoiding for instance isolation top contacts in zone close to the most critical circuitry) is common practice to limit the electron concentration and the lateral transfer of electrons to the tubs.

Without any special precaution the collected current will depend on the substrate resistivity, on-die distance from the injecting point and the area of the collecting epi.pocket (see Fig. 5.49).

Q_1 emitter going above V_{dd} vice versa forward bias the junction p-body/n-drain injecting holes that, from the subcollector scatter through the depleted region, are, as a consequence, transported to the substrate.

Although n^+ buried layer and sinker partially inhibits this holes transport, a fraction can penetrate these regions and cause the substrate potential locally to rise above the nominal ground potential (peak substrate voltage can rise up to tens of volts).

This can cause the N-tubs nominally biased to ground (for instance the collector of a saturated npn transistors) to act as the emitter of a lateral npn transistor that, forward biased, inject electrons into other epi-pockets.

Hardwiring the isolation walls to GND can, in this case, avoid substrate potential to rise as a consequence of holes injections but, for the reasons stated previously, it better activates Q_1 and increases the lateral diffusion of electrons currents (see Fig. 5.50; the asterisk, for each curve, shows where the substrate contact has been placed).

A substrate contact, cancelling in fact the natural decay of the injected current and better activating the nearby npn transistor increases the local amount of collected current.

What mitigates T_1 magnifies the effectiveness of T_2 and makes the management of parasitic currents in H bridge structures difficult to be handled.

Among all the different ways, intended to control these currents, a well-known method concerns surrounding the power element with epi-tubs connected to GND or V_{dd} creating, in such a way, a preferential collector that, conveying most of the current, acts as a barrier preventing the current to be injected in other epi-pockets.

Further improvement can be obtained by using the structure of Fig. 5.51 that, is usually quite efficient against lateral diffusion of current induced by N-tubs going below GND.

T_1 represents the parasitic transistor close to the power stage, T_2 an equivalent component close to critical epi-tubs.

T_1 turning on and saturating keeps T_2 off decreasing this last component collector current.

As a matter of fact, since substantially lay-out dependent, all these effects are difficult to be simulated and controlled just by design techniques; very often known by expert designers these still represent the major cause of failures and circuit redesign in Power I_{Cs}.

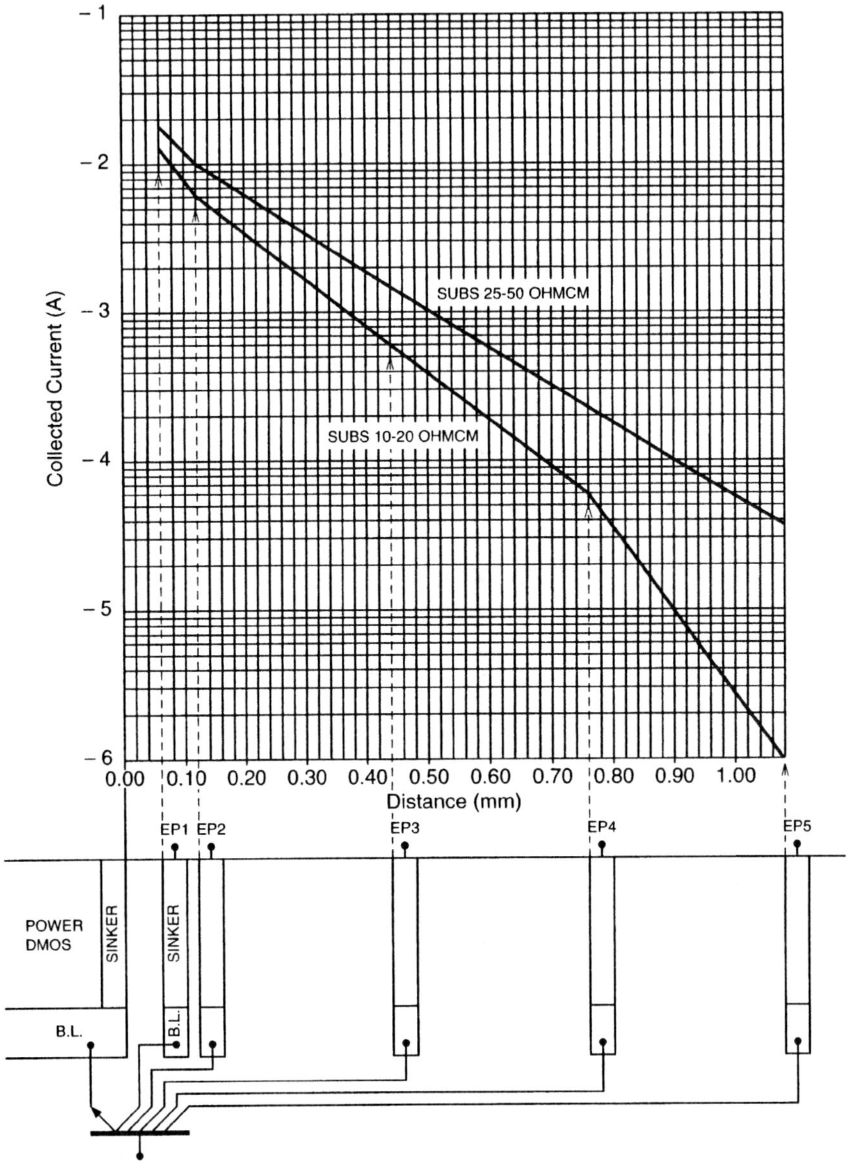

Fig. 5.49 Distributed npn transistors effects: backside substrate contact tied to GND; epi collecting pockets area = 0.15 mm², injected current 1 A

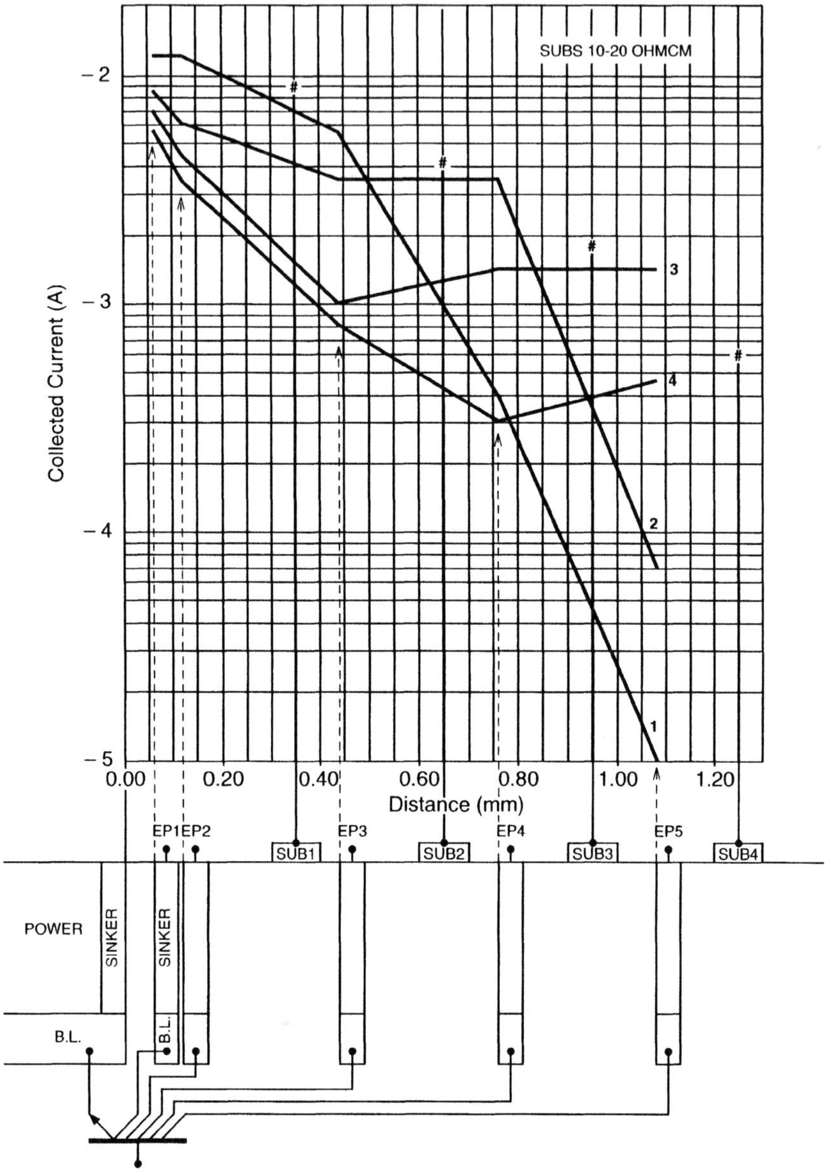

Fig. 5.50 Effect of on-die substrate contacts: collected current V_s distance

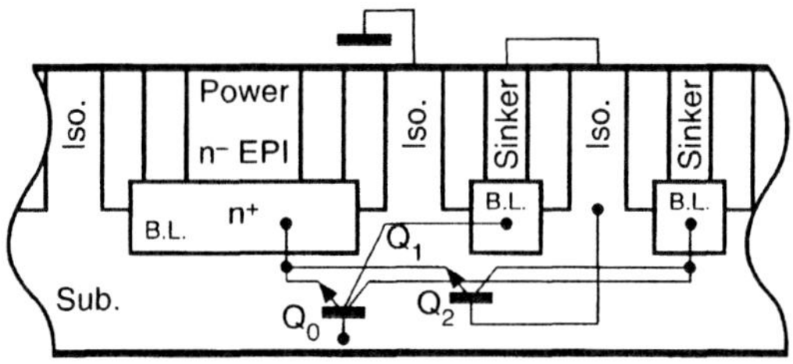

Fig. 5.51 Barrier against electrons parasitic injection

References

1. Muller RS, Kamins TI, Device Electronics for Integrated Circuits, pp 314-317, John Wiley & Sons
2. Cini C, Diazzi C, Rossi D (1985) An High Efficiency, Mixed Technology Motor Driver IC's, pp 267–277, Proceeding of PCI
3. Murari B (1993) Mixed Power IC Technologies and Small System Integration, invited paper ESSCIRC
4. Antonietti P, Power Integrated Circuits: Physics, Design, Applications, pp 4.13–4.17, McGraw-Hill
5. Chavez W, Kanner J (1993) Reverse Battery Protection: A New Niche for Bipolar Transistors, pp 9–17 PCIM
6. Grant DA, Gowar J, Power Mosfets: Theory and Applications, pp 84–91 135–138, John Wiley & Sons
7. Danz G (1994) High-voltage IC's Simplify Half-Bridge Driver Design, pp 135–142 EDN
8. Cini C, Diazzi C, Rossi D (1986) High Frequency, High Current Switching Regulator in Bipolar, CMOS, DMOS mixed technology, pp 62–73, Proceeding of PCI
9. Andreini A, Contiero C, Galbiati P, Storti S (1987) Design of High Side Driver in Multipower BCD and Vipower Technologies, Proceeding of IEDM, Washington DC
10. Paparo M, Javurek W, Smith III WH & Pietrobom G (1994) A very Low Noise, uP Interfaced, Multiple High Side Driver with Self Diagnostics for Car Radio Application, 940256 proceeding of SAE 1994
11. Storti S, Consilieri F, Paparo M (1988) 30 A, 30 V DMOS Motor Controller and Driver, IEEE Journal of Solid State Circuit, vol 23
12. Pietrobon G, Rossi D (1989) A Reconfigurable DMOS Control Chip for an Electronic Typewriter, pp 200–201 Proceeding of ISSCC
13. Brasca G & Botti E (1992) A 100V/100W Monolithic Power Audio Amplifier in Mixed Bipolar-MOS Technology, ICCE
14. Villa FF (1984) La rottura secondaria nei circuiti integrati di potenza, Elettronica e Telecomunicazioni, No 3, 1984
15. Fay G (1986) Current-Mirror FETs Cut Cost and Sensing Losses, pp 193 200, EDN
16. Widlar R (1991) Controlling Substrate Currents in Junction-Isolated IC's, pp 1090 1097, IEEE J Solid State, vol 26, No 8

CHAPTER 6

Motion Control

R. Gariboldi

6.1 Introduction

Motion control is one of the branches in the electronic systems where the circuit evolution and the progress in technology have already moved together.

At the end of the seventies, the bipolar technology was applied for the first time to the integration onto the same chip of signal and power parts in the emerging motion control applications.

In the eighties, the mixed technologies (as BCD: Bipolar - CMOS - DMOS) brought a significant breakthrough in the art of power integration.

Finally, in the last few years, a new generation of ultra-efficient motion control ICs has been designed integrating onto the same chip complex smart power systems including many different power stages with the associated control circuitry (COMBO chips).

In fact, IC design and technology are now so closely related that they are virtually no longer separable. The application of mixed technologies in the design of motion control ICs makes it possible to dissipate very low power at reduced costs since the devices can be assembled in very compact multipin packages hence heatsinks, cooling fans and so on can be either of small size or better, totally avoided.

In this chapter, after the presentation of basic topics such as voltage overdrive generators, power output stages and the feedback control theory, we will describe some new architectures of smart power ICs dedicated to drive stepping motors, DC motors, brushless motors and voice coils.

Several circuits will be derived from hard disk drives which represent today the leading force of smart power integration, showing also how the versatility of mixed technologies allows the optimization of each subcircuit within the system.

A fundamental problem always present in motion control ICs during current recirculation on intrinsic flyback diodes of DMOS transistors when driving inductive loads, is the activation of parasitics. If not properly controlled and managed these can lead to circuit malfunctioning (latch-up) or, even worse, to permanent device damage [1, 2, 3]. Special protection barriers are introduced to reduce this effect [4, 5]. This aspect is considered in detail in chapter 5.

Finally some practical examples of smart power ICs will be described.

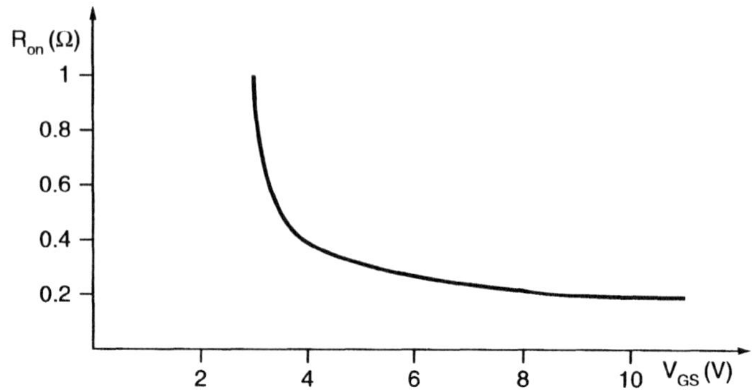

Fig. 6.1 Power DMOS on-resistance versus gate-source voltage

6.2 Voltage Overdrive Generator

A typical characteristic of the on-resistance R_{on} of a power DMOS transistor versus gate-source voltage is reported in Fig. 6.1.

This curve shows that to drive the DMOS in full linear conduction and to minimize its on-resistance, it is necessary to apply a gate source voltage of about 10 V (called voltage overdrive).

Hence, for high-side drivers, an auxiliary voltage always 10 V higher than the drain supply voltage and in tracking with it, is required.

We will see several techniques used to realize this function.

6.2.1 Bootstrapping Capacitor

In general, this technique consists of charging at the voltage overdrive a large value capacitor (tenths of nanofarads) which has stored enough energy to quickly charge the gate input capacitance of the power DMOS.

This is automatically achieved by the switching action of the power device itself.

If we consider the circuit configuration in Fig. 6.2, where two power devices are present, when M_1 is on, capacitor C_B will be charged at the required overdrive voltage V_{OD} (minus the voltage drop on diode D). This can be set by an internal bias regulator or it can be another external supply.

As soon as M_1 is turned off and M_2 on, the voltage on capacitor C_B is applied to the internal drive circuitry of M_2 (top driver) so that the output voltage V_O starts rising while the bootstrapping capacitor will keep the appropriate gate-source voltage. Diode D blocks the discharging path toward the bias regulator and it can be integrated.

The top driver circuit has an applied voltage equal to V_{OD} but must withstand a voltage equal to V_S plus V_{OD} with respect to the substrate.

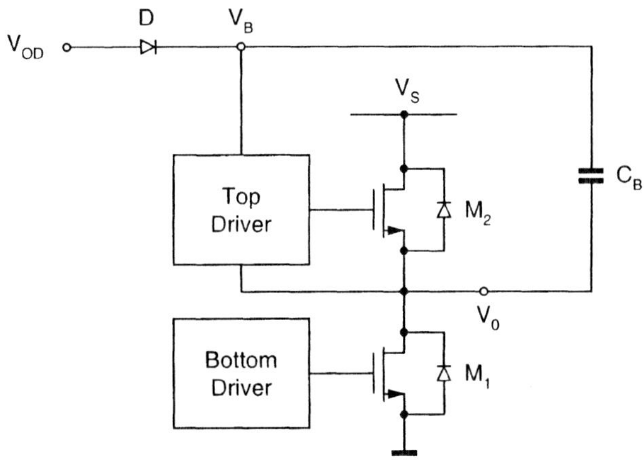

Fig. 6.2 Bootstrapping capacitor voltage overdrive generator

The design requirement of C_B is to provide adequate energy to the driving circuit without decreasing its voltage when operating at minimum frequency.

This technique allows very fast switching times but has the disadvantage of requiring a large off-chip capacitor and a dedicated pin for each high side DMOS.

6.2.2 Central Charge Pump

In complex smart power ICs, the bootstrap solution is very expensive because it requires one pin and one external capacitor per high-side driver and hence a larger package.

These drawbacks have been overcome by designing a unique voltage overdrive generator which is able to deliver sufficient energy to supply all high-side switches thus creating a supply bus to which all high-side drivers are connected.

There are two possibilities:

a) if diodes D_1, D_2 are integrated, three pins and two external capacitors are necessary (see Fig. 6.3);
b) if D_1, D_2 are external, two pins and two external capacitors are required.

Basically the circuit includes an internal CMOS oscillator operating at high frequency (up to several MHz) supplied by the required voltage overdrive V_{OD}. The oscillator drives the pump capacitor C_P (typically tenths of nanofarads) that in turn does not charge directly an internal DMOS gate but a large (hundreds of nanofarads) storage capacitor C_S.

In steady-state the voltage V_{CP} is given by:

$$V_{CP} = V_S + V_{OD} - V_{drop} \tag{6.1}$$

where V_{drop} is the average total voltage drop defined, with good approximation, as:

Fig. 6.3 Central charge pump voltage overdrive generator

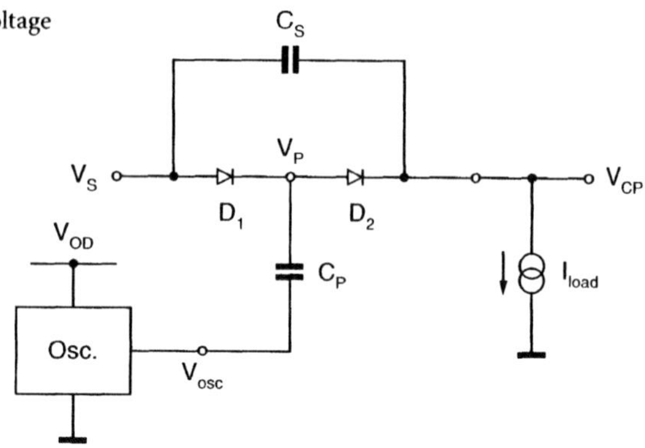

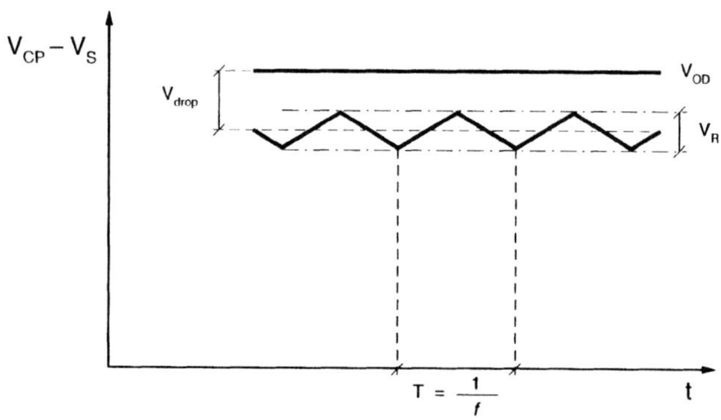

Fig. 6.4 Steady-state voltage waveform of a loaded central charge pump

$$V_{drop} = 2V_D + \frac{I}{f \cdot C_P} \tag{6.2}$$

with V_D diode voltage drop, I load current and f oscillation frequency.

Voltage V_{CP} has also a superimposed ripple, whose peak to peak value is:

$$V_R = \frac{I}{f \cdot C_S} \tag{6.3}$$

Figure 6.4 shows the voltage overdrive waveform.

6.2.3
Low Drop Central Charge Pump

As equation 6.2 clearly shows, the total voltage drop is strongly affected by the diode drops. This is a significant drawback when operating at low voltage from battery supplied systems.

Fig. 6.5 Low-drop central charge pump

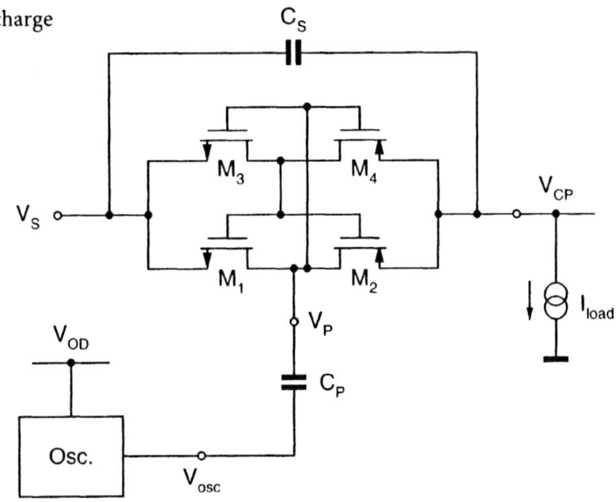

A circuit, which substitutes diode drops with CMOS on-resistance switches, has been proposed and discussed in detail by Gariboldi and Pulvirenti in [4].

The basic structure is shown in Fig. 6.5.

Essentially M_1, M_2, M_3, M_4 (with body shorted to source) are CMOS transistors operating as a latch triggered by the signal delivered by pump capacitor C_P.

With this configuration the voltage drop V_{drop} can be written as:

$$V_{drop} = \left(2R_{on} + \frac{1}{f \cdot C_P}\right) \cdot I \qquad (6.4)$$

where R_{on} is the on-resistance of transistors M_1 and M_2 assumed equal. Thus, by properly sizing the switches, a very low drop-out can be obtained.

6.2.4 Step-Up Regulator

In applications where high current from the voltage overdrive generator is required, a step-up regulator can be used.

This solution is expensive since it requires an inductor, some large value low ESR external capacitors and the internal circuitry is more complex.

Furthermore, to minimize interference and noise, special attention must be given to carefully separate this block within the device from the surrounding circuits by special protection barriers.

However, with this approach it is possible to obtain another voltage supply with current capability as high as desired also starting at very low voltage.

A simple and widely used self-oscillating step-up regulator operating in discontinuous current mode is shown in Fig. 6.6.

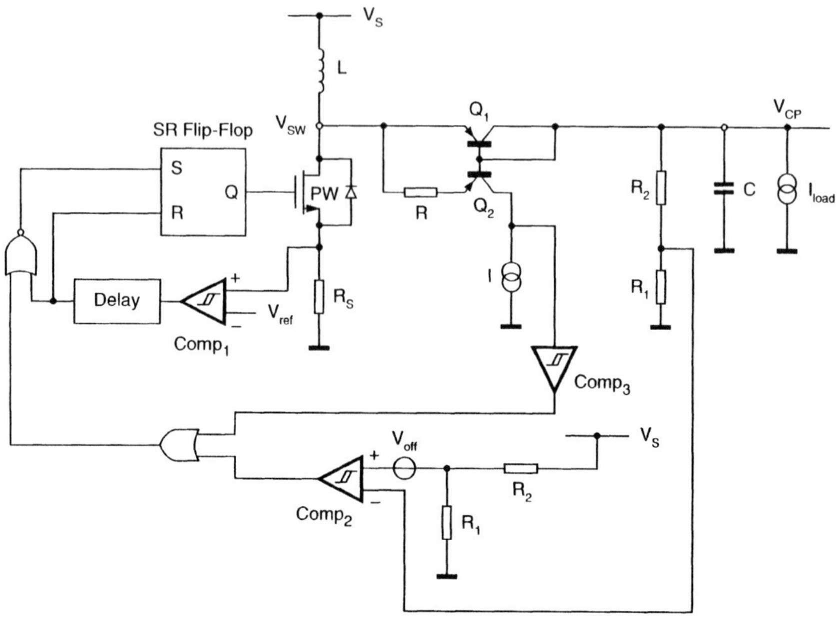

Fig. 6.6 Step-up converter voltage overdrive generator

The output voltage V_{CP} is given by the relationship:

$$V_{CP} = V_S + V_{off} \cdot \left(1 + \frac{R_2}{R_1}\right) \tag{6.5}$$

where V_{off} is a fixed voltage, that together with resistors R_1, R_2 sets the overdrive.

The operation is as follows: when the power DMOS transistor PW is on, current flows through the inductor L and energy is stored.

When the voltage on the sensing resistor R_S is higher than the reference voltage V_{ref}, that is, when the maximum current peak is reached, comparator $Comp_1$ commutates giving a reset signal to the SR flip-flop and turning off PW.

In this condition, the current in the inductor L goes into the diode-connected transistor Q_1 to the load. This holds until both comparators $Comp_2$ and $Comp_3$ change state. In fact, $Comp_3$ toggles when the output voltage V_{CP} becomes lower than the value set by equation 6.5, while $Comp_2$ guarantees the discontinuous operation since it commutates only when the sensing current in Q_2, proportional to the current flowing in the inductor, is lower than the very small internal current reference I.

When both conditions are satisfied a set signal is delivered to the flip-flop so that PW turns on again.

More details about this class of DC-DC converters can be found in Chapter 7.

6.3 Power Driving Stages

The switching characteristics of a DMOS transistor depend on the gate current which determines how fast the device capacitances are charged and discharged.

We will analyze low-side drivers where the source of the DMOS is connected to ground and high-side drivers which are more complicated because they make use of power limited auxiliary voltage supplies for proper overdriving.

In many applications (for instance hard disk drives), it is of utmost importance to control the output voltage slope for EMI reduction. Examples of such circuits will also be illustrated.

6.3.1 Low-Side Driver

A simple circuit used for fast switching transients is shown in Fig. 6.7.

It consists of an inverter chain preceded by a level shifter (made of M_1, M_2, M_3, M_4) that receives an input low voltage CMOS logic signal and transforms it in a logic signal swinging from ground to the voltage V_{reg} (that is internally generated and is about 10 V).

Transistors M_5, M_7, M_8, and M_6, M_9 are tapered and, together with current booster npn Q_1, supply the input capacitance of the power DMOS with adequate drive for fast on/off switching times.

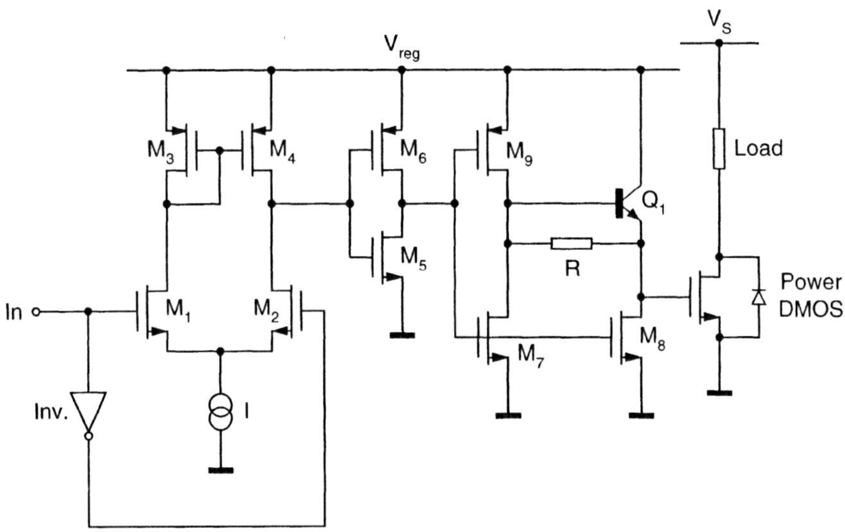

Fig. 6.7 Fast switching low-side driver

6.3.2
High-Side Driver

The simplest solution is the bootstrapping capacitor as shown in Sec. 6.2.1.

However for COMBO chips, where many independent high side drivers are present, this solution becomes too expensive because, as explained above, a number of external capacitors and pins equal to the high-side power DMOS is necessary. Thus a solution with a common voltage overdrive generator is much more common.

We will see a high-side driver circuit that, while minimizing the current consumption from the overdrive generator, allows fast switching times (see Fig. 6.8).

The circuit operation is as follows.

When the input logic signal In is low, M_9 and M_{10} are off, while M_{11} is on and sinks a small current I_L to keep Q_1 in saturation and consequently the power DMOS off.

As soon as In goes high M_{11} is turned off while current is flowing through M_9, M_7 and M_8 turns off Q_1, so the power DMOS input capacitor is charged by M_{10}, M_1, current mirror M_3 – M_4 and diode D by a high current pulse I_H of time duration T_d.

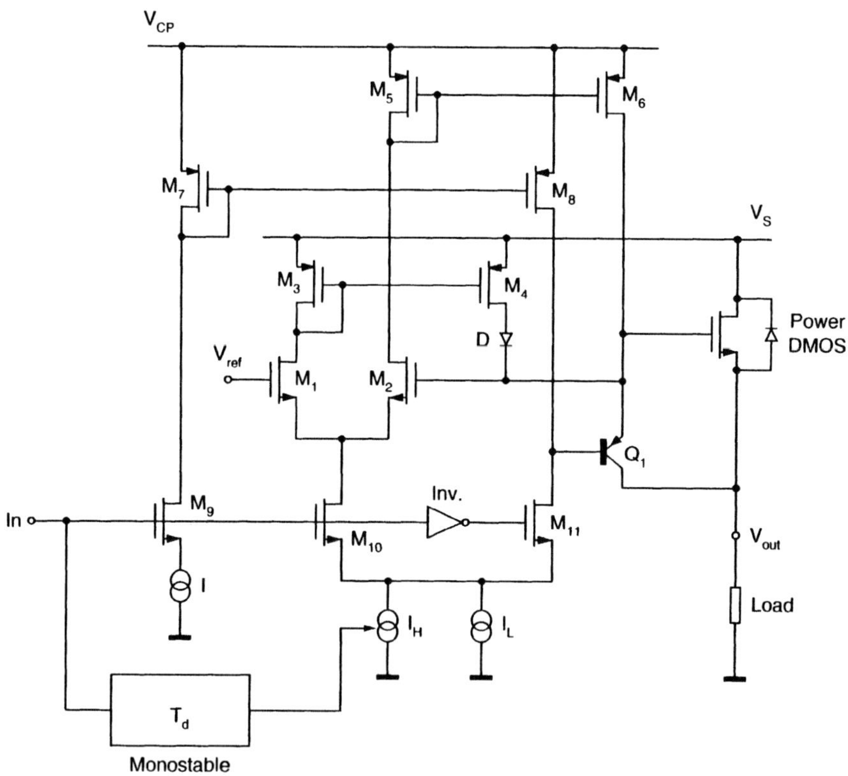

Fig. 6.8 Fast switching high-side driver

When the gate potential of the power DMOS reaches the value V_{ref}, M_1 turns off and M_2 turns on so that now the gate is charged by M_2 and current mirror M_5-M_6.

This means that M_1, M_2 operate like a comparator and current is sunk from the voltage overdrive generator only when necessary.

When the power DMOS is fully on, only a small current I_L just enough to overcome the gate leakage, is absorbed from V_{CP}. Diode D blocks the current path toward V_S through the body diode of M_4 when the gate voltage of the power DMOS becomes higher than V_S.

The same timing procedure is repeated when the power DMOS is turned off: as soon as In goes low the current through M_{11} is I_H for the time T_d needed to quickly discharge the gate equivalent capacitor, afterward, only a small current I_L is maintained active.

Another advantage of this circuit is that the input capacitor of the power DMOS is discharged by Q_1 to the load further increasing efficiency.

6.3.3
Low-Side Slew Rate Control

The basic concept is to use an integrator as shown in Fig. 6.9, where A is a unity gain buffer.

The output voltage slope SR will be:

$$SR = \frac{I}{C} \tag{6.6}$$

Capacitor C defines the dominant pole of the loop. Furthermore the low output impedance of the buffer kills the pole generated by the DMOS input equivalent capacitance improving loop stability with inductive loads.

A simple circuit implementation is shown in Fig. 6.10.

It consists of a classic AB push-pull output stage [6] made of Q_1, Q_2, Q_3, Q_4, Q_5, with the quiescent current set by I_Q.

Fig. 6.9 Basic circuit for low-side slew rate control

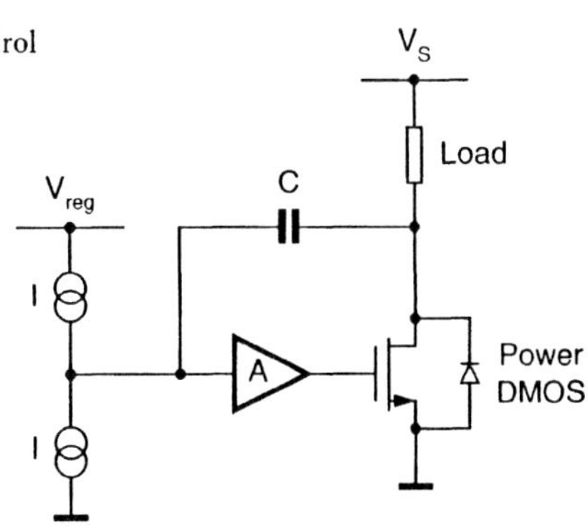

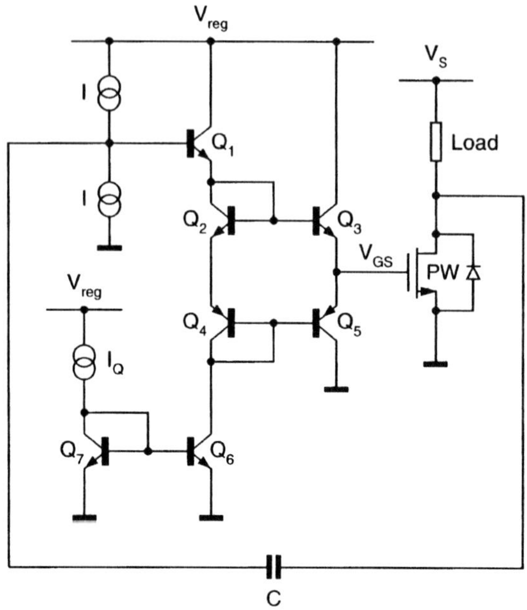

Fig. 6.10 Circuit implementation of a low-side driver with slew rate control

6.3.4
High-Side Slew Rate Control

The simplest method to control the output voltage slew rate for this source follower configuration, is to take advantage of the input capacitance C_{in} of the power DMOS (see Fig. 6.11, where the intrinsic capacitances of the power PW are explicitly indicated as C_{GS} and C_{GD}).

In fact, the capacitance C_{in} during rise and fall times, when the power PW is in the saturation region, can be considered rather constant and equal to the gate-drain capacitance C_{GD}.

Consequently, if the gate node of the power DMOS is charged and discharged with constant current sources I_C and I_D, the slew rate is given by:

$$SR = \frac{I_{C,D}}{C_{in}} \tag{6.7}$$

Of course, to drive the DMOS transistor PW fully on, current generator I_C must be connected to the voltage overdrive generator V_{CP}. However, this basic circuit has a drawback.

In fact, if I_D were constant, a propagation delay time t_d would be present on the output voltage V_{out}, during the fall time (see Fig. 6.12).

To avoid this delay, current I_D must be boosted until the gate voltage V_G goes lower than V_{out}, that is, until the voltage overdrive has been zeroed.

A circuit implementation of this concept is shown in Fig. 6.13.

Fig. 6.11 High-side slew rate control with constant current sources

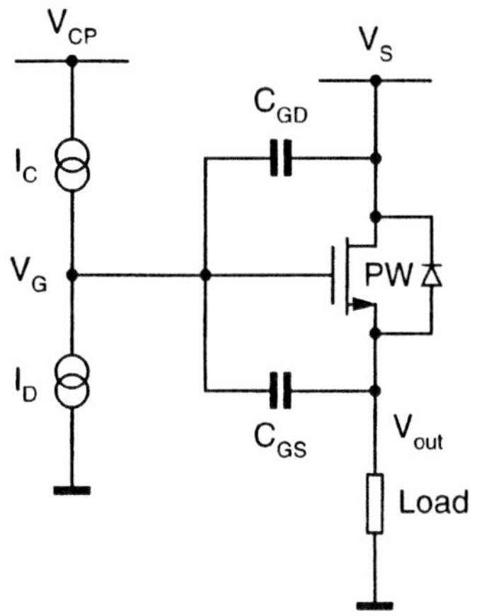

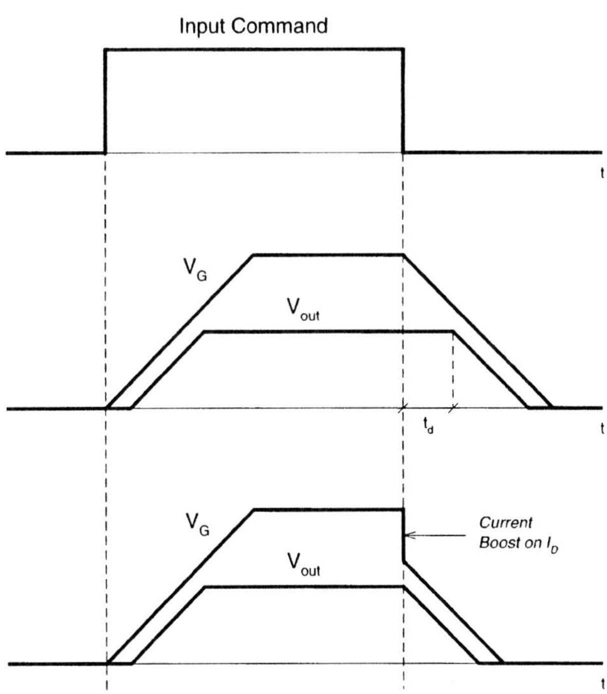

Fig. 6.12 Waveforms of a high-side driver slew rate control showing how the boost on the discharging current avoids the propagation delay time

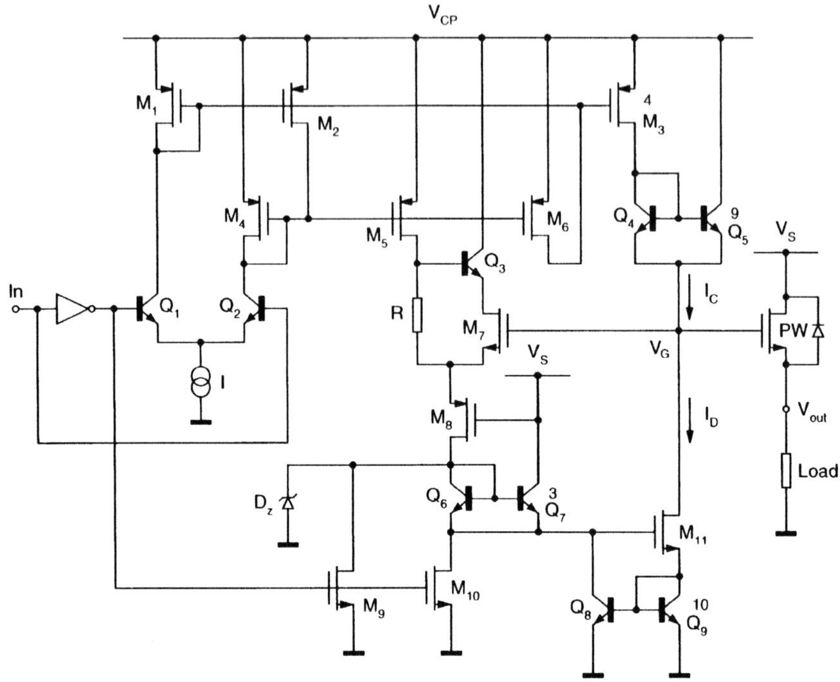

Fig. 6.13 Circuit schematics of a high-side driver slew rate control with constant current sources

If In = 0, Q_1 is on, and the input capacitor of PW is charged with the current:

$$I_C = 40 \cdot I \tag{6.8}$$

because of the current multiplication of mirrors $M_1 - M_3$ and $Q_4 - Q_5$.

The slew rate will be:

$$SR_{rise} = \frac{40 \cdot I}{C_{in}} \tag{6.9}$$

At the same time M_2 holds off $M_4 - M_5 - M_6$ while M_9 and M_{10} keep off $Q_6 - Q_7 - Q_8 - Q_9 - M_{11}$. This happens during the rise time and the gate voltage will arrive at the voltage V_{CP} (minus the diode drop of Q_4) supplying the necessary voltage overdrive to PW.

When In = 1, Q_2 turns on and M_5 delivers the current I to the base of Q_3 that, multiplied by its current gain h_{feQ3}, sources a high current through M_7. This current, cascoded by M_8 and multiplied by 40 with current mirrors $Q_6 - Q_7$ and $Q_8 - Q_9 - M_{11}$ quickly discharges the gate capacitor of PW giving the boost effect on the discharge current I_D.

This condition holds until M_7 is on, that is, until the gate voltage V_G of PW with respect to ground is about:

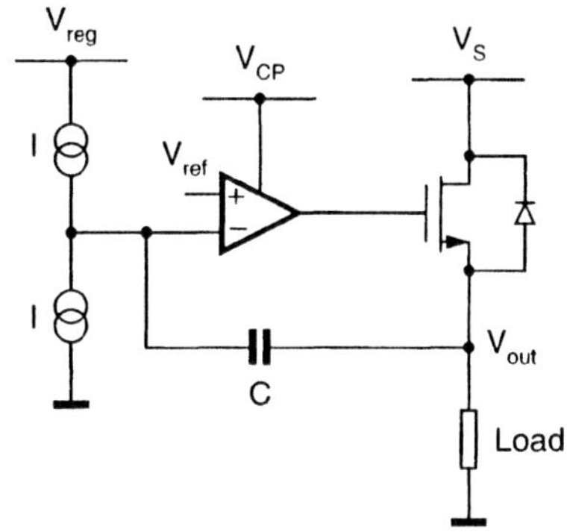

Fig. 6.14 Basic circuit for an accurate slew rate control of a high-side driver

$$V_G = V_S + V_{thM7} + V_{thM8} \tag{6.10}$$

where V_{thM7}, V_{thM8} are the threshold voltages of M_7, M_8 and V_S is the supply voltage. Afterwards, M_8 is always on but the current supplied by M_5 is no longer multiplied by the gain of Q_3, so the output voltage slew rate will be again:

$$SR_{fall} = \frac{40 \cdot I}{C_{in}} \tag{6.11}$$

To summarize:

$$I_D = 40 \cdot h_{feQ_3} \cdot I \quad \text{for} \quad V_G \geq V_S + V_{thM7} + V_{thM8} \tag{6.12}$$

$$I_D = 40 \cdot I \quad \text{for} \quad V_G \leq V_S + V_{thM7} + V_{thM8} \tag{6.13}$$

Furthermore in this situation, M_6 actively turns off the mirror $M_1-M_2-M_3$ while D_Z limits the voltage on the gate of M_{11}.

This circuit allows to obtain a good matching between rise and fall times and propagation delays: a jitter of ± 30 ns maximum has been measured between input and output signals for a 3 μs pulse width.

An accurate slew rate control can in general be achieved also for high-side drivers using an integrator as shown in Fig. 6.14.

However, this circuit is more sophisticated because the op. amp. has to be designed with low current consumption as it must be supplied by the voltage overdrive generator.

A circuit implementation which follows this approach, is described in detail in [7].

6.4 Motion Control Fundamentals

In this section, we will review at first a circuit topology very often used in motion control: the full (or H) bridge.

The second basic topic that will be reminded is the feedback theory and its application to motor driving.

6.4.1
The Full Bridge Configuration

The switching power device configuration shown in Fig. 6.15 is called a full (or H) bridge [8].

This configuration is often used to control speed, position and torque of stepping, DC, brushless motors and voice coils.

It takes a DC supply voltage V_S and provides a four quadrant control to a load connected between two pairs of power switching transistors. Because the DMOS are reversible devices, they allow current to flow bidirectionally, so the voltage across the load and the direction of current can be of either polarity. If the load is a motor, the sense of rotation will be reversed.

There are two ways of driving an inductive load:

1) PWM (pulse width modulation) bipolar voltage switching: diametrically opposite pair switches (A, D and B, C) are driven on and off simultaneously.
   If the time where A and D are on, is equal to the time where B and C are on, that is the duty cycle D_C is 50 %, the average output voltage and current across the load will be zero (Fig. 6.16 a).
   As the B, C pair locked conduction is increased (and correspondingly the conduction time of A, D decreased), so $D_C \leq 50$ %, an average current will flow through the load from Out_2 to Out_1 (Fig. 6.16 b).
   Conversely when the conduction time of A, D is increased and $D_C \geq 50$ %, the average current through the load will be from Out_1 to Out_2 (Fig. 6.16 c). Hence, only one signal, that is the duty cycle, is necessary to modulate both amplitude and direction of the current.
2) PWM with unipolar voltage switching: the voltage of the output terminal of one leg of the full bridge is held stationary, while the average voltage of the opposite leg is varied by a duty cycle of a pulse width modulated signal. This kind of operation is shown in Fig. 6.17 a, b, c, where one upper transistor of

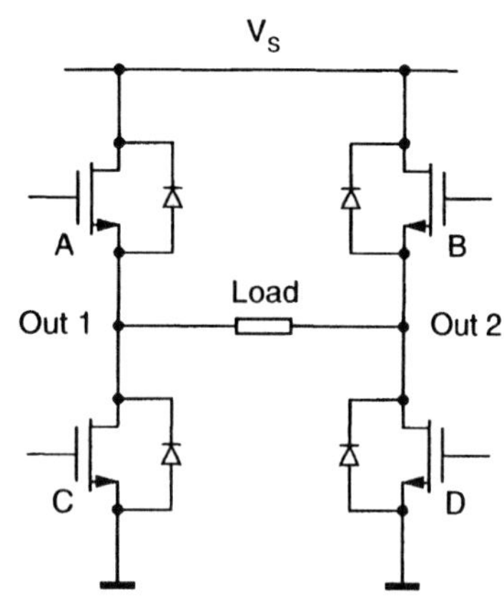

Fig. 6.15 The full bridge configuration

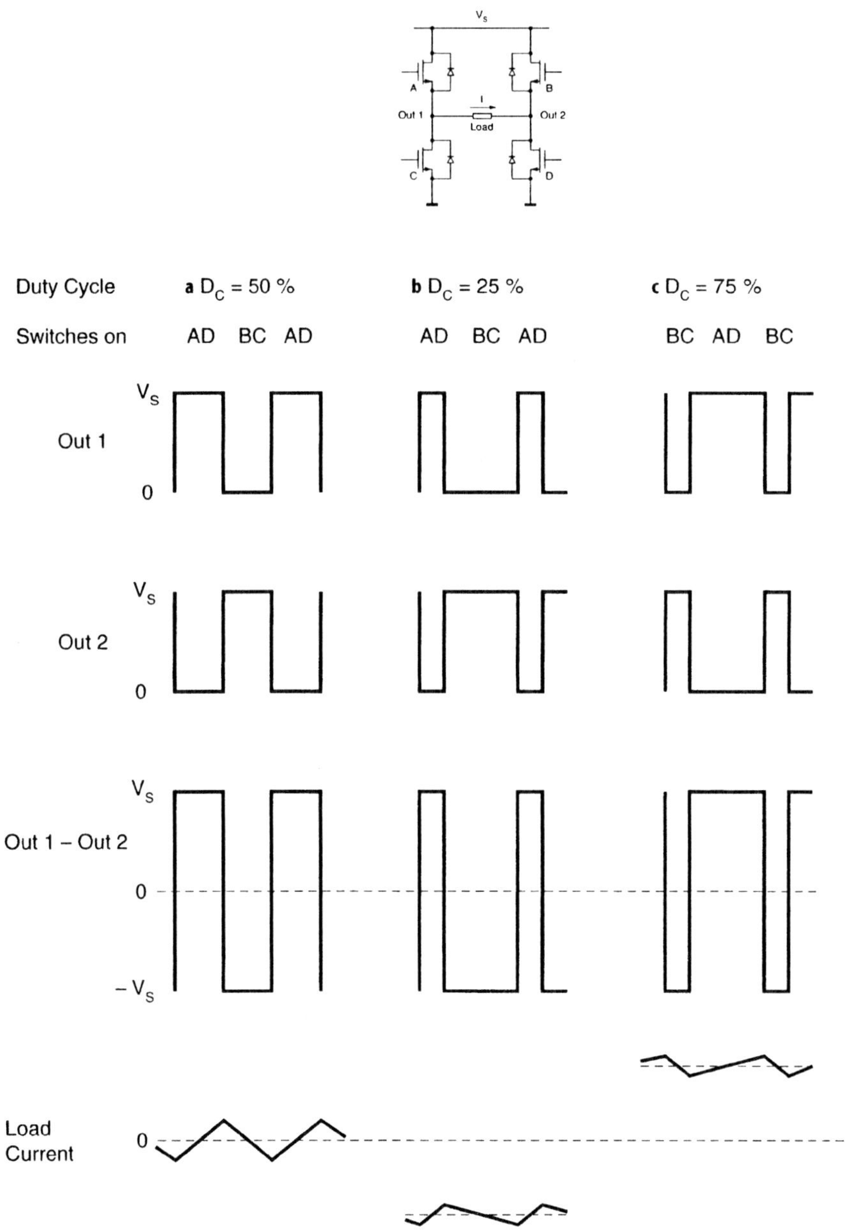

Fig. 6.16 Idealized waveforms of PWM bipolar voltage switching: **a** $D_C = 50\%$, **b** $D_C = 25\%$, **c** $D_C = 75\%$

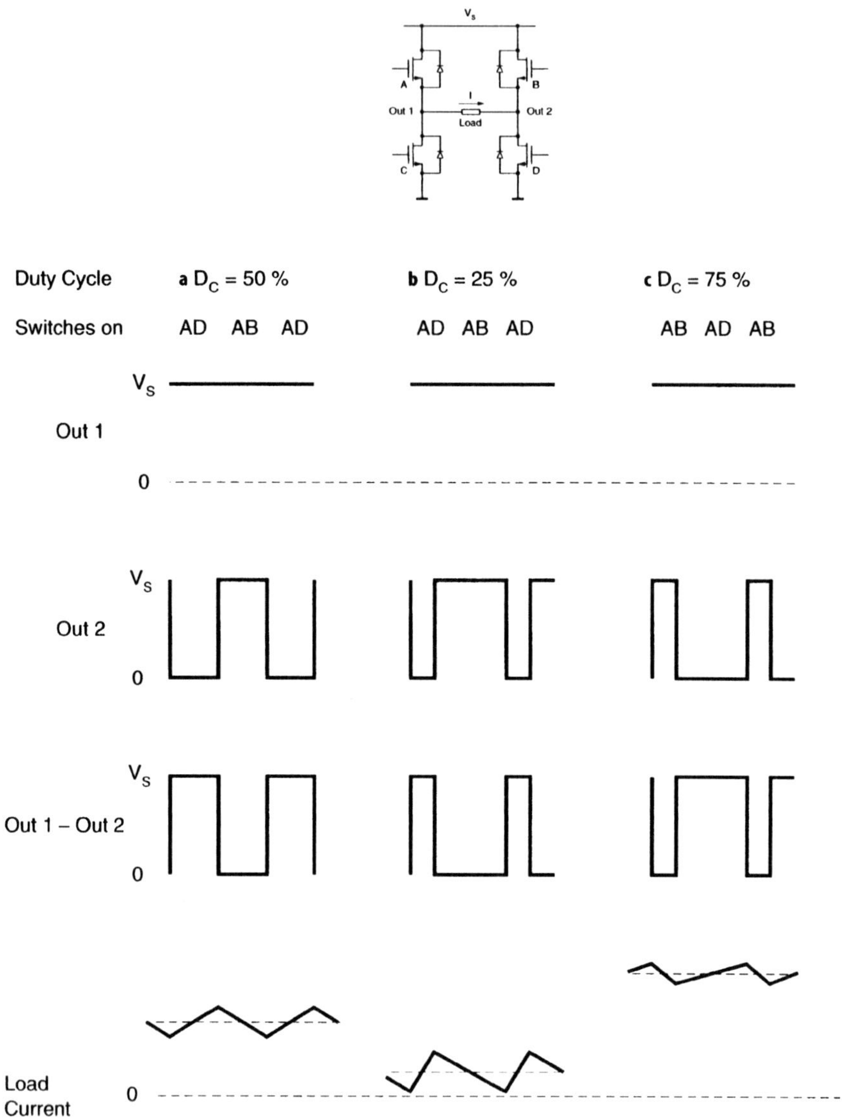

Fig. 6.17 Idealized waveforms of PWM unipolar voltage switching: **a** $D_C = 50\%$, **b** $D_C = 25\%$, **c** $D_C = 75\%$

one leg is always held on (in the figure is the high-side transistor A), while the other leg is chopping.
With this topology, the current reversal can only be accomplished by exchanging the leg activation, while the current magnitude depends on the duty cycle: in this case we need two different signals.

In both cases, a masking time (also called dead or blanking time) is always introduced to avoid cross-conduction and high current shoot-through between top and bottom transistors of the same leg (A, C and B, D).

For example, with PWM bipolar voltage switching, when A and D are turned off, it is necessary to wait for the dead time (typically around 100 nsec) before turning B and C on: during this period, there will be current recirculation to the supply rail through the flyback diodes of transistors B and C. However, after the dead time, B and C are switched on and the recirculation current is then commutated to the transistors themselves, thus increasing the circuit efficiency because of the lower voltage drop. This effect is known as active recirculation.

In PWM unipolar voltage switching, active recirculation can be applied to one transistor of the chopping leg.

6.4.2 Open-Loop and Closed-Loop Motor Control

Motion control can be performed in open loop or closed loop [9].

An open-loop system is shown in Fig. 6.18.

With this control the output will follow the input as long as the system variables are constant. Any change in loop, amplifier gain or any other variable will cause a variation of the output.

Although open loop control is adequate for many applications, the trend toward better speed regulation have led to closed loop control (servo system).

In its elementary form, a closed-loop control consists of an amplifier, a motor and a sensor as a feedback element (Fig. 6.19).

The amplifier drives the motor in such a way to keep the feedback variable detected by the sensor equal to the imposed reference. This means that the output variable is essentially insensitive to variations in electrical parameters or motor load conditions.

However, for these closed-loop servo systems overall stability and transient response are of primary importance and special attention must be paid to their evaluation.

Stability can be studied with the usual methods as root focus techniques or Bode plots. The amplifier can operate in linear or switching mode. Linear ampli-

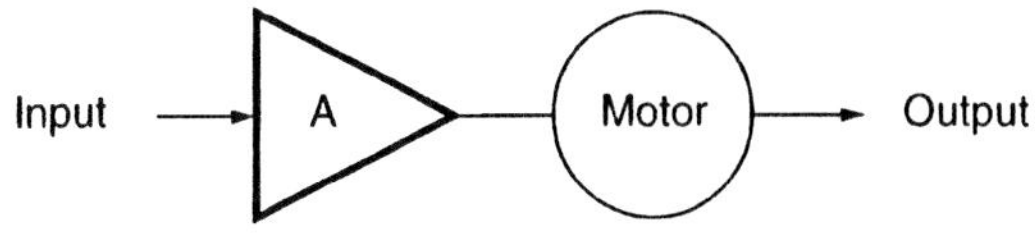

Fig. 6.18 Open-loop motor control

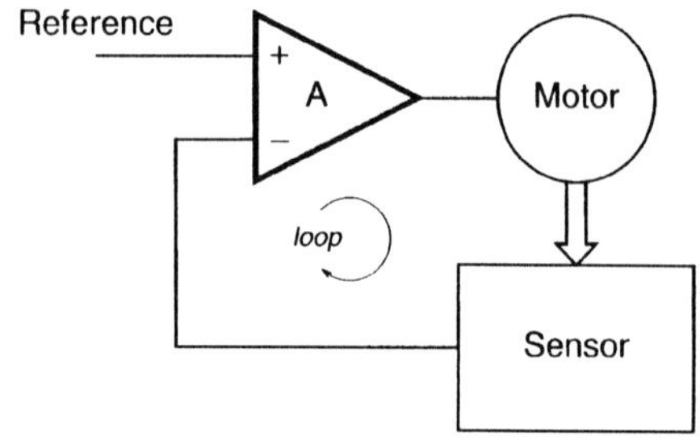

Fig. 6.19 Closed-loop motor control

fiers are used in high performance servo systems (for example voice coil drives) but have the disadvantage of high power dissipation.

Switching amplifiers offer high efficiency but are more difficult to implement and are noisy.

The simple example shown in Fig. 6.19 is just one of the many possible feedback techniques. In general, there are many feedback loops according to the different sensed variables. In the following, we will describe the different possibilities:

1) position: motor angular position must be controlled for example, when locking the shaft in a desired position. Sensing can be performed by Hall effect elements [10] (magnetic sensing) or photodiode encoder (optical sensing).
2) speed: the motor angular velocity must track a defined speed profile. The transducer can be directly a tachometer (which delivers an output voltage proportional to the shaft speed) or indirect by exploiting Hall effect elements.
3) torque: motor torque (and hence motor current) is to be controlled. This is usually an inner loop where a maximum current limit is set thus realizing the protection of the IC against overloads or short circuits and the protection of the motor by limiting the maximum torque.
   In fact, during start-up (fast acceleration) the current is generally several times higher than the continous steady-state current because the back electromotive force (BEMF) is zero since the motor is at stall and the rotor must win the static friction. This is immediately reflected in a fundamental design requirement: the power DMOS transistors of the output stage must be sized in order to achieve the minimum start-up current at the minimum supply voltage in order to optimize the dynamic response.
   The motor current can be sensed by measuring the voltage drop in a sensing resistor in series with the load or by current mirror techniques, taking a small current proportional to that flowing in the power DMOS transistors. The last solution minimizes the power dissipation but requires a higher circuit complexity and is less precise.
4) multiple variables: many different variables are sensed and closed in feedback loops (for instance, position and current, position and speed and so on). Many examples of multiple loop control will be described in the following.

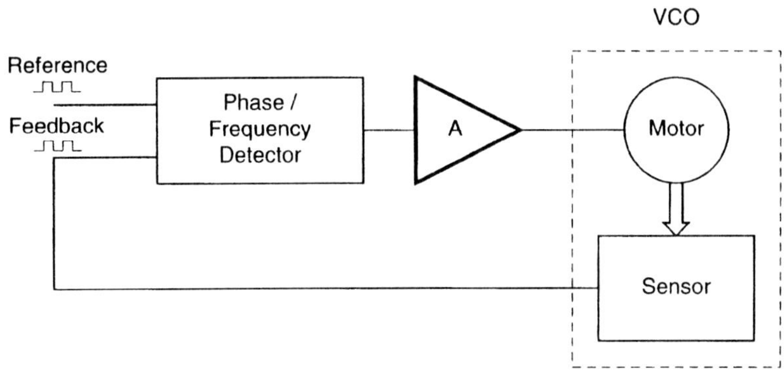

Fig. 6.20 Phase-locked loop motor control

The feedback configurations described above have permitted to improve significantly the system performance.

However, since the needs for precise speed control are continually increasing in new applications (for instance, in hard disk drives), phase-locked servo loops have been designed and fully integrated. Fig. 6.20 shows a block diagram of a phase-locked loop motor speed control where the combination of the motor and the sensor behaves as a voltage controlled oscillator (VCO).

When the system is phase-locked, the motor velocity follows the frequency of the input independently of motor, load and other parameter variations. Because the phase frequency detector compares not only the frequencies of the reference and feedback signals, but also their phases, the system settles at zero velocity error, so the speed regulation is as good as the command frequency.

The integrator present in the structure makes the DC gain virtually infinite which is desirable for high precision, but stability must be carefully investigated and compensation networks introduced.

In this chapter we will define closed loop systems those where position or speed are sensed and controlled.

6.5 Stepping Motor Drives

6.5.1 The Stepping Motor

The stepping motor consists of a rotating permanent magnet (rotor) surrounded by stator poles carrying the windings (see Fig. 6.21).

The concept behind the stepping motor operation is that the rotor aligns itself with the field generated by current applied to stator windings. Physically, the torque is generated because curved magnetic lines have the tendency to become as short and straight as possible like elastic strings while minimizing the

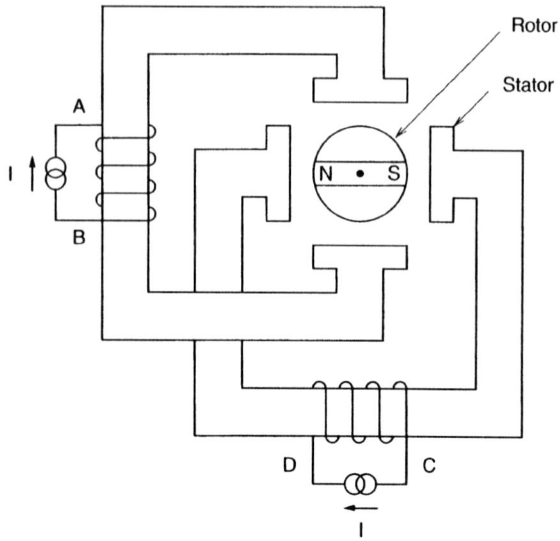

Fig. 6.21 Basic structure of a stepping motor

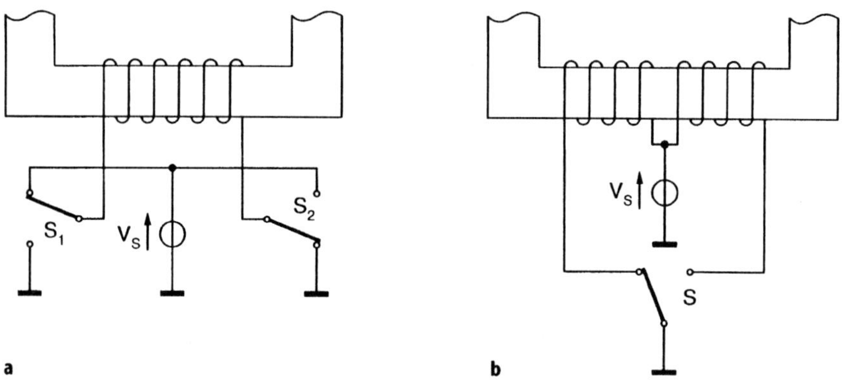

Fig. 6.22 Winding excitation current in a stepping motor: **a** bipolar mode; **b** unipolar mode

magnetic reluctance (this phenomenon is known as Maxwell stress [11]). By switching current to the various stator windings in a regular cyclic manner, the rotor will follow, by continually seeking alignment with the energized poles.

For a motor of this type, there are three possible drive sequences. The first is to energize the windings by changing the current polarity in the sequence AB, CD, BA, DC (BA means that the winding AB is energized but in the opposite sense). This sequence is known as "one phase on" full-step or wavedrive mode. Only one phase is energized at any given moment.

The second possibility is to energize both phases together so that the rotor always aligns itself between two pole positions. Called "two phase on" full step, this mode is the normal drive sequence for a stepping motor and gives the highest torque.

The third option is to energize at first only one phase, then two phases, then again only one, etc., so that the motor moves in half-step increments.

This sequence, known as half-step mode, halves the effective step angle of the motor but gives a less regular torque. For rotation in the opposite direction (counter clockwise), the same three drive sequences are used except, of course, that the order is reversed.

The stepping motor simplified in Fig. 6.21 has four stator poles resulting in a step angle of 90 degrees (45 degrees if the half-step sequence is used). Real motors have many more poles to reduce the step angle to a few degrees, but the number of windings and the drive sequences are unchanged.

Current in the windings can be delivered in two different ways (Fig. 6.22a, b).

a) by a single coil and two switches (bipolar mode).
b) by two coils and one switch (unipolar mode).

The advantage of the bipolar connection is that only one winding is present on the magnetic core, but it requires two switches.

The unipolar connection has the advantage of having only one switch but to get the same number of coils as the bipolar, the wires must be thinner so the winding presents a higher ohmic resistance decreasing efficiency. In fact, for the same motor power dissipation (and so for the same current), the bipolar mode offers a torque 40 % higher.

6.5.2 Stepping Motor Driving Circuits

A simple drive system for a stepping motor can be realized by supplying the phases according to a sequence of external pulses: it is expected that the stepping motor is able to follow every pulse. This operating mode, which is called "open loop drive", is interesting and widely used in applications of speed and control positions because expensive sensors and feedback loops are avoided. However, the performance of a stepping motor is limited under this kind of control. For example, a stepping motor driven in the open loop mode may fail to follow a pulse command when the frequency of the pulse train is too high or the inertial load too heavy: the motor motion can be oscillatory.

The performance of a stepping motor can be greatly improved by employing position and/or speed feedback to define the proper phase switching sequence at proper timings. This type of control, which is called "closed-loop drive", requires some kind of position sensor to detect the rotor position (for instance, a photodiode encoder).

The closed loop control is advantageous over the open loop control not only because a stop failure never occurs, but also because the motion is much quicker and smoother.

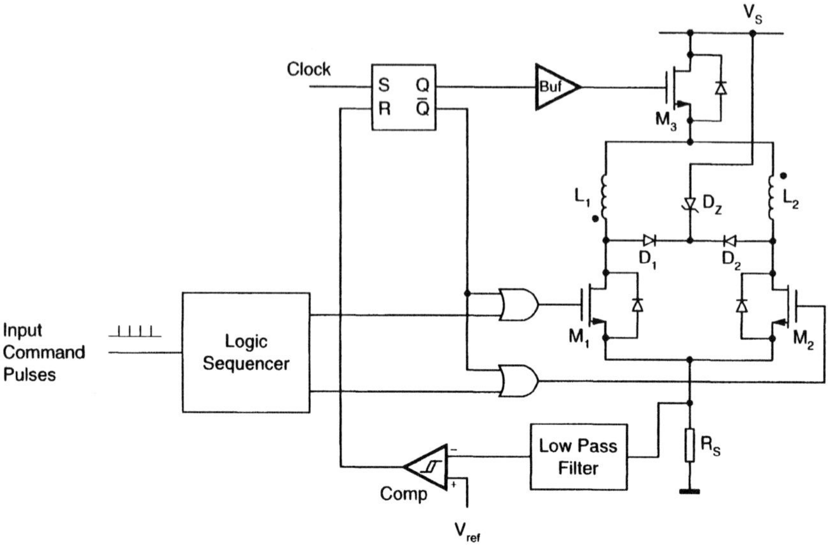

Fig. 6.23 Open-loop unipolar current control

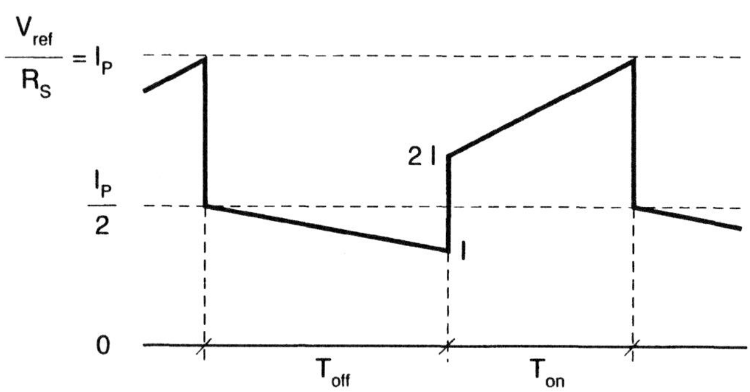

Fig. 6.24 Current waveform in the inductor L_1 of Fig. 6.23

Anyway, for both driving possibilities, it is necessary to limit the motor power dissipation by a current control within each winding. The best solution to accomplish this goal is to use a PWM control scheme. With this approach the only losses are caused by the on-resistance of the switches and the windings series resistance. Hence efficiency, intended as the ratio between output mechanical energy and input electric energy, is optimized.

A simple integrated current control for an open loop drive in a unipolar driving scheme is shown in Fig. 6.23.

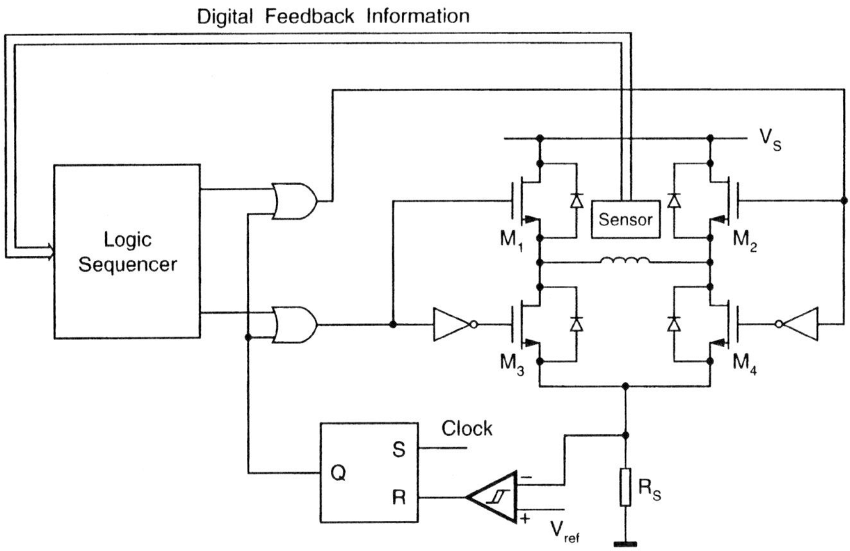

Fig. 6.25 Closed-loop bipolar current control

With M_1, M_3 on and M_2 off, the current in L_1 rises up to the peak:

$$I_P = \frac{V_{ref}}{R_S} \tag{6.14}$$

where V_{ref} is the threshold voltage of comparator Comp and R_S is the sensing resistor. At this point the comparator output resets the flip-flop and M_3 switches off. When M_3 is off, the current recirculation path is through L_2, L_1, M_1 and the flyback diode of M_2. It can also be shown that just after M_3 turns off, if the number of turns of L_1 is equal to that of L_2 and supposing also ideal the coupling between them, the current drops to $I_P/2$.

Since the DMOS is a perfectly reversible device, in order to additionally reduce the dissipation, M_2 is also driven on, M_3 remains off (PWM off time) up to a synchronization clock pulse which sets on again the flip-flop (active recirculation, see Sec. 6.4.1). The low-pass filter has the function of increasing the system noise immunity while D_1, D_2, D_Z clamp the voltage on the drain of M_1 and M_2 avoiding dangerous spikes. The winding current behaviour in L_1 is shown in Fig. 6.24.

An integrated current control for a bipolar stepping motor in a closed-loop configuration is reported in Fig. 6.25.

The chopping current control is performed in the same way as unipolar motors, but there is a position sensor that is used to supply feedback pulses to close the loop. The logic sequencer can also be controlled by a microprocessor programmed to execute the best timing to move the rotor from one position to another.

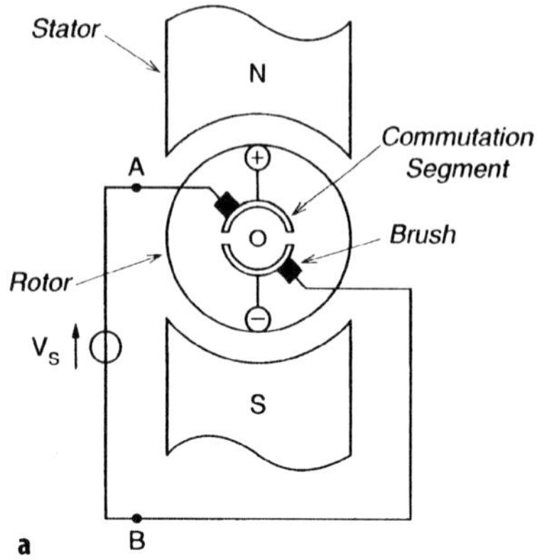

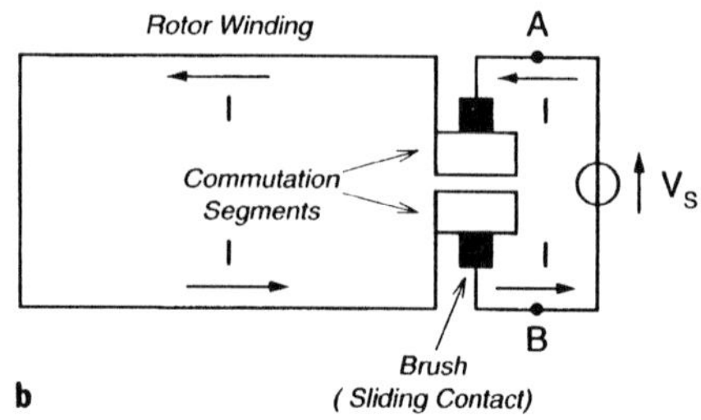

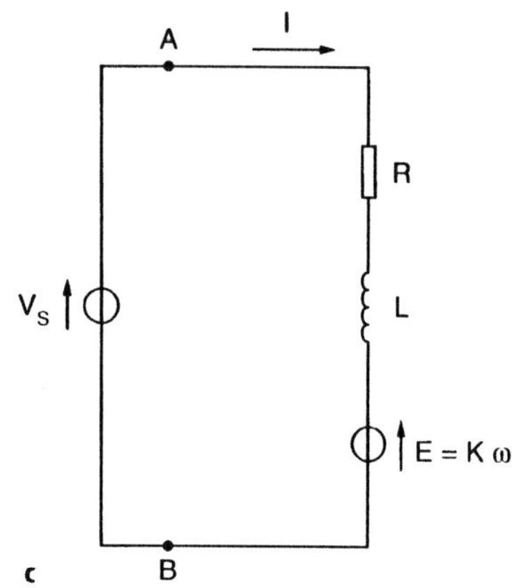

Fig. 6.26 The DC Motor: **a** front view; **b** side view; **c** electric equivalent circuit

6.6 DC Motor Drives

6.6.1 The DC Motor

An elementary DC motor is composed of three main parts: a stator generating a magnetic field (for example a permanent magnet), a rotor which carries the armature windings and some type of sliding contact arrangement (carbon brushes) to introduce current into the moving rotor winding (Fig. 6.26 a, b).

The rotor winding is continuous without beginning or end, and it is connected to the commutation segments which are insulated from each other and rotating with the shaft.

The current commutation into the rotor winding to provide continuous rotation is mechanically performed by the stationary brushes which make contact between the commutation segments (and hence the rotor windings) and a DC voltage source that supplies voltage and current.

The electric equivalent circuit of the motor is reported in Fig. 6.26 c where R and L are the series resistance and inductance of the rotor winding while E is the Back Electromotive Force (BEMF) induced on the rotor conductors during rotation and proportional to the angular speed.

6.6.2
DC Motor Driving Circuits

DC motors can be driven in open-loop or closed-loop configurations.

Open-loop control is used whenever an exact speed control is not essential. Applications of this type are found for example in the automotive industry for driving pumps, fans, wipers and power window lifts.

The circuit shown in Fig. 6.27 is an example of an open-loop drive where the motor current is not controlled in feedback but depends on a control voltage V_C which sets the duty cycle D_C in a PWM unipolar voltage switching H-bridge (see Sec. 6.4.2).

When M_1 is on, M_3 is off, while M_2 and M_4 are chopping according to the PWM control.

The concept is to generate a sawtooth waveform with a fixed frequency relaxation oscillator by charging with current I_{osc} up to the threshold voltage V_{ref} and discharging quickly through N_1 to ground, the capacitor C_{osc}. This ramp is compared with the control voltage V_C, which can be programmed from ground to V_{ref}. The output voltage V_{PWM} sets the duty cycle D_C of the PWM control and hence the motor current I_M, according to the expression:

$$I_M = \frac{V_S - E}{R_{onM1} + R_{onM4} + R} \cdot D_C = \frac{V_S - E}{R_{onM1} + R_{onM4} + R} \cdot \frac{V_C}{V_{ref}} \tag{6.15}$$

The significant waveforms of the circuit are reported in Fig. 6.28.

The basic circuit of an open-loop drive with PWM current feedback control and the relevant waveforms are shown in Fig. 6.29 a, b.

In this case, the voltage on the sensing resistor R_S (proportional to motor current and torque) is the feedback signal that is compared with a reference voltage which, in turn, defines the duty cycle of the PWM control. This low-value

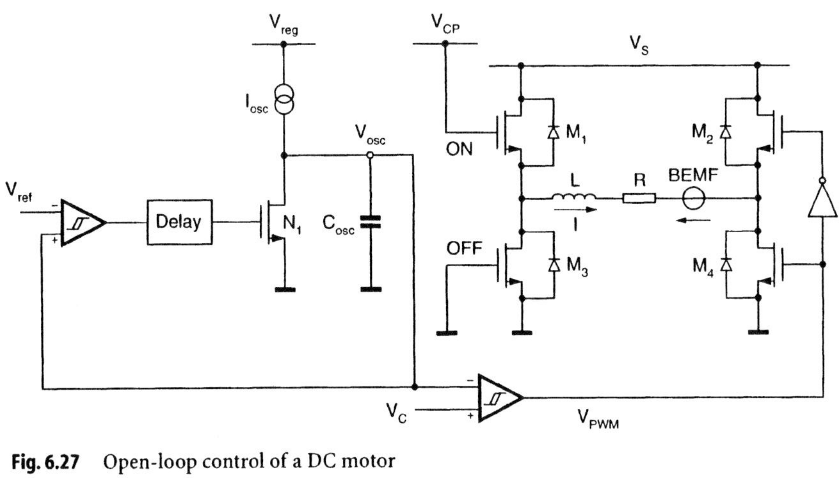

Fig. 6.27 Open-loop control of a DC motor

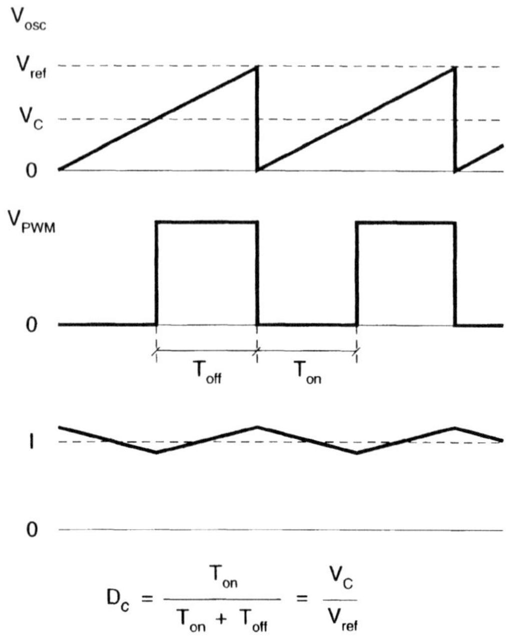

Fig. 6.28 Waveforms of the circuit of Fig. 6.27

resistor creates a usable voltage proportional to the motor current without appreciable power losses.

The current ripple can be reduced by increasing the clock frequency. This can be easily achieved in a mixed technology by exploiting CMOS logic and fast switching power DMOS transistors.

With this loop configuration the current does not depend on motor characteristics and load, electrical parameters or voltage supply, but only on the reference voltage V_{ref} and the sensing resistor R_S:

$$I = \frac{V_{ref}}{R_S} \tag{6.16}$$

Closed-loop control in DC motors can be accomplished in many different ways as reported in [12]. To illustrate the structure of these systems, we will describe here an integrated solution which allows to obtain high-positioning accuracy (comparable to stepping motors) but using a low-cost optical encoder.

The block diagram is shown in Fig. 6.30.

This solution does not require high precision mechanism such as conventional tacho-converters, since it exploits the slope between crossovers in the optical encoder signals elaborated by the speed detector block. So the speed conversion is independent of the encoder waveform shape.

Two control loop operating modes are used: speed control and position control.

At the beginning of a positioning action the system operates in speed control mode. The microcontroller (CPU) applies a speed profile to the DAC, that by

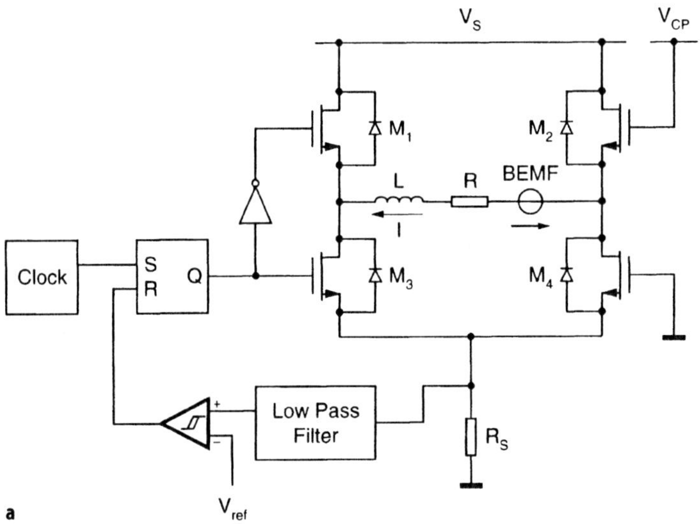

a

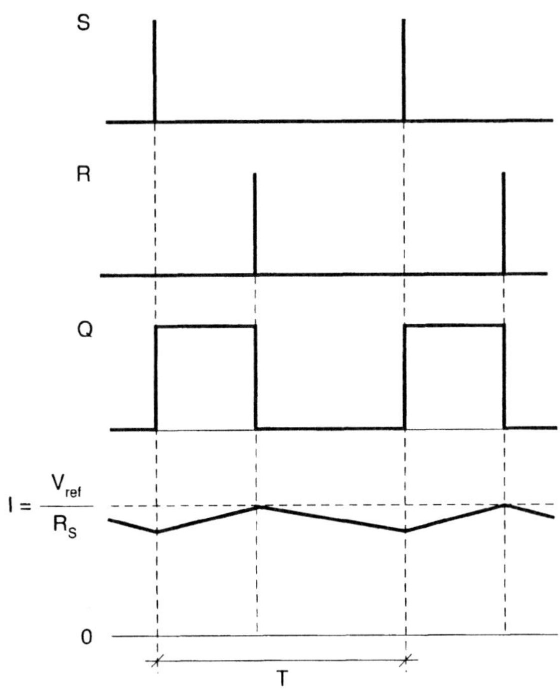

b

Fig. 6.29 Open-loop DC motor with PWM current control: **a** circuit implementation; **b** relevant waveforms

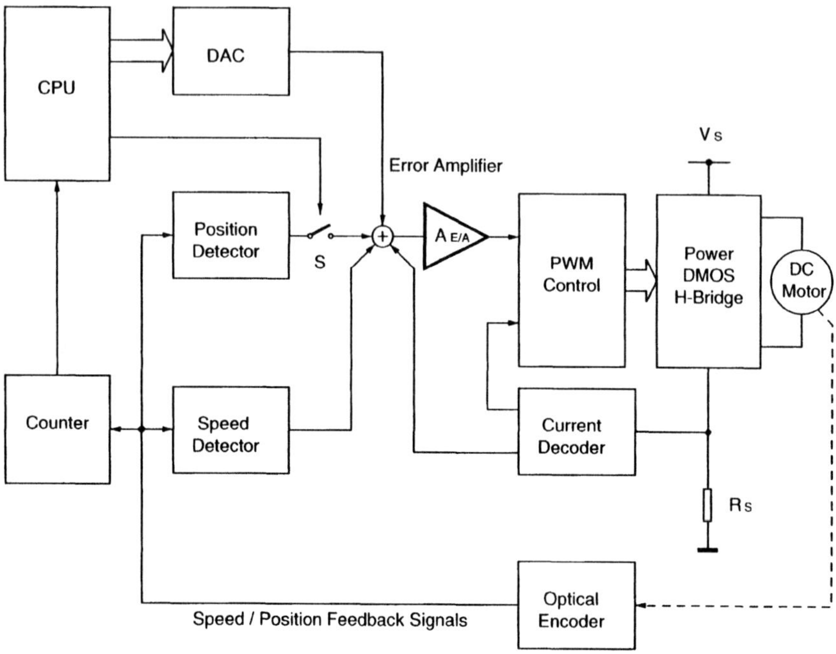

Fig. 6.30 Block diagram of an integrated control of a DC motor including a main velocity loop and a final positioning loop

means of the error amplifier, drives in PWM bipolar mode the power DMOS output stage.

Since the motor was at stall, the current rises rapidly accelerating the motor to the desired speed which is maintained by the optical encoder feedback signal. At the same time, the counter counts the pulses coming from the speed sensor to determine the travelled distance. As soon as the target position is approached, the microcontroller modifies the speed profile decelerating the motor step by step.

When the final position is very close, the microcontroller closes switch S and enables the position loop by connecting the sensor output directly to the error amplifier, thus forcing the motor to stop and hold the position corresponding to a zero-crossing of the sensor signal. Feedback is provided by the sensing resistor R_S that delivers a voltage proportional to motor current to the current decoder block. This information is also used to set the maximum current limit.

It can be underlined that for this circuit the transconductance gain is set by R_S, but the maximum current limit can be defined independently, allowing great flexibility in the applications.

This combination of closed-loop speed control, pulse counting and closed-loop position control gives very fast and precise positioning.

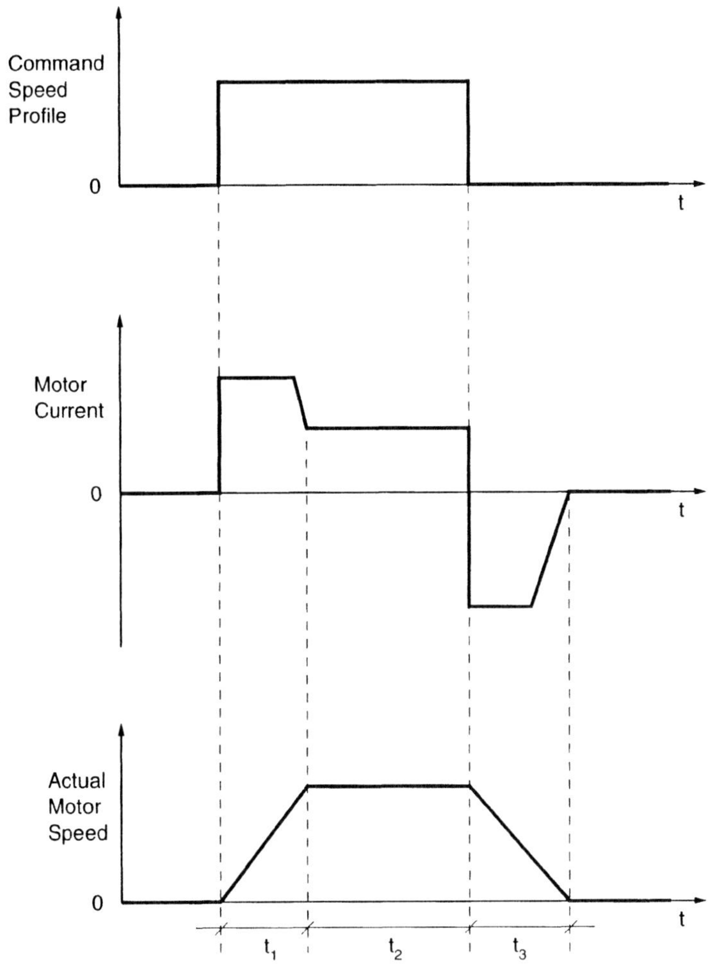

Fig. 6.31 Waveforms of command speed profile (DAC output), motor current and actual motor speed of an elementary positioning operation

Figure 6.31 shows the main waveforms of an elementary positioning operation, that, as is evidenced, can be split in four different phases:

a) acceleration: in t_1 the speed loop is saturated and the current goes at a maximum defined by the current limit;
b) constant speed: in t_2 the speed feedback assumes control and the current is set by the transconductance gain of the loop;
c) deceleration: during t_3 the current reaches a maximum in the reverse direction again defined by the current limit;
d) positioning: after t_3 the positioning loop becomes active.

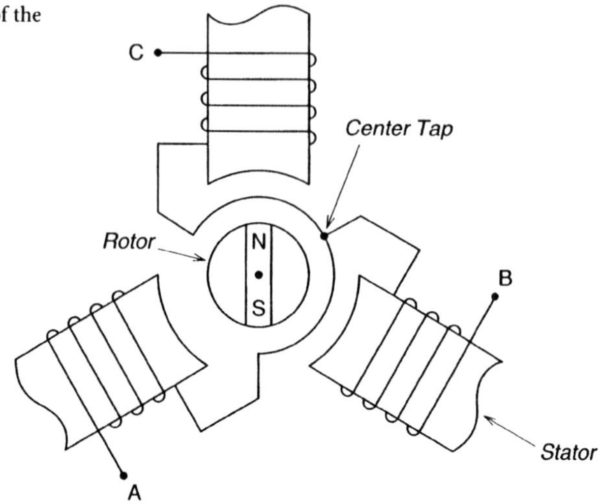

Fig. 6.32 Basic structure of the brushless motor

6.7 Brushless Motor Drives

6.7.1 The Brushless Motor

The brushless motor consists of a permanent magnet rotor "chasing" sequentially switched stator poles (Fig. 6.32).

Practically, the stator is supplied by a balanced set of three-phase currents. This generates a rotating magnetic field that drives the rotor synchronously.

The great advantage of this structure compared to DC motors is that the commutation of stator poles to produce rotation is accomplished by switching electronic devices that substitute the brushes and commutator system.

Hence the poor reliability of conventional mechanical brushes and commutators and radio frequency noise are eliminated by performing the commutation (current reversal) electronically. Another advantage is that the active conductors are on the stator, so the heat removal is easier.

Virtually all brushless motors are of either two-phase or three-phase design.

6.7.2 Brushless Motor Driving Circuits

The most popular driving configuration for a three-phase brushless motor is shown in Fig. 6.33.

With this stator winding arrangement each phase is oriented 120 degrees from the other. The six transistors are connected to the end points of each stator leg and thus form a three-phase full wave motor control (see waveforms in

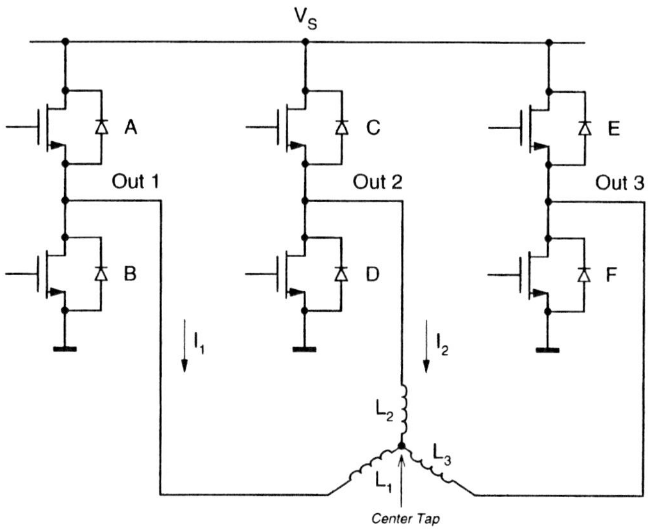

Fig. 6.33 Driving configuration of a three-phase brushless motor

Fig. 6.34 referred to an inductive load where flyback voltage spikes above the supply and underground are also shown).

From the basic operation described above, we can realize of a fundamental difference between DC, stepping and brushless motors: DC motors require only a voltage supply for their operation while stepping and brushless motors need a voltage plus a proper driving sequence of current pulses to excite the stator windings.

The concepts of feedback control developed for stepping and DC motors apply to brushless motors as well. We will see two examples derived from high performance smart power ICs for hard disk drive applications where multiple control loops are present. For this application, stability of angular speed of the spindle motor and hence of the hard disk, is of utmost importance.

a) Transconductance loop (Fig. 6.35).

In this circuit the motor current is determined by a control voltage V_C. The voltage across a sensing resistor R_S connected in series with the source of all lower DMOS transistors provides the feedback signal that is utilized to regulate the motor current. This one is defined by the expression:

$$I_{motor} = \frac{V_C}{A_S \cdot R_S} \tag{6.17}$$

where A_S is the voltage gain of the sense amplifier. So the current is regulated by a linear transconductance loop that drives the lower DMOS transistors.

The control is passed to each lower DMOS in succession by the multiplexer following the commutation sequence imposed by the Hall cells by the control logic.

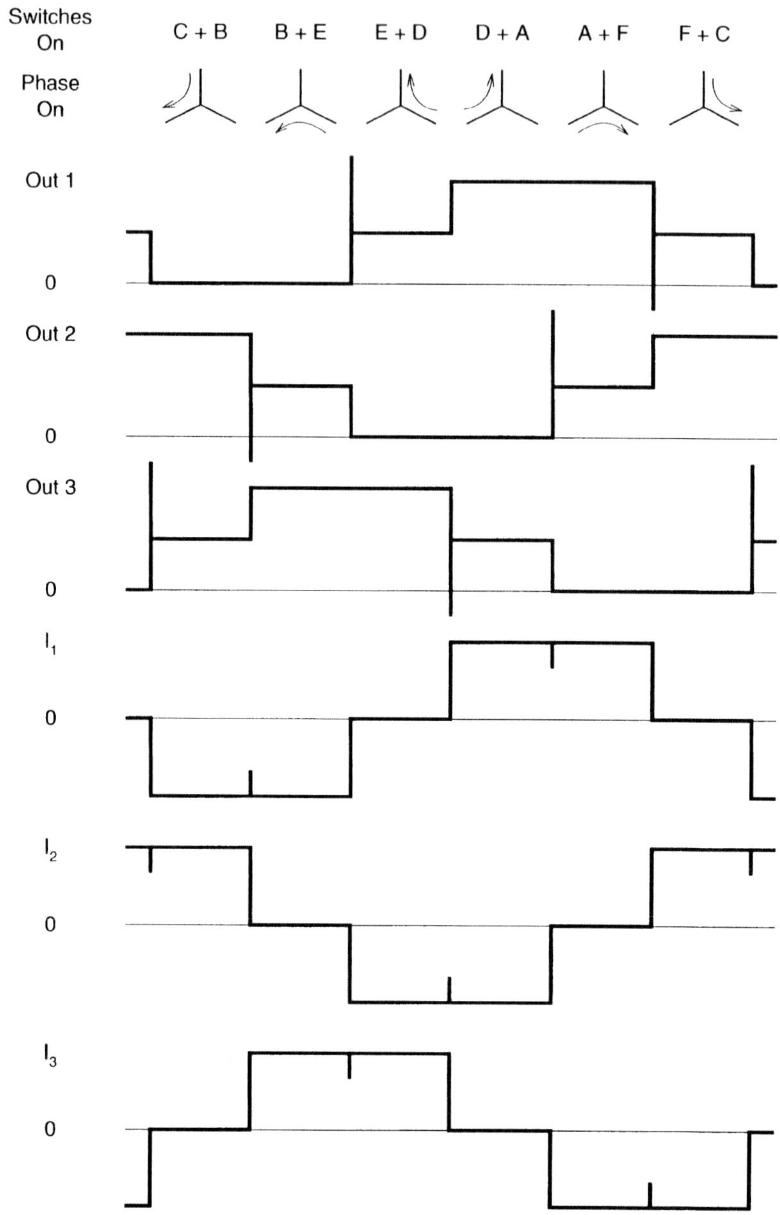

Fig. 6.34 Voltage and current waveforms of a three-phase brushless motor with inductive load

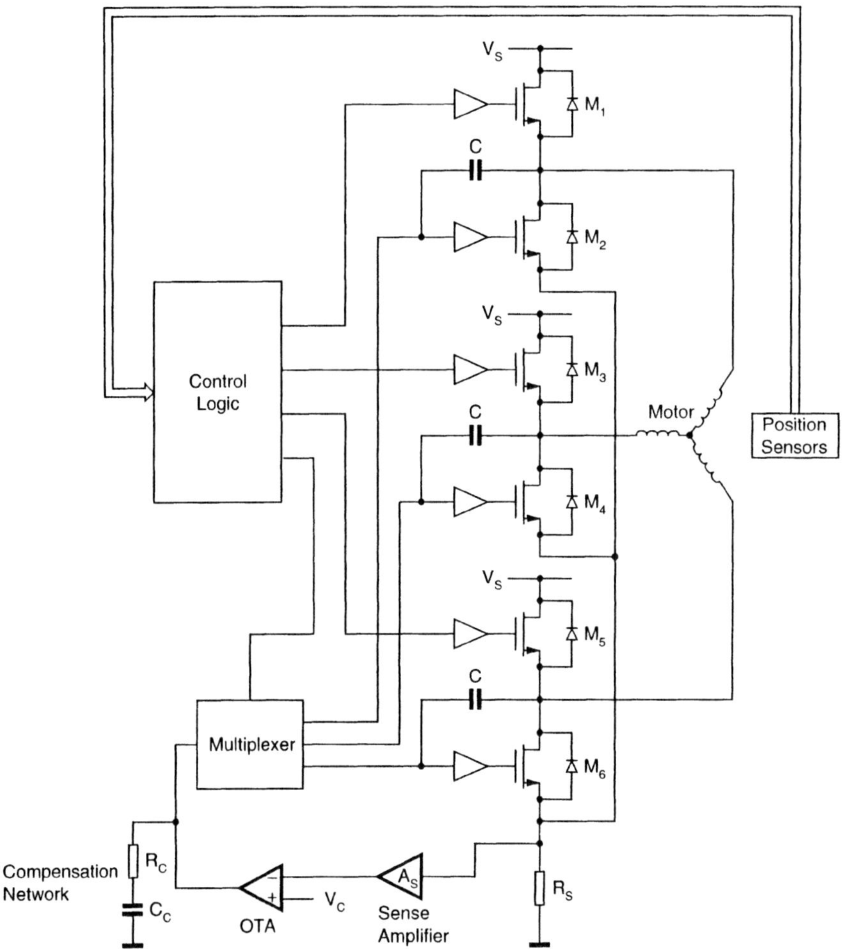

Fig. 6.35 Transconductance conversion closed-loop control

To avoid recirculation of the current flowing in the motor coils when each phase is switched off, the turn-off slew rate of the upper and lower drives is controlled. In this way, the current in the switched-off coil will decrease to zero, while the total current in the motor is kept constant by the transconductance loop.

This solution allows to improve the transient response and to minimize the noise and EMI during the commutations. Furthermore, snubber networks on the load are eliminated [13].

The requested design features of the operational transconductance amplifier (OTA) are: low input offset voltage, high output resistance R_o and large bandwidth.

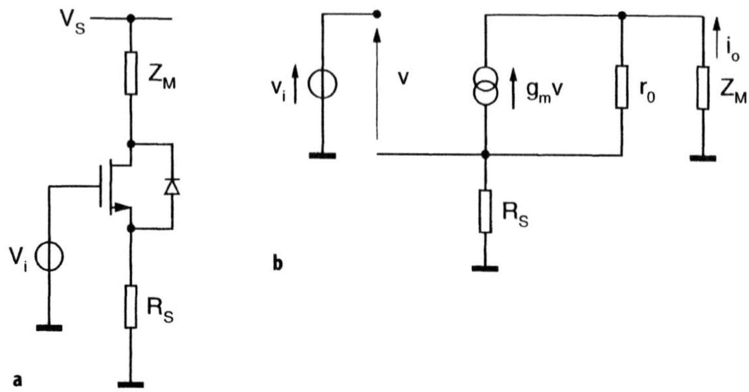

Fig. 6.36 Simplified lower drive: **a** DC circuit; **b** AC equivalent circuit

High output resistance is required to make the off-chip compensation network independent of the OTA output characteristics.

Large bandwidth is necessary to avoid interference with the other poles and in particular with those introduced by the motor. These poles should be out of the loop but in fact they are present because of the limited output resistance r_o of the lower DMOS transistors.

This can be easily understood if we consider the simplified lower drive circuit in Fig. 6.36a. The transfer function can be calculated from its AC equivalent circuit (Fig. 6.36b) as:

$$\frac{i_o}{v_i} = \frac{g_m}{1 + g_m \cdot R_S + \frac{Z_M}{r_o} + \frac{R_S}{r_o}} \tag{6.18}$$

where Z_M is the equivalent motor impedance, g_m is the DMOS transconductance, and R_S is the sensing resistance.

This relationship demonstrates that only if r_o is very high the motor poles do not enter into the loop.

Furthermore, to kill the poles coming from the intrinsic capacitances of the DMOS, its gate is driven by a buffer stage with a very low output impedance (see Sec. 6.3.3).

The overall loop compensation and the optimization of the transient response are realized by R_C and C which introduce a pole and a zero respectively given by:

$$f_p = \frac{1}{2\pi R_o C_C}\,;\; f_z = \frac{1}{2\pi R_C C_C} \tag{6.19}$$

b) Phase-locked loop control [14, 15].

The circuit discussed above is simple but since the lower power DMOS transistors operate in saturation, the power dissipation is not optimized. Efficiency

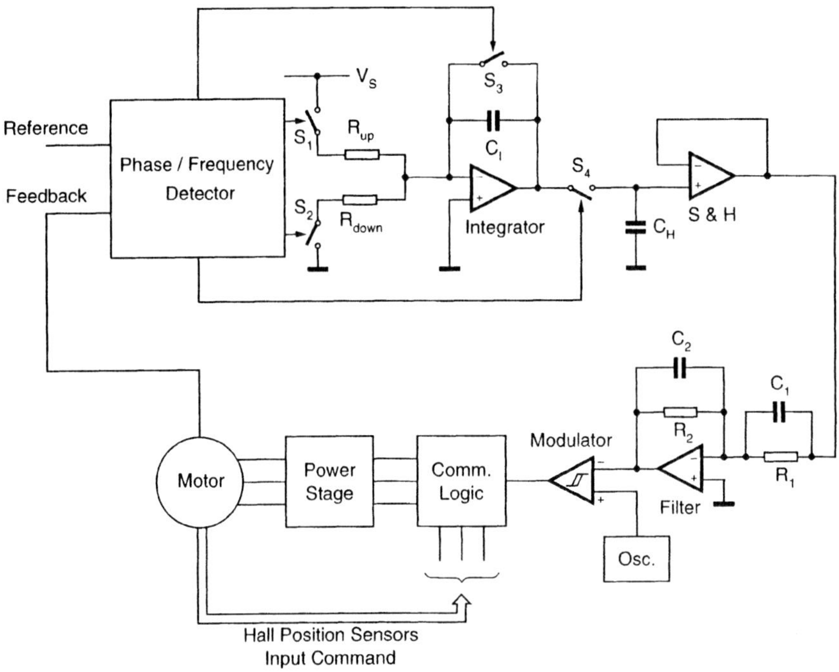

Fig. 6.37 Block diagram of an integrated phase-locked loop motor control

and accuracy can be improved using digital techniques combined with high precision analog functions: this approach leads to a circuit solution particularly suitable of integration using mixed technologies.

An example of an integrated control loop used in hard disk drives and allowing for a speed stability better than 0.03 % is shown in Fig. 6.37. In fact, since the disk data storage density increases, so does the precision required in controlling the speed of the spindle motor for data synchronization.

Essentially, the circuit works as follows: the controller compares a reference frequency and phase to a feedback frequency and phase signal related to the motor speed and generates a proportional repetitive pulse width which drives the power stage into the motor (PWM control).

The reference phase/frequency used as control loop error detection and speed synchronization control is a crystal-derived clock frequency corresponding to the "once around" rate of the motor.

The phase/frequency error is determined through time comparison of occurrence of the feedback signal coming from the motor relative to the quartz crystal oscillator reference. Depending on whether the phase of feedback signal is advanced or delayed, the error signal will be used to close one of two analog switches S_1 and S_2 either from V_S to one side of resistor R_{up} or from ground to one side of R_{down}.

The other side of these resistors is fed into a summing node of an op.amp. (Integrator). A capacitor across the op.amp. (along with R_{up} and R_{down}) forms the time interval of integration. A second op.amp. (S & H), that is configured as a unity gain buffer, in combination with capacitor C_H, acts as a hold network to maintain a stable voltage to the loop filter between phase error measurements.

Time integration will occur for the period of the phase error. The integrator output ramps positive or negative (relative to a system reference voltage) during the error interval. At the end of the error interval, the integrator charge is forwarded to the hold capacitor C_H by closing switch S_4 in series with it.

When sufficient time for charge transfer has elapsed, the switch is opened and the integrator capacitor C_I is shorted through switch S_3 until the next phase error measurement begins.

The output of op.amp. S & H is a buffered sample and hold voltage and becomes the loop filter input. This consists of an op.amp. (Filter) which is configured as an active filter whose time constant is set by R_1, R_2 and C_1, C_2 to provide the control loop with the desired dynamic response and optimum stability.

The output of the filter is fed into the modulator that is simply a comparator. It compares the voltage coming from the loop filter and proportional to the phase error with a fixed frequency triangle waveform, generating a pulse width modulation (PWM) signal that is delivered to the commutation logic circuitry, where, by the power stage, it enables application of voltage to selected motor windings defined by the Hall effect sensors for controlled time intervals.

6.7.3 Sensorless Commutation

We have seen so far that the input sequence of pulses in a brushless motor has been obtained by some kind of position transducer (for instance, Hall effects sensors).

However, as brushless motors get smaller there is less space available. Furthermore, Hall effect sensors are also sensitive to temperature variations and noise always present in motor applications. All of these factors reduce motor reliability. Last but not least Hall effect sensors are expensive and consume power which is strongly undesirable in battery driven systems.

In the last few years a new concept has emerged. It consists of using the Back Electromotive Force (BEMF) as position sensor feedback signal [16]. To understand the working principle, we will refer to the basic motor structure shown in Fig. 6.38, where for better comprehension, the rotor is the external permanent magnet and the stator is the inner coil.

Considering Faraday's Law [17]:

$$V = -\frac{d\phi}{dt} = \text{BEMF} \tag{6.20}$$

and since the flux through the coil depends on the angular position, we get the BEMF induced in the coil as shown in Fig. 6.39.

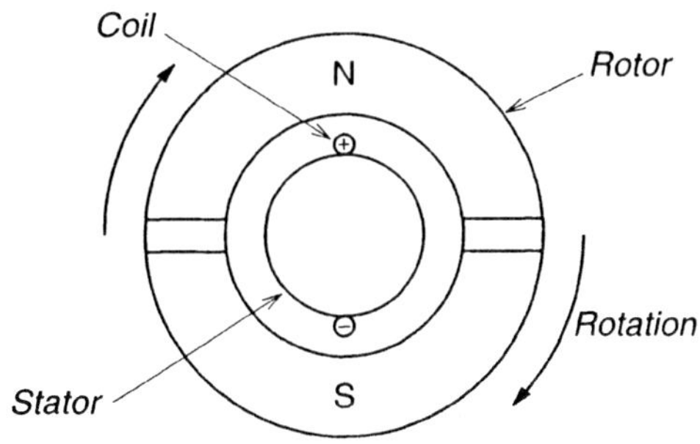

Fig. 6.38 Simplified single-phase brushless motor

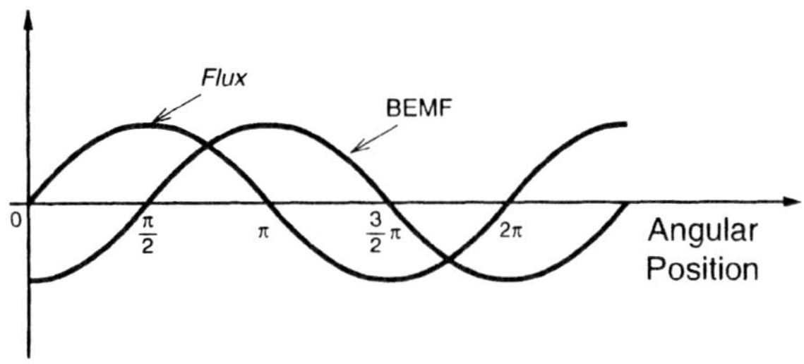

Fig. 6.39 Flux and BEMF induced in a coil according to Faraday's Law

As appears, the BEMF zero-crossing is an indication of the rotor angular position.

This feature can be practically implemented in three-phase motors by sensing the BEMF in the non-driven coil relative to the center tap (for applications where the center tap is not available, three resistors connected in a star configuration can be used).

Two major drawbacks of sensorless commutation are backward rotation at start-up and self-synchronization of a moving rotor at any speed. However thanks to the adoption of fully integrated digital techniques, these problems have been solved giving 100 % of reliability.

The basic problem lies in the discrimination between BEMF and commutation noise in the position detection process. Since low-pass filtering is not acceptable due to both cost/space/effectiveness requirements and excessive phase delay introduced, a digital masking technique of the noise has been implemented.

The block diagram of the system is shown in Fig. 6.40. The operation can be summarized as follows: the period between zero crossings of BEMF is measured by a counter and, at each transition, the content is stored and utilized to generate a digital delay (which minimizes the torque ripple) and a noise masking time. At the end of the digital delay, the phase commutation is performed by the sequencer and the masking time is initiated to protect the BEMF detection circuit from the noise generated by the commutations which might otherwise provide false information. Both times are calculated as proportional to the

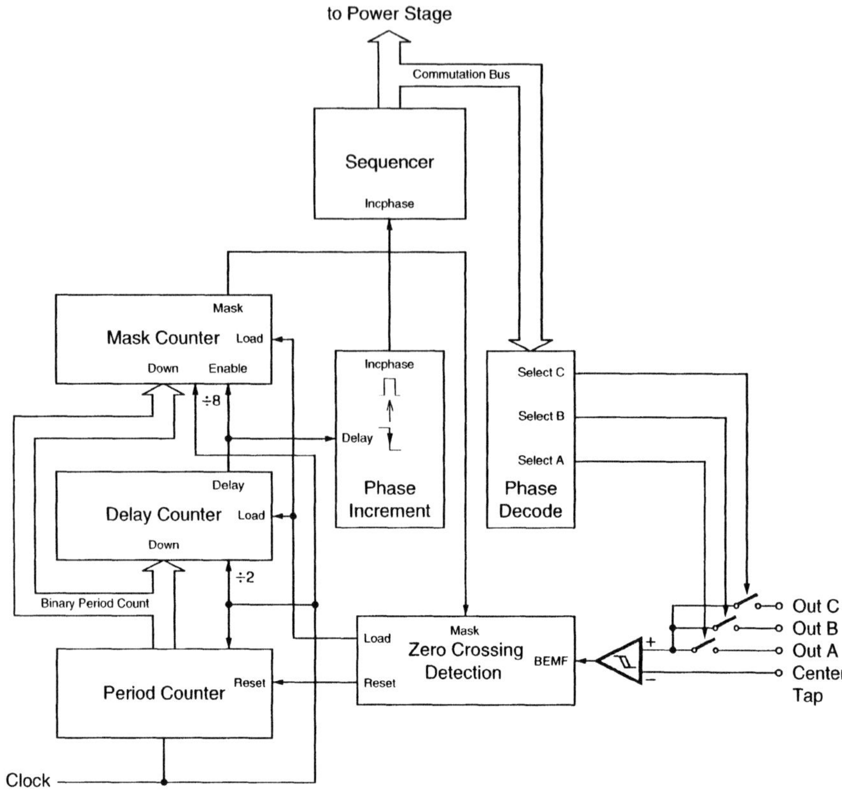

Fig. 6.40 Block diagram of a fully integrated digital circuit to generate the driving sequence of a three-phase brushless motor by sensing the BEMF zero-crossing in the floating coil

period quantities, thus maintaining an electrical angle relationship which is independent of the speed at which the system is operated. Accuracy of delays and jitter performance are therefore only limited by the clock frequency of the counters.

6.7.4
Motor Braking

When a motor drives a load that is essentially inertial in nature, a long coasting period may be required when the motor is turned off. This flywheel effect can be undesirable for certain applications, such as hard disk drives.

Slowdown by mechanical braking often proves to be both awkward and costly. In systems employing speed control feedback, an error signal is generated when there is a difference between feedback information and the information provided by the reference. Braking is achieved by decrementing the reference value, thus causing the central system to seek a new equilibrium.

Dynamic braking is a widely used technique in which the motor is operated as a generator after being disconnected from its source of electrical power. When a machine is functioning as a motor, it develops a counter EMF; when the same machine operates as a generator, it develops a counter torque. Therefore, if the armature terminals of a motor are immediately shorted following turn off, a very rapid deceleration of the rotating shaft will be achieved.

In disk drive the braking function is needed to avoid damage to the heads and magnetic media. Furthermore, it must be performed without depending on the system power supply. Fig. 6.41 shows a complete disk drive system architecture including voice coil and spindle motor [18].

When the brake function is started, all six power DMOS of the spindle motor are turned off. Hence the outputs are tristated and the motor will continue its rotation supplying a BEMF proportional to its speed. The synchronous rectification provided by the flyback diodes of the DMOS themselves gives the power to supply the voice coil driving circuitry for a time sufficient to park the heads (see Sec. 6.8.4).

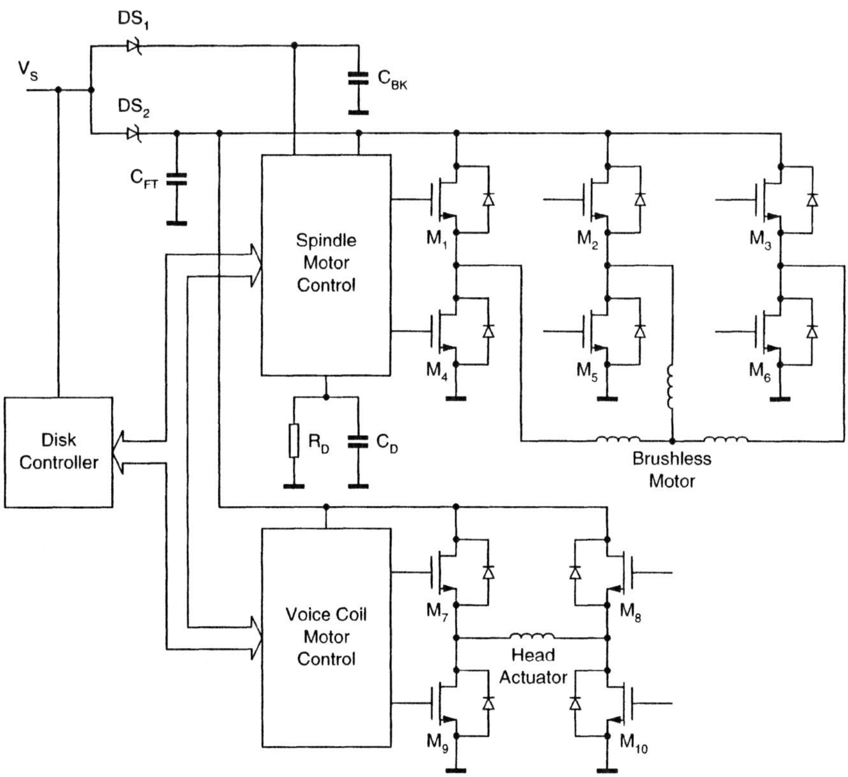

Fig. 6.41 Hard disk drive system architecture

As soon as the heads have been parked, the motor can be braked stopping its rotation in a very short time by turning on all lower DMOS and keeping off the upper DMOS transistors. The energy required to turn and keep on the lower DMOS and the parking circuitry, is taken from an external capacitor C_{BK}. This expedient and the decoupling Schottky diodes DS_1, DS_2 allow the braking procedure even if the supply voltage has been removed. Capacitor C_{FT} is the usual filtering capacitor on the voltage supply while resistor R_D and capacitor C_D set the time necessary to park the heads after which the motor is braked.

A practical circuit implementation of the function described above is shown in Fig. 6.42, where only one motor phase is shown for clearness.

M_{1PW} and M_{2PW} are the power DMOS transistors of the output power stage. During normal operation when the brake command is low, capacitors C_{FT}, C_{BK} are charged very close to the supply voltage V_S, capacitor C_D is charged at the voltage V_Z minus the base-emitter voltage of transistor Q_2 while M_1 is off and M_2, M_3 are fully on. So M_4, M_5, M_7, M_8 are also off and M_{1PW}, M_{2PW} operate normally.

When the brake command is pulled high or V_S is removed, Q_1 is forced in saturation and M_1 turns on both M_4 and M_5.

Since M_6 was previously on, M_{1PW} and M_{2PW} are turned off and the output is tristated.

Hence the energy delivered by the BEMF of the rotating motor is used to supply the voice coil drive and to park the heads. At the same time capacitor C_D

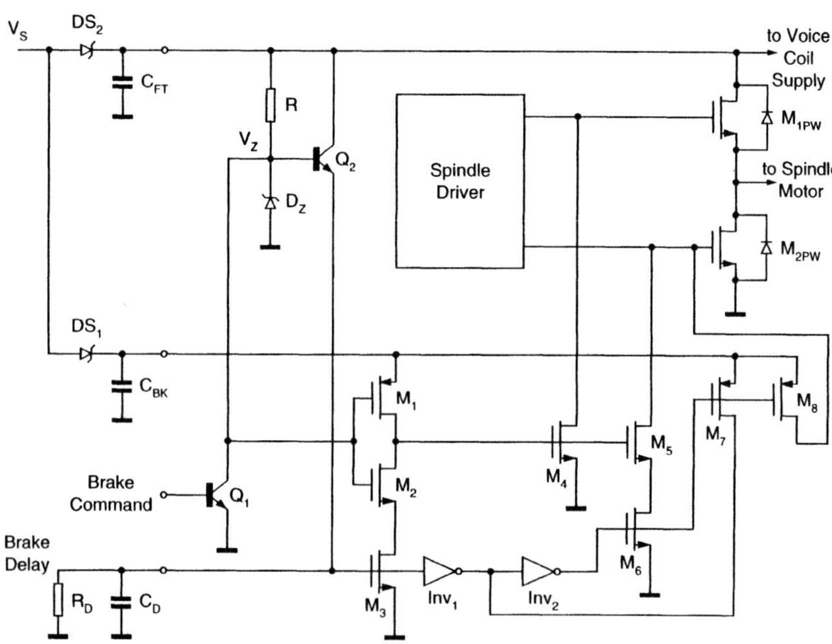

Fig. 6.42 Circuit implementation of the braking function

discharges exponentially on resistor R_D so that when the voltage goes lower than the threshold of inverter Inv_1, M_6 turns off and M_7, M_8 are turned on. This means that M_{1PW} is confirmed off but M_{2PW} turns on by the action of M_8 (M_7 provides only a positive feedback to speed up the transition). During this condition all three lower DMOS are fully on, thanks to the energy delivered by capacitor C_{BK}, braking quickly the motor.

6.8 Voice Coil Motor Drives

Voice coil actuators are so called because they work on principles originally used in radio loudspeakers [19].

These electromagnetic limited motion devices are used in linear and rotary motion applications exclusively in closed loop where tight servo control is of utmost importance. The voice coil motor drive works essentially on the same principle as a DC motor drive (see Sec. 6.6.2).

Basically the driving circuit implements the transconductance amplifier function:

$$I = G \cdot V_{in} \tag{6.21}$$

where a DC voltage V_{in} determines, through a given transconductance G, the coil current I (see Fig. 6.43).

6.8.1 The Voice Coil Motor

The voice coil is basically a linear motor. Just as in standard acoustic speaker, current in a moving coil produces a magnetic field which pushes against a permanent field in the stationary magnet assembly.

Because the moving force is proportional to the amount of current in the coil, a smooth, easily controlled linear movement is produced. The direction of movement is defined by the direction of current flow in the coil.

In its simplest form, a linear voice coil motor consists of a tubular coil located within a radially oriented magnetic field (Fig. 6.44a, b).

The permanent magnets facing the coil are arranged of the same polarity. An inner core of ferromagnetic material and connected to the permanent magnets, closes the magnetic circuit.

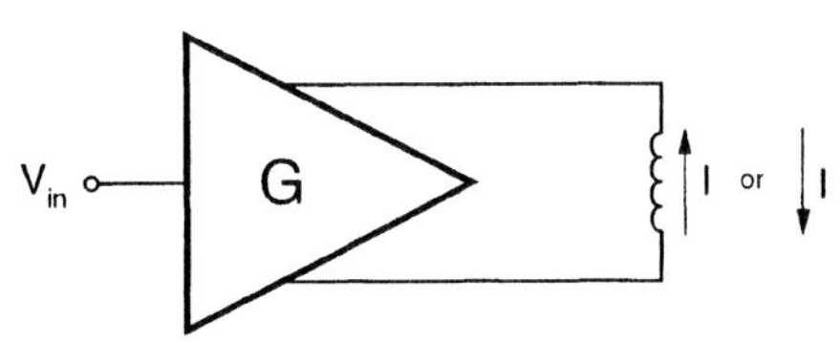

Fig. 6.43 Basic voice coil transconductance loop

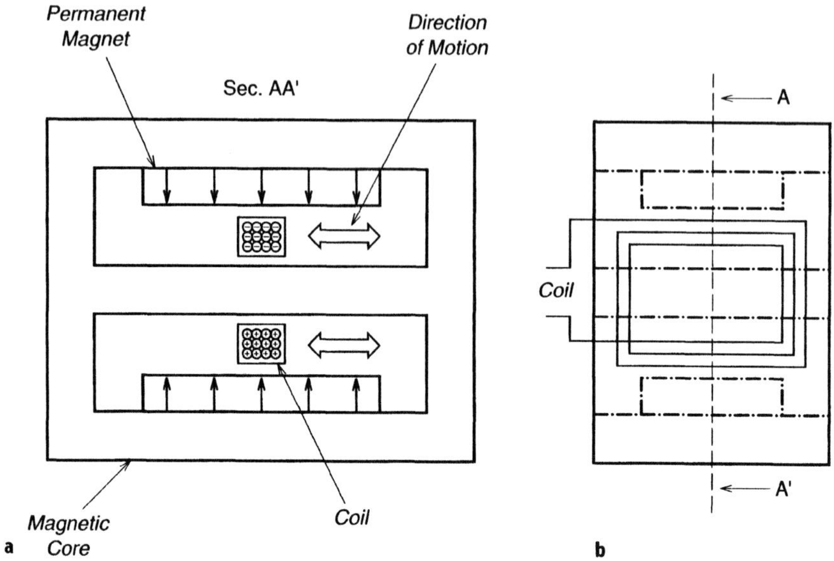

Fig. 6.44 Linear voice coil motor: **a** front view; **b** side view

Since the magnetic field within the air gap between the magnets and the coil can be kept very constant, the force versus the linear distance is extremely flat: the force degradation at the extremes of the motion can be kept below 5 %.

There are two different voice coil motor designs. They are categorized as rotary and linear actuators. These designs both use the same type of voice coil motor but with different mechanical methods of translating the motor force into coil movement.

A rotary actuator is shown in Fig. 6.45.
Here the coil pivots around a fixed fulcrum so ratings are in units of torque instead of force.

The rotary voice coil motor finds wide application in hard disk drives. As is known a magnetic disk for mass storage is divided in circular and concentric tracks carrying the bits of information.

The function of the voice coil motor is to move the read/write head on the desired track according to a specified velocity profile.

6.8.2 Class B Voice Coil Drives

DMOS full bridge drives (with the current controlled in linear mode by the low-side transistors) allow for the realization of extremely low power (class B) systems which do not require snubber networks for load compensation, yet maintaining high accuracy on gain and offset parameters.

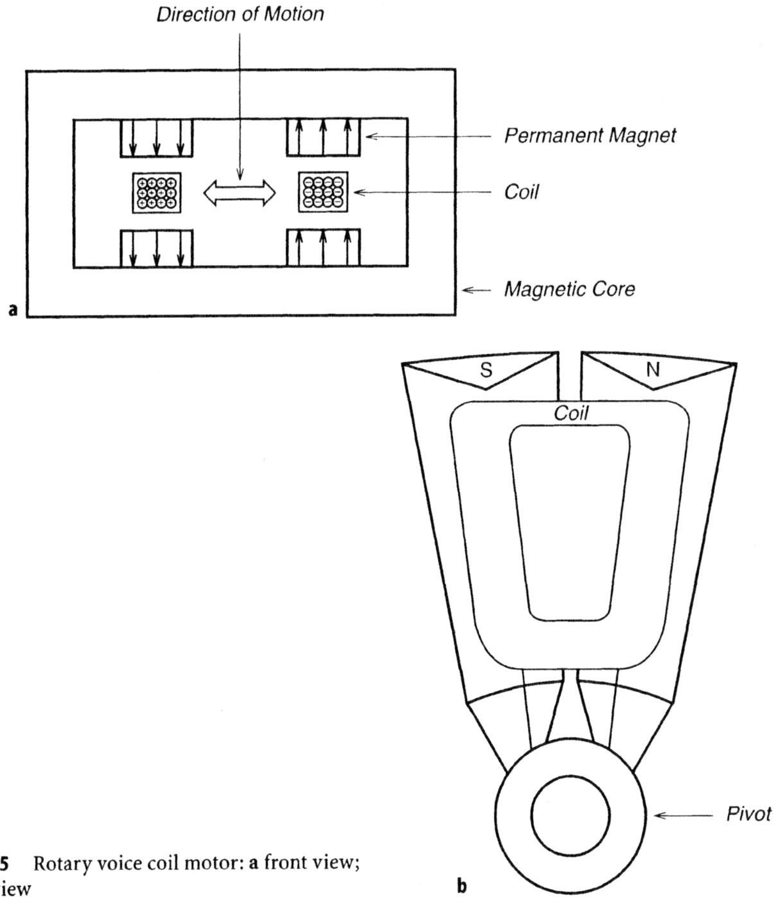

Fig. 6.45 Rotary voice coil motor: **a** front view; **b** top view

Figure 6.46 shows the principle of operation of such a driver.

This circuit accomplishes a full swing of the output voltage from rail to rail of the supply voltage V_S with current capability ranging from 1 to 2 A with a bandwidth of 30 KHz on inductive loads.

The loop transconductance G is given by:

$$G = \frac{1}{R_S} \tag{6.22}$$

and can be programmed by the user with the external sensing resistor R_S.

An important application of high performance voice coil servo-systems is found in hard disk drives for the positioning of the read/write head on the desired track of the disk.

There are basically two operating modes of the head actuator in this application: seeking and tracking.

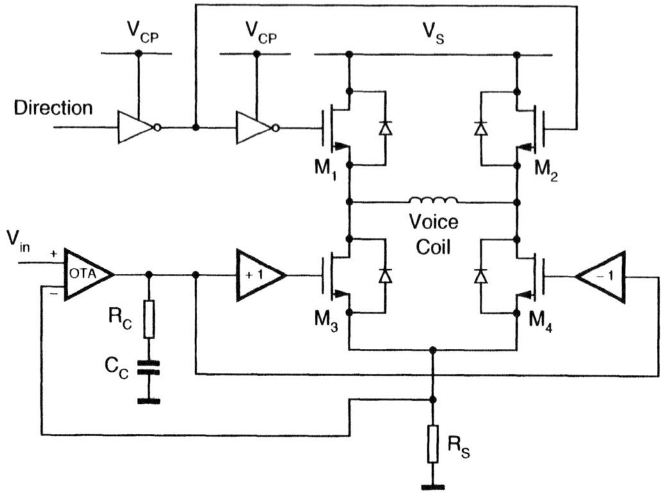

Fig. 6.46 Class B voice coil drive

During seeking, the disk controller receives an instruction from the microprocessor to either retrieve or write data from another track and the actuator input is adjusted to force an accelerating current in the load. The actuator input is the voltage V_{in} which represents the desired head velocity or trajectory at that particular position and moment. About midway from the target track, the control system will instruct the head to decelerate by reversing the direction of current in the coil.

The head actuator will then decelerate towards the destination track where the coil receives no current at all, but it is just floating above the hard disk by hydrodynamic air bearing.

A typical seeking action takes about 10 to 20 ms.

During tracking, the head actuator makes only small adjustments in position to compensate for head drifts or to move on vicinity tracks so only a momentary small sinking or sourcing current is needed.

A typical tracking action takes about 200 to 400 μsec.

The head actuator for an optimum seeking action must move from one track to another in the shortest time possible with minimum power dissipation and avoiding mechanical vibrations. So it is of utmost importance to select a velocity profile with an optimum trade-off between these three parameters. The most common velocity profiles are the triangular and trapezoidal.

The triangular profile is used for intermediate distance seeking, while the trapezoidal is used for long-distance seeking. Fig. 6.47 shows the idealized waveforms of the so-called "bang-bang" seeking where the arrival on the destination track is performed with zero velocity.

To obtain such velocity profile it is necessary that the current in the coil is driven to the maximum in one direction (constant acceleration) for half the

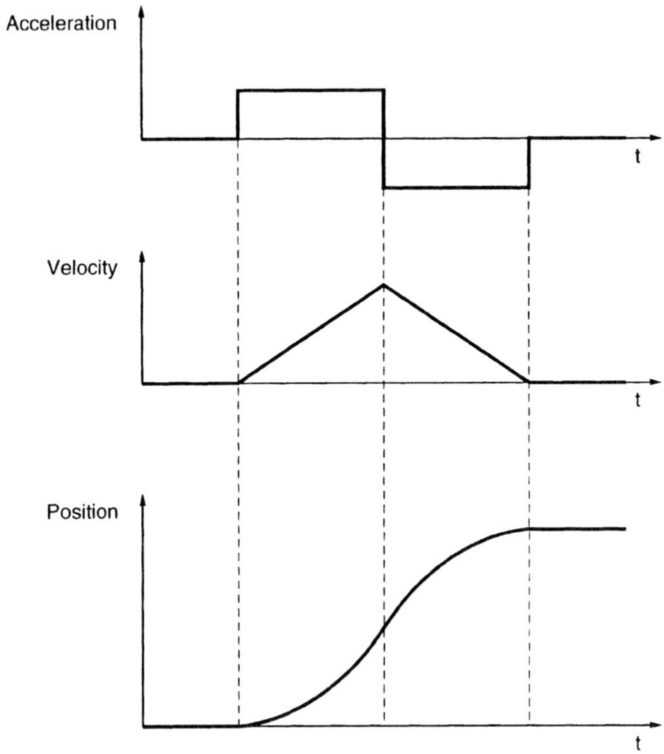

Fig. 6.47 Acceleration, velocity and position of the read/write head during the "bang-bang" seeking

distance and then is driven in the reverse direction (constant deceleration) for the remaining distance.

According to Faraday's Law, the movement of the coil in a magnetic field will induce a BEMF which acts to resist to the movement and is proportional to the velocity: this means that BEMF and speed will have the same profile.

Each seeking or tracking operation is performed in five different actions: acceleration, flyback, active recirculation, deceleration and rapid decay [20] (see Fig. 6.48).

During acceleration (see Fig. 6.48 a) one diagonal pair of DMOS transistors is turned on. Since the sensing voltage V is zero, the transconductance loop is completely unbalanced and both transistors are in full linear conduction.

At the beginning the BEMF is zero and develops with a mechanical time constant $\tau_m \approx 20$ msec much longer than the electrical time constant $\tau_e \approx 50$ µsec.

Hence the current in the coil will have a peak given by:

$$I_{peak} = \frac{V_S}{2\,R_{on} + R_L + R_S} \tag{6.23}$$

where R_{on} is the on-resistance of the DMOS transistors supposed equal, R_L is the coil resistance and R_S is the sensing resistance.

After the peak the current will decrease as the BEMF builds up with time constant τ_m.

When the head has reached half distance from the arrival track, all transistors are turned off and kept off for a duration of time (dead time) sufficient to avoid cross-conduction between top and bottom transistors.

In this condition, since the coil has stored energy, there is recirculation above the supply rail and below ground thanks to the intrinsic diodes of M_2 and M_3 called D_2, D_3 in Fig. 6.48 b.

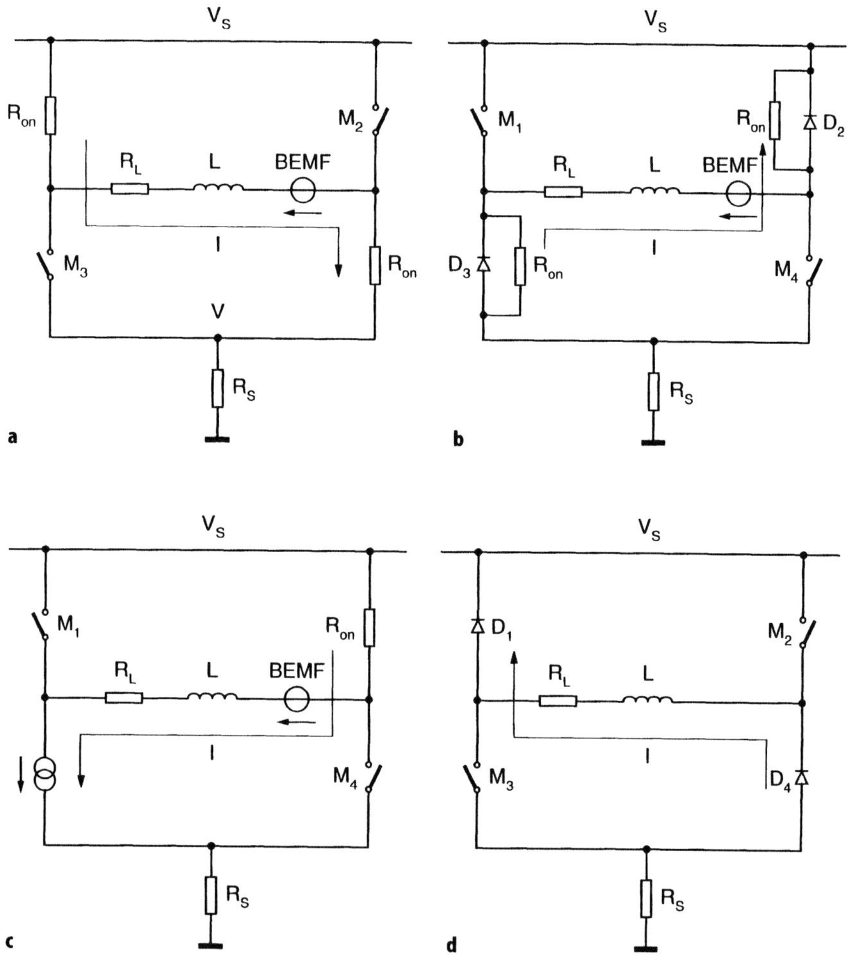

Fig. 6.48 Equivalent circuit of voice coil motor and head actuator full bridge during a seeking event: **a** acceleration; **b** flyback/active recirculation; **c** deceleration; **d** rapid decay

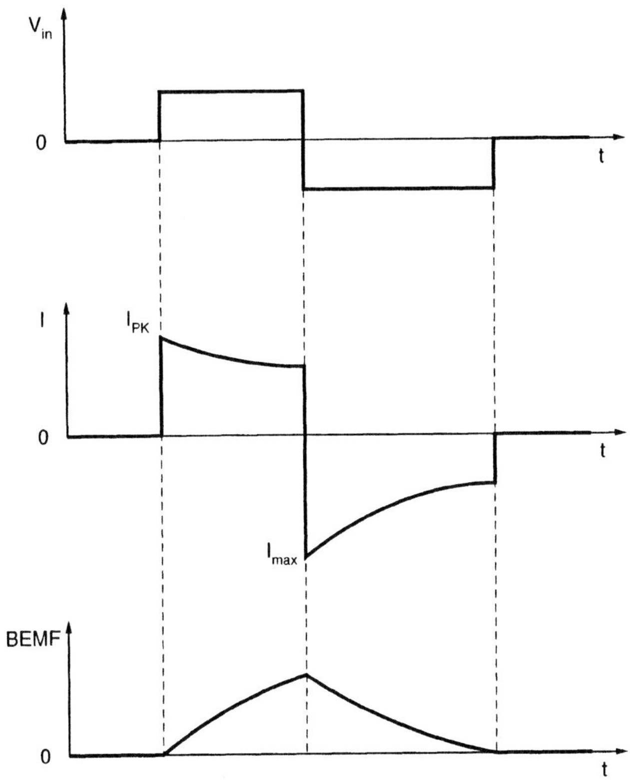

Fig. 6.49 Input command voltage, current in the coil and BEMF waveforms for a "bang-bang" seeking

After the dead time M_2 and M_3 are turned on, thus shunting their own intrinsic diodes and carrying the recirculation current. This effect is known as active recirculation (see also Sec. 6.4.1).

When recirculation is over, the current changes direction and the head decelerates while the motor is braking (refer to Fig. 6.48 c).

Since this phenomenon occurs with the electrical time constant τ_e, the BEMF can be assumed constant and adds up to the supply voltage giving the maximum peak current just after the current reversal:

$$I_m = \frac{V_S + BEMF}{2\,R_{on} + R_L + R_S} \tag{6.24}$$

Well after this peak the servo system resumes control and profiles a deceleration sequence optimized for a given disk drive: in this phase M_2 works as a switch while M_3 operates as a current generator whose drain-source voltage is defined by the transconductance loop.

At the time where the BEMF is zero, the head will be positioned over its target track and the input command voltage V_{in} of the transconductance loop will be forced to zero.

However, the energy remaining in the inductor must be dissipated before starting the on-track operation. This is accomplished during rapid decay when all transistors are off and allow current recirculation through the intrinsic diodes of M_1 and M_4 (D_1, D_4 in Fig. 6.48d).

The relevant current and voltage waveforms of the circuit are shown in Fig. 6.49.

6.8.3 Class AB Operation

The major drawback of class B operation discussed above, is the intrinsic non-linearity present in the transfer characteristic of the power output stage when the bridge is balanced and the load current is zero. This reflects in an undetermined region for the actuator position.

This problem can be solved using a bridge configuration with the output power stage working in class AB with a well controlled quiescent current. This approach requires more electronics but allows maximum precision in the positioning of the head actuator.

The circuit is in general described in Fig. 6.50.

After straightforward calculations the transfer function can be written as:

$$G = \frac{I_L}{V_{in}} = \frac{R_2}{R_1 R_S A_S} \cdot \frac{-G_{loop}}{1 + G_{loop}} \qquad (6.25)$$

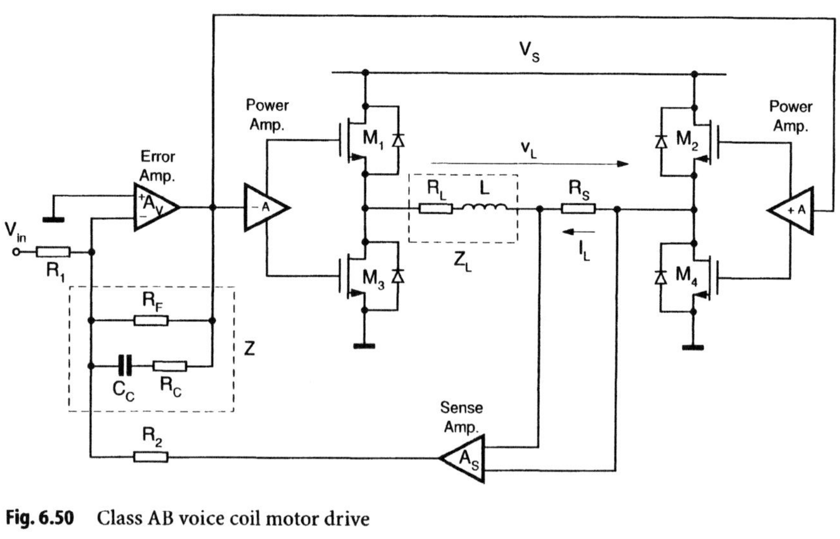

Fig. 6.50 Class AB voice coil motor drive

where R_2, R_1 are external resistors, R_S is the sensing resistance in series with the load, A_S is the sense amplifier gain and G_{loop} is the open-loop gain which is in turn given by:

$$|G_{loop}| = + \frac{2\,A\,A_S\,R_S\,Z}{R_2\,(R_S + Z_L)} \tag{6.26}$$

In this expression A is the output power amplifier gain working in class AB (see chapter 10 for the circuit description), Z represents the impedance in feedback with the error amplifier A_V and Z_L is the equivalent impedance of the motor.

At low frequencies assuming typical parameters as: $A_S = 10$, $A = 16$, $R_S = 0.2\ \Omega$, $R_1 = R_2 = 1\ K\Omega$, $Z \approx R_F = 1\ M\Omega$, $Z_L \approx R_L = 7\ \Omega$, we have:

$$|G_{loop}| = 78.9\ dB$$

This means that the transconductance expression (6.25) can be simplified to:

$$|G| = \frac{R_2}{R_1\,R_S\,A_S} \tag{6.27}$$

Using the numbers reported above it will be:

$$G = 0.5\ A/V$$

To look into the system stability, it is necessary to evaluate the overall transfer function. This can be easily done calculating the transfer function of each sub-circuit.

For this reason Equation 6.26 can be rewritten in the form:

$$|G_{loop}| = A_{E/A} \cdot A_{PW} \cdot A_S \cdot A_L \tag{6.28}$$

where $A_{E/A}$ represents the closed-loop gain of the error amplifier A_V. So it will be:

$$A_{E/A} = \frac{Z}{R_2} = \frac{R_F}{R_2} \cdot \frac{1 + sR_C\,C_C}{1 + s\,(R_F + R_C)\,C_C} \tag{6.29}$$

Moreover A_{PW} is the full output power op.amp. gain equal to:

$$A_{PW} = 2A \tag{6.30}$$

while A_S is the sense amplifier gain and A_L is the attenuation of the load given by:

$$A_L = \frac{R_S}{R_S + Z_L} = \frac{R_S}{R_S + R + sL} \tag{6.31}$$

The frequency dependence of the power op.amp. A and of the sense amplifier A_S can, at first analysis, be neglected since they introduce poles and zeroes at frequencies much higher than the bandwidth.

The first condition to check for overall system stability is that each amplifier of the loop is stable. If this condition holds the loop frequency compensation and stability will be achieved by the appropriate dimensioning of resistor R_C and capacitor C_C in feedback with the error amplifier A_V.

To compensate the loop with a large bandwidth the idea is to introduce a low frequency dominant pole and a zero that compensates the pole introduced by the motor.

Specifically the frequency of the motor pole, according to Equation 6.31, is:

$$f = \frac{R_S + R_L}{2\pi L} \tag{6.32}$$

For instance with $R_S = 0.2\ \Omega$, $R_L = 7\ \Omega$ and $L = 1.2$ mH, we have:

$f \approx 1$ KHz

From Equation 6.29 the error amplifier introduces one zero and one pole given by:

$$f_z = \frac{1}{2\pi R_C C_C} \tag{6.33}$$

$$f_P = \frac{1}{2\pi\,(R_F + R_C)\,C_C} \tag{6.34}$$

Hence, we will set:

$f = f_z$

or

$$\frac{R_S + R_L}{2\pi L} = \frac{1}{2\pi R_C C_C} \tag{6.35}$$

Resistor R_F is used in practice to set the DC maximum loop gain since it results from Equations 6.28, 6,29:

$$|G_{loop,DC}| = \frac{R_F}{R_2} \cdot A_{PW} \cdot A_S \cdot A_L \tag{6.36}$$

Afterwards the dominant pole frequency f_P will be chosen such that the overall bandwidth is not influenced by the poles introduced by A_{PW} and A_S (in any case, higher than 100 KHz).

Finally, R_C and C_C can be calculated from the system of Equations 6.34 and 6.35.

For the system described above, we have obtained f = 1.5 Hz with a bandwidth of about 25 KHz.

6.8.4 Parking

Another important function easily integrated in mixed signal power ICs for hard disk drives together with the voice coil drive, is the head retract (parking).

This function has to be performed following a specific logic command or when the supply voltage is going down.

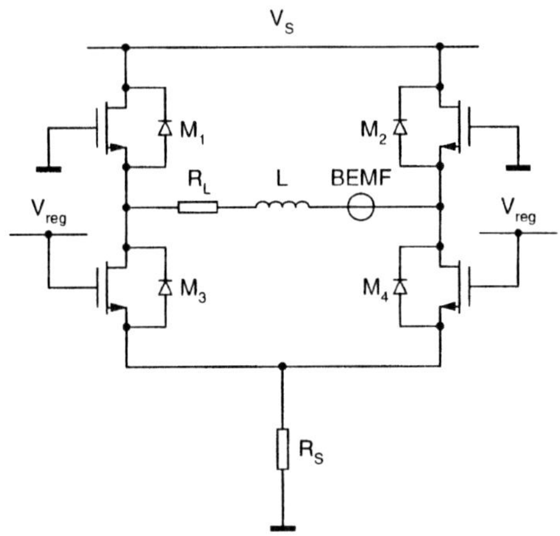

Fig. 6.51 Braking action before parking

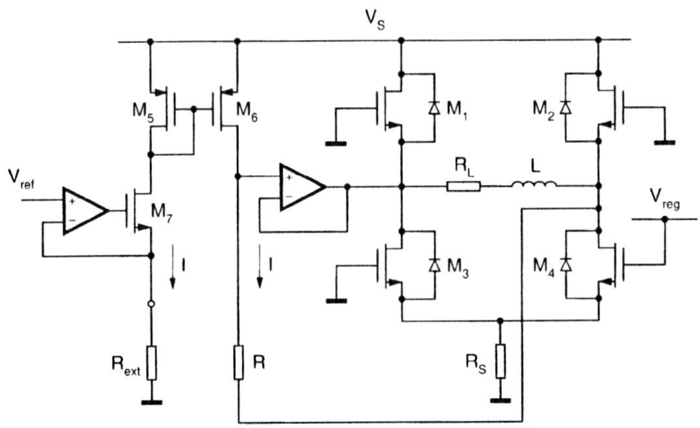

Fig. 6.52 Circuit implementation of the parking function

In any case, the head must have sufficient energy (delivered by the BEMF of the spinning spindle motor) to move back to the parking position within about 200 msec, after which the spindle motor will start its braking action (see Sec. 6.7.4).

The parking function is executed in two phases:

1) the voice coil is braked and stopped discharging the BEMF by turning on M_3 and M_4 as shown in Fig. 6.51;
2) afterwards a constant and controlled current is maintained in the coil (typically 20 mA) to prevent the risk of over-acceleration and damage which results if the head hits the stop. This is accomplished turning off M_3, turning fully on M_4 and applying a voltage externally programmable on the coil.

Figure 6.52 shows a practical circuit to implement this function.

The current I is set externally to the value:

$$I = \frac{V_{ref}}{R_{ext}} \qquad (6.37)$$

and is mirrored by $M_5 - M_6$, so that the voltage applied on the coil will be:

$$V_{park} = R \cdot I = \frac{R}{R_{ext}} \cdot V_{ref} \qquad (6.38)$$

thus allowing a constant velocity head retract.

6.9 Applications

As previously outlined, the integration onto the same chip of signal circuitry and power elements driving inductive loads, creates problems of parasitics. Hence, special layout techniques and protection barriers must be introduced to reduce the interference between these sections. A careful layout floorplanning is thus mandatory for the device operation: in some cases the layout is even more important than the design itself. However, sometimes the chip floorplanning is dictated by other requirements (for example, the pin configuration or the frame of the package) so the best trade-off must be found in each case by a detailed analysis of the circuits involved (that is why we often talk of "art of power integration").

We will see in the following some practical examples of complex multi-function smart power ICs (COMBO chips) addressed to motion control in many different applications.

In all these devices, the utilization of mixed technologies has increased the system performance and efficiency allowing the integration on the same chip of many different power functions combined with analog and digital circuits, hence shrinking physical size, improving reliability and, at the same time, reducing cost [21].

6.9.1 Fax Driver

This smart power IC is used in the new generation fax machines and includes 16 power DMOS transistors plus all the electronics necessary for motor current control.

The device can be directly interfaced with a microprocessor and serial or parallel data loading are possible. The chip photograph is shown in Fig. 6.53.

The power section of this device is able to drive a unipolar stepping motor for the scanner, a unipolar stepping motor for the printer paper feed and a full bridge for driving a DC motor for the paper cutter.

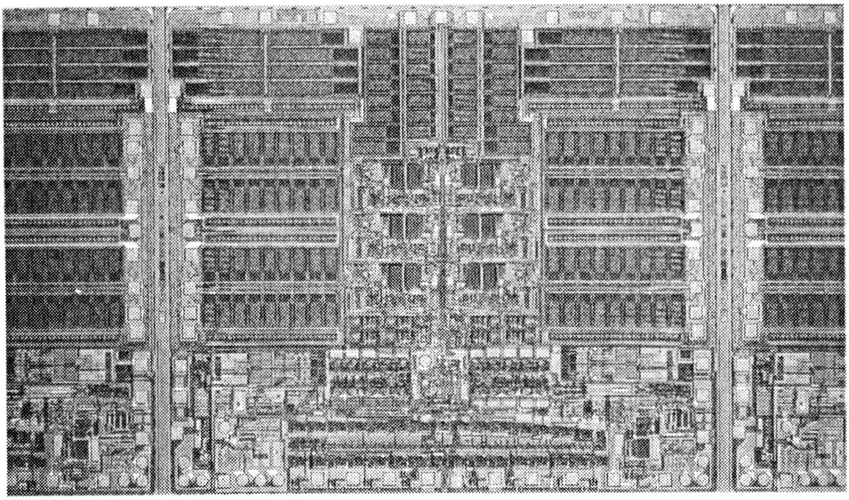

Fig. 6.53 The photograph of a smart power IC for a fax machine, integrating three motor control/drive circuits

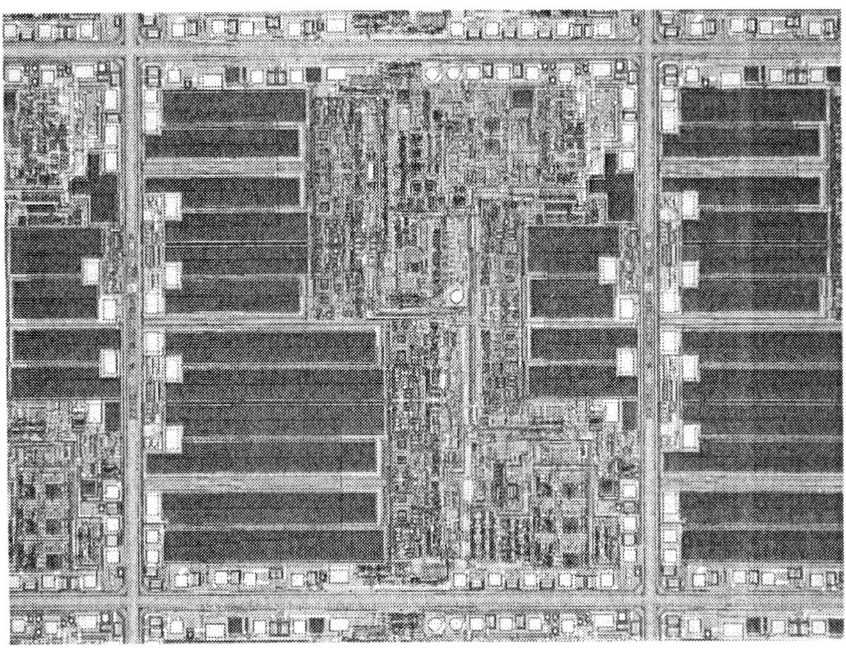

Fig. 6.54 Combo chip for VCR applications including a double three-phase motor drive and a DC motor drive

6.9.2
VCR Driver

This device consists of a double three-phase brushless motor drive plus a full bridge DC motor drive [22].

The first three-phase motor drive is utilized for the Drum motor control, that is, the motor which carries the video heads.

The second motor drive is intended for the Capstan motor whose function is to move the video tape in the cassette deck. Detection of the motor position is carried out by three comparators coming from Hall effect sensors.

The Load motor section includes also a full bridge DC motor drive used to load/unload the cassette and to move the tape close to the video heads.

The IC is also equipped with service circuitry for auxiliary control functions and supplies. The microphotograph of the chip is shown in Fig. 6.54.

6.9.3
Hard Disk Drives

Hard disk drives have played a fundamental role in leading the development of the high level of integration and very complex smart power ICs, because the space available in 2.5 inch and smaller drives is extremely limited.

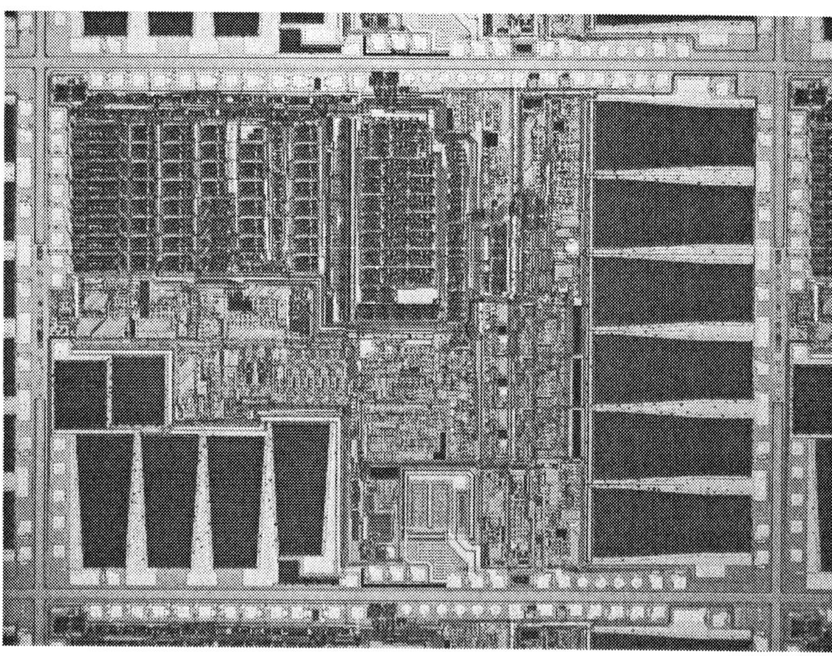

Fig. 6.55 Smart power IC integrating all the circuits needed to control and drive the spindle motor and the voice coil motor

An example of such devices is a single smart power IC that includes all the circuits required to control and drive both the spindle motor and the head positioning actuator (the voice coil) [23].

This very complex chip (see Fig. 6.55), which contains 12 power DMOS transistors and more than 10 000 other transistors, connects directly to the bus of an eight-bit microprocessor and drives directly the three-phase sensorless brushless motor that rotates the disk spindle and a voice-coil servo-positioning system that controls the read/write head.

Because of the low power dissipation this device is packaged in a 64 lead plastic quad flat package and operates without heatsink.

A linear voice coil motor drive IC with externally programmable transconductance gain allowing high precision in the head positioning system, is shown in Fig. 6.56.

The power output stage is composed of two power amplifiers operating in class AB with four DMOS transistors having on-resistance of 0.25 Ω maximum closed in H-bridge configuration with an output current capability up to 3 A.

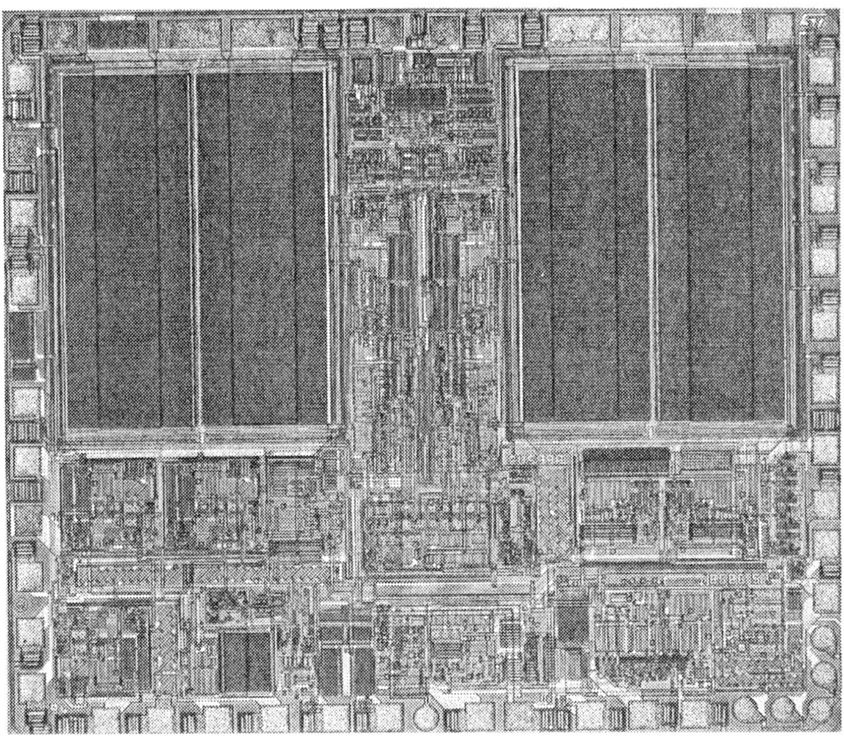

Fig. 6.56 A voice coil motor driver IC with class AB output stage allowing high precision in the head positioning system

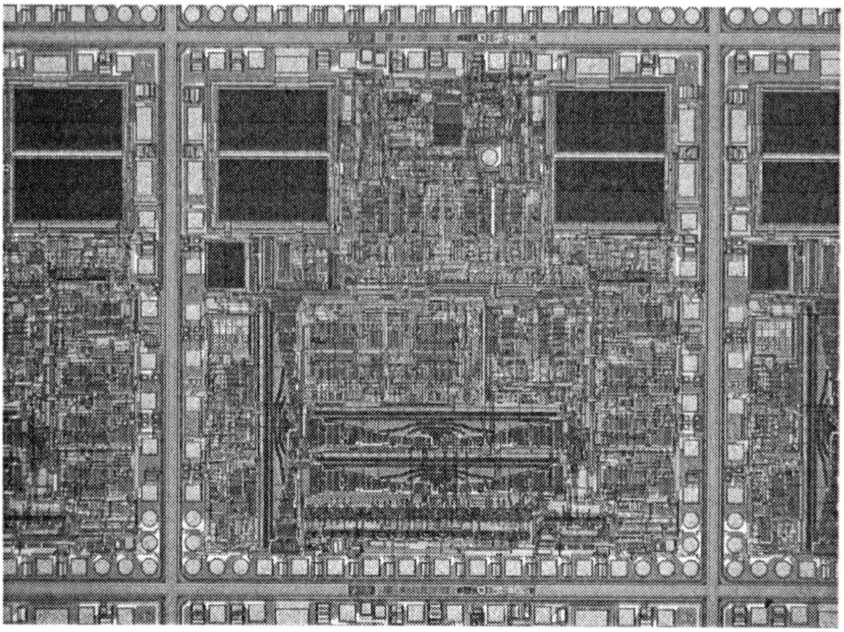

Fig. 6.57 Voice coil motor controller including three 8-bit DACs

The device implements automatic brake and parking of the head actuator performed by a command or when a failure condition is detected by on-chip power supply voltage monitors. It also includes a complete set of service blocks like power on reset, enable, overtemperature warning output and overtemperature protection. A regulated 5 V output for external usage is also provided together with a gate driver output which has the function of driving an external MOSFET that disconnects the power supply.

Figure 6.57 shows a novel voice coil motor controller incorporating three 8-bit DACs and offering significant improvements in terms of cost, external component count, board space and performance in the 3.5 inch disk area.

The device functionality can be summarized as follows:

- 8 bit microprocessor interface;
- three 8-bit DACs with 5 μsec conversion time;
- transconductance loop which contains all the blocks for the current loop control. This includes sense amplifier, error amplifier with gain switch and power amplifier composed of a full bridge DMOS transistor with on-resistance of 0.4 Ω maximum operating in class AB with current capability up to 2 A DC, 2.5 A peak;
- auxiliary blocks such as park function with velocity limit, thermal protection and power monitor functions.

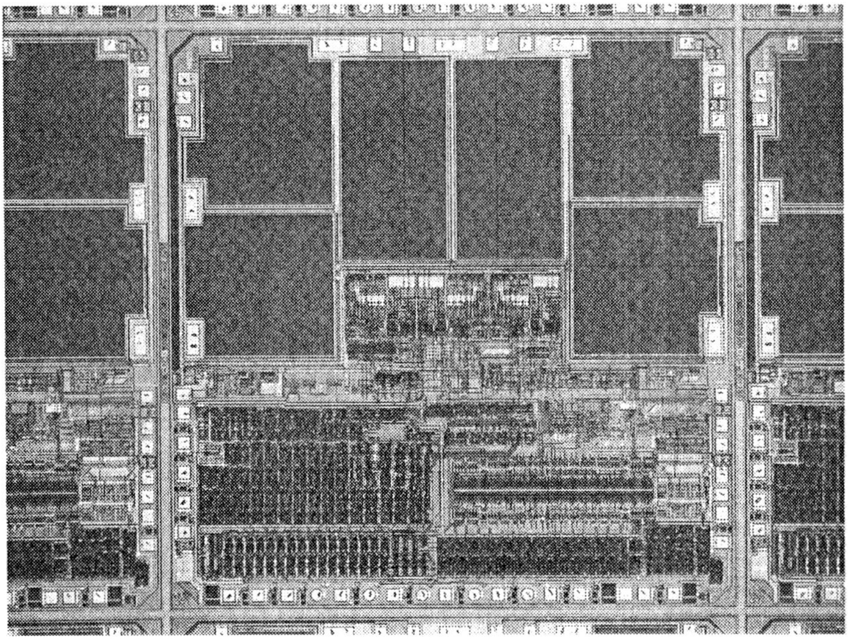

Fig. 6.58 Sensorless spindle motor controller IC with phase-locked loop control

An example of a smart power sensorless spindle motor controller IC, is shown in Fig. 6.58.

This device, used to commutate and speed control a three-phase, 8 poles brushless motor operating at 3600 and 5400 RPM, has the following features:

- sensorless commutation by sensing the BEMF of the undriven motor terminal and speed control via a precision digital phase-locked loop;
- stand-alone operation or an external PMOS transistor can be connected to the output bridge for higher power applications. In this configuration the internal DMOS transistors are sequenced in full conduction state, while the external PMOS is the linear control element;
- the motor current limit is externally programmable;
- programmable functions such as phase switching timing optimization for motor efficiency, autostart or microprocessor supervised start-up and slew rate control;
- output stage with six power DMOS transistors with on-resistance of 0.33 Ω maximum and capable to deliver 3 A DC, 5 A peak;
- serial port is included minimizing the number of I/O pins;
- energy recovery mode for head retraction (parking) followed by dynamic braking;
- stuck rotor and backward rotation detection;
- thermal shutdown.

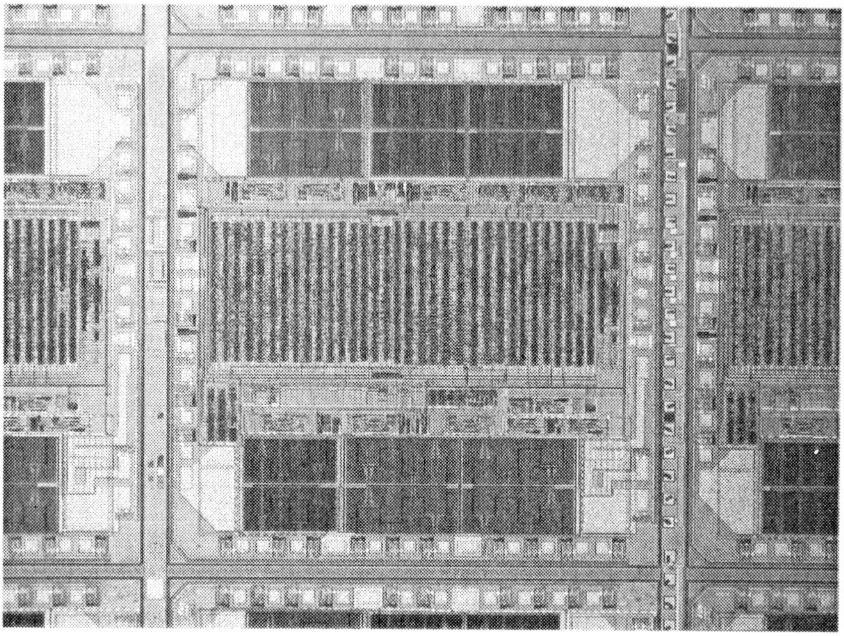

Fig. 6.59 A BiCMOS three-phase sensorless brushless motor drive for low voltage applications

BCD (Bipolar, CMOS, DMOS) is not, however, always the best solution in this application field. A BiCMOS technology has allowed the development of a three-phase sensorless brushless motor drive operating on 3 V supplies for battery-operated portable systems [24].

Rail to rail output voltage swing, mandatory at low voltage, has been achieved using a CMOS power stage having low-side n-channel and high-side low-threshold p-channel power transistors (see Fig. 6.59). The chip implementation also combines digital design automation with hand-crafted analog.

References

1. Murari B (1978) Power Integrated Circuits: Problems, Trade-offs and Solutions. IEEE J Solid-State Circuits, vol 13, no 3, pp 307–319
2. Widlar RJ, Yamatake M (1988) A Monolithic Power Op Amp. IEEE J Solid-State Circuits, vol 23, no 2, pp 527–535
3. Widlar RJ (1991) Controlling Substrate Currents in Junction-Isolated ICs. IEEE J Solid-State Circuits, vol 26, no 8, pp 1090–1097
4. Gariboldi R, Pulvirenti F (1994) A Monolithic Quad Line Driver for Industrial Applications. IEEE J Solid-State Circuits, vol 29, no 8, pp 957–962
5. United States Patent (1984) no 4,466,011
6. Sedra AS, Smith KC (1987) Microelectronics Circuits. HRW, second ed
7. Gariboldi R, Pulvirenti F (1995) A 70 mΩ Intelligent High Side-Switch with Full Diagnostics. Proceedings of the 21th ESSCIRC (Lille)

8. Mohan N, Undeland TM, Robbins WP (1989) Power Electronics: Converters, Applications and Design. J Wiley & Sons
9. Electro-Craft Corporation (1980) DC Motors-Speed Controls-Servo Systems. Fifth ed Electro-Craft Corporation, Hopkins, MN
10. Baltes HP, Popovic RS (1986) Integrated Semiconductor Magnetic Field Sensors. Proceedings of the IEEE, vol 74, no 8, pp 1107–1132
11. Kenjo T (1984) Stepping Motors and Their Microprocessor Control. Clarendon Press, Oxford
12. SGS-Thomson Microelectronics (1992) Designers' Guide to Power Products. 2nd ed AN 380, pp 215–229
13. Carobolante F, King AJ (1991) Spindle Motor Drive IC Designed in BCD Technology with Improved Transient Characteristics. Proceedings of the 20th Annual Symposium. Incremental Motion Control Systems and Devices
14. Cameron S, Carobolante F (1992) Speed Control Techniques for Hard Disk Drive Spindle Motors Drivers. Proceedings of the 21th Annual Symposium. Incremental Motion Control Systems and Devices
15. Best RE (1984) Phase locked loops. Mc Graw Hill
16. Peters D, Harth J (1993) ICs Provide Control for Sensorless DC Motors. EDN
17. Jackson JD (1975) Classical Electrodynamics. J Wiley & Sons
18. Williams RK, Chang A, Concklin B, Pham G (1994) A Fully Integrated HDD Power IC with Novel Head Retract Feature. Proc of the 6th Int Symp on Power Semiconductor Devices and Power ICs, Davos, Switzerland
19. Black B, Lopez M, Morcos A (1993) Basics of Voice Coil Actuators. PCIM pp 44–46
20. Williams RK, Chang A, Cornell ME, Concklin B (1991) Design and Operation of a Fully Integrated BiC/DMOS Head Actuator PIC for Computer Hard-Disk Drives. IEEE Transactions on Electron Devices, vol 38, no 7, pp 1590–1599
21. Murari B (1993) Mixed Power IC Technologies and Small System Integration, presented at the 19th European Solid-State Circuits Conference, Sevilla, Spain
22. Bertolini L, Gleim G, Heizmann F (1994) Combined Smart Power IC for VCR with PWM Controlled Motor Voltage. IEEE Transactions on Consumer Electronics, vol 40, no 3, pp 377–386
23. Cavezzale M, Nessi M (1993) Smart Power for Advanced Disk Drives. PCIM pp 230–236
24. Nessi M, Fucili G (1993) BiCMOS Smart Power Chip for Disk Drives Operates on 3 V Supplies. Electronic Engineering, pp 43–44

CHAPTER 7

Switching Regulators

M. R. Borghi

7.1 Introduction

Switching converters have caused a revolution in the supply technique field, allowing a drastic reduction of the power supply encumber and consequently a decrease in the weight and volume of the equipment where they are used.

In off-line power supplies, switching converters with a small high frequency transformer have replaced the encumbering off-line transformers.

In post-regulators, high efficiency switching converters have substituted the more dissipative linear regulators, increasing the delivered power in the same space, or reducing the dimensions.

For example, notebooks in average include six power supplies. It would be difficult to obtain such light portable computers without using switching techniques.

The switching power supply field is very extensive. It includes AC-AC, AC-DC, DC-AC and DC-DC regulators, whether voltage increasing or decreasing, inverting and not. Every type of regulator that makes an energy conversion using switching techniques belongs to this field. They provide an energy storage in reactive components and its subsequent transfer to other reactive components, by opening and closing switch transistors.

The most diffused switching regulators are AC-DC and DC-DC converters, that have a lot of commercial uses.

The AC-DC converters are in all equipment supplied by the 50 (or 60) Hz electrical network but including components and circuits that operate with DC voltages, often lower than 50 V for user protection.

The DC-DC conversion is necessary, for example, in equipment fed by batteries, whose voltages are converted in the various voltages required by the supplied devices. DC-DC converters are also used to make transformer post-regulation where the secondary side voltage is different or less precise than the voltage required by part of the supplied circuits.

These switching converters are usually closed-loop regulators where the controlled parameter (voltage or current) is measured in a continuous way or by sampling, and is compared with a reference. The error signal is converted in a digital signal that drives the switches.

In semiconductor technologies a lot of different controller circuits for switching regulators have been developed. Moreover for low and medium power applications, up to 100 W, the power integrated technologies have allowed the integration of the switches on the same controller chip.

The most used switching regulators can be classified into three categories depending on the kind of reactive components involved: the Single Ended Inductor circuits, the Diode Capacitor and the Transformer Coupled ones.

This classification is very schematic and makes it difficult to include some kinds of converters, but it is suitable enough to classify the most diffused ones. The Transformer Coupled circuits include the most part of the off-line converters and it is dealt with in Chap. 8. The Diode-Capacitor ones are very low current regulators and in power integrated converters they are used as voltage multipliers to create the right supply for the main regulator. Some of them are described in Chaps. 5 and 6.

The Single Ended Inductor circuits for DC-DC converters are those more frequently made with power integrated technologies, consequently, they are the main subject of this section. In these regulators, the first reactive component is an inductor that stores magnetic energy and transfers part of it to a load while another part is converted into electrostatic energy in a capacitor and stored for a following load supply.

In the first chapters, their basic topologies are described as concerns the different inductor conduction modes and the various regulation and control methods, with an outline to the problem to find a reliable and manageable model to study their dynamic behaviour. In the last chapters, the main characteristics of the integrated switching regulators are shown using as an example a Buck converter.

7.2 Main Topologies of Integrated Switching Regulators

Thanks to their versatility, Single Ended Inductor circuits have developed a lot of different configurations. Herewith the basic topologies and the operation modes of switching regulators are summarized [1].

7.2.1 Buck Converter: Step-Down Voltage Regulator

This regulator, shown in Fig. 7.1a, converts a higher input voltage into a lower regulated voltage. Figure 7.2 shows its main waveforms in steady state. To simplify, on the following the inductor charge and discharge are approximated with linear shapes because the error, compared to the real exponential waves, is generally below 1%.

The switch gate is driven by a T-period square wave, that allows three possible states for the network, shown in Figs. 7.1b, c, d. Three intervals of time must be considered:

T_{ON}: switch-on time,
T_D: diode-on time,
T_I: idle time, where neither the switch nor the diode conduct.

Fig. 7.1 Buck Converter

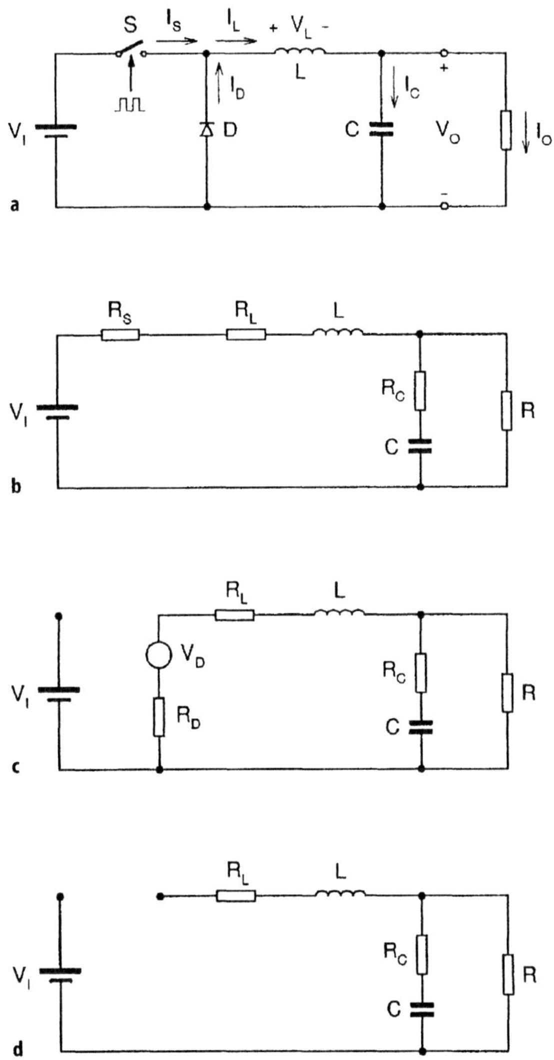

They are related by the following relationships:

$T_{OFF} = T_D + T_I$: switch-off time (7.1)

$T = T_{ON} + T_D + T_I$ (7.2)

The ratio between the switch-on time and the switching period is called duty cycle:

$$D = \frac{T_{ON}}{T} \tag{7.3}$$

Fig. 7.2 Buck Converter Main Waveforms

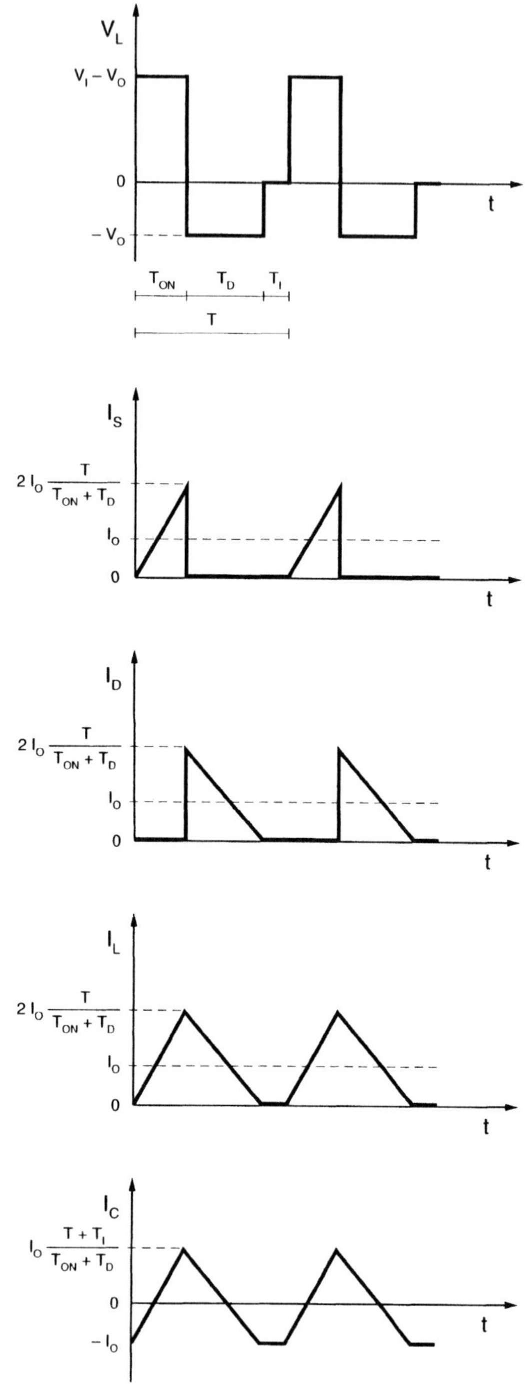

These Figures also include the network parasitic resistances. The reactive parasitic components are negligible in steady-state analysis for usual switching frequencies from some tens of kHz to a few hundreds of kHz. However, they are important to evaluate the regulator efficiency. These considerations are valid for all the converter topologies described in this chapter.

In the switch conduction time Ton (Fig. 7.1b) the inductor current increases. As far as it is lower than the load current I_O the lacking current is supplied by the capacitor C. When the inductor current exceeds I_O the extracurrent goes into the capacitor that stores energy.

In the switch-off time T_{OFF} the inductor discharges through the freewheeling diode D during T_D (Fig. 7.1c), while in the idle time T_I no current flows in the inductor and the load is supplied by the capacitor (Fig. 7.1d).

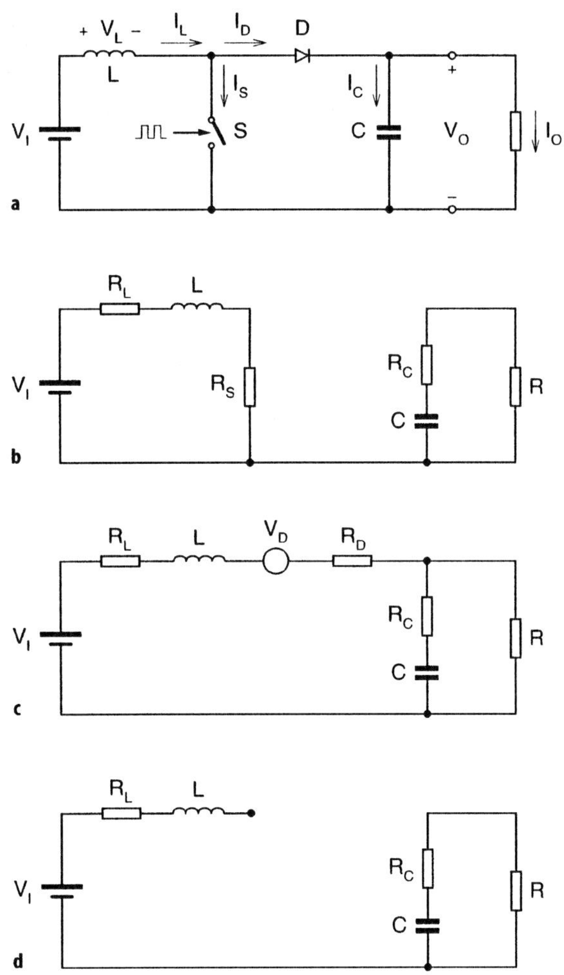

Fig. 7.3 Boost Converter

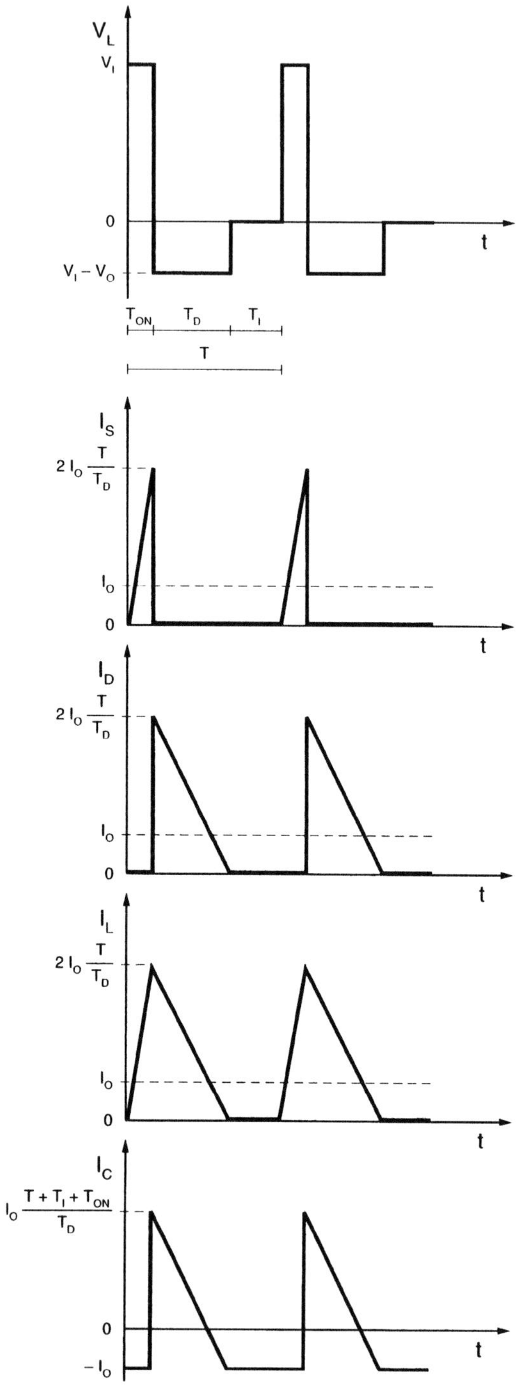

Fig. 7.4 Boost Converter Main Waveforms

The capacitor value must be big enough to keep the output voltage constant, that is, to make negligible the voltage ripple due to its current variation. This ripple depends both on the capacitor and on the capacitor parasitic resistance, usually of some tens of mΩ.

7.2.2 Boost Converter: Step-Up Voltage Regulator

A Boost Converter is shown in Fig. 7.3a and its main waveforms in the steady state are plotted in Fig. 7.4. It converts a lower input voltage into a higher regulated output voltage. As in the Buck Converter, there are three possible network states, shown in Figs. 7.3b, c, d.

During the switch conduction time T_{ON}, the inductor stores energy and the load current is supplied by the capacitor. During the switch-off time this energy is transferred to the output through the diode D during T_D, and in the idle time T_I the load is again fed by the capacitor.

In a Buck Converter, the inductor supplies the output both in T_{ON} and T_D; in a step-up regulator it is connected to the output only in T_D. This means that the energy stored in a Boost regulator inductor, and therefore its current, is higher, with the same output current, as can be seen comparing Figs. 7.2 and 7.4.

7.2.3 Buck-Boost Converter: Polarity Inverting Voltage Regulator

Figure 7.5a, b, c, d show the converter, and its three possible states, while in Fig. 7.6 its main steady-state waveforms are reported. When the switch is on the inductor stores energy and the capacitor feeds the load. During the switch-off time the inductor supplies the output for the time T_D. During T_I the inductor is fully discharged and it cannot invert the current polarity because the diode D opposes, hence the load is again supplied only by the capacitor.

As in a Boost converter the inductor is connected to the output only during its freewheeling time, but the current flows in the opposite direction, thus the regulated voltage has an inverted polarity compared to the input voltage.

7.3 Inductor Continuous and Discontinuous Conduction Modes

Whatever the topology, a converter works in Discontinuous Conduction Mode (DCM) if the inductor is fully discharged at the end of the freewheeling time, as in Figs. 7.2, 7.4, 7.6. It operates in Continuous Conduction Mode (CCM) if the coil keeps part of its magnetic energy stored and the inductor current never goes to zero.

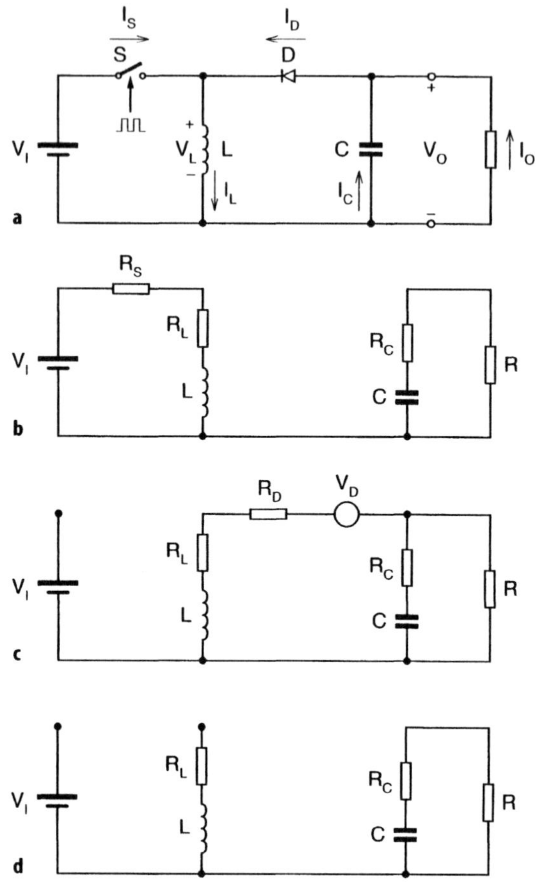

Fig. 7.5 Buck-Boost Converter

In CCM the idle time T_I is equal to zero, and the freewheeling time T_D is the total switch off time T_{OFF}, as can be seen from Eqs. 7.1, 7.2:

$$T_I = 0$$
$$T_{OFF} = T_D$$
$$T = T_{ON} + T_{OFF}$$

Figures 7.7 and 7.8 show the comparison between inductor current waveforms in DCM (Figs. 7.7a, 7.8a) and CCM (Figs. 7.7b, 7.8b) for a Buck (Fig. 7.7) and a Boost (Fig. 7.8) regulator. A Buck-Boost converter has waveforms similar to those of a Boost one.

Keeping the same switching period T the Discontinuous Mode requires a smaller inductor, but the ripple current is bigger for the same load current. Thus the voltage ripple using the same output capacitor is bigger. In a DCM Buck converter the inductor peak current is at least double than the output current. In a DCM Boost converter it is again bigger and it depends on the ratio between

Fig. 7.6 Buck-Boost Converter Main Waveforms

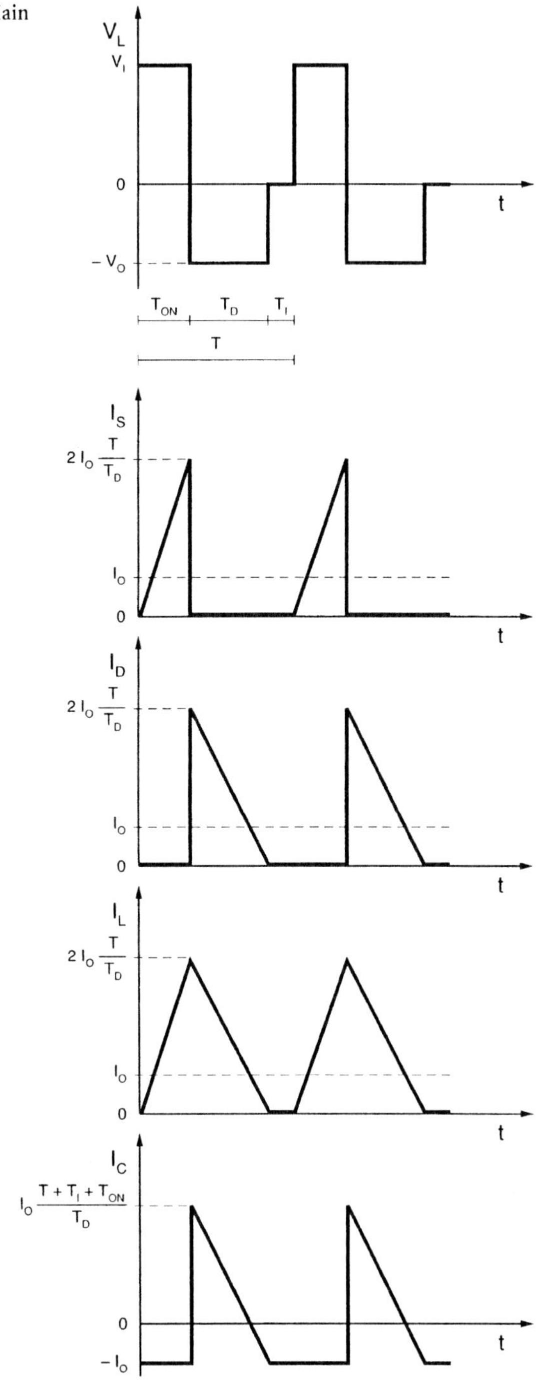

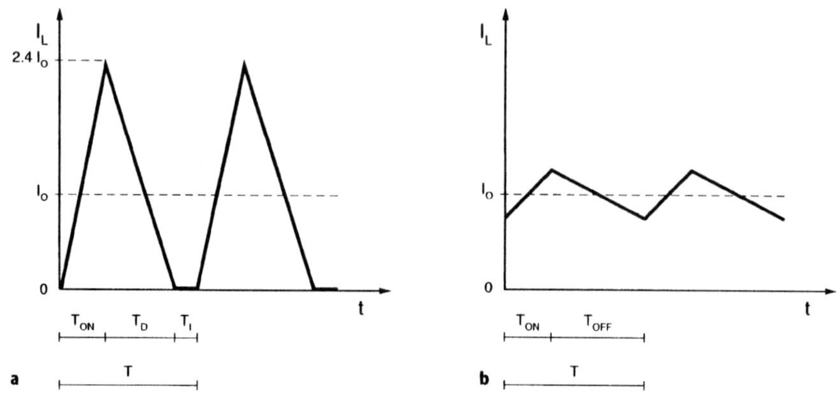

Fig. 7.7 Buck Converter: DCM and CCM Inductor Current

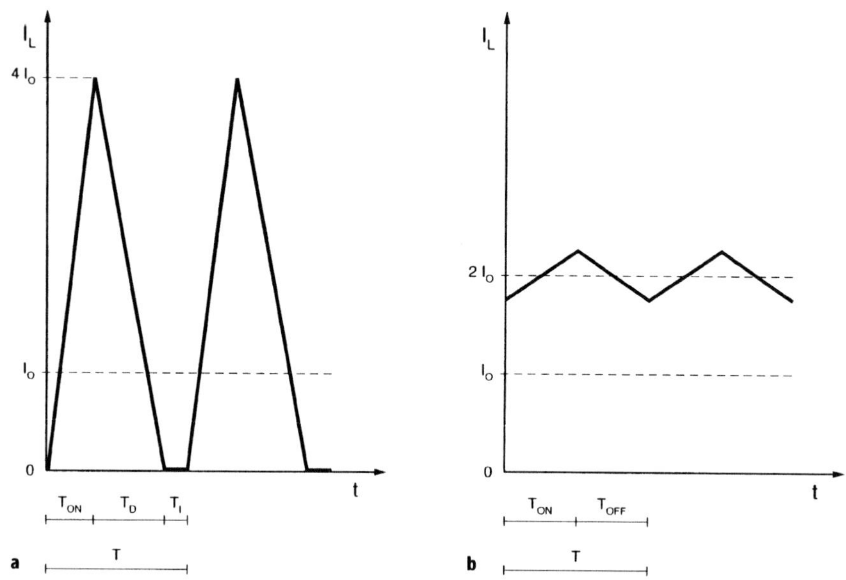

Fig. 7.8 Boost Converter: DCM and CCM Inductor Current

the diode on-time and the period T. With T_D equal to T/2, as shown in Fig. 7.8, the inductor current is four times higher than the load current.

Regulators usually operate in Continuous Mode at heavy output currents, higher than 0.5 ÷ 1 A, in Discontinuous Mode at light loads.

A converter working in Discontinuous Mode has a faster closed-loop response to heavy loads and line voltage variations, due to the complete inductor discharge that makes the system insensitive to its previous cycle.

The DCM Control to Output Gain has a single-pole low frequency characteristic. The CCM Gain has a two poles response due to the LC filter. In Boost and Buck-Boost topologies, it also presents a right half plane zero that delays their closed-loop responses.

We will now describe the output voltage versus input voltage relationships in the steady state for the three main topologies, both in DCM and in CCM. These relationships are evaluated neglecting parasitic resistances and diode drops. The CCM relationships are finally revised keeping them in mind. The DCM relations with parasitic resistances can be found following the same method.

From the previous equations duty cycle expressions are evaluated. Neglecting parasitic effects, in Continuous Mode, the duty cycle depends only on input and output voltages (Eqs. 7.6, 7.15, 7.25). In Discontinuous Mode it depends also on the output load (Eqs. 7.12, 7.20, 7.30).

7.3.1 Buck Converter CCM

The main waveforms are shown in Fig. 7.9.

To obtain the relationship between the output voltage (V_O) and the input voltage (V_I) we have to consider that in steady state, once that any transition due to load or input voltage variation has ended, the inductor DC current is constant and the inductor current is the same for any time nT. Thus the current increase during T_{ON} must be equal to current decrease during T_{OFF}. Keeping the linear shape approximation as mentioned in paragraph 7.2.1, this means:

$$\Delta I_{L,T_{ON}} = \Delta I_{L,T_{OFF}}$$

$$\Delta I_{L,T_{ON}} = \frac{1}{L}(V_I - V_O)\,T_{ON}$$

$$\Delta I_{L,T_{OFF}} = \frac{1}{L}V_O\,T_{OFF}$$

$$\frac{1}{L}(V_I - V_O)\,T_{ON} = \frac{1}{L}V_O\,T_{OFF} \tag{7.4}$$

$$V_O = V_I\,\frac{T_{ON}}{T_{ON} + T_{OFF}}$$

From Eq. 7.3 it can be written

$$V_O = V_I\,D \tag{7.5}$$

or:

$$D = \frac{V_O}{V_I} \tag{7.6}$$

Fig. 7.9 Buck Converter: CCM Main Waveforms

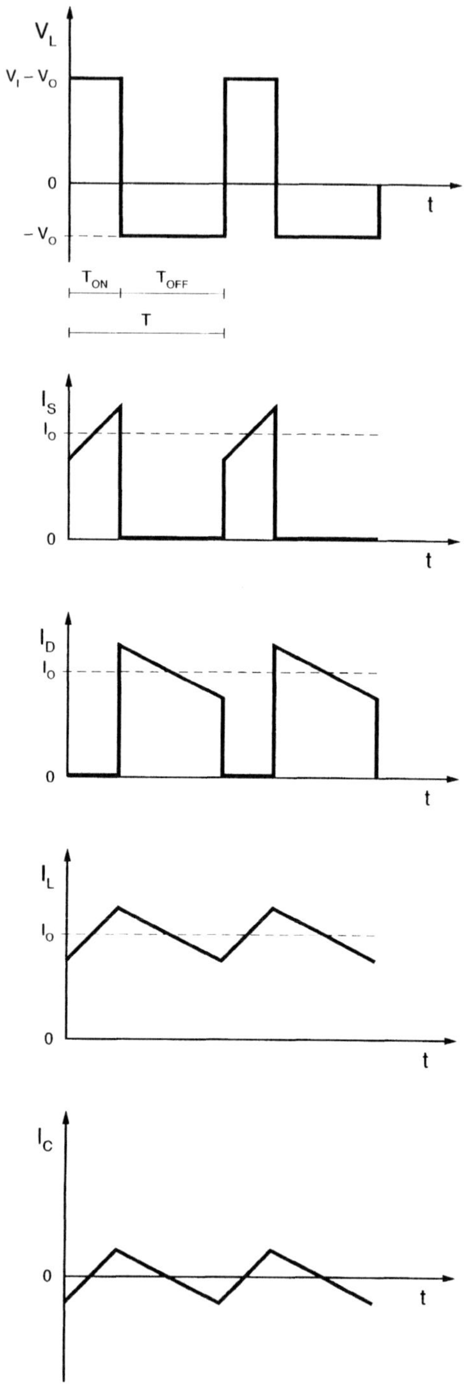

Considering the parasitic resistances and the diode drop, the voltages across the inductor changes and relation 7.4 becomes:

$$\frac{V_I - V_O - (R_S + R_L)I_O}{L} T_{ON} = \frac{V_O + V_D + (R_D + R_L)I_O}{L} T_{OFF}$$

$$V_O = V_I D - V_D(1 - D) - (R_S + R_L)I_O D - (R_D + R_L)I_O(1 - D) \tag{7.7}$$

$$D = \frac{V_O + V_D + (R_D + R_L)I_O}{V_I + V_D + (R_D - R_S)I_O} \tag{7.8}$$

Equations 7.7 and 7.8 have been found considering the parasitic resistances voltage drops due only to the average inductor current, instead of to the real current waveforms. This approximation neglects the drop on the capacitor parasitic resistance and cannot be acceptable any more at high duty cycles when V_I is very close to V_O.

7.3.2 Buck Converter – DCM

Steady-state main waveforms are shown in Fig. 7.2.

Relation 7.4 is still valid substituting T_{OFF} with T_D:

$$\frac{1}{L}(V_I - V_O)T_{ON} = \frac{1}{L} V_O T_D \tag{7.9}$$

But in DCM there is a third variable T_I because $T = T_{ON} + T_D + T_I$, hence we need a second equation that is obtained considering the inductor average current equal to the load current I_O. An equivalent load resistance can be obtained for which $I_O = V_O/R$. Thus we can write:

$$\frac{V_O}{R} = \frac{(T_{ON} + T_D)T_D V_O}{2LT} \tag{7.10}$$

Combining Eqs. 7.9 and 7.10, the output voltage versus input voltage relationship can be obtained:

$$V_O = \frac{2V_I}{1 + \left(1 + \frac{8L}{RT}\frac{1}{D^2}\right)^{\frac{1}{2}}} \tag{7.11}$$

The duty cycle D can now be explained as:

$$D = \left(\frac{2L}{RT}\right)^{\frac{1}{2}} \frac{V_O}{[V_I(V_I - V_O)]^{\frac{1}{2}}} \tag{7.12}$$

To have a Discontinuous Conduction Mode, it must be that:

$$T_{ON} + T_D < T$$

From Eqs. 7.9 and 7.10 the DCM condition becomes:

$$\frac{2L}{RT} < 1 - \frac{V_O}{V_I}$$

Fig. 7.10 Boost Converter: CCM Main Waveforms

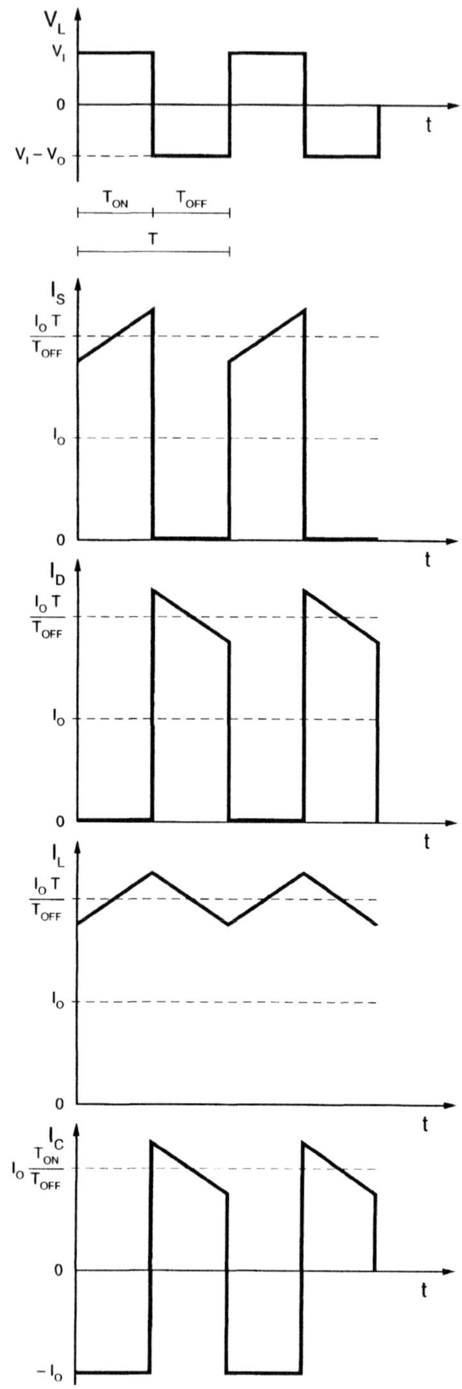

7.3.3
Boost Converter – CCM

The steady-state waveforms are shown in Fig. 7.10. Equalizing the inductor current increase to its decrease in each period, we can find the relationship between V_O and V_I:

$$\frac{V_I}{L} T_{ON} = \frac{V_O - V_I}{L} T_{OFF} \tag{7.13}$$

$$V_O = V_I \frac{T_{ON} + T_{OFF}}{T_{OFF}}$$

that can be written as:

$$V_O = V_I \frac{1}{1 - D} \tag{7.14}$$

or:

$$D = \frac{V_O - V_I}{V_O} \tag{7.15}$$

Adding the parasitic drops due to the average inductor current:

$$\frac{V_I - (R_S + R_L)I_O T/T_{OFF}}{L} T_{ON} = \frac{V_O - V_I + V_D + (R_D + R_L)I_O T/T_{OFF}}{L} T_{OFF}$$

that becomes:

$$(V_O + V_D)D^2 + [V_I - 2V_O - 2V_D + (R_S - R_D)I_O]D + V_O + V_D - V_I + (R_D + R_L)I_O = 0 \tag{7.16}$$

From Eq. 7.16 it appears that the parasitic resistances make a CCM Boost converter a second degree system. For a given output voltage there are two possible duty cycles, one stable, D_1, and another one, D_2, unstable, as shown in Fig. 7.11 by the qualitative waveforms of V_O versus D.

Figure 7.11 shows that the maximum output voltage that can be obtained for a given input voltage is limited due to parasitic resistances. This limit does not appear in Eq. 7.14. The maximum output voltage decreases when the load current rises because also the parasitic drops increase.

During a Boost regulator start-up, the converter tries to reach the maximum value of the duty cycle, moving towards the diagram falling slope, with the risk that the output capacitor will not be charged. To avoid this a soft start circuit is integrated, that brings the converter into its steady state slowly increasing the value of duty cycle.

The risk of moving from a stable to an unstable duty cycle also exists for regulators working at high duty cycles when some large transient appears. To avoid this, the regulators can keep the maximum duty cycle limited around $90 \div 95$ percent of the period.

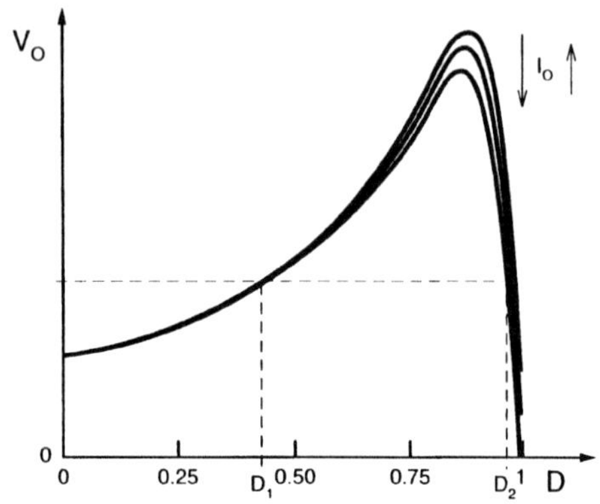

Fig. 7.11 Boost Converter: Output Voltage versus Duty Cycle Qualitative Diagram

7.3.4 Boost Converter – DCM

Referring to Fig. 7.4, Eq. 7.13 becomes

$$\frac{V_I}{L} T_{ON} = \frac{V_O - V_I}{L} T_D \tag{7.17}$$

The second relationship is found equalizing the charge delivered by the inductor to the output in T_D to the total charge supplied to the load in the switching period T:

$$\frac{V_O}{R} = \frac{V_I T_{ON} T_D}{2LT} \tag{7.18}$$

From Eqs. 7.17 and 7.18 we have:

$$V_o = \frac{V_I}{2}\left[1 + \left(1 + 2D^2 \frac{RT}{L}\right)^{\frac{1}{2}}\right] \tag{7.19}$$

$$D = \left(\frac{2L}{RT}\right)^{\frac{1}{2}} \frac{[(V_o - V_I)V_o]^{\frac{1}{2}}}{V_I} \tag{7.20}$$

To have a Discontinuous Conduction Mode it must be that

$$T_{ON} + T_D < T$$

T_{ON} and T_D can be calculated from relations 7.17 and 7.18 to find the DCM condition:

$$\frac{2L}{RT} < (V_O - V_I)\frac{V_I^2}{V_O^3} \tag{7.21}$$

Even considering parasitic effects in a DCM Boost converter there is only one possible duty cycle for each output voltage.

Fig. 7.12 Buck-Boost Converter: CCM Main Waveforms

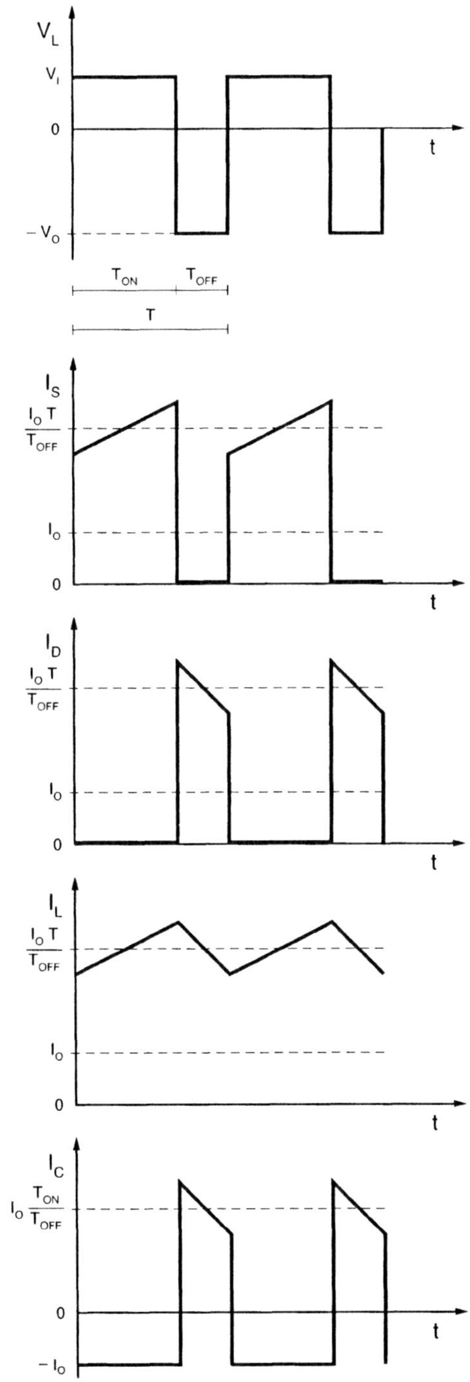

7.3.5
Buck-Boost Converter – CCM

CCM steady state waveforms are shown in Fig. 7.12.

As in the other converters the inductor current rise during T_{ON} is equal to its decrease in T_{OFF}, that means:

$$V_I \frac{T_{ON}}{L} = - V_o \frac{T_{OFF}}{L} \tag{7.22}$$

and the output versus input voltage relation is

$$V_o = - V_I \frac{T_{ON}}{T_{OFF}} \tag{7.23}$$

that can be written as

$$V_o = - V_I \frac{D}{1 - D} \tag{7.24}$$

and

$$D = \frac{V_o}{V_o - V_I} \tag{7.25}$$

As shown in relationship 7.23, if the switch-on time is shorter than its off-time the converter works in step-down mode. When T_{ON} is longer than T_{OFF} it operates in step-up mode.

Taking into account the parasitic resistances Eq. 7.22 becomes:

$$\frac{[V_I - (R_S + R_L)I_oT/T_{OFF}]T_{ON}}{L} = \frac{[(R_D + R_L)I_oT/T_{OFF} - V_o + V_D]T_{OFF}}{L}$$

and we obtain:

$$(V_o - V_D - V_I)D^2 + [V_I - 2V_o + 2V_D + (R_D - R_S)I_o]D + V_o - V_D - (R_D + R_L)I_o = 0 \tag{7.26}$$

From Eq. 7.26 it appears that a Buck-Boost converter is a second degree system. The output voltage versus duty cycle qualitative diagram is shown in Fig. 7.13. There are two possible duty cycles for each output voltage: one stable and another one unstable. They are valid the same considerations made for the Boost Converter.

7.3.6
Buck-Boost Converter – DCM

The steady-state waveforms are shown in Fig. 7.6. The relationship 7.22 becomes:

$$V_I \frac{T_{ON}}{L} = - V_o \frac{T_D}{L} \tag{7.27}$$

and the average inductor current during T_D is equal to the load current.

Fig. 7.13 Buck-Boost Converter: Output Voltage versus Duty Cycle Qualitative Diagram

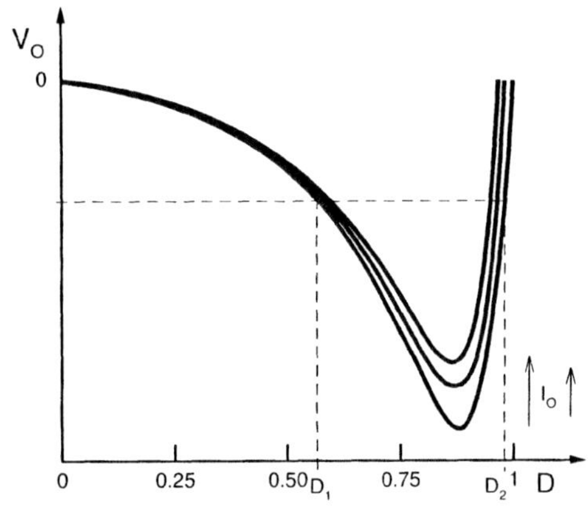

$$\frac{1}{2}\frac{V_O T_D^2}{LT} = \frac{V_O}{R} \tag{7.28}$$

from Eqs. 7.27 and 7.28 the relationship between output and input is obtained:

$$V_o = -V_I D\left(\frac{RT}{2L}\right)^{\frac{1}{2}} \tag{7.29}$$

and the duty cycle D can be expressed as:

$$D = \left(\frac{2L}{RT}\right)^{\frac{1}{2}}\frac{V_O}{V_I} \tag{7.30}$$

The Discontinuous Conduction Mode condition is:

$$\frac{2L}{RT} < \left(\frac{V_I}{V_I - V_O}\right)^2 \tag{7.31}$$

7.4 Regulation Methods

Some converters called Pulse Frequency Modulators (PFM), regulate the output varying the switching frequency. They usually have a fixed on or off time. Yet, at present, the most used regulators are fixed frequency ones that regulate the output changing the ratio between the switch on and off times. They are called Pulse Width Modulators (PWM).

There are advantages and drawbacks for both PFM and PWM. But usually PWM are preferred because the fixed frequency simplifies the filtering and the shielding of radio-frequency and electromagnetic interferences, especially if more regulators are used in the same equipment.

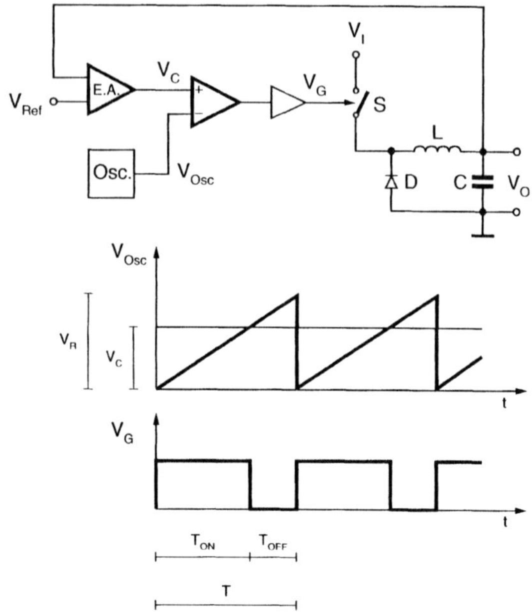

Fig. 7.14 Voltage Mode Control

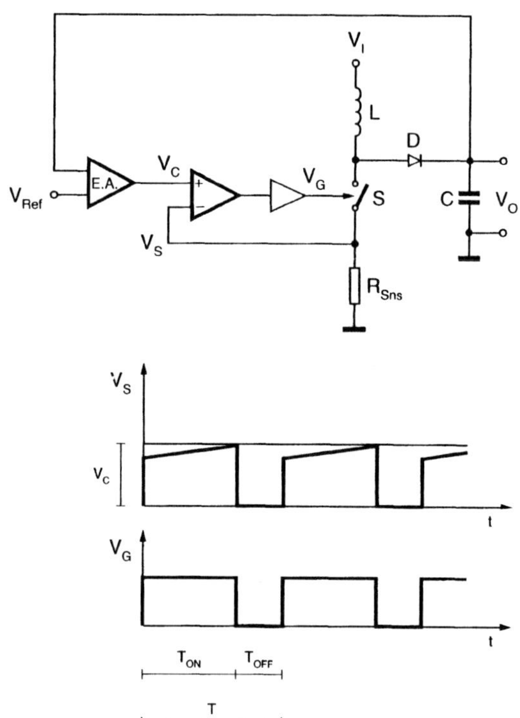

Fig. 7.15 Current Mode Control

7.4.1
Closed Loop Control Methods

PWM and PFM regulators are usually closed loop systems with negative feedback, in order to have good dynamic responses and line and load performances. The regulation is made by a comparison between a control signal and a ramp signal, that generates a logic drive for the switch.

The control signal is the difference between the regulated output and a reference. This difference can be more or less amplified and filtered, depending on the regulator design. The filter can be applied to the error amplifier input or on its output or between input and output.

There are also power supplies, both PFM and PWM, where the switch logic drive is made directly by the comparison between the regulated signal and the reference.

If the ramp signal is generated by an oscillator, as shown in Fig. 7.14, the control is called Voltage Mode.

The ramp signal can be a voltage proportional to the inductor current, as in Fig. 7.15. This control is called Current Mode. It closes a second feedback loop inside the voltage loop. The inductor current is detected either by a resistor, as shown in Fig. 7.15 or by a transformer.

7.4.1.1
Current Mode Control

The Control to Output Gain of a Continuous Conduction Mode regulator with a Current Mode Control is without the inductor pole and becomes a single-pole diagram, at low frequencies, with the advantage of a faster transient response.

It has the drawback of a bigger load effect on the steady-state output voltage, because the ramp signal load dependence is corrected by the voltage loop with an error amplifier input offset.

In Current Mode Control the ramp slope depends on the input voltage, as shown for example by relationship 7.4. An input voltage change modifies this slope and the ramp intersection with control voltage, providing a voltage feed-forward that anticipates the input voltage variation effect, modifying the duty cycle with a consequent good line regulation.

The most used Current Mode Control, shown in Fig. 7.15, is a Peak Current Mode because the power switch is turned off when the inductor peak current crosses the control voltage V_C. In common regulators, where the purpose is the output current control, this method is not so effective in Boost and Buck-Boost topologies, where the difference between the inductor peak current and the load current can be high as shown in Figs. 7.4, 7.6, 7.10, 7.12. It is more effective in a Buck converter where the inductor is at the output, and also here the difference between the peak and the average current must be corrected by the voltage loop with an offset between the error amplifier inputs that depends on the inductor current ripple and the error amplifier gain.

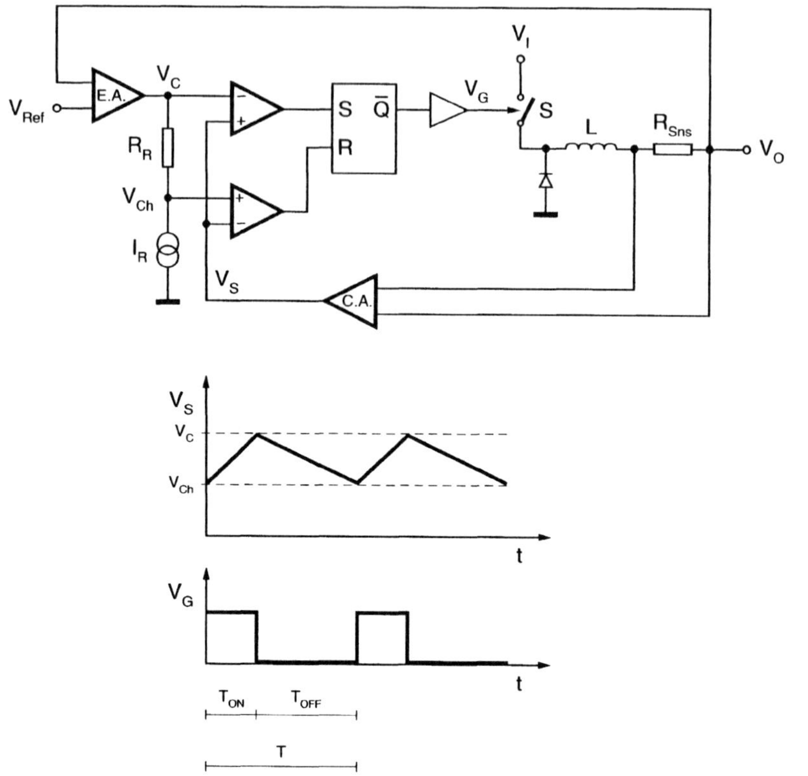

Fig. 7.16 Hysteretic Current Mode

A Peak Current Mode is unstable for a duty cycle exceeding 0.5, resulting in subharmonic operating mode that can be eliminated by adding to the sampled inductor ramp a slope of suitable value, usually equal to the inductor current downslope, solution called "Slope Compensation".

Another problem is the extreme noise sensitivity because the current ramp is usually small if compared to the current level programmed by the control voltage. A little noise easily changes the power switch turn off, moving the system into subharmonic oscillations.

For these drawbacks different kinds of Current Mode Control methods have been introduced, like the Average Current Mode described in Chap. 8 and the Hysteretic Current Mode. The latter, shown in Fig. 7.16, compares inductor current with two voltages: the first is the control voltage and the second is obtained by subtracting a fixed voltage to the control voltage. This solves the instability problems, but requires a full period inductor current control instead of a current sampling during T_{ON}, that is simple to introduce into an integrated regulator by monitoring the power switch current.

The Hysteretic Current Mode is a variable frequency control, but it can be changed into a constant-frequency one with few additional circuits as described in [2].

7.4.1.2
Voltage Feedforward Control

In Voltage Mode Control the oscillator ramp is fixed, and it has a slow response to input voltage changes, the regulator reacts only after their effect detection through the feedback path.

This problem is solved with an oscillator ramp variable with the input voltage. This improved Voltage Control is called "Voltage Feedforward Control".

The optimal feedforward compensation for the described converter topologies can be found as referred in [3] by equalizing to zero the output voltage derivative versus the input voltage, called "Audio-Susceptibility".

Considering Fig. 7.14, with the origin on the ramp minimum voltage, it can be written:

$$D = \frac{T_{ON}}{T} = \frac{V_C}{V_R} \tag{7.32}$$

where V_C = control voltage and V_R = peak ramp voltage.

Substituting on relations 7.5, 7.14, 7.24 they become:

$$V_o = V_I \frac{V_C}{V_R} \tag{7.33}$$

$$V_o = +V_I \frac{V_R}{V_R - V_C} \tag{7.34}$$

$$V_o = -V_I \frac{V_C}{V_R - V_C} \tag{7.35}$$

Deriving and equalizing to zero we obtain the optimal feedforward for the three topologies:

Buck: $$V_R = \alpha V_I \tag{7.36}$$

Boost: $$V_R = -\frac{V_C}{\frac{V_I}{V_\beta} - 1} \tag{7.37}$$

V_β = arbitrary constant, $V_\beta > V_I$

Buck-Boost: $$V_R = V_C + \gamma V_I \tag{7.38}$$

Combining 7.36 ÷ 7.38 with 7.33 ÷ 7.35, the output to input relationships become:

$$V_o = \frac{V_C}{\alpha} \tag{7.39}$$

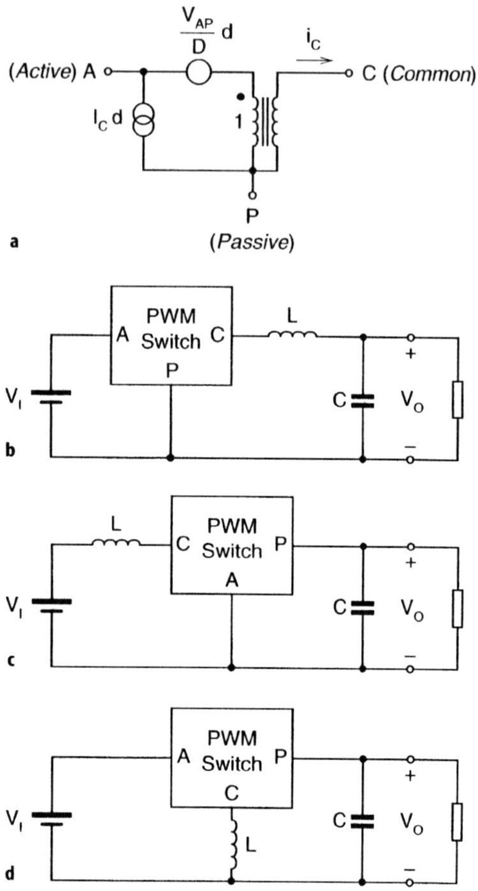

Fig. 7.17 a: PWM Switch Model. V_{AP} is the voltage between active and passive terminals, Ic is the current in the common terminal, ic and d are the small signal variations of Ic and D. The three basic Converters: Buck **b**, Boost **c**, Buck-Boost **d**, with PWM Switch Model **a** inserted

$$V_O = V_\beta \tag{7.40}$$

$$V_O = \pm \frac{V_C}{\gamma} \tag{7.41}$$

and the Control to Output Gain: $G_C = \frac{dV_o}{dV_C}$ becomes

Buck $$G_C = \frac{1}{\alpha} \tag{7.42}$$

Boost $$G_C = 0 \tag{7.43}$$

Buck-Boost $$G_C = \frac{1}{\gamma} \tag{7.44}$$

In Buck and Buck-Boost topologies the feedforward facilities the control loop compensation network design because the Control to Output Gain becomes independent of the input voltage. In Boost converter this gain becomes nil and the optimal feedforward cannot be used. In Boost regulators only a partial voltage feedforward can be used.

7.4.2 An Outline of Dynamic Behaviour Analysis of Switching Regulators

The analysis of the behaviour of a system means the formulation of a mathematical model that reproduces with accuracy the connections among the system variables and their time derivatives to predict the steady working point and the small signal characteristics. Besides, it is desirable to find a linear mathematical model, with differential equations having time invariant coefficients, that can be solved using the Laplace transform.

This method, as referred to in [4], transforms differential equations into simpler solution algebraic ones, and relates differential equations solution with the harmonic analysis method, that is important both for its direct translation in the experimental measurements field and for studying the system dynamic behaviour with the circuit designers' usual methods, Bode plots, with computer standard simulators such as Spice.

In switching regulators the signals that describe system status, the inductor current and the output capacitor voltage, are time variable but continuous signals. The problem is the non-linear operation mode of the switching regulator itself. For this reason different linear mathematical models have been studied and proposed, for example [5, 6, 7], to try to reproduce in a reliable mode its dynamic behaviour. A converter behaviour prediction up to one half the switching frequency seems to be adequate because according to the sampling theory it is not possible to transfer information at frequencies higher than one half the sampling frequency as shown in [8].

The critical test for these models is the ability to predict Current Mode Control instabilities with their subharmonic and chaotic oscillations.

In 1990 a method has been published [6], based on the identification of a three-terminal non-linear component, called "PWM Switch" that includes the total regulator non-linearity of some PWM converters, including the three basic ones described in paragraph 7.2.

After determination of its equivalent electrical circuit this can be substituted to the converter active and passive switches in the regulator circuit. According to [6] it is possible to obtain its DC and small signal characteristics and to perform with accuracy its closed-loop frequency analysis using standard simulators without the use of special purpose programs. An example is shown in [9].

Figure 7.17 shows the equivalent DC small signal model of the PWM Switch and the three basic converters including this model.

According to [7] these methods of analysis based on this Switch Model can be applied not only to PWM converters but also to PFM ones.

In closed-loop regulators, having determined the Control to Output Gain, the error amplifier and the compensation network can be designed to obtain the closed-loop transfer function that satisfy the desired characteristics: line and load regulation, and a good dynamic response together with loop stability.

7.5 Main Characteristics of Integrated Switching Regulators

In the following paragraphs we will describe an example of an integrated Buck Regulator, with its features, its circuits and their evolution in recent years. The same circuit with little changes can be used in other converter topologies.

7.5.1 PWM, Voltage Feedforward, Buck Regulator Core

Figure 7.18 shows a regulator block diagram with the power integrated circuit and the external components. The IC includes a constant temperature voltage reference V_{ref}, using Zener diodes or more frequently a Bandgap voltage generator. The reference voltage is usually trimmed during wafer testing to guarantee tolerances of a few percent.

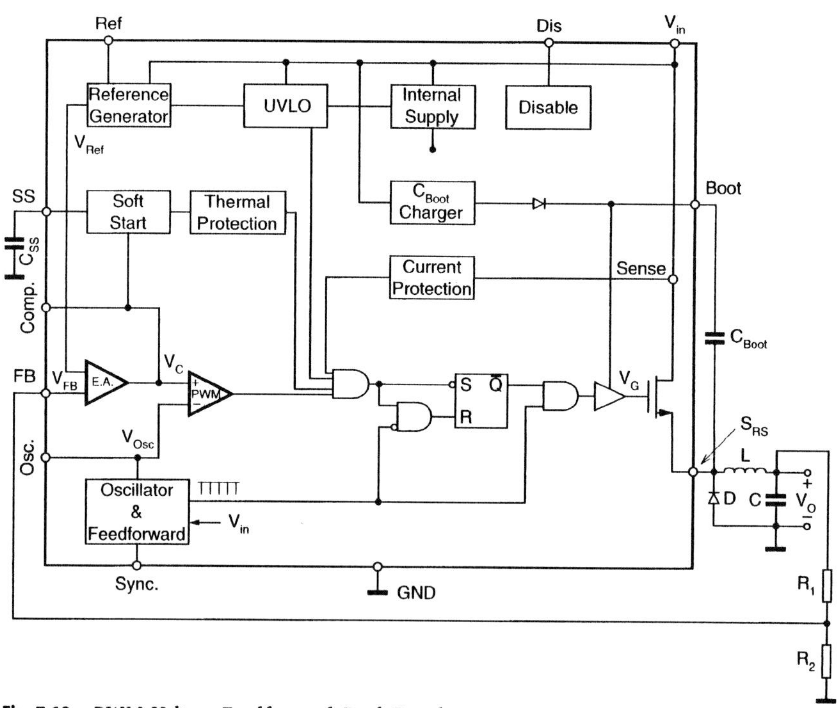

Fig. 7.18 PWM, Voltage Feedforward, Buck Regulator

The new generation of regulators for the supply of very low voltage logics and microprocessors with very strict supply voltage tolerances, can also have the reference trimmed after packaging to guarantee precision better than 1%, including temperature dependence.

The regulated output voltage V_O is fedback through a divider to the negative input of an operational amplifier, V_{FB}, whose non inverting input is connected to the reference.

The amplified error signal is the control signal V_C, that is compared with the oscillator ramp V_{osc}. When the two signals become equal the comparator output changes state and the switch is turned off. The switch off is latched at comparator output to avoid power transistor multiple switching due to the noise generated by its turn off.

The oscillator features a voltage feedforward technique with the ramp slope proportional to the input voltage. When V_I varies the slope changes to keep a constant "volt second" product on the inductor, and the duty cycle immediately varies with a minimum error amplifier output shift.

A clock pulse from the oscillator resets the PWM latch, that switches on the power transistor. The latch signal goes to a floating power stage that drives the switch.

When the driving stage is supplied by a bootstrap capacitor (see Chaps. 5 and 6), as that shown in Fig. 7.18, the clock forces a minimum switch-off time in each period, that limits the regulator maximum duty cycle, to recharge the Bootstrap capacitor. If different driving techniques are used the power transistor can be on all the period. This happens, for example, when the input voltage has a minimum value close to the regulated output voltage.

7.5.2
Regulator Protections and Features

At the regulator start-up the output capacitor is discharged, the error amplifier is unbalanced and the system runs with a duty cycle limited only by current protection up to the fully capacitor charge, with an overstress on the power switch. The delay of the filtered error amplifier in moving its output to the right duty cycle level and the inductor extracharge can then cause an output voltage overshoot.

To avoid these problems a soft start function is usually integrated. It controls the start-up duty cycle for a monotonically increase of the output voltage, without stressing the power switch.

An integrated power regulator often includes some protection circuits, such as current limitations, thermal and overvoltage protections. One example of a thermal protection is described in Chap. 5.

In general it includes power management circuits, as an "Under Voltage Lockout" of the power switch when the input voltage and the internally generated supply voltages and references are below their minimum value. This to avoid dangerous regulator behaviours when these voltages cannot guarantee the right circuit operations.

Sometimes, circuits that provide external flags when the input and/or the output voltages are below minimum levels are integrated.

The device can also have an input or an output or an input-output synchronization pin, to work as the "master" regulator or as a "slave" one, in equipment where more converters have to operate together at the same switching frequency.

Most of the last generation regulators also include an inhibit function to switch off the power transistor and sometimes to put all the IC in "Sleep Mode" with an extremely low current consumption: some tens of μA, decreasing to a few μA in the next generation devices.

Differently from the most part of integrated circuits for other application fields that are supplied by well-controlled input voltages, a power supply integrated circuit is fed by input voltages that are unregulated or with a very poor regulation. It also must be feasible for a wide input voltages range. This situation obliges to integrate voltage regulators dedicated to internal circuits. First of all, to increase their total line rejection, but also to provide them the right supply voltage required by their operation or by the technology. And, when input voltage becomes very high, to keep the current consumption low, without increasing the chip area with high value resistances and high voltage components.

In regulators with bootstrap capacitor there is a 10 V voltage generator to charge this capacitor with the voltage that optimizes the power DMOS switch resistance R_{DSON}. When there is a pin available to filter this voltage generator output, it can also be used to supply the internal signal circuits. This is more conveniently done by using two separate chip pads connected to the pin with two different wires, to avoid voltage spikes on signal parts, due to the bootstrap capacitor charge and to the diode recovery. If a filter pin is not available it is better to add a specific signal supply voltage.

7.5.3 Error Amplifier and Compensation Network

A regulator compensation network is designed to have enough gain at low frequencies to satisfy regulation accuracy requirements and line rejection. Some bandwidth is required for good load transient responses. The cross-over frequency is usually well below one half the switching frequency.

The CCM Voltage Mode Buck Regulator loop gain can be written as:

$$T(s) = A(s)\frac{V_I}{V_R}\frac{(1 + s/\omega_z)}{1 + 2\delta s/\omega_n + (s/\omega_n)^2} \tag{7.45}$$

where A(s) is the error amplifier and compensation network voltage gain, including a resistor divider between the regulated output and the error amplifier feedback input when $V_o > V_{ref}$.

$$\frac{V_I}{V_R}\frac{(1 + s/\omega_z)}{1 + 2\delta s/\omega_n + (s/\omega_n)^2} = \text{Control to Output Gain}$$

$$\omega_Z = \frac{1}{R_C C}$$

$$\omega_n = \frac{1}{\sqrt{LC}} \frac{1}{\sqrt{1 + R_C/R}}$$

$$\delta = \frac{L + C\,R\,R_C}{2\sqrt{RLC\,(R + R_C)}}$$

and, as previously defined:

$$R = \frac{V_O}{I_O}$$

R_C is the output capacitor equivalent serie parasitic resistance (ESR).

The loop gain relationship 7.45 neglects inductor and switch parasitic resistances, diode drop and regulator propagation delays but is enough for a first approximation analysis.

The Control to Output Gain corner frequencies are

$$f_z = \frac{1}{2\pi R_C C}$$

$$f_n = \frac{1}{2\pi\sqrt{LC}} \frac{1}{\sqrt{1 + R_C/R}}$$

The Control to Output Gain depends on V_I. It includes the typical frequency response of a second order LC filter with a natural frequency and a damping factor dependent on load resistance and capacitor ESR. The ESR presence introduces a zero, too.

Figure 7.19 a shows the simplified schematic of the error amplifier. It is a transconductance operational amplifier whose basic characteristics are:

- $G_M = 4$ mA/V
- $R_O = 2.5$ Mohm
- $Av_O = 80$ dB
- I source/sink = 200 μA
- Input Bias Current = 0.3 μA
- $V_{ref} = 5.1$ V

Its open loop gain has two poles: $f_{p1} = 20$ kHz and $f_{p2} = 20$ MHz. In the overall regulator loop compensation the 20 MHz pole can be neglected because it is a frequency well below the zero dB axis and the error amplifier can be schematized by the equivalent circuit shown in Fig. 7.19 b, and its voltage gain is:

$$A_{EA}(s) = G_M \frac{R_O}{1 + sR_O C_O}$$

with $C_O = 3$ pF.

Thanks to its high output impedance, the regulator compensation network can be inserted between the error amplifier output and ground. Looking at Fig. 7.20 a, the total voltage gain becomes:

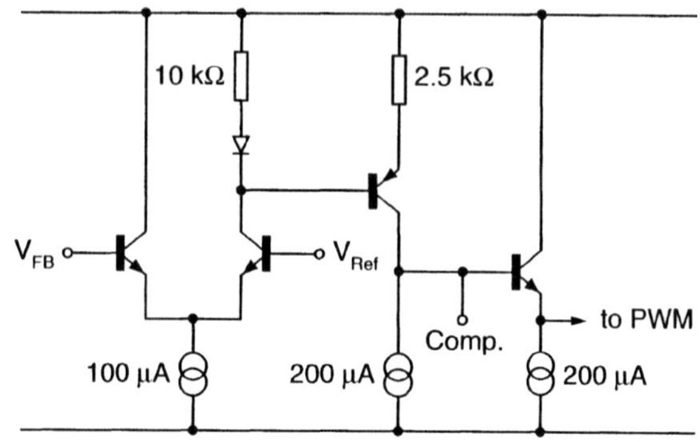

a

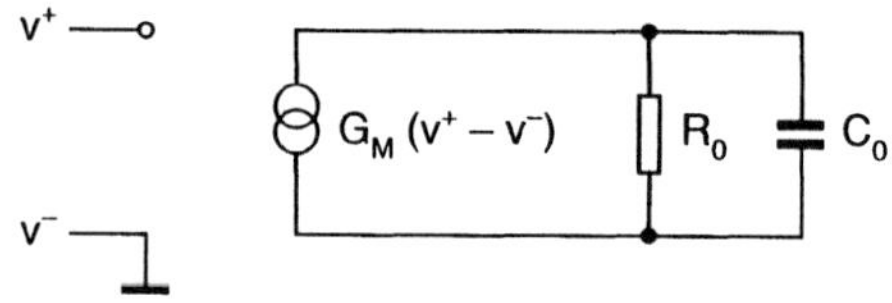

b

Fig. 7.19 Error Amplifier Circuit **a** and its Low Frequency Equivalent Circuit **b**

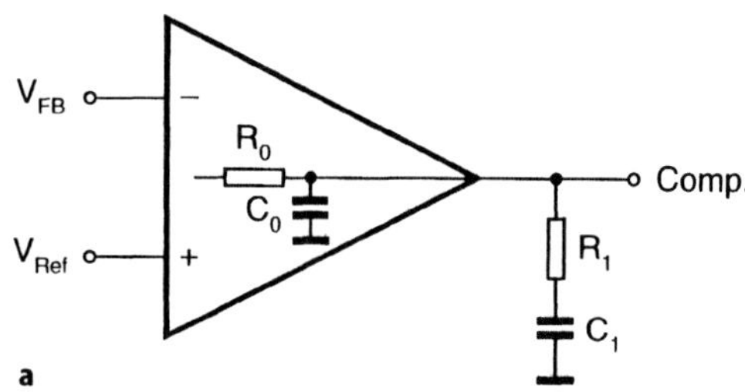

a

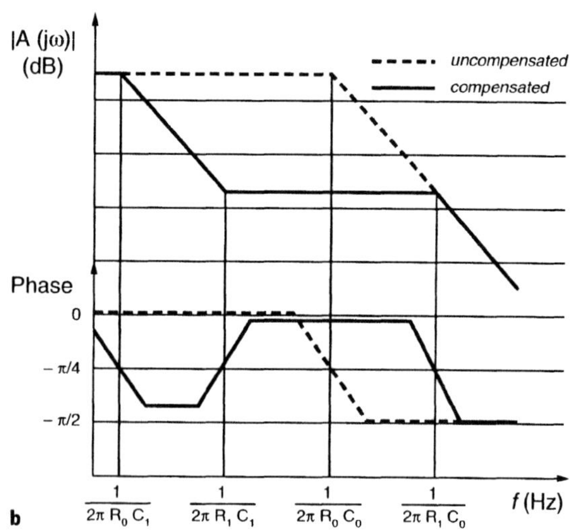

b

Fig. 7.20 Compensated Error Amplifier **a** with its Bode Plots **b**

$$A(s) = G_M \frac{R_O(1 + sR_1C_1)}{s^2R_OC_OR_1C_1 + s(R_OC_1 + R_OC_O + R_1C_1) + 1}$$

with $R_1 < R_O$ and $C_1 > C_O$. The Bode plots are shown in Fig. 7.20b. There is a low frequency pole:

$$f_{p1} = \frac{1}{2\pi R_O C_1},$$

a zero:

$$f_{z1} = \frac{1}{2\pi R_1 C_1},$$

that is placed at the output filter frequency f_n.

The second pole:

$$f_{p2} = \frac{1}{2\pi R_1 C_O}$$

can be neglected because it is a high frequency pole.

This kind of compensation introduces a dominant pole f_{p1} at a frequency lower than the LC filter frequency and compensates its two poles with a new zero f_{z1} and with the converter zero f_z due to filter capacitor ESR.

This could seem a risky method because of the ESR spread, but it is frequently used because of its simplicity. A qualitative loop gain plot is shown in Fig. 7.21.

For regulated voltages higher than V_{ref}, the presence of a feedback resistor divider introduces the second compensation zero (Fig. 7.22):

$$f_{z2} = \frac{1}{2\pi R_2 C_2}$$

Fig. 7.21 Qualitative Loop Gain Plot

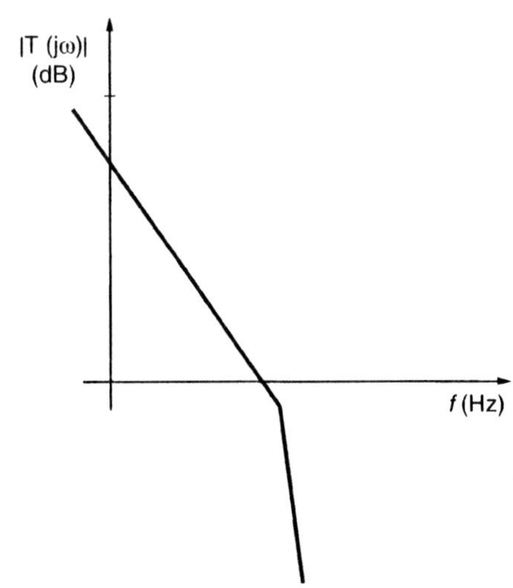

This network also adds a second pole:

$$f_{p2} = \frac{1}{2\pi R_P C_2}$$

with $R_P = R_2//R_3$, that must be put in correspondence with ESR zero.

Another example of compensation network is shown in Fig. 7.23, whose transfer function is:

$$A(s) = \frac{(1 + sR_1 C_1)(1 + sR_2 C_2)}{sR_1 C_1 \left(1 + s\frac{C_1}{G_M}\right)}$$

This network introduces a low frequency pole and two zeros to compensate the second order filter poles.

This compensation network has a second pole:

$$f_{p2} = \frac{G_M}{2\pi C_1}$$

due to the high error amplifier input impedence.

Fig. 7.22 Compensation Network for Output Voltage Higher than V_{REF}

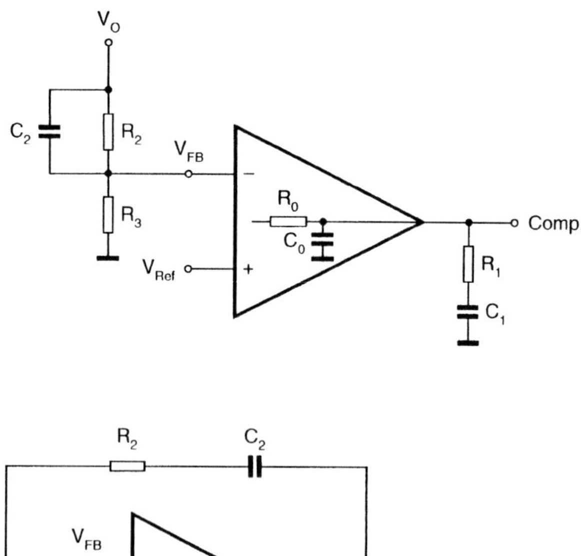

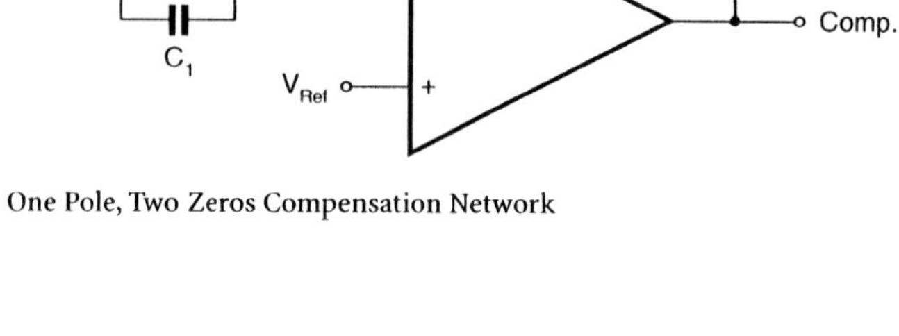

Fig. 7.23 One Pole, Two Zeros Compensation Network

The capacitance C_1 must have a low value to shift this pole at high frequency. This is possible choosing a high value for R_1, compatibly with the offset introduced by this resistance with the error amplifier input bias current.

The error amplifier sink and source currents fix the minimum load impedence that can be driven without limiting the output voltage swing. High values of these currents are important for fast load transient responses, because they allow high output voltage slew rates. Mostly a high sink current would be desirable to quickly decrease the duty cycle, for example, when a load disconnection happens, avoiding regulated output overvoltages. But the asymmetry between source and sink current must not be high in large output ripple application, because it can give an offset voltage shift with an error on the regulated output.

7.5.4 Oscillator and Feedforward Compensation

Figure 7.24 shows the waveforms of an oscillator circuit. The peak-to-peak ramp voltage depends on V_I, to implement a feedforward compensation of input voltage variation that reduces the error amplifier output changement to regulate the duty cycle. The ideal ramp voltage would be that shown by relation 7.36: $V_R = \alpha V_I$, the real one obtained by Fig. 7.24 circuit is:

$$V_R = \frac{V_I - 9V_{BE}}{9} \tag{7.46}$$

for input voltages between 15 V and 45 V.

To keep constant the frequency when V_I varies, the oscillator capacitor is charged with a current depending on the input voltage:

$$I_{CH} = \frac{V_I - 9V_{BE}}{R_{OSC}}$$

The oscillator discharge current is constant and its value is 20 mA. Oscillator charge and discharge times are:

$$T_{CH} = \frac{V_R}{I_{CH}} C_{OSC} = \frac{R_{OSC}C_{OSC}}{9}$$

$$T_{DISCH} = \frac{V_R}{20\ \text{mA}} C_{OSC}$$

and switching frequency is:

$$f_s = \frac{1}{\dfrac{R_{OSC}C_{OSC}}{9} + T_{DISCH}}$$

Switching frequency is practically independent of input voltage because the only dependence is through T_{DISCH}, that is negligible compared to charge time, as shown in Fig. 7.24.

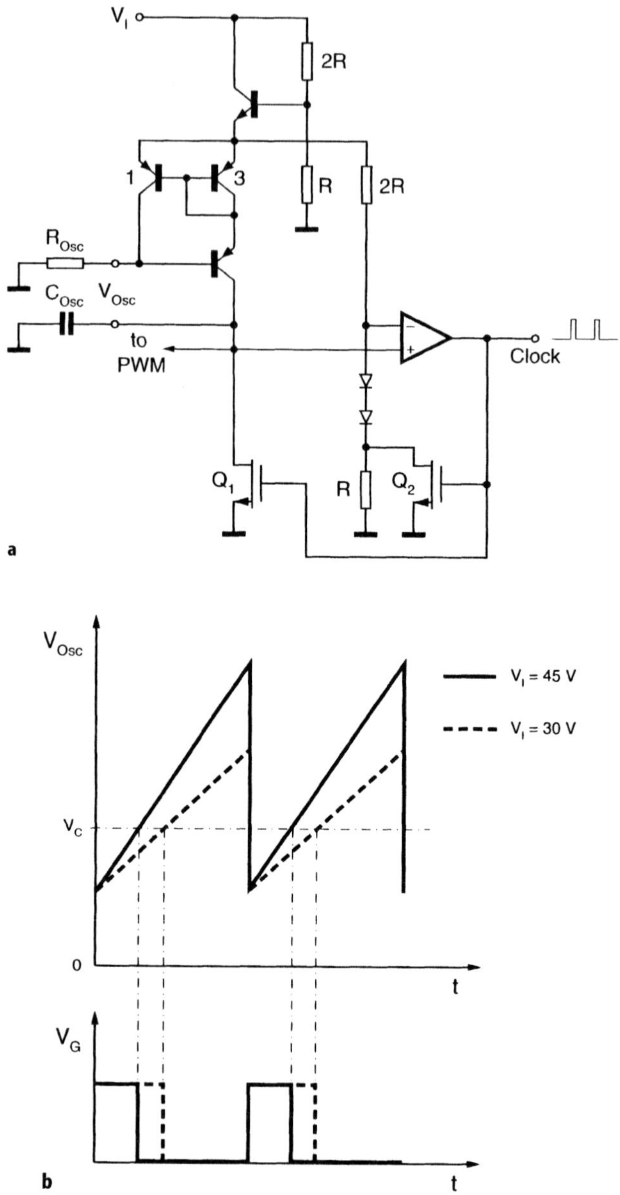

Fig. 7.24 First Example of Oscillator with Feedforward Compensation **a** and its Waveforms **b**

Substituting the V_R expression 7.46 in the loop gain relation 7.45, we find:

$$T(s) = A(s) \frac{9}{1 - 9\dfrac{V_{BE}}{V_I}} \frac{1 + s/\omega_z}{1 + 2\delta s/\omega_n + (s/\omega_n)^2} \tag{7.47}$$

The strong loop gain input voltage dependence in Eq. 7.45 makes it difficult to design an optimized compensation network. An input voltage change from 20 V to 30 V causes a 50 percent gain variation. With the same input voltage change, the voltage feedforward Control Loop Gain of Eq. 7.47 gives only a 10 percent gain variation. On small signal input voltage variation the loop gain is practically constant.

Considering the total input voltage range of 15 V – 45 V, Eq. 7.47 gives only a 40 percent gain decrease instead of the 300 percent increase as in Eq. 7.46.

Another oscillator example is that shown in Fig. 7.25, where the ramp voltage is:

$$V_R = \frac{V_I - 1}{6} \tag{7.48}$$

for an input voltage range between 8 V and 55 V.

Here the oscillator capacitor is charged directly by an external resistance connected to V_I. The ramp peak voltage is always low compared to V_I.

Charge and discharge times are:

$$T_{CH} = R_{OSC} C_{OSC} \ln \frac{6}{5}$$

$$T_{DISCH} \cong K C_{OSC}$$

$$K \cong 100 \quad [\Omega]$$

and the switching frequency is:

$$f_S = \frac{1}{R_{OSC} C_{OSC} \ln \dfrac{6}{5} + K C_{OSC}}$$

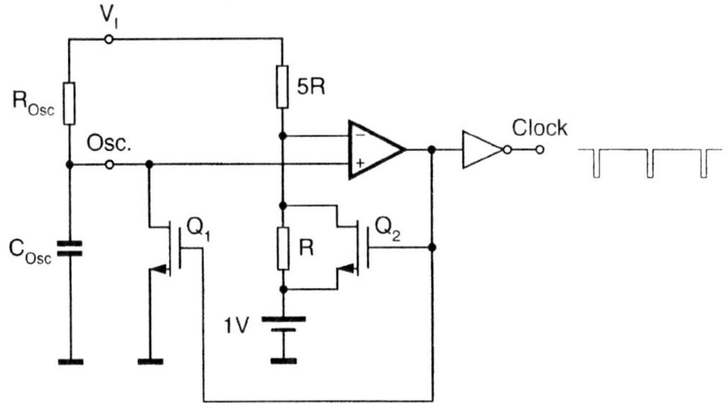

Fig. 7.25 Second Example of Oscillator with Feedforward Compensation

Substituting the ramp expression 7.48 in the regulator loop gain 7.45 we obtain:

$$T(s) = A(s)\frac{6}{1 - \frac{1}{V_I}}\frac{1 + s/\omega_p}{1 + 2\delta s/\omega_n + (s/\omega_n)^2}$$

with a loop gain variation of about 11 percent for the entire V_I range.

The error amplifier output range should be larger than the oscillator ramp to swing from zero to the maximum value of the duty cycle, but only slightly larger to avoid delays. Also using a feedforward technique this is possible simply by limiting the error amplifier output range with a clamp that increases proportionally to the input voltage as the oscillator ramp.

Another technique to keep the loop gain independent of input voltage is considered in [10]. This technique has not a feedforward effect because the oscillator does not change its slope changing its peak value, but has different slopes on its

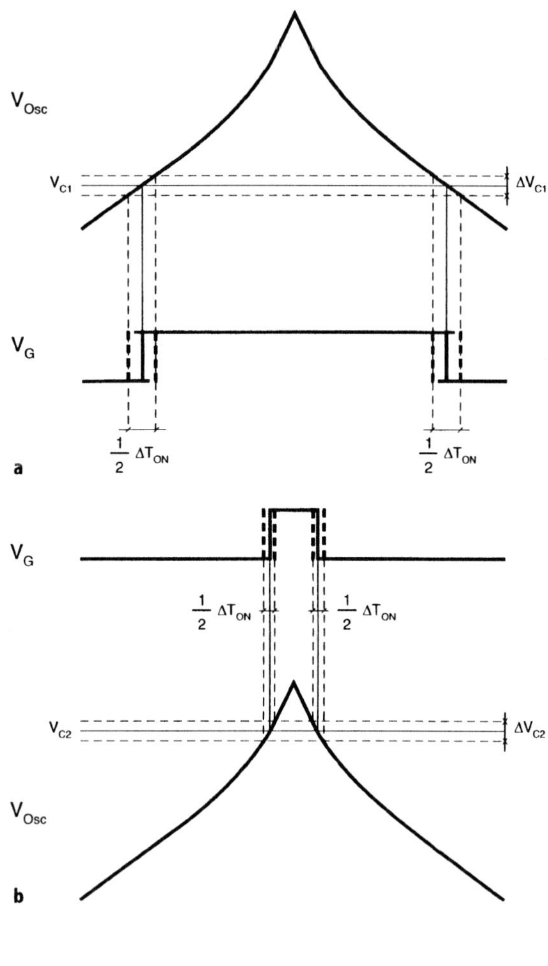

Fig. 7.26 Oscillator Waveform to have a Loop Gain Independent of Input Voltage

waveform, as shown in Fig. 7.26. It is linear only on its lower part, where the slope is minimum. It becomes exponential and its slope progressively increases towards the top.

The slope of the line tangential to each waveform point is inversely proportional to the duty cycle change due to a small control voltage variation.

For low input voltages the duty cycle is high and the control voltage intersects the oscillator waveform where the duty cycle variation is maximum.

At high input voltages duty cycle is low and its change with the control voltage is minimum.

This means having a ramp voltage proportional to V_I in Eq. 7.45, and, at least at first approximation, a loop gain independent of V_I.

7.5.5 PWM Latch

In noisy regulator ambient it is better to latch the switch off command, as shown in Fig. 7.18. The reset signal is a clock synchronous with the oscillator. In the examples shown in Figs. 7.24 and 7.25 it is generated by the oscillator discharge.

The regulator can need a duty cycle value equal to zero, thus if an off-command remains on PWM output during the clock time the flip-flop cannot be reset and the switch does not turn on.

This can be risky in regulators working with small duty cycles, because the PWM comparator allows the turn-on of the switch at the end of the oscillator discharge, when also the clock pulse to reset PWM latch ends. To avoid that the regulator does not reset the switch-on and that it works at one half the switching frequency, the clock pulse end at flip flop input is delayed or, with circuital arrangements, the oscillator discharge end is anticipated at PWM input.

Current mode regulators have a similar problem when they work near the maximum duty cycle, because the clock reset pulse must be longer than the output stage switch-off delay, plus the PWM comparator delay time in changing its status after the power transistor turn-off.

7.5.6 Power Stage

The output stage with power switch and its driving circuitry is very important to obtain good performances at high switching frequencies.

BCD processes DMOS transistors are well suited for regulator switches. They can withstand high voltages and high currents without second breakdown problems, DMOS have good R_{DSON} with low gate charge and fast gate charge paths for high switching speed.

BCD processes availability of both p and n high voltage components is very important to optimize the driving stage.

Driving techniques are various. Figure 7.27 shows one example of a power stage circuit. A Darlington NPN pair Q_1, Q_2, supplied by a current generator Q_5,

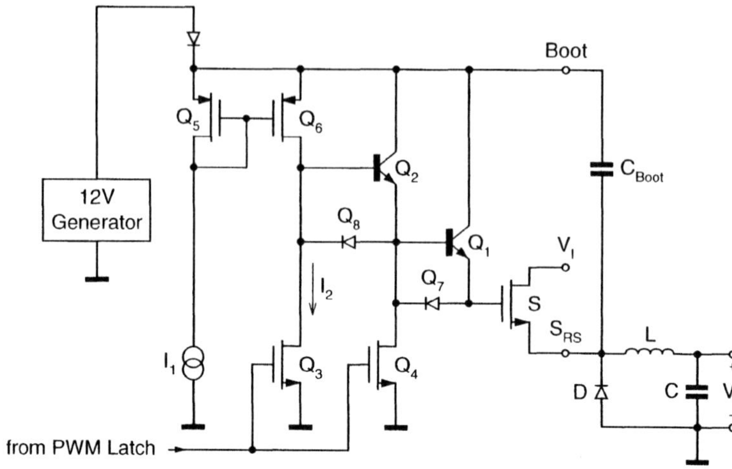

Fig. 7.27 First Example of Power Switch Driving Stage

Q_6, turns on the power DMOS S. When the switch S is on, the charge required from CBOOT is only the P-channel diode bias current I_1.

A DMOS pair Q_3, Q_4 switches off the power transistor through diodes Q_7 and Q_8, to interdict first of all, the transistors Q_2 and Q_1. When the switch S is off, the driving stage has a current consumption equal to $I_1 + I_2$, that does not discharge the bootstrap capacitor, because it is supplied by the internal regulator while the load current flows in the freewheeling diode.

Figure 7.28 shows the L4970 switching regulator photograph. The technology is BCD of the first generation. The power switch is a 13 mm², 60 V VDMOS, with a 100 mΩ R_{DSON} for a 10 A load current. It has the driving stage shown in Fig. 7.27 with $I_1 = 400\ \mu A$, $I_2 = 2.1$ mA, Q_1 and Q_2 areas 45 and 125 times the minimum NPN area. The total driving area is 1 mm². Rise and fall times have a typical value of 50 ns.

This regulator, designed in the mid-eighties has excellent features compared to regulators of that time made with bipolar power stages. This circuit can supply a load current of 10 A. Figure 7.29 a shows a switch drop-out comparison between the bipolar regulator L296 and the BCD one L4970 versus their load current. At a junction temperature of 125 °C, the first one has a 2.1 V drop with 4 A, the second a 0.9 V, and a 1.65 V drop with 10 A. Figure 7.29 b compares their power dissipation related to the input voltage V_I: it is evident the improvement.

The switch-off technique shown in Fig. 7.27 pulls down the power transistor gate to ground. It is an extremely fast turn-off method because the gate source voltage becomes negative, but it has two main drawbacks: it wastes the gate charge and disturbs the ground path with high current spikes. To avoid signal ground pollution, a specific power ground pad can be provided.

For very low threshold power MOS using this technique there could be the risk of turning on again the switch when its source becomes negative compared

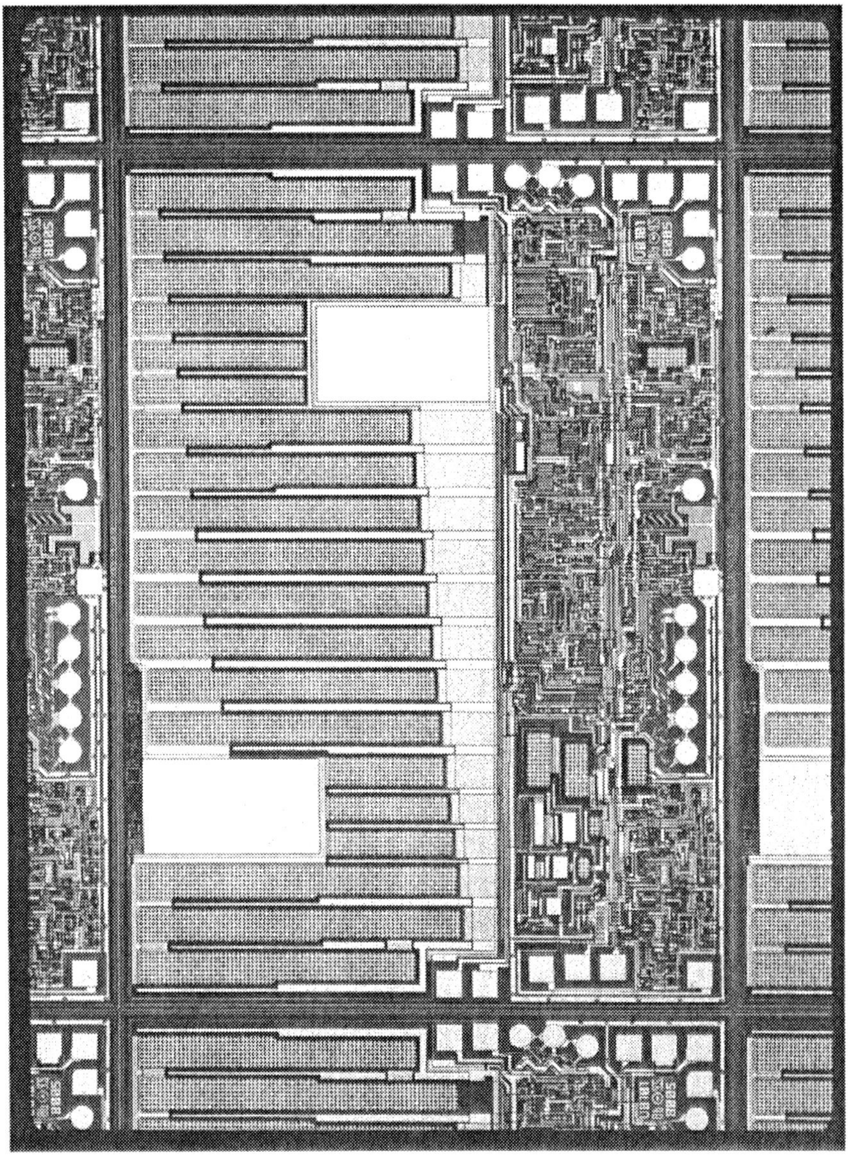

Fig. 7.28 L4970 Switching Regulator Photograph

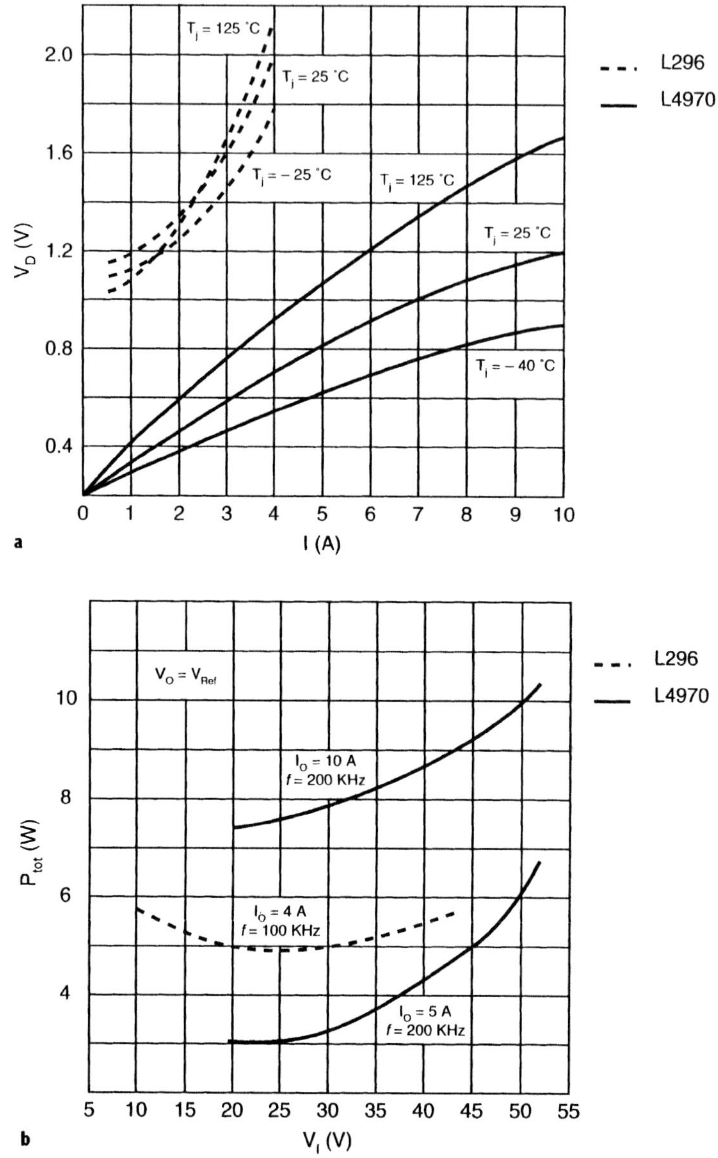

Fig. 7.29 Power Switch Dropout Comparison **a** and Power Dissipation versus Input Voltage Comparison **b** between L296 and L4970

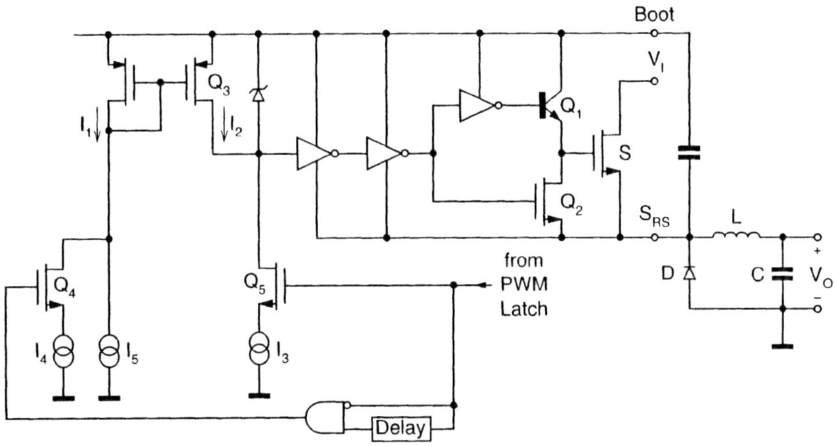

Fig. 7.30 Second Example of Power Switch Driving Stage

to ground. This is not the situation of the L4970 regulator that has a 3 V threshold VDMOS.

Figure 7.30 shows a driving stage where the power DMOS is switched off by a transistor Q_2 inserted between its gate and source. In this way, the gate charge is not dissipated to ground but delivered to the output.

The power DMOS driving signal is amplified by some inverters and the PWM signal is applied to a small DMOS transistor Q_5 with little ground disturbance. The power switch turns on when Q_5 switches on and it is switched off by a P-channel current generator Q_3.

In this way, the driving signal to the floating stage first inverter is opposite compared to the power DMOS source edge and the rise and fall of the switch source confirm the driving command. If the first inverter driving polarity was the same as that of power DMOS source, oscillations could appear on the switching edges.

When transistor S is on, the bootstrap capacitor supplies a current $I_I + I_3$. For a fast switch-off, the currents I_1 and I_2 increase for a short time. After this current pulse a little current I_1 flows. This is important to avoid the discharging of the bootstrap capacitor C_{BOOT} at light loads, when power switch remains off for long times, and its source goes to the output voltage and the bootstrap voltage moves together with source. In this situation, it is not the voltage generator, but C_{BOOT} itself that supplies the off-current I_1.

This kind of output stage is used, for example, in the L4973 switching regulator, where the current required from C_{BOOT} during T_{ON} is $I_1 + I_3 = 400\ \mu A$, the switch-off current pulse $I_1 + I_2$ is equal to 800 μA for 100 ns, and T_{OFF} current consumption I_1 is 20 μA.

The device L4973 is made using BCD II technology. The switch transistor is a 60 V VDMOS with a R_{DSON} of 150 mΩ that can supply 3.5 A. The components supplied between C_{BOOT} and the switch source are 12 V logic MOS, plus an NPN driver Q_1.

N-channel power switch thresholds are usually low related to the necessary gate-source voltage to optimize their R_{DSON}. Turn-on is not critical because the switch threshold voltage is when the transistor Q_1 collector emitter voltage is high and its current is maximum. Therefore its dimension can be small. In new generation BCD, a P-channel transistor can be used instead.

The drawback of the driving stage shown in Fig. 7.30 consists in the need of a very low switch-off MOS Q_2 resistance, because it limits the final DMOS gate discharge current and with low thresholds, its switch-off time. For the same reason this output stage configuration needs a very low switch-gate resistance, too.

In regulators that need a very high efficiency even for light loads of some mA, the driving signals are latched by a flip-flop inserted between C_{BOOT} and the power transistor source. There are current pulses only to switch the regulator and there is no current consumption after switching. They are made with the same techniques used in very high voltage converters described in Chap. 8.

The choice between more complex energy saving circuits or simpler driving solution obviously depends on the regulator application.

The bipolar technology regulators had epitaxial pockets connected to the power stages nodes that reach a negative voltage during freewheeling time, therefore they had to be protected against the parasitic npn elements described in Chap. 5.

In BCD technologies the availability of different kinds of components allows to not connect epi pockets to the switch source, thus this kind of parasitic effect can be eliminated.

Unluckily very fast high current switchings have increased high frequency substrate noise, as described in [11] and [12]. This phenomenon disturbs the devices because of the capacitive coupling with epi pockets, when they become more sensitive to noise as their operative current level decreases.

7.5.7
Power Switch Gate Supply Voltage

High-side DMOS needs a gate supply voltage that is 10 V higher than V_I. This auxiliary supply can be provided using a step-up, a diode capacitor voltage multiplier or a bootstrap capacitor, as described in Chaps. 5 and 6.

In low input voltage regulators working at low switching frequencies the last two techniques can be combined.

At high frequencies, diode-capacitor multipliers could not be able to provide, with a reasonable area and efficiency, the charge lost by the bootstrap capacitor to supply the gate. In this case a step-up solution becomes necessary. The same happens when a chip includes more than one regulator.

With a step-up converter, the driving stage voltage must be clamped at the source switch, as shown in the simplified schematic in Fig. 7.31, and not to exceed its gate-source maximum voltage.

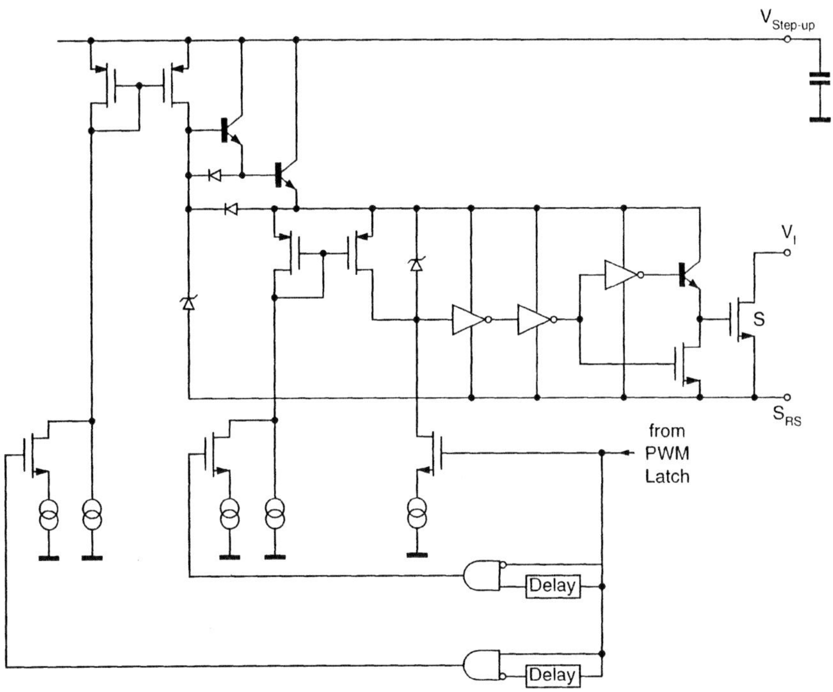

Fig. 7.31 Power Switch Driving Stage with a Step-up Supply

When the bootstrap technique is used, the regulator needs a long enough freewheeling time T_D to recharge the bootstrap capacitor through the diode. This puts a limit to the minimum load current when the regulator works at high switching frequency, because at light loads the regulator works in Discontinuous Mode with very short T_{ON} and T_D.

To solve this problem suitable circuits can be implemented. An internal switch can be used to provide the necessary recharge when the bootstrap voltage becomes too low, or the regulator can work at lower frequencies, in Pulse Skipping Mode as described in paragraph 7.6.3, with efficiency improvement.

7.5.8
Current Protection

The Buck Converter, with its power switch connected to the input voltage, can be well protected against dangerous overcurrents by opening the switch.

Voltage Mode regulators need current limitation circuits. One example is shown in Fig. 7.32. Current sensing is made on a metal resistor R_{Sns} in series with a DMOS switch partition, to minimize the sensing resistor contribution to switch drop and power dissipation. However the Q_2 area cannot be much smaller

Fig. 7.32 Example of Current Protection Circuit

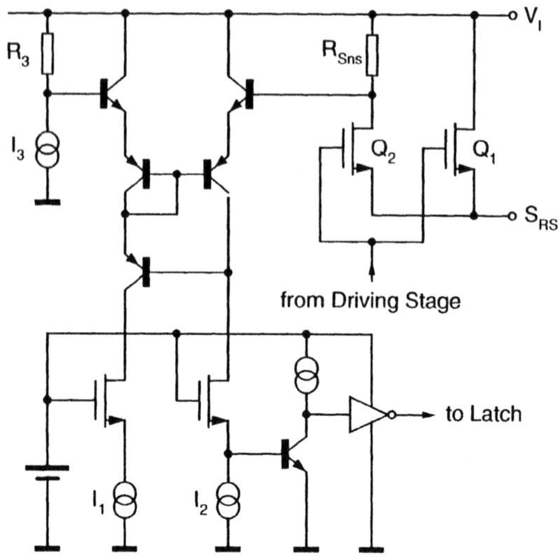

than the Q_1 area to have a good match between the two DMOS parts. Q_2 R_{DSON}, R_2, can be, for example, 8 times Q_1 R_{DSON}, R_1, and R_{Sns} 1.3 times Q_1 R_{DSON}. The current limit threshold is:

$$I_{CC} = \frac{R_3 I_3}{R_{Sns}} \left(\frac{9R_1 + R_{Sns}}{R_1} \right)$$

Current limitation often needs an input filter or a logic output mask to blank unwanted turn-off due to current spikes when the power transistor is switched on.

Current limitation and other protection outputs are usually connected to the PWM flip-flop and their switch-off signals are latched.

To implement current limitation it is also possible to sense directly the drop between DMOS source and drain, without a sensing resistor, but masking the current protection output signal in the switch-off time, when this drop becomes high. For low input voltages, it is simple to use this method, but when they are high, the circuit needs high voltage switches to protect the comparator sensing input and it becomes more complex.

Whatever the current protection circuit, after the DMOS switches off, any current information disappears. In the next cycle the power transistor turns on again and immediately switches off if the fault is still present.

A Current Mode regulator that senses current in series with the switch, or on the switch itself, has an intrinsic "pulse-by-pulse" current limitation, whose threshold depends on the error amplifier maximum output voltage.

Both kinds of regulators, Current or Voltage Mode, to protect themselves from heavy short circuit conditions could need a further protection.

During short circuit the output voltage becomes very low. Looking back at Eq. 7.4 the inductor current discharge slope is proportional to V_o, and for low

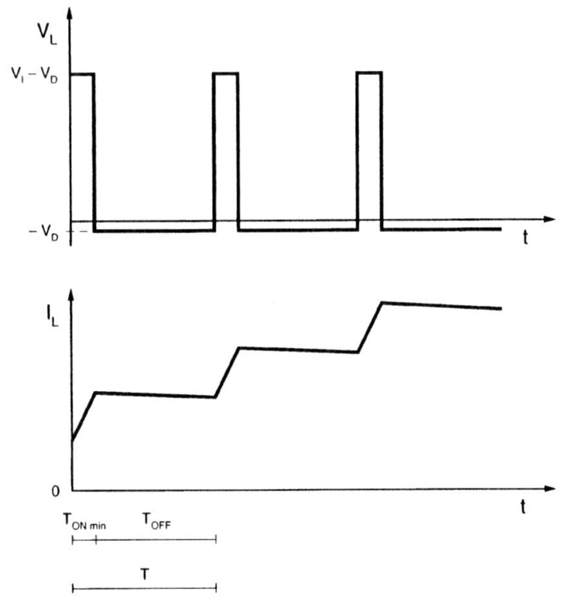

Fig. 7.33 Pulse by Pulse Current Protection with an Heavy Short Circuit

V_O this slope becomes very small. The charge slope is proportional to $V_I - V_O$, for low V_O becomes high.

In this case to turn on the power switch in every period could result in a continuous output current increase as shown in Fig. 7.33, up to a value well higher than the protection threshold. The situation becomes more and more critical with input voltage increase.

We will consider an extreme case of a nil residual output voltage, with discharge slope depending only on diode drop and neglecting other parasitic effects. As it can be seen from Eq. 7.8, with $V_I = 40$ V, at a switching frequency of 200 kHz, to keep a constant short circuit current a minimum on time equal to:

$$T_{ON} = \frac{0.5}{40} \cdot 5 \cdot 10^{-6} = 63 \text{ ns}$$

is needed. This time includes current protection and output stage propagation delays and can be difficult to obtain, also because of switch on leading edge blanking.

The Current Mode regulators that control the inductor current in all the switching period, by a sensing resistor between the inductor and the regulated output, or in the ground path, do not have this problem. Because the power switches cannot turn on as long as inductor current discharges to the current limit threshold. But they require at least one additional pin and they have a larger dissipated power.

Without a full period current control it is possible to protect regulators against heavy short circuit in different ways, used both in Voltage and Current Mode regulators.

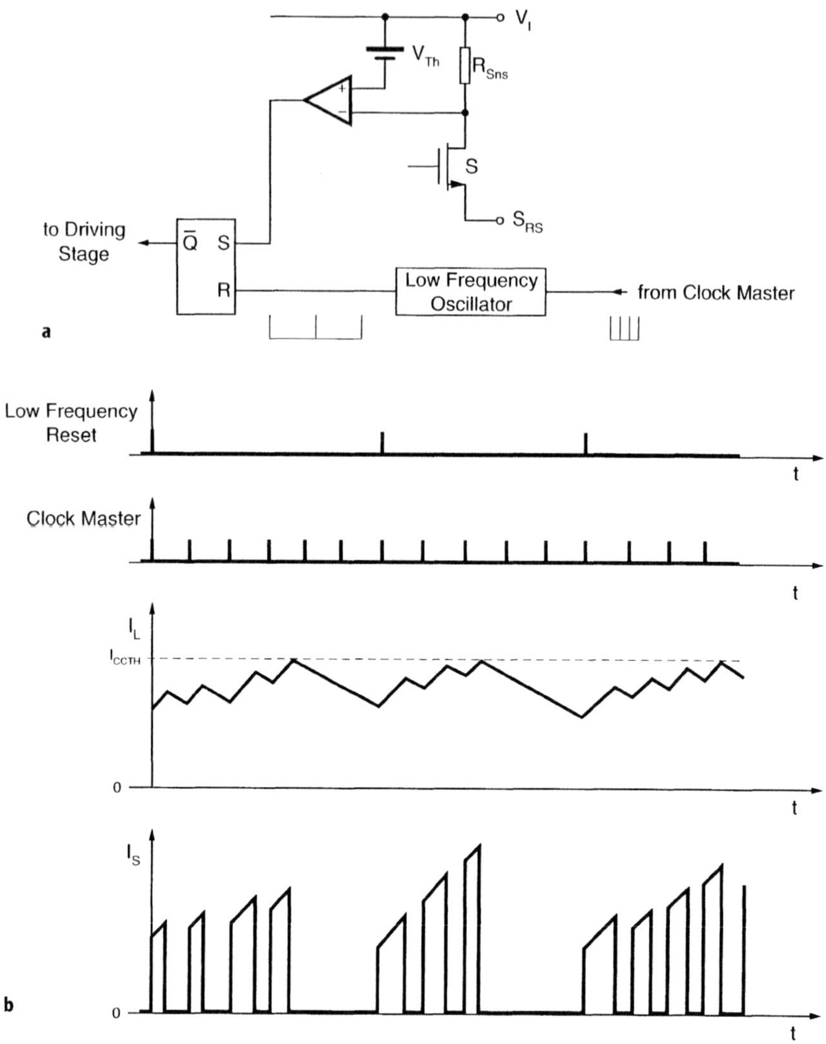

Fig. 7.34 Short Circuit Protection with Switching Frequency Decreasing **a** and its Waveforms **b**

The first one is to reduce the switching frequency when the current limit threshold is reached. The method is shown in Fig. 7.34: when the comparator senses an overcurrent, it sets the flip-flop, that switches off the power transistor and keeps it off until a reset clock pulse comes from an internal low frequency oscillator, different from the main one.

After the reset pulse, the regulator restarts to switch at its nominal switching frequency. If the overload condition is still present, the protection cycle is repeated.

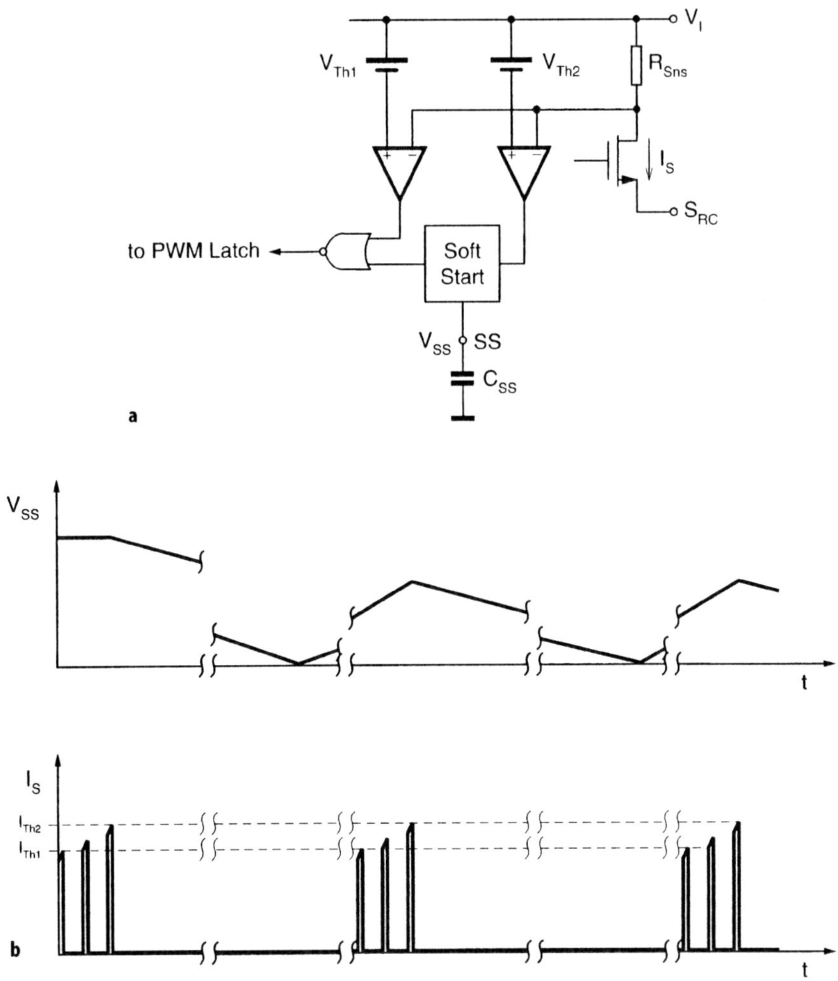

Fig. 7.35 Hiccup Current Protection **a**, with its Waveforms **b**

This kind of protection requires that the low frequency current ripple on the maximum value of the load current does not exceed the current limitation threshold, to not lose the possibility of returning to the correct value of the switching frequency after an overload.

A second protection method shown in Fig. 7.35 is called "Hiccup" current limitation. It is usually added to a pulse-by-pulse current protection. When an overload happens the last one reduces the switch-on time till its minimum, depending on transmission delays. If the peak current still increases and reaches the threshold of the second comparator, it sets the soft start latch, that discharges the soft start capacitor with a current generator and keeps in the off-state the regulator power switch as long as it is fully discharged and a new soft start phase begins.

If the short circuit is still present the regulator works at very low frequency, whose value depends on soft start slopes.

7.5.9
Soft Start

As explained in paragraph 7.5.2, Buck regulators usually need a soft start up to charge the output capacitor. In Fig. 7.36 a soft start circuit is shown together with its waveforms. The soft start capacitor C_{SS} is charged with a constant current up to the generator voltage. In the charge time the error amplifier output voltage, V_C, is clamped by a PNP transistor Q_1 to a voltage that is one base emitter voltage higher than the soft start pin.

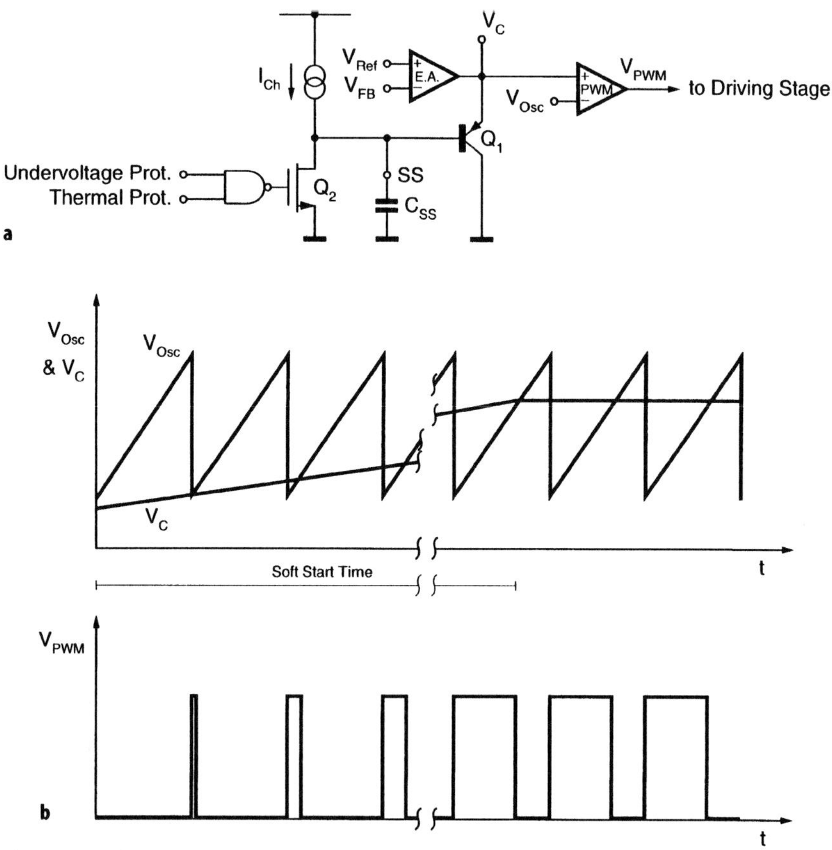

Fig. 7.36 Soft Start Circuit a and its Waveforms b

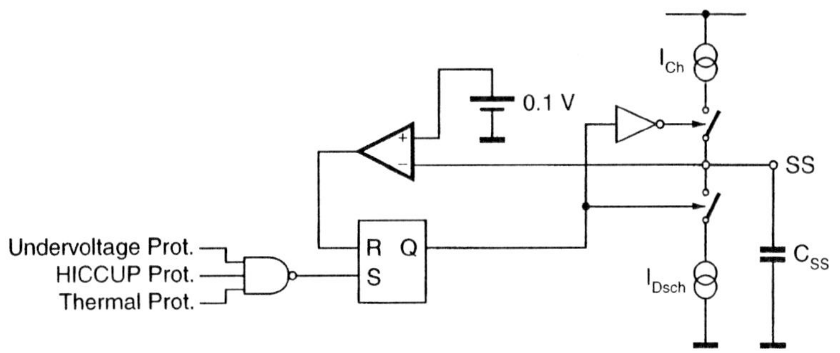

Fig. 7.37 Soft Start Circuit with Latched Discharge

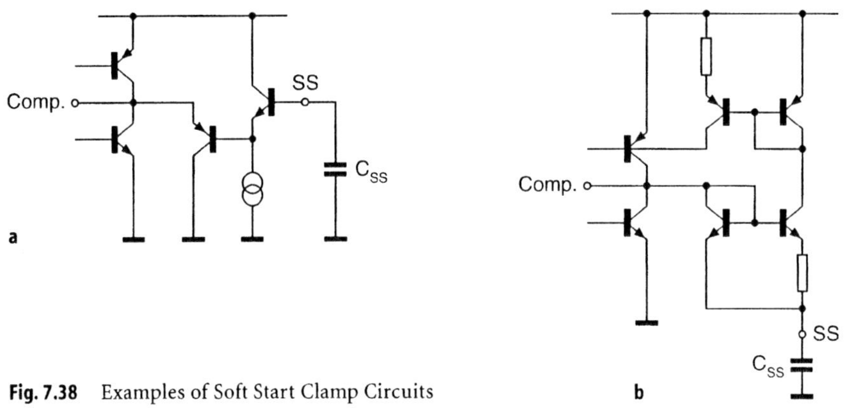

Fig. 7.38 Examples of Soft Start Clamp Circuits

When V_C crosses the oscillator ramp, The PWM comparator begins to switch on the power transistor, slowly increasing its on-time. It reaches its nominal duty cycle and the soft start capacitor ends its charging process without influencing error amplifier output.

If a fault happens as an undervoltage or a thermal problem, C_{SS} is quickly discharged by transistor Q_2. When the fault signal ends the C_{SS} charging process re-starts.

Sometimes, the soft start capacitor is discharged slowly, for example, by a current generator as shown in Fig. 7.37. To use the C_{SS} fall time to implement some protections, as the hiccup current limitation described in paragraph 7.5.8. In this case, the fault signal is latched to guarantee the C_{SS} complete discharge if it disappears.

If the error amplifier output has a high sink current, the solution with a PNP clamp cannot be adopted because the PNP base current becomes comparable with the soft start charge current. A PNP Darlington pair usually gives too much voltage drop if compared with the minimum oscillator voltage. In this case, different clamp solutions are adopted, as those shown in Fig. 7.38.

Alternatively, if possible, C_{SS} clamps an error amplifier internal node, with the advantage that the regulator compensation capacitor is discharged by the error amplifier output source current instead of by the soft start through its clamp circuit.

This solution also allows the error amplifier output voltage to have a useful level lower than one V_{BE}.

Another possible soft start solution is to use the pulse-by-pulse current limitation, described in paragraph 7.5.8, to keep the duty cycle low, by a slow increase of its threshold up to its nominal value. But this method can be adopted with success only if the load is applied after the output voltage rising to its final value, especially if the load is more similar to a current generator, independent of the output voltage, than to a resistor, because the regulator cannot charge the output capacitor until the inductor current is higher than the load current.

7.6 Switching Regulators Efficiency

High efficiency is one of the most important features of a switching regulator. Power is lost partly on the chip and partly on the external components. All the parasitic resistances give their contribution to dissipation, but the main causes are the diode and the switch. In this paragraph the simplified expression of their power losses in a Buck Converter are examined. Firstly we assume that ripple current is negligible related to average inductor current.

On both switch and diode we have the conduction losses:

$$P_{S1} = R_S I_O^2 D$$

$$P_{D1} = V_D I_O (1 - D)$$

On the switch there are the switching losses due to its rise and fall times, T_R and T_F:

$$P_{S2} = \frac{V_I I_O}{2} (T_R + T_F) f_s$$

and the power losses due to gate charge, Q_G:

$$P_{S3} = V_I Q_G f_s$$

Some charge is also lost on the diode, for its activation and recovery. Another power loss is due to the regulator circuit current consumption, I_B:

$$P_B = V_I I_B$$

The contribution of the various dissipation factors to the total regulator efficiency is different in the case of heavy or light load currents.

7.6.1 Heavy Load Currents

In regulators working at high load currents, to decrease the power dissipation is important to minimize the DMOS switch resistance, mostly at high duty cycles.

To improve efficiency they also need to reduce the switching times by decreasing the parasitic capacitance and by optimizing the paths to DMOS cell gates.

The reduction of switching losses is very important in high input voltage regulators and to increase the frequency.

The gate charge losses become significant when the frequency increases. In order to decrease them is necessary to reduce the gate capacitor and to use high efficiency methods to charge the gate.

BCD technology evolution has allowed to reduce all these various contributions to power dissipation. The availability of components with good behaviour at very small currents, together with new circuital techniques, has also decreased the dissipation due to chip current consumption. In ten years it has been reduced from some tens of mA to a few mA in high input voltage regulators and to less than 1 mA in the low voltage ones.

Efficiency improvement has made it possible to move from power packages such as multiwatt, with big heatsinks, to smaller dual in-line ones with a few cm^2 of copper and finally to surface mounting packages.

The table in Fig. 7.39 shows a power loss comparison among three similar 3.5 A, high voltage, Buck regulators. The first L296 is a bipolar one designed in the eighties. The second and third, L4974 and L4973, are BCD first and second generation chips and power losses have been reduced as shown, allowing the described package improvement.

A further significant efficiency improvement can be obtained by substituting the external diode with a low resistance synchronous rectifier power switch S_2, as shown in Fig. 7.40.

Regulator circuitry becomes more complex because the synchronous transistor S_2 needs a driving circuit. The power stage is relatively simple because it is related to ground.

The S_2 switch turn-off transistor must have a very low resistance, not only for fast switching off, but also to avoid that S_2 goes on again when its drain rises. This because the charge current of S_2 drain-gate capacitor goes to ground

HIGH VOLTAGE BUCK REGULATORS EVOLUTION

DEVICE	TECHNOLOGY	f_S (KHz)	V_I (V)	V_O (V)	I_O (A)	P_{tot} (W)	PACKAGES
L296	HVO	100	35	5	3.5	4.5	MULTIWATT
L4974	BCD I	100	35	5	3.5	2.6	DIP 16+2+2
L4973	BCD II	100	35	5	3.5	1.3	SOP 12+4+4 DIP 12+3+3

Fig. 7.39 High Voltage Regulators: Power Dissipation Comparison

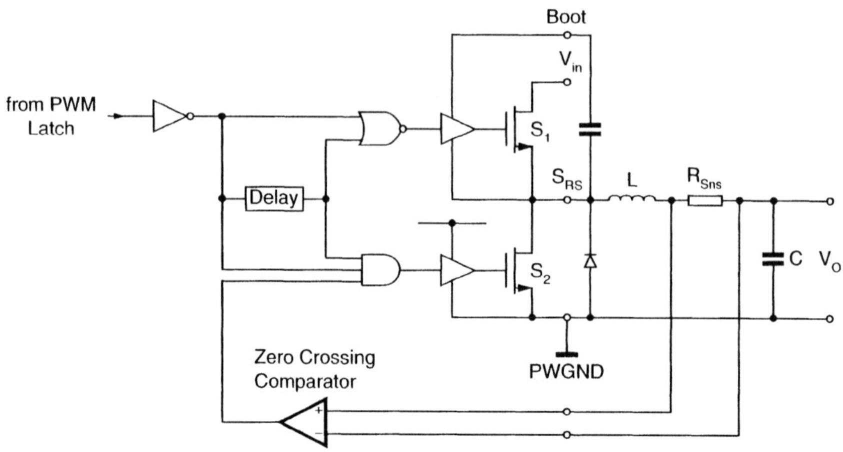

Fig. 7.40 Buck Regulator with Synchronous Rectifier

through the turn-off transistor and it must not cause a gate voltage increase high enough to switch on again S_2.

When a synchronous rectifier is used, it is necessary to insert delays between each power transistor switch-off and the other one switch-on to avoid cross-conduction currents. For this reason, switch S_2 has a Schottky diode in parallel, to conduct the inductor current during these delay times.

If the regulator can go in Discontinuous Conduction Mode, the synchronous rectifier MOS must be switched off as soon as its current goes to zero. This is necessary to avoid inverting the inductor current and discharging the output.

Attention must be paid to the chip layout design because to integrate a synchronous DMOS rectifier means to quickly switch a big epi pocket from a high input voltage to a negative one.

7.6.2 Integrated Double Regulator

Thanks to efficiency improvements, it is possible to integrate more than one regulator into the same chip. One example is the L4985 double regulator whose photograph is shown in Fig. 7.41. It is a BCDII, 19 mm^2 chip. Its block diagram is shown in Fig. 7.42. It integrates a 3 A Buck regulator with a 70 mΩ, 30 V LDMOS switch, that also has a gate driving section for an external synchronous rectifier MOS. The second regulator is a 5 W Boost Converter. It has a 1 Ω, 40 V VDMOS switch connected to ground, that can also be used in topologies with a transformer.

The IC input voltage range is 4.5 V – 22 V. It has an internal low power step-up converter to provide the necessary supply voltages to circuitry. The device includes many of the features described in the previous paragraphs.

Fig. 7.41 Integrated Double Regulator

One application board is at a switching frequency of 80 kHz, with $V_I = 14$ V, $I_{01} = 3$ A, $V_{01} = 5$ V for the Buck converter with a synchronous rectifier, $I_{02} = 0.21$ A, $V_{02} = 24$ V, for the Boost one, with a total output power of 20 W. The measured efficiency is 92% with 1.6 W dissipation. The IC is available in So20 package.

7.6.3 Light Load Currents

In portable equipment, such as notebooks and cellular phones, high efficiency is necessary with very low currents levels, too. Because they must operate for very long periods at minimum loads without discharging the batteries.

In this situation conduction and switching losses are no longer dominant, while gate charge losses and circuit bias currents become more significant.

To keep the efficiency high at low load currents, the regulator does not work in Discontinuous Mode at its nominal frequency but switches at a very low frequency, in Pulse Skipping Mode, also called "Burst Mode", whose waveforms are shown in Fig. 7.43.

Usually the Pulse Skipping frequency is not completely free, because the power transistor switching on can remain synchronous with the main oscillator clock.

In the Pulse Skipping Mode the switch and diode currents during T_{ON} and T_D are bigger than those in DCM, but the idle time T_I is so high that their contribu-

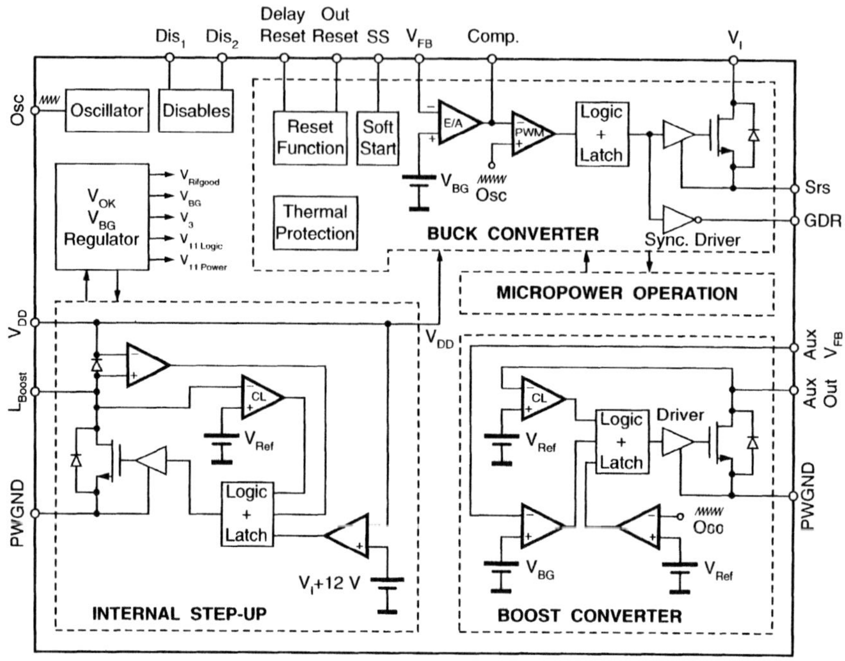

Fig. 7.42 L4985 Block Diagram

tion remains negligible. All switching losses, including those of the transistor gate and diode are also negligible.

Now the way to keep the efficiency high is to reduce as low as possible the circuit current consumption, sometimes also disabling the less important circuits when loads are light. Today, the regulators for portable equipment have some hundreds of μA bias currents, while the new generation would have less than one hundred μA.

Actually, working in Burst Mode, efficiencies higher than 80 % on a 5 V, 10 mA output can be obtained.

One method for operating in Pulse Skipping Mode is to use a comparator with a threshold higher than the nominal regulated voltage. When the load current decreases the output voltage rises over this threshold. If this happens, the error amplifier and PWM stage are disabled and this comparator with a little hysteresis regulates the output voltage. When the switch current increases over a threshold value the system goes back to the PWM mode.

It is possible to move from PWM to Burst Mode also sensing the regulator on-time. When it becomes smaller than a reference value, the Pulse Skipping Mode begins. When the pulse frequency increases over a threshold, the regulator goes back to PWM mode.

Another way of forcing the regulator to switch at low frequency when load is light is to keep on the power switch till when its current is lower than a given

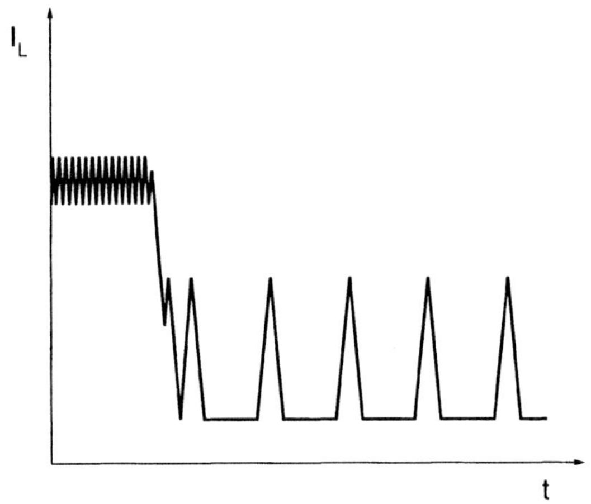

Fig. 7.43 Inductor current from PWM to Pulse Skipping Operation Mode

threshold, independently of the error amplifier status. This is practically a constant T_{ON} operation mode at light loads.

References

1. Burgio C, Cammelli P (1985) La Dinamica dei Regolatori Commutati a Modulazione di Durata. Politecnico di Milano
2. Anunciada V, Silva MM (1990) New Constant-Frequency Current Mode Control for Power Converters, Stable for all values of Duty Ratio, and Usable in All Four Quadrants. IEEE Transaction on Industrial Electronics, vol 37, no 4
3. Calderoni L, Pinola L, Varoli V (1992) Optimal Feed-Forward Compensation for PWM DC/DC Converters with Linear and Quadratic Convertion Ratio. IEEE Transactions on Power Electronics, vol 7, no 2
4. Marro G (1982) Controlli Automatici. Zanichelli
5. Brown AR, Middlebrook RD (1981) Sampled – Data Modeling of Switching Regulators. Proceedings of the IEEE Power Electronics Specialist Conference
6. Vorperian V (1990) Simplified Analysis of PWM Converters Using Model of PWM Switch. Part 1: Continuous Conduction Mode. IEEE Transactions on Aerospace and Electronic Systems, vol 26, no 3
7. Ridley RB (1991) A New, Continuous-Time Model for Current-Mode Control. IEEE Transactions on Power Electronics, vol 6, no 2
8. Gregorian R, Temes GC (1986) Analog Mos Integrated Circuits for Signal Processing. John Wiley & Sons
9. Shrager A (1993) High Voltage Current Mode Controlled TWT Power Supply Design, Simulation and Implementation. Official Proceedings of the Twenty-Seventh International Power Convertion Conference
10. Scolio J (1988) New Switching Regulator's Unique Compensation Scheme Assures Design Simplicity. Official Proceedings of the Sixteenth International PCI Conference
11. Rainal AJ (1994) Eliminating Inductive Noise of External Chip Interconnections. IEEE Journal of Solid State Circuits, vol 29, no 2
12. Su DK, Loinaz MJ, Masui S, Woley BA (1993) Experimental Results and Modeling Techniques for Substrate Noise in Mixed-Signal Integrated Circuits. IEEE Journal of Solid State Circuits, vol 28, no 4

CHAPTER 8

High Voltage Integrated Circuits for Off-Line Power Applications

C. Diazzi

8.1 Introduction

Power integrated circuits have made an enormous progress in the last 10 years, thanks to the impetus given by the introduction of switching techniques for the control of power (low dissipation), and the availability of BCD (Bipolar, CMOS, DMOS) mixed technologies.

BCD technologies allowed high integration density on the signal digital/analog control sections, together with the implementation of the power DMOS components on the same chip, with its well known characteristics of high efficiency and ruggedness. This opened new application fields for the so called SMART POWER devices, and allowed the integration of power levels never reached before with classical Bipolar IC technologies.

Nevertheless "OFF-LINE APPLICATIONS" have been, up to a few years ago, the domain of discrete devices due to the cost of high voltage integration. This chapter illustrates how recent technological progress allowed to obtain ICs with breakdown voltages in excess of 600 V, using standard junction isolated techniques, which made it possible for ICs to become competitive even in those high voltage applications traditionally reserved to power discrete (Bipolar and MOS), and passive components (transformers).

The "off-line environment" is described, together with the consequent demand, in terms of voltage capability, on the silicon according to the application. Some examples are given of high voltage integrated circuits for off-line applications: switch-mode power supplies, power factor controllers, motor drivers, lighting. In particular, circuital techniques addressed to high efficiency, high voltage driving are described.

8.2 Technological Introduction

The possibility of high voltage integration became possible with the introduction of the DMOS component as the element able to withstand high voltage, together with low voltage Bipolar and CMOS devices for the implementation of analog and digital control sections on the same chip. Typically three types of isolation techniques are used for high voltage IC implementation: Junction isolation, Self-isolation, Dielectric isolation. The first two techniques are today the most utilized by the industry due to their more economical approach, as to the Dielectric isolation approach, even if the technological development could change the above situation in the near future, especially for very high voltage technologies from 300 V to 700 V.

Self-isolation is typical of MOS processes, where the isolation between components is automatically obtained by circuit layout and no purposeful junctions are provided. Such approach is very simple and economical; the trade-off is the limited flexibility in terms of devices available and in terms of architectural implementation of the circuit. Typically such approach leads to CD (CMOS-DMOS) processes, where the CMOS is used for the control sections and the DMOS is the high voltage power component.

Junction isolation is a well known technique extensively used in the production of Bipolar ICs, where reverse biased PN junctions are used to separate components. This is a flexible approach in which Bipolar, CMOS, DMOS devices can be implemented on the same chip (BCD technologies). From the architectural point of view of the circuit there are much less constraints compared to self-isolation techniques, both for the control sections, and for the power sections, allowing the implementation of flexible power stages: low-side, high-side, multiple power stages,... Both self-isolation and junction isolation approaches have the limitation of parasitic interactions between components like substrate injection, latch-up, dV/dt.

From the point of view of current flow in the power device three approaches are possible: vertical, lateral-vertical, lateral.

Vertical current flow is typical of discrete devices, and the utilization of this approach in integrated form is suitable for very high current applications ($\geq$10 A). In this case the configuration for power stages is limited to one or more power devices with common drain.

Vertical-lateral current flow is the most extensively used in integrated circuits with voltage rating from 20 V to 250 V, and with average current rating (1 A – 10 A). There is no limitation about the topology of the power stages that can be implemented.

Lateral current flow is typical of high voltage integration (300 V to 700 V) and it is suitable for current levels up to a few Amps. The topology of the output power stage is limited to one or more source grounded power devices.

For a detailed analysis of the above subjects refer to chapter 1, where ICs technologies are treated, and to references [1, 2].

8.2.1 IC Technologies for Off-Line Applications

Off-line applications require, as will be discussed later from an application point of view, voltage capability in the range of 350 V to 700 V.

The standard approach to withstand such high voltage values, using junction isolation techniques, would require very thick epitaxial layers with the consequence of waste of silicon area due to the amount of lateral diffusion. The solution to this problem was given by the RESURF (Reduced Surface Field)

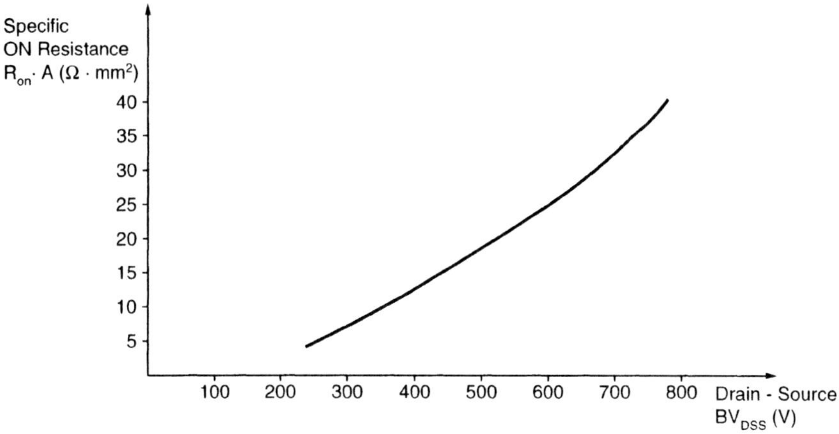

Fig. 8.1 $R_{on} \cdot$ Area comparison for LDMOS in H.V. BCD technology

approach [3] that allows to withstand high voltage in very thin epitaxial layers ($\leq 10\ \mu m$). The RESURF approach implies lateral current flow in the high voltage DMOS (Lateral DMOS) and the limitation to source grounded topology.

A figure of merit for power density in integrated and discrete components is given by the product $R_{on} \cdot$ Area expressed in $\Omega \cdot mm^2$. Figure 8.1 shows $R_{on} \cdot$ Area as a function of Drain-Source Breakdown Voltage, BV_{DSS} for LDMOS implemented in H.V. RESURF BCD technologies.

As can be seen from the figure the $R_{on} \cdot$ Area for 600 V devices is in the range of $25\ \Omega \cdot mm^2$.

Usually ICs for off-line applications are distinguished into:

- HVPIC which integrate high voltage power DMOS in the output stage to drive directly the load.
- HVIC which integrate the high voltage sections (level shifters, drivers) necessary to interface external high voltage power devices.

The following table shows a summary of the main characteristics of today's integrated technologies for "Off-line" applications.

As the table shows the two basic technologies available are different in terms of process architecture complexity. CDMOS process allows the integration of CMOS control together with high voltage source grounded Lateral DMOS. Typical applications are HVPIC where the control section does not require accurate or sophisticated control. It is important to notice that the low voltage components can be used only referred to ground, so that it is not possible to implement low voltage control or low voltage drivers floating from ground to the maximum voltage.

BCDMOS process is more expensive, but it gives much more flexibility as far as the implementation of the low voltage control section is concerned. The high voltage DMOS has the same limitations and characteristics, being a Lateral

Table 1

Process	Isolation	μLitogr.	N. Masks	Low volt. Comp.	H. V. Comp.	$R_{on} \cdot A$ H. V. Comp.	H. V. Floating Drivers
CDMOS	Self isolated	2 μm	10 levels	CMOS 15 V	LDMOS Gnd Sour. BV_{DSS} 300 – 700 V	25 Ωmm² at 600 V	NO
BCDMOS	Junct. Isol.	2 μm	15 levels	CMOS 15 V BIPOLAR 25 V	LDMOS Gnd Sour. LIGBT BV_{DSS} 300 – 700 V	LDMOS 25 Ωmm² at 600 V	Yes

DMOS in Source grounded configuration. The extra number of masks that are used to implement the junction isolation and the bipolar sections, also allows the implementation of Lateral IGBTs that are convenient in terms of power density integration (4 to 5 times), with the trade-off of a lower switching speed. Finally, with BCD technology it is also possible to realize low voltage sections able to float from ground to the maximum BV voltage, giving the maximum flexibility for the implementation of HVPIC and HVIC, and allowing the best choice in terms of system and power partitioning between integrated and discrete. As regards the flexibility for the output power stage topology, both processes have the limitation of source grounded power devices, so that it is not possible to implement source follower, half bridge, full bridge topologies. Such limitation is overcome by dielectric isolation that gives the maximum flexibility avoiding any interaction between components.

8.3 Off-Line Environment

In many applications a load has to be fed with a current of variable frequency and amplitude, with energy supplied directly from the mains voltage. If the load must be supplied from the mains, the mains voltage is rectified, filtered and supplies a DC rail. The load is then connected to the DC rail via a "switch topology", that can be realized by one or more power devices, that have the function of controlling the current, and hence the power through the load. Such applications, in which the load is directly fed by the rectified mains voltage, are called "Off-line" applications, and include:

- Electronic ballasts for low pressure and high pressure lamps
- Switch- mode power supplies
- Motor drives for DC and AC motors

Figure 8.2 shows the typical configuration for off-line applications.

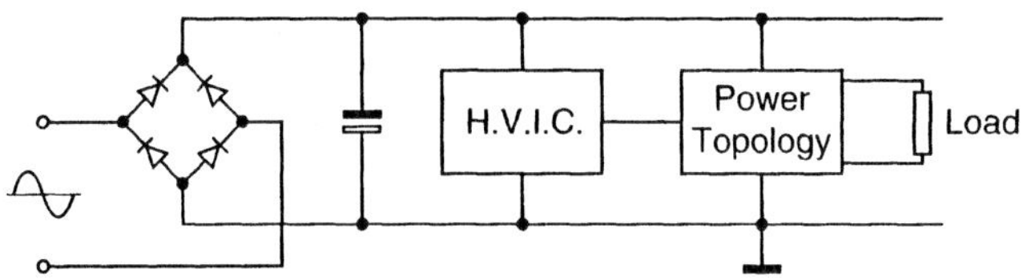

Fig. 8.2 Off-line Configuration

The monophase sinusoidal input voltage RMS value is in the range from 110 V typical for USA and Japan, to 220 V for Europe (240 V UK), and 277 V for industrial USA.

The electricity distribution companies supply sinusoidal voltage at a frequency of 50 Hz in Europe and 60 Hz in USA; the voltage is practically very close to the pure sinusoidal form at the terminals of the alternators, but the same is no longer true for its waveform at the user's terminals, since the electrical energy undergoes a certain number of influences between its generation and its supply to the user. Different types of irregularities are present in the voltage supplied to consumers:

- slow changes in voltage
- frequency variation
- harmonics of the fundamental mains frequency
- sudden changes in voltage (due to switching of loads)
- dc components
- voltage dips
- spikes or surges
- HF signals

These irregularities find their origin, not only in the network itself (during faults, switching, operation of automatic devices), but also in the use of many appliances that produce voltage change, harmonics, unbalances and dc components. Under these conditions, the characteristics of voltage irregularities cannot be expressed in deterministic terms, but rather in probabilistic terms; this means that the distributor cannot guarantee that at a given point in the system and at a given time, the nominal values are not exceeded.

We will discuss separately the issue of harmonic distortion that is today regulated by law, and that gives origin to the active power factor control in which ICs are becoming widely used.

Here we will dedicate our attention to the issue of spikes or surges that conditioned the voltage capability of the devices applied to off-line systems [4, 5].

8.3.1 Spikes Caused by Transient Phenomena

Usually in LV distribution this type of disturbance has its origin in operations such as the blowing of fuses, the opening of switches, the operation of contactors for motors or for household electrical appliances, etc. [5]. As an example, statistical measurements carried out in Europe on an overhead network supplying domestic consumers, showed the following distribution amplitude of spikes:

- 2% with an amplitude between 600 and 800 V (up to 1200 V during blowing of a fuse)
- 10% with an amplitude between 400 and 600 V
- 88% with an amplitude between 200 and 400 V

With regard to the duration, the distribution of spikes was as follows:

- 14% with a duration less than 1 μs
- 85% with a duration between 1 μs and 10 μs
- 1% with a duration between 10 μs and 100 μs
- less than 0.1% with a duration greater than 100 μs

As concerns occurrence in terms of absolute values we can give the following data:

Domestic consumers: 2 spikes/day ≥100 V; 1 spike/2 months ≥1000 V.
Industrial consumers: 20 spikes/day ≥100 V; 2,5 spikes/2 months ≥1000 V.

The energy contained in these pulses cannot be specified; it depends on the source of the spikes and can vary from 1 mJ to some Joules.

Although at the point of production many disturbances have fronts with very short rise times, by travelling through the system they are attenuated by the impedances and may become oscillatory nature. In Europe, the IEC recommendations are usually followed, which specify a voltage transient of the following characteristics: $V_{peak} = 1$ kV, rise time = 1.2 μs, half value time = 50 μs, series resistance = 5 ohm and energy content of 2 Joule.

Due to the existence of high energy voltage spikes, electronic systems which are directly supplied from the mains require a voltage limiting configuration. With reference to Fig. 8.2 in which the typical off-line configuration is shown, we can see that the buffer energy storage filtering capacitor is the main element that provides voltage limitation. It must be noted that the voltage limitation is a function of the value of the capacitor and of its serial impedance. The voltage increase across the storage capacitor due to transients will be much higher in a miniaturized system where a small capacitor is used.

Table 2 shows the result of a test in which the load after the rectifier and filter capacitor is a resistor, with different values of filter capacitors. The voltage transient recommended by IEC was used. Figure 8.3 shows the test setup configuration.

The result of the test showed that the resistive load does not considerably change the shape of the transient coming from the mains. No significative differences were found for the peak overvoltage due to different power ratings. In fact the voltage peak value is defined by the voltage divider made of the impedance of the capacitor and the line impedance.

The minimum breakdown voltage requirement is determined by the maximum sine wave amplitude, with superimposed voltage spike amplitude present on the device after the rectifier across the filter capacitor. In the hypothesis of low

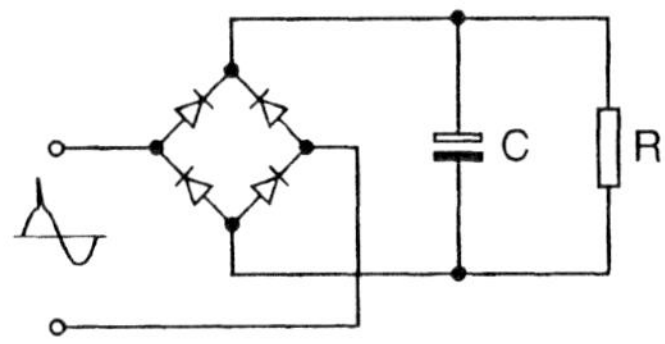

Fig. 8.3 Test set-up to evaluate the effect of voltage transient

Table 2 Overvoltage on capacitor-resistive load for R = 380 Ω (580 W)

Capacitor Type	Rs mΩ (100 Hz)	Zs mΩ (10 Khz)	Spike 200 V	Spike 400 V	Spike 600 V	Spike 800 V	Spike 1000 V
47 μF	2700	1500	23 V	52 V	85 V	120 V	150 V
100 μF	950	400	12 V	20 V	30 V	50 V	60 V
220 μF	580	350	8 V	18 V	30 V	40 V	50 V

Table 3 BV Requirements for off-line systems in presence of 1000 V mains surge

V mains RMS	V mains peak + 20%	Spike on C filter	BV Requirement
Europe 220 V	372 V	150 V	522 V
UK 240 V	406 V	150 V	556 V
USA 110 V	186 V	150 V	336 V
USA Indust. 277 V	470 V	150 V	620 V

power systems, with small values and low quality filtering capacitors, in the worst condition a superimposed 150 V spike can be considered, to guarantee a reliable operation of the system even in correspondance with a 1000 V main surge. Table 3 reports the Breakdown Voltages (BV) to which the device is submitted as a function of the value of the mains voltage supplying the system.

From the above table it is evident that to cover off-line applications for mains voltages ranging from 110 V to 277 V_{RMS}, integration processes able to withstand voltage between 350 V and 700 V are required.

In some applications, if a guarantee must be provided to survive to surges higher than 1000 V, active clamping devices as Voltage Dependent Resistors (VDR) are used. Such devices are able to clamp high levels of power, being the dissipation distributed all over the bulk of the device; the trade-off is extra cost and the reliability of the overall system that is defined by the reliability of the VDR.

8.4 Mains Harmonic Distortion and Power Factor Correction

In Fig. 8.2, the typical configuration to supply electronic systems from the mains is shown. The presence of the storage capacitor to obtain a filtered DC voltage

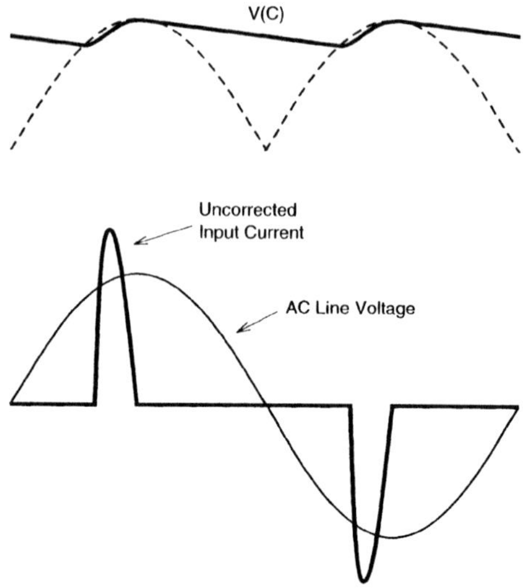

Fig. 8.4 AC line input voltage and current for uncorrected AC/DC conversion

from the AC line is responsible for peaks in the current drawn from the mains, causing harmonic distortion. Figure 8.4 shows typical input voltage and current waveforms in the application circuit of Fig 8.2.

The simple rectifying circuit draws power from the line when the instantaneous AC voltage exceeds the capacitor voltage. Because power is taken in proximity of the line voltage peaks, the resulting spikes of current are extremely non sinusoidal with high armonic content. This results in poor power factor, and the apparent input power is much higher than the real power. Power factors of 0.5 to 0.7 are common.

8.4.1 Power Factor: Definitions

Power factor is a measure of how close the apparent power is to the real power.

"Apparent Power" is defined as the product of rms voltage and rms current:

$$\text{Apparent Power} = P_A = V_{rms} \cdot I_{rms} \tag{8.1}$$

This quantity however is not the power utilized by the load, or the "Real Power":

$$\text{Real Power} = P_R = \frac{1}{T}\int_0^T v(t) \cdot i(t)dt \tag{8.2}$$

If the applied voltage is sinusoidal and the circuit is linear we have that equation 8.2 is:

$$\text{Real Power} = P_R = V_{rms} \cdot I_{rms} \cdot \cos\alpha \tag{8.3}$$

where α is the phase angle between the sinusoidal voltage and current.

Power factor is by definition the ratio between Real Power and Apparent Power:

$$\text{Power Factor} = P_F = \frac{\text{Real power}}{\text{Apparent power}} \qquad (8.4)$$

From the definitions it follows that $P_F = \cos\alpha$ for a linear load, and that $P_F = 1$ for a resistive load.

In the general case represented by the waveforms of Fig. 8.4, the input current waveform is rich in unwanted harmonics of the line frequency that do not contribute to deliver the real power to the load.

Due to the high distortion of the input current the apparent power is much higher than the real power, and this situation is responsible for extra heating in the distribution wires and can cause a premature activation of the circuit breakers. In such situation equation 8.4 can be written as:

$$\text{Power Factor} = P_F = \frac{I_F}{I_{RMS}} \cos\alpha \qquad (8.5)$$

where I_F is the rms value of the fundamental frequency component of the input current, I_{rms} is the rms value of the total input current, and α is the phase angle between voltage and current fundamental components. I_{rms} is the square root of the sum of the squares of the fundamental plus all harmonic components of the input current waveform:

$$I_{RMS} = \sqrt{(I_F)^2 + (I_2)^2 + (I_3)^2 + \ldots + (I_N)^2} \qquad (8.6)$$

Equations 8.5 and 8.6 state the power factor in terms of a distortion factor and of a phase angle. Usually in switch mode controls the phase angle is small and the dominant factor is waveform distortion.

8.4.2
Power Factor: The Sstandards

The rapid increase of electronic systems connected to the mains (electronic ballasts, switch mode power supplies and motor controls) are causing serious problems to the utilities for energy distribution, forcing the governments to introduce standards and limitations to the harmonic content of the current drawn by electronic systems, according to different classes of instrumentation. Electrotechnical Commission defined the standards for power factor correction IEC555; European standards are defined in EN60555 [6].

According to these standards electronics equipment is classified into four classes A, B, C, D, as shown in Table 4.

The different classes specify the maximum content for even and odd harmonics of the input current. The specifications are referred to the absolute current values, independently from the power of the equipment, or in terms of percent of the fundamental. As an example Table 5 reports the maximum level of 3rd harmonic allowed for the four equipment classes.

Table 4

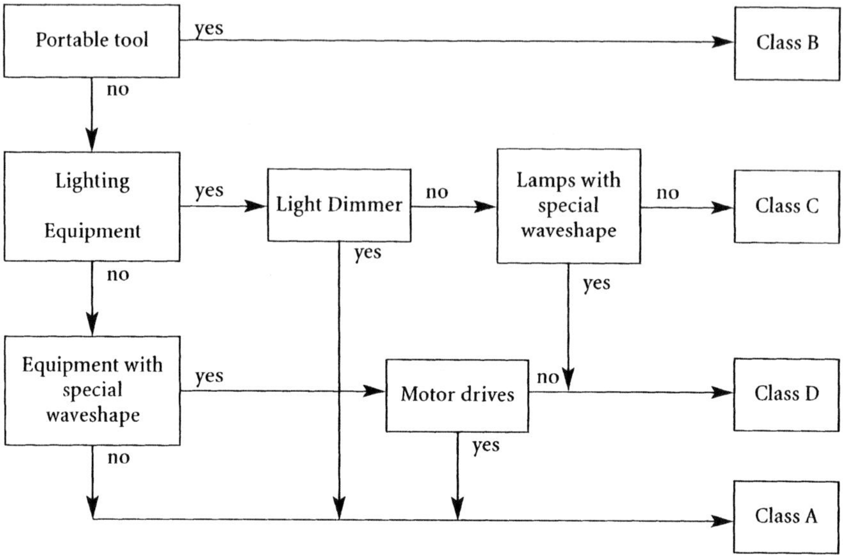

Table 5

	Class A	Class B	Class C	Class D
Maximum 3rd Harmonic	2.30 A	3.45 A	30%	3.6 mA/W or 1.08 A

8.4.3
ICs for Power Factor Correction

Power Factor Correction (PFC) can be achieved with either a passive or active input circuit. Passive circuits usually contain a combination of large capacitors, inductors and rectifiers that operate at the AC line frequency. Active circuits incorporate high frequency switching conversion for power processing. They are smaller, lighter in weight, and more efficient than a passive circuit that yields similar results. Different topologies are utilized to perform power factor correction. Figure 8.5 shows some possibilities with the main characteristics and trade-off.

The Boost topology is today the most popular due to its simplicity and cost effectiveness. Beyond the characteristics listed in the above table, it must be noted that the inductor in series with the input reduces high frequency noise, and allows a reduction in the cost and volume of the RFI input filter [7]. The volume of the inductor is also reduced due to the fact that only a part of the

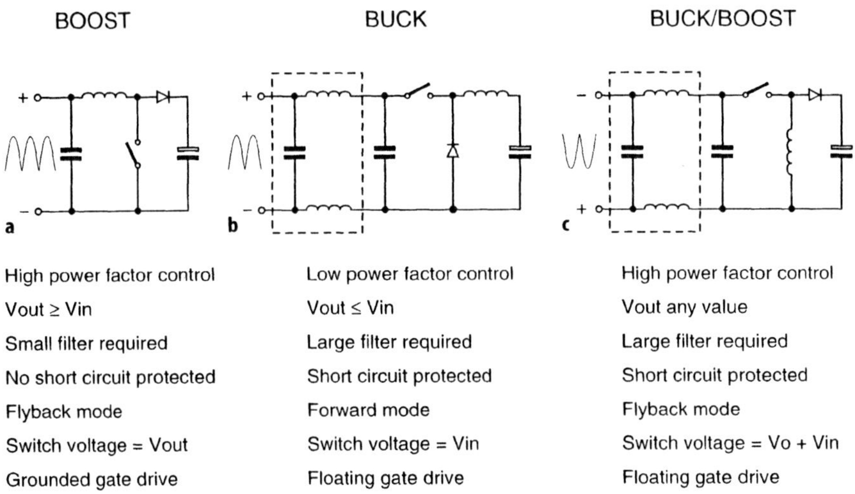

Fig. 8.5 Basic Topologies for Power Factor Correction

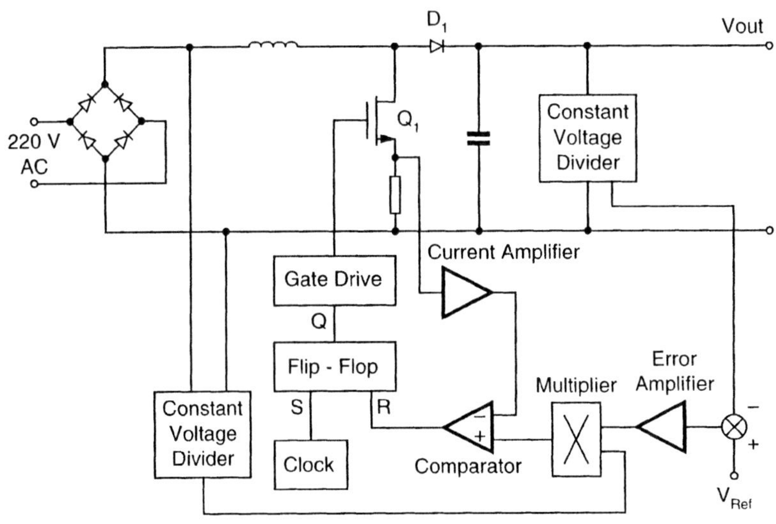

Fig. 8.6 Block Diagram of Boost topology for Power Factor Correction

energy transferred to the output is stored, being the output supplied by the mains during inductor demagnetization. Due to the above characteristics of good performances and low system cost, ICs that perform the function necessary for Boost topology power factor correction were developed. Figure 8.6 shows the block diagram of a simple solution based on Boost topology.

With the control performed by this circuit, any complex load can be made to appear resistive to the AC line, reducing significantly the harmonic current content.

The PFC circuit must generate a constant DC output voltage, and draw a sinusoidal input current in phase with the sinusoidal input voltage. The operation of the PFC boost converter can be summarized as follows. The AC mains voltage is rectified by a diode bridge and the rectified voltage delivered to the boost converter. The boost converter section, using a switching technique, boosts the rectified input voltage to a DC controlled output voltage V_{out}. The power section consists of a boost inductor, a controlled power switch, a boost diode, an output capacitor and the control circuit. The selection of the power components is mainly related to the parameters of the application such as: Input mains voltage, Output power, DC Output voltage, Switching frequency. The system control contains two loops: an "external" voltage loop and an "inner" current loop. The output DC high voltage is reduced by a voltage divider and compared with an accurate voltage reference; the difference between the two is supplied to the input of an error amplifier. The output error voltage is not used to control directly the gate drive of the output power DMOS, but is fed to the input of an analog multiplier. The other input of the multiplier is a fraction of the input rectified voltage. The multiplier output voltage is therefore:

$$V_{mult.out} = V_{error} \cdot k\, V_{peak} \sin \omega t \tag{8.7}$$

with $K < 1$, where $V_{peak} \sin \omega t$ is the input AC voltage. The $V_{mult.out}$ is the signal delivered as the reference of the current control loop. The current control loop is realized by a comparator that receives on the plus input the $V_{mult.out}$ signal, and on the minus input the current signal sensed on the source of the power DMOS. When the peak current exceeds the $V_{mult.out}$ signal the power DMOS is switched off by the control logic and is switched on by the clock signal. In this way the $V_{mult.out}$ signal, whose shape is the same as the rectified AC voltage and whose peak is proportional to the voltage loop error signal, forces the input main current to be sinusoidal and in phase with the input voltage, and with an amplitude that can sustain the output power and voltage.

The clock signal that defines the turn on time of the power DMOS, can be delivered by a fixed frequency oscillator, or defined by the moment in which the inductor current falls to a defined level. In the first case, we have a fixed frequency control; in the second case, a variable frequency control.

Figure 8.7 shows, in the case of variable frequency control, the inductor current and DMOS gate drive voltage, and turn on of the power DMOS which occurs at zero inductor current.

Variable frequency is the simplest control method to achieve power factor correction; it is widely used for low power applications (30 W to 150 W).

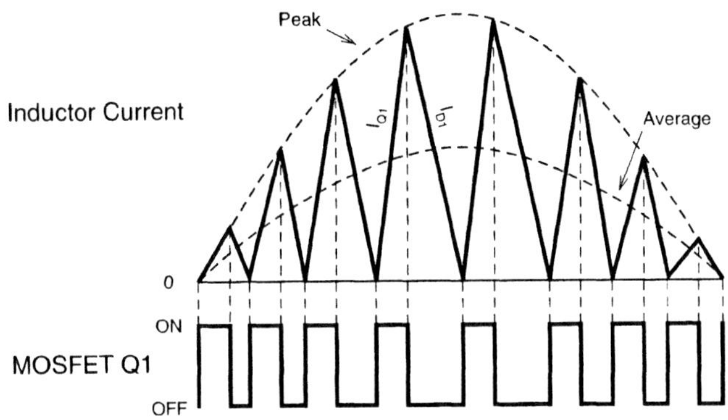

Fig. 8.7 Inductor current and DMOS Gate Voltage Waveforms for variable frequency control

Frequency changes as a function of the load power and input voltage, so that variable frequency is used in applications where the power is constant and the supply voltage is defined. For higher power levels and/or universal mains applications fixed frequency control is preferable. The Boost configuration increases the output voltage to the maximum value of the input voltage. Typically the output regulated voltage has a value of 385 V DC for 220 V_{rms} applications. Power factors from 0.97 to 0.99 can be achieved, with a conversion efficiency higher than 90%. Semiconductor components manufacturers developed in the last years control ICs to perform power factor correction both at variable frequency, both at constant frequency. PFC ICs on the market are generally controller devices, implementing the analog control and driving sections on one chip, while the power device is usually a discrete DMOS with a voltage capability in the range of 800 V. From the technical point of view it would be possible to integrate into a HVPIC the complete solution, at least for low power, low-end applications, but the solution of ICs used as a PFC controller and a discrete power DMOS for the output stage is today a standard approach, because many semiconductor companies are providing the controllers compatible with each other, so probably this approach will continue to be the most utilized in the future as well. Controller ICs utilize Bipolar or BCD technologies, and the voltage capability of such devices ranges from 20 V to 35 V. The pin count for PFC controllers is from 8 pins for the simplest solutions to 20 pins. The 8 pins solution is the most widely used for low power (35 W to 60 W) and fixed mains; it is generally used for Lighting applications. The 20 pins solution is suitable for more sophisticated applications, at higher power, fixed frequency, universal mains, and finds its typical application in preregulators for power supplies.

Figure 8.8 shows the block diagram of a PFC control IC, 8 pins solution [8, 9].

The 8 pins device performs an hysteretic current control in transition mode (inductor current reaches zero and then the power device is switched on) at variable frequency. The functional behaviour of the 8 pins device is the same described in Fig. 8.6. The blocks performing the basic functions are the Voltage

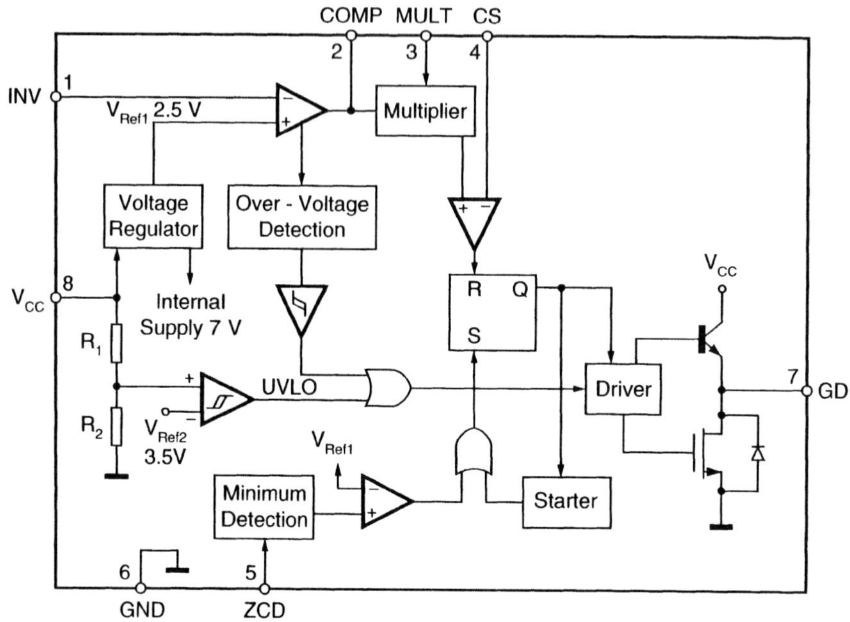

Fig. 8.8 SGS-THOMSON L6560, 8 pins PFC controller. Block diagram

Reference, Undervoltage Lockout, Error Amplifier, Multiplier, Peak Current Comparator, Driver and Control Logic. A Starter is provided to give the first pulses to the power device; after that, the power device is turned off when the peak current reaches the reference value present at the output of the Multiplier, and turns on again when the current in the power device is zero, sensed via the ZCD input on the secondary winding of a transformer, coupled with the main Boost inductor. Another block is the Overvoltage detection that provides a fast protection against output overvoltages that may be present during transients. Figure 8.9 shows the application diagram of L6560 IC.

Figure 8.10 shows the block diagram of a PFC control IC, 20 pins solution [8, 9].

The basic difference compared to the 8 pins solution is that L4981 implements a PWM average current control using feedforward line regulation and fixed frequency switching. The oscillator simultaneously turns on the power switch and starts the ramp of the PWM current control. The average inductor current is compared with the current reference by means of the current error amplifier which operates as an integrator by means of external compensation, allowing the circuit to accurately follow the current reference and limiting the current loop bandwidth for good PWM operation. The current reference is obtained, as in the previous case, by sine wave modulating the error voltage of the voltage control loop. In every Power Factor Corrector circuit it is mandatory that the loop bandwidth is very small (below 20 Hz), otherwise the output voltage ripple would be amplified by the error amplifier and multiplied by the input rectified sinewave,

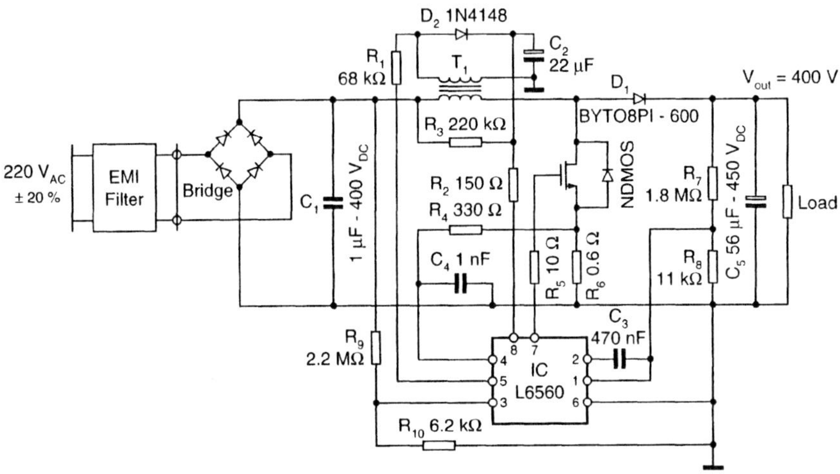

Fig. 8.9 L6560 Application circuit

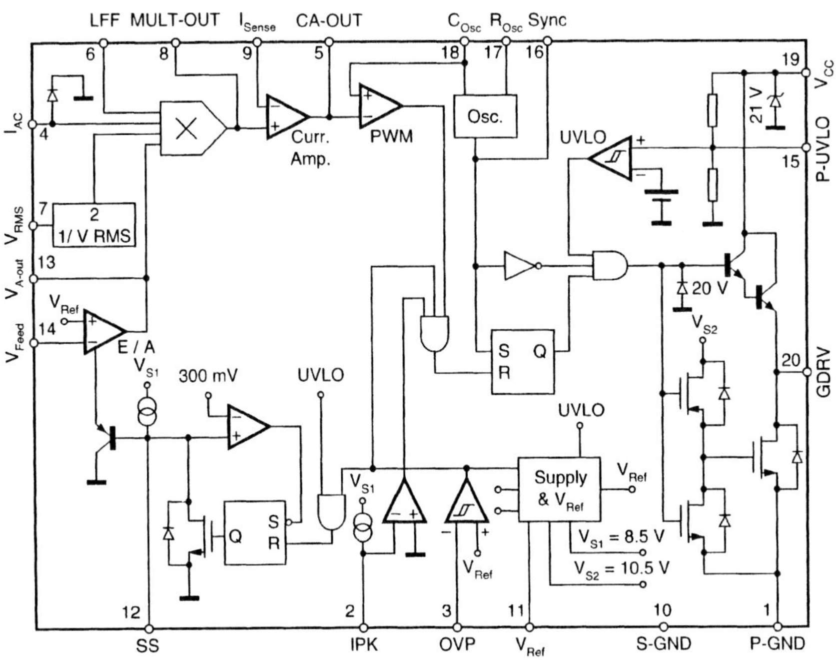

Fig. 8.10 SGS-THOMSON L4981, 20 pins PFC controller. Block diagram

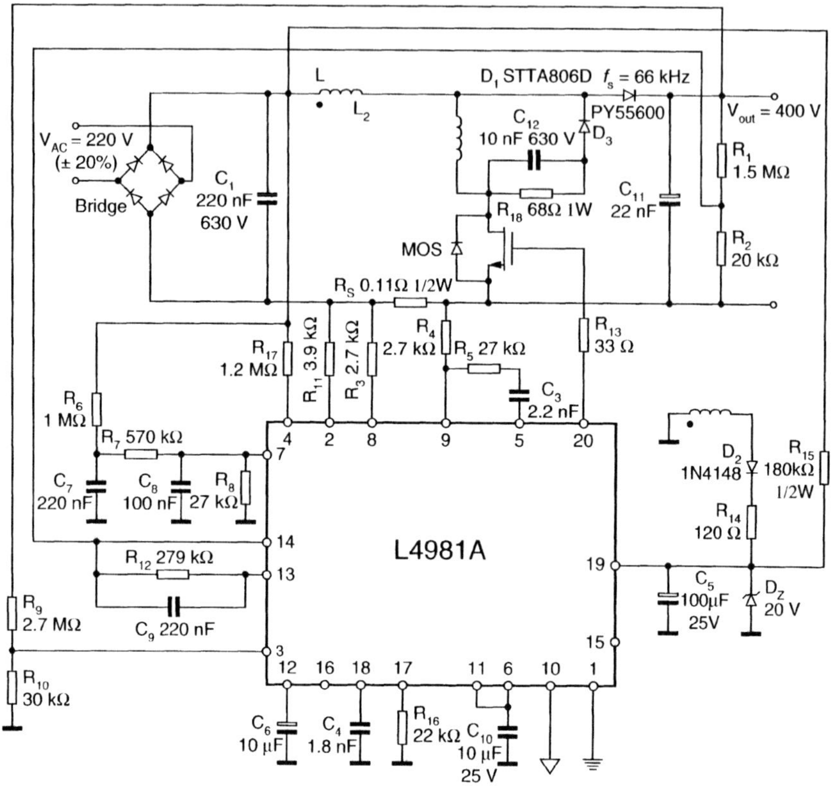

Fig. 8.11 L4981 Application Schematic

giving as a result a reference current with a contribution of twice the mains frequency, causing current distortion. The relatively slow control loop is unable to react to rapid line or load changes causing output overshoots. Line and load feedforward functions are a good improvement as they help the control circuit to respond quickly to input/output variations. If the mains voltage rises by a factor of two, for a constant load, the input current and the reference current should fall to one half. To keep the same error voltage signal requires dividing the sine wave modulating signal by four. So, the voltage loop bandwidth is kept constant and voltage overshoot are avoided with dynamic loads. Extra functions provided are Overvoltage Protection with hysteresis, Pulse by Pulse Current Limiting, Synchronization function, Under Voltage Lockout with hysteresis and programmable turn on threshold. Besides that, an optional version of the device provides a switching frequency modulation capability for noise reduction. The switching noise energy can be distributed by varying the switching frequency thus reducing the peak amplitude of noise voltage. This means that the input filter attenuation required to meet VDE standards on EMI is lower and the filter size can be reduced. Figure 8.11 shows the application schematic of L4981 IC.

8.5 Off-Line Power Supply Topologies

One of the most utilized electronic circuits directly supplied from the mains is the power supply, for energy conversion from the mains voltage to a lower DC level. There are two approaches to reach the target: one is to use a low frequency transformer (50 or 60 Hz according to the mains frequency) to lower the mains voltage to a second level, not stabilized, and then to regulate from the second level with a switch mode power supply to the requested value, for example the 5 V needed for microprocessor supply; the second one is to convert directly from the rectified mains voltage to the desired value. The first approach may be useful if an intermediate voltage is needed with no requirements for stabilization (for instance to supply motors with no need for speed control) and has the advantage of lower voltage electronics, but the disadvantage of bulky low frequency transformers at the input. The second approach is presently the most utilized due to its compactness and higher efficiency and the availability of high voltage electronics, and is known as "OFF-LINE" supply. Different topologies are available for OFF-LINE power supplies, and they can be distinguished according to the phase in which energy is transferred to the output, and to the configuration of the power devices. A short description of the different topologies will now be presented.

The most utilized converter topologies can be divided into two categories: FLYBACK and FORWARD converter [10]. In the FLYBACK converter energy is stored into an inductor during the part of the period in which the power device is in conduction ("ON" state), and transferred to the load during the part of the period in which the power device is not conducting ("OFF" state). In the FORWARD converter, energy is stored into an inductor and contemporarly delivered to the load during the part of the period in which the power device is conducting, and transferred to the load during the part of the period in which the power device is off. By controlling the "ON" time, it is possible to control the energy transfer and therefore the output voltage. The reader should refer to the specialized literature and to the bibliography for a detailed analysis of voltage and current waveforms and working equations of the converter topologies.

In Fig. 8.12 a summary of the mostly utilized off-line power supply topologies, together with the required maximum voltage rating of the power device is reported. Flyback configurations are utilized in the low power range from 10 W to 200 W. Forward configurations are utilized for output power up to 250 W in single transistor configuration, and up to 500 W in double switch structures. Push-pull, Half/Full Bridge configurations are used for higher power levels from 500 W to 2 KW.

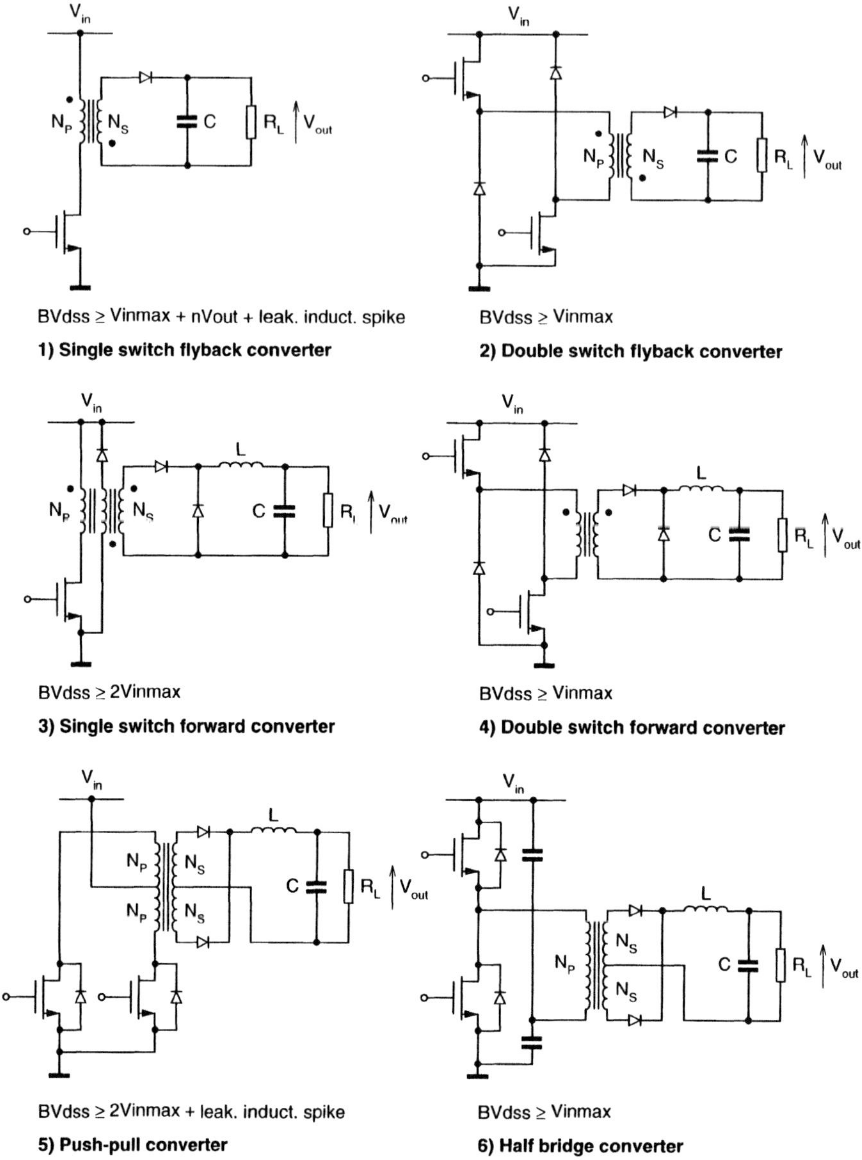

Fig. 8.12 Off-line Power Supply Topologies

8.5.1
ICs for Off-Line Power Supplies

As it can be seen from the above topologies, the simplest solution consists of a single switch, source grounded, that is utilized for the lower power range. The load on the drain is inductive and this causes a drain to source voltage higher than the input voltage. On the other hand, half bridge or full bridge configurations intrinsically limit the voltage to the same value of the input voltage, but require more power devices and high side floating drivers for the source follower devices. ICs are widely utilized to control power supplies. The most popular ICs utilized up to now are low voltage controller ICs, with voltage capability in the range from 20 V to 35 V, that provide the control functions and the driving output current capability for the power devices. The driving to high side power devices is provided via transformer coupling.

High voltage ICs used to perform power supplies function are today becoming a reality, and are already present on the market. Completely monolithical solutions are dedicated to low power flyback/forward configurations where only one power device is needed; this is due to economical considerations and technological limitations. On the other hand, high voltage ICs can be utilized in this field just for controlling and high-side driving of discrete power devices, as will be described in a following section dedicated to high voltage driving ICs. In this case they can be applied to whatever output power topology.

We want to focus our attention here to High Voltage Power ICs for power supply applications, in which also the power device is integrated. Such approach must be competitive with the classical solution in which a low voltage controller is utilized, driving a discrete power device. To be competitive an integrated process using a reduced number of masks is needed, so that usually CMOS-DMOS self isolated processes are utilized. Power dissipation and chip size is also an important feature, limiting the available output power in the range of 100 W. As an example a family of power ICs developed by POWER INTEGRATIONS, INC., for off-line power supply implementation, will now be described. Figure 8.13 shows the functional block diagram of the so called TOPSwitch™ devices family. It implements, with three pins, the functions necessary for an off-line switched mode control system. A high voltage N-channel power MOSFET (700 V) with gate driver, a voltage mode PWM controller with integrated 100 KHz oscillator, a high voltage start-up bias circuit, a bandgap voltage reference, a bias shunt regulator/error amplifier for loop compensation and a fault protection circuitry, including overtemperature protection are integrated into a single chip. TOPSwitch is a self biased and protected linear control current to duty cycle converter device with open drain output. During normal operation, the power MOSFET duty cycle linearly decreases with increasing CONTROL pin current. To implement all the required control, bias, and protection functions, the DRAIN and CONTROL pins perform several functions. For a detailed functional description the reader should refer to POWER INTEGRATIONS, INC. application notes and datasheets.

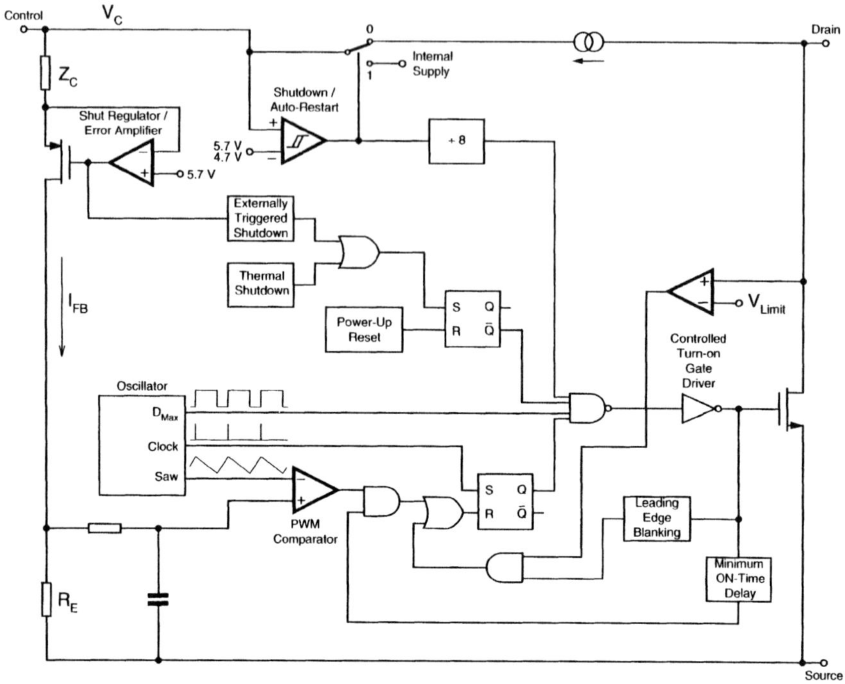

Fig. 8.13 POWER INTEGRATIONS, INC. TOPSwitch functional block diagram. (by courtesy of POWER INTEGRATIONS)

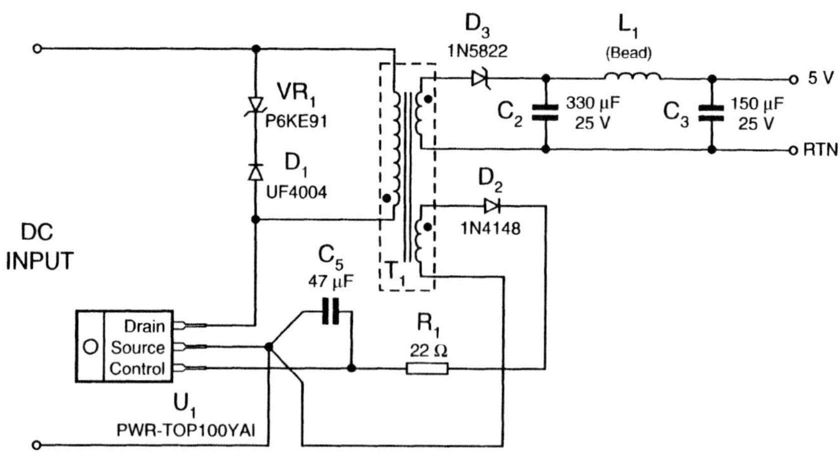

Fig. 8.14 Schematic Diagram of a 5 V, 5 W Bias Supply utilizing the PWR-TOP200. (by courtesy of POWER INTEGRATIONS)

The devices can be used to implement Buck, Boost, Flyback or Forward topology. Figure 8.14 shows a typical application schematic of a 5 V, 5 W Bias supply.

8.6
High Voltage ICs for Control and Driving

In paragraph 8.2.1 it was already said that ICs for off-line applications can be distinguished into HVPICs integrating the power devices, and HVICs integrating the high voltage sections (level shifters, drivers) to interface external high voltage power devices. In this paragraph we will focus the attention on HVICs for high voltage driving, giving examples of the circuital techniques utilized, and describing some applications. HVICs can be considered composed of two parts: a low voltage section integrating the analog and digital control required for the application, and eventually the low voltage driver for a source grounded power transistor, and the High Voltage section integrating the stages required for the level shifting of the digital signal from ground to the floating source of the upper power DMOS. Figure 8.15 shows the partitioning of the chip.

Typical application fields for HVICs are Electronic ballasts, Switched mode power supplies, DC and AC motor drives [11, 13].

The most important requirements for such H.V. drivers are:

- Switching frequency in the range of hundreds of kHz
- Intrinsic protection against cross-conduction
- Very high dV/dt capability especially for inductive load applications (5 V/ns to 10 V/ns)
- Supply of the necessary gate drive current to the external power MOS
- Very low current consumption of the overall circuit for an easy and economical supply from the mains
- Voltage capability in the range of 600 V to 700 V

Fig. 8.15
HVIC Chip partitioning

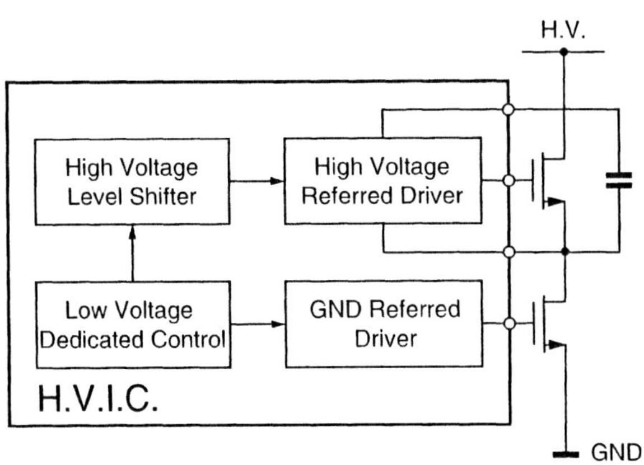

The discrete solutions for high voltage driving are implemented with pulse transformers, which show limitations or trade-offs compared to the performances achievable with integrated solutions, the more evident being cost, manufacturability, reproducibility of system performances considering the overall spreads of power devices and of load. Description of standard driving techniques using pulse transformers, optocouplers or discretes can be found in ref. [12] of the bibliography.

8.6.1
IC Techniques for High-Side, High Voltage Driving

Referring to Fig. 8.15, the main purpose of the high voltage section is the efficient driving of the Gate Source terminals of the upper source follower device. The driving circuit must be able to satisfy the following requirements:

- to be referred to the Source of the power DMOS
- to be able to override the supply voltage value during the ON state, due to the fact that $V_{gs} \geq 10$ V is needed to guarantee a good conduction in the DMOS channel.

The most utilized techniques for high-side driving are Capacitive Bootstrap, Charge Pump and Capacitive Coupling, or a mix of the above techniques to get the best of the achievable perfomances. Driving circuits have already been discussed in chapter 5 of this book. The Capacitive Bootstrap technique, that is the most utilized for Off-Line applications driving will be here analyzed and discussed in detail.

8.6.2
Capacitive Bootstrap Driving

Capacitive Bootstrap driving is based on the generation of a voltage ($\cong 12$ V) across a capacitor, that has the function of a floating supply for the driver, with the negative terminal referred to the Source of the power device.The Bootstrap Capacitive action is shown in Fig. 8.16.

Three phases can be distinguished as:

Phase 1: Refresh of + 12 V Across Cboot

In the first phase, C_{boot} is charged to + 12 V through the diode and the turn on of the two devices to ground. In the simple schematic of Fig. 8.16, Q_1 drives the switch off of the upper DMOS and Q_2 provides the path to ground for the charging current.

Phase 2: Bootstrap Action

During this phase the logic circuits drive the turn on of the upper DMOS by switching off Q_1 and Q_2. While the DMOS is turning on, its Source potential is rising toward + HV. Bootstrap capacitor keeps the voltage constant across itself, so that the driver is supplied at HV + 12 V. The Bootstrap diode is reverse biased and decouples the Driver supply voltage from the + 12 V voltage.

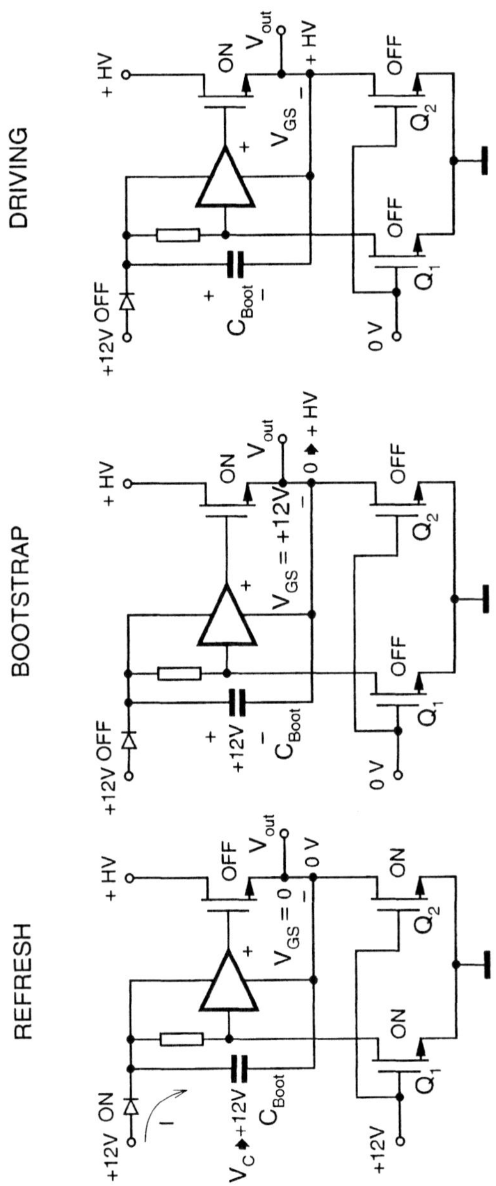

Fig. 8.16 Capacitive Bootstrap Driving

Phase 3: Driving

It is the phase in which the upper DMOS is on and drives the load, and it is the continuation of phase 2. While the DMOS is in the on phase C_{boot} discharges due to leakage and to the current consumption of the driving circuit.

It is evident from the above description that the Capacitive Bootstrap technique does not allow an indefinite on state of the upper DMOS due to the discharge of Bootstrap Capacitor, so that it can be used only if the system is working at a known switching frequency. Such a technique has the advantage of being relatively simple, allowing high switching speed and requiring a limited number of high voltage components. It is the most utilized in switching high voltage drivers. The schematics shown in Fig. 8.16 are very simple but have some limitations and cannot be used in these form in the real case. The next paragraph describes in more detail a real driving and level shifter circuit.

8.6.3 High Voltage Level Shifter

The most important feature, for efficiently driving a high side high voltage MOSFET, is low power dissipation in performing the level shifting function. The circuit of Fig. 8.17 shows the same level shifter discussed in Fig. 8.16, in which the drain to substrate parasitic capacitance is considered.

To avoid power dissipation and discharge of the bootstrap capacitor, the phase of the level shifter is such that the upper POWERMOS is on when the H.V. DMOS level shifter is off. In this way the logic signal is supplied to the upper buffer via resistor R_1 in which no current is flowing, and which is forcing a high level at the buffer input. Such a solution is good considering only the DC conditions, but shows heavy limitations if transient behaviour is examined. During the turn on transition of the upper POWERMOS, while the POWERMOS is off, the DMOS level shifter is on, and $C_{D\text{-}SUB}$ is discharged. At turn on, the DMOS

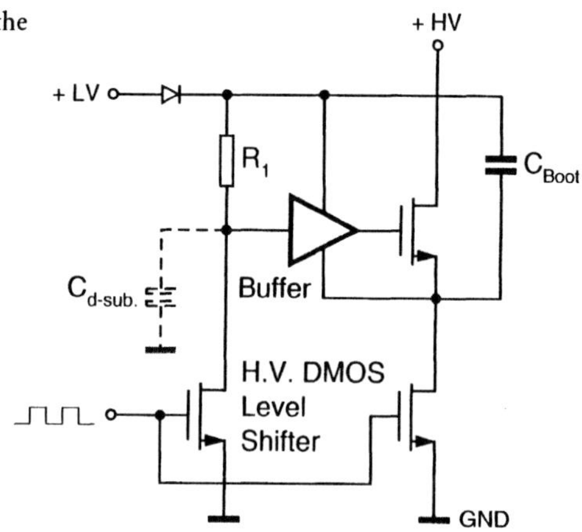

Fig. 8.17 Simple level shifter showing the Drain-Substrate parasitic capacitance

level shifter is turned off, and the voltage at the buffer input increases towards the buffer threshold with a time constant defined by $R_1 \cdot C_{D\text{-}SUB}$. When the buffer threshold voltage is reached the output of the buffer is turning on the POWERMOS that increases the source voltage, driving the load, with its switching speed. The bootstrap capacitor keeps its voltage constant, and thus induces the same dV/dt of the source at the supply node of the buffer. At this point the input buffer node decreases its relative potential because the capacitance $C_{D\text{-}SUB}$ must be charged via the R_1 resistor, and it is not possible to provide a time constant smaller than the switching speed of the output POWERMOS. Due to the above phenomena the driving buffer is changing its state, trying to switch off the POWERMOS, and the output slew rate decreases and the buffer turns on again. In other words there is a kind of negative feedback, between the output switching node and the input driving signal, that does not allow a good switching behaviour for the POWERMOS. Considering the turn off transition the situation is the same, but less critical because the parasitic capacitor is discharged by the current provided by the DMOS level shifter. It is also clear that it is not possible

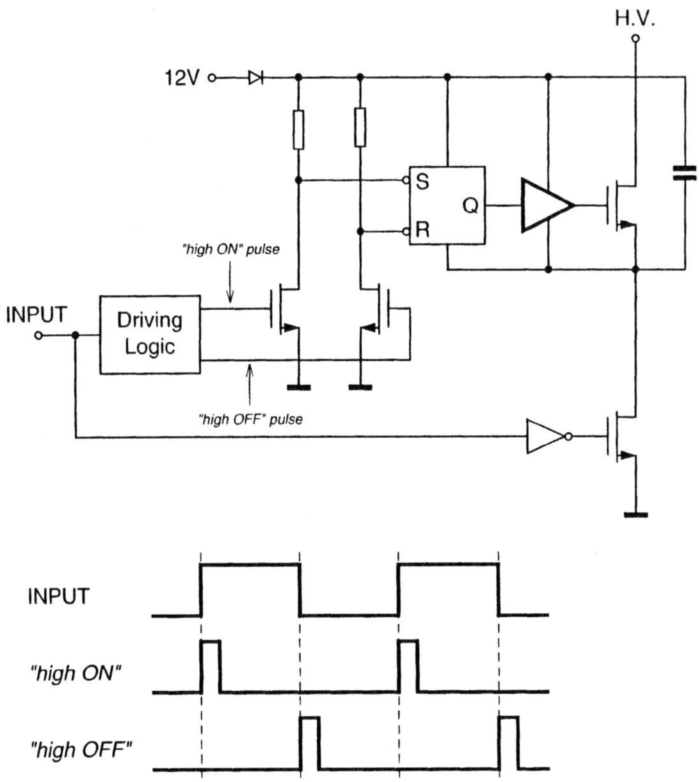

Fig. 8.18 Level shifter implemented with two H.V. DMOS

to change the working phases of the above level shifter, because in this case the on state of the POWERDMOS would be induced by a DC on state of the DMOS level shifter that would discharge the bootstrap capacitor. Due to the above considerations, the level shifter stage is typically realized with two H. V. DMOS providing "on" and "off" pulses that are latched in the floating driver structure (Fig. 8.18).

The circuit of Fig. 8.18 works in the following way. The digital signal at the input of the level shifter logic defines which one of the two POWERMOS devices is on or off, for instance level "0" means lower POWERMOS on and upper POWERMOS off and viceversa. The level shifter logic generates two pulses, whose duration is in the range of 200 ns, one at the occurrence of the rising edge of the digital signal and one at the occurrence of the falling edge of the digital signal. Such pulses are applied to the gates of the two H.V. DMOS and are transferred via current pulses to the resistors connected to the drains and latched into the driver. The solution presented overcomes the problems explained previously, because the signal is transferred by a low impedance and the pulse technique assures a negligible power dissipation. Typical values are 2 KΩ for the drain resistors and current pulses of about 5mA. At switching frequency of 100 KHz, the average current consumption due to the 200 ns current pulses is:

$$I_{AV.} = I_{peak} \cdot \frac{\text{Pulse width}}{\text{Period}} \cdot 2 = 200\ \mu A$$

The power dissipation with a 300 V rail would in this case be $P_D = 60$ mW.

8.6.4
Shoot-Through Protection for Half Bridge Topologies

The logic integrated in the H.V. drivers usually provides protection against shoot-through or contemporary conduction of the output power stage. The circuit assures that some predefined time has elapsed before turning on one of the two power devices after the turn off of the other power device. It is very simple to provide such function with CMOS logic: one example of which is shown in Fig. 8.19.

In the circuit of Fig. 8.19 two signals are generated from the IN signal: Drive Low and Drive High. It is assumed that the "High" level of the signal drives the POWERMOS transistor on and viceversa. As shown in the figure, a delay t_D is generated between the switching off of one power device and the switching on of the other one. The delay is generated by the RC time constant and by the threshold of the Inverter 2. The dead time that provides the shoot-through protection can also be programmed by the utilizer by an external resistor that defines a current level for charging or discharging the delay capacitor. Such approach is very flexible and is one of the reasons that make the IC drivers preferable compared to discrete solutions both for flexibility and for reliability of the solution.

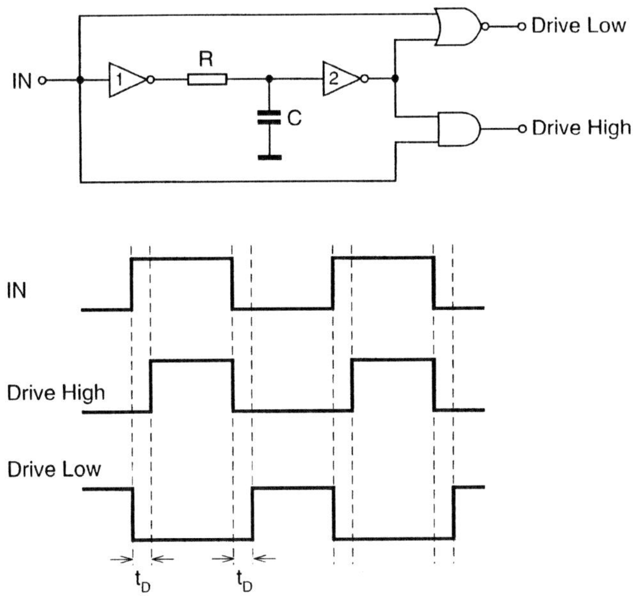

Fig. 8.19 Digital circuit for shoot-through protection of the output power stage

8.6.5 Buffer Implementation in H.V. Drivers

The buffer that drives the gates of the POWERMOS devices must provide the peak current required to charge and discharge the input capacitance at the required speed, and the gate to source voltage. The most common solution utilizes CMOS drivers that have the advantage of being simple, and technologically compatible with high voltage integration. HVICs for driving available today have an output current capability in the range from 100 mA to 2 A, allowing a very wide spectrum of applications. From the technological point of view, it should be noted that the bootstrap capacitor driving techniques allow high voltage driving utilizing low voltage buffers. The high voltage components required are DMOS for level shifting and a bootstrap diode. The buffer is a low voltage device, but it must be able to float up and beyond the maximum supply voltage, so that it must be integrated into a well able to float and to withstand the maximum process voltage (see paragraph 8.2.1).

8.6.6 Bootstrap Diode Integration

The bootstrap diode function was described in the paragraph 8.6.2 in which the bootstrap capacitor driving is explained. The diode must be able to withstand

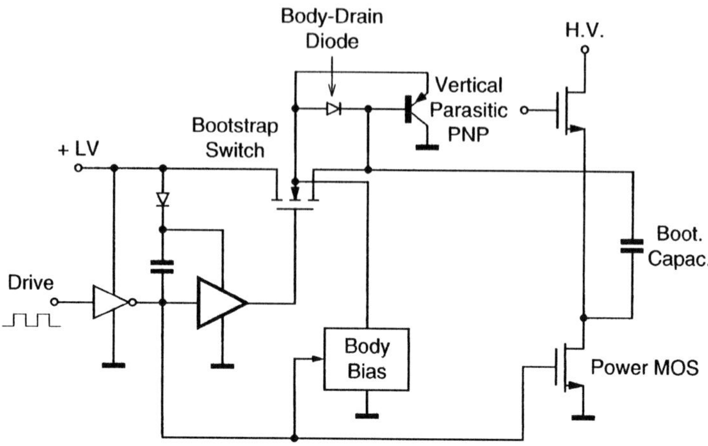

Fig. 8.20 Implementation of the bootstrap diode function with a gate driven DMOS

the maximum voltage of the process. In most of the HVICs available today the bootstrap diode is not integrated and a discrete component is used. From the technological point of view the diode cannot be integrated easily in a junction-isolated process, due to the associated parasitic structures. For instance, since the diode must withstand a high voltage, it must be a base-collector diode that has an associated vertical parasitic pnp when working in forward conduction mode. The bootstrap diode can however be integrated in the form of a gate driven DMOS synchronized with the logic driving signals and this solution is utilized in some of SGS-THOMSON devices that will be described later. An example is shown in Fig. 8.20.

In Fig. 8.20 is shown the vertical parasitic pnp associated to the body-drain diode of the DMOS. The DMOS has the source connected to + LV supply and the drain is connected to positive termination of the bootstrap capacitor; the gate is driven by an integrated bootstrap capacitor driver able to drive the gate higher than the source of the DMOS when it is driven on. The value of the integrated bootstrap capacitor is in the range of 20 pF. The DMOS used as a bootstrap switch is driven in synchronization with the lower POWERMOS (both are on at the same time to charge the external bootstrap capacitor). It must be noted that, during the charging phases of bootstrap capacitor, the DMOS conducts with a current flowing from source to drain, because the point connected to the positive terminal of the bootstrap capacitor is the drain that has therefore to withstand the high voltage. An extra circuital block, called "Body bias", has the function of biasing the body of the DMOS at the proper potential during the commutation phases to avoid the turn on of parasitic junctions like the vertical pnp shown in the figure. This polarization can be easily provided because the DMOS in the integrated form is a lateral device and its body is accessible. The integration of a simple bootstrap diode function can be obtained with some circuital complexity, but it is worth in those applications where room saving on the application printed circuit board is mandatory; moreover the integrated solution, even if complex, is competitive in terms of cost with the discrete diode.

8.6.7
Generation of a Low Voltage Supply in Off-Line Applications

We have seen that in high voltage applications there is anyway a considerable part of the circuit, dedicated to the control and the driving of the ground referred POWERMOS, that must be supplied at low voltage, in the range from 12 V to 15 V. The generation of the low voltage supply is a critical part of the application, because it must be taken from the high voltage rail (300 V) in an efficient way, to avoid power dissipation, and with the most economical solutions. The HVIC plays an important role in the way the low voltage supply is generated. Usually the generation of the low voltage supply starts directly from the high voltage rail, supplying a capacitor, connected in parallel to the low voltage supply, through a resistor. During the first phase the system is in stand by condition, not switching, and the current consumption of the IC must be as low as possible to avoid power dissipation in the resistor connected to the high voltage rail. However the IC must in any case be able to perform some basic functions during this phase, that is to sense the increasing low voltage supply, to decide when to start the switching of the system when the low voltage supply has reached a defined value, and to be able to control with its logic circuits and drivers the off state of the output power devices. Such functions are provided by the so called Under Voltage Lock Out circuit, that is a comparator with some hysteresis, which senses the low voltage supply, whose output controls the status of the system. During the start up phase the current consumption of the IC must be lower than 200 μA, and be guaranteed in the worst case over the applicative temperature range, to keep low the power dissipation on the supply resistor, that remains always connected to the mains. Once the supply voltage has reached the startup value, the system starts switching, and increases its current consumption, therefore the supply voltage must be provided by a non dissipative, low impedance supply generator derived by the output. The generation of the low voltage supply after startup can be done in different ways. One possibility is to provide the low voltage via a secondary winding, put on the main inductor or transformer supplying the load, according to the application. Examples of this approach are given in Fig. 8.9 and 8.11 of this chapter where the PFC application circuits are shown. Another way of generating the low voltage supply after start up is by capacitive supply, directly from output. This is a simple and convenient method, that is also very efficient when driving inductive loads. Fig. 8.21 shows the basic circuit.

The capacitor C_1 is connected with one terminal to the switching output, and with the other terminal to C_2 via diode D_2, and to ground via diode D_1. When the output voltage V_0 of the half bridge is switching from GND to + HV a charge is flowing to C_2, the capacitor across which the low voltage supply is generated, via C_1 and D_2 that is forward biased. When the output half bridge voltage is switching from + HV to GND the diode D_2 decouples C_2 from C_1 and the diode D_1 is

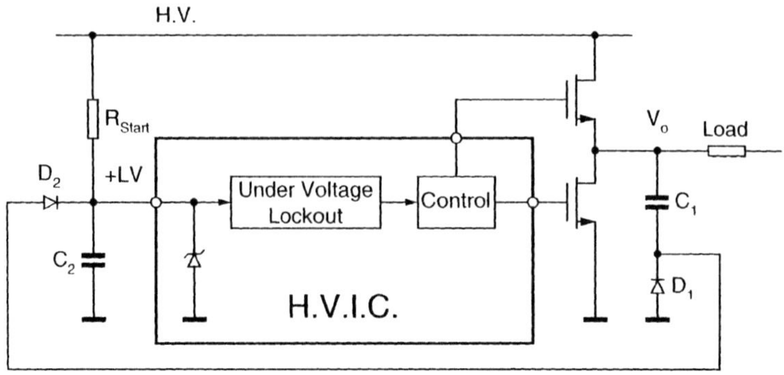

Fig. 8.21 Example of low voltage supply generation via capacitive supply

forward biased and provides the current path for the discharge of C_1. Also in this case a start resistor is needed that charges the C_2 capacitor before the half bridge starts commutating. It must be noted that, when driving inductive loads, the output is switching from low to high and from high to low forced from the output inductor, while the two power MOS are off due to the shoot-through dead time protection. This means that the circuit described utilizes a charge pump technique in which a charge is taken from the output load and delivered to C_2 via C_1 and then given back to the output when discharching C_1 via diode D_1. For this reason the power dissipated to generate the low voltage supply is practically very low. The only high voltage components needed for the low voltage supply generator are the start resistor and capacitor C_1. This charge pump circuit has some application limits as far as the supply current capability is considered; the supply generator can in fact be seen as a voltage generator whose value is + HV with an impedance in series, whose value is $\frac{1}{FC_1}$ where F is the switching frequency. The value of C_1 is usually chosen in the range of 500 pF (it is responsible for the output voltage slew rate and its value cannot be too high), so that in the hypothesis of a 300 V DC rail and of a switching frequency of 100 Khz the current capability from the charge pump supply is of 15 mA. This is the maximum current that the IC can sink, including the output average current needed to drive the gates of the POWERMOS, so that the current consumption of the IC to perform the control functions must be very low, usually in the range of a few mA. The value of capacitor C_2 is chosen according to the maximum acceptable ripple on the + LV supply and according to the peak current that must be delivered by the drivers. The resistor R_{start} can in some cases be integrated into the HVIC in the form of an active device (lateral DMOS). With this solution the start switch can be switched off after start up, allowing for less stringent requirements for start up current consumption that in case of use of the resistor must be very low ($\leq$200 μA). The IC must include also a shunt regulator (a Zener diode) to avoid that, in situations of low current consumption, the charge pump supply increases the value of the voltage + LV up to dangerous values.

8.7 Examples of HVICs for Lighting Applications

In this paragraph some examples of HVICs for lighting applications will be presented. Figure 8.22 shows the actual trend in electronic ballasts which consists of the substitution of POWER BIPOLARS output stages in self-resonant configuration using pulse transformers for the base driving, with integrated control and drivers adapted to a POWERMOS output stage.

The IC solution reduces the component count, avoids pulse transformers and improves the control characteristics that can be achieved. A practical example is shown in Fig. 8.23 where the application of the L6569 device is presented. It is a half bridge driver, rated at 600 V, whose frequency is programmable by external RC.

The internal block diagram of the device is shown in Fig. 8.24, where it is possible to see the basic blocks described in the previous paragraphs.

The lamp ballast electronics must generate, for a good lamp start and burning, the following sequence:

- Preheat of electrodes
- Ignition
- Burning

During preheat time the circuit controls the current flowing through the electrodes for a defined time to prepare a good ignition, which is the next phase. Preheat phase is important for a long life of the lamp, because the electrodes must be at a certain temperature before the thermoionic emission takes place. During ignition time the voltage across the lamp is increased up to the value (1 kV) at which a discharge occurs, thus providing the lamp turn on. Normal burning is the steady state condition in which the circuit must ensure a good power control for a constant light output. The electrical characteristic of a gas discharge lamp is shown in Fig. 8.25. As it can be seen, after the ignition voltage the lamp works on a negative impedance characteristic. For a stable working point the lamp must be driven by a current generator, provided by the series inductor.

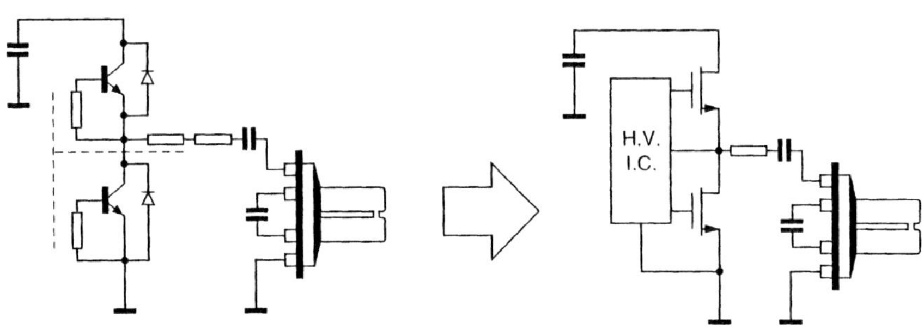

Fig. 8.22 Trend in electronic ballasts

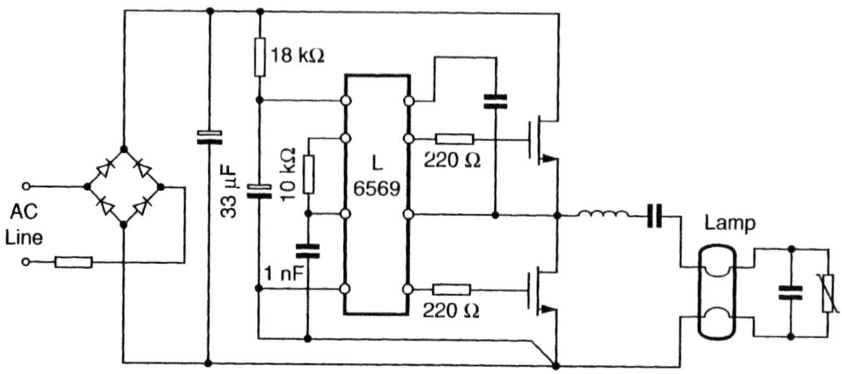

Fig. 8.23 L6569 Lamp Ballast application

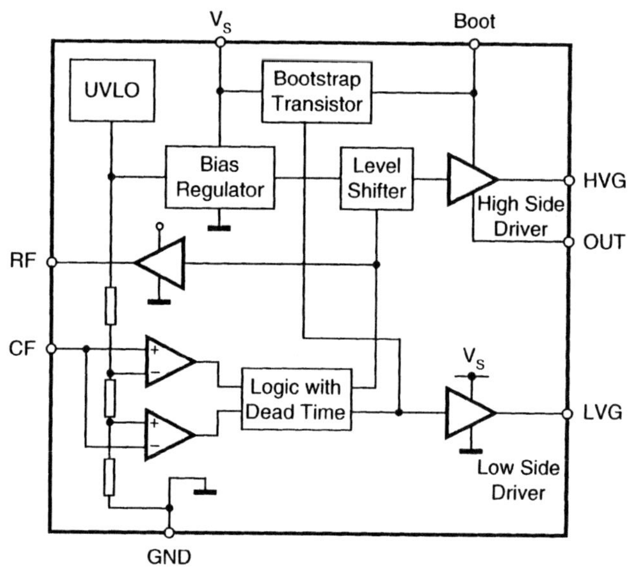

Fig. 8.24 L6569 Block Diagram

The above described characteristics can be obtained with a frequency control system that defines the working point on a resonant load. The resonant load consists of the series inductor, the capacitor in parallel to the lamp and the lamp itself that behaves as a resistor when on, and as an open circuit when off. In the application example of Fig. 8.23 the frequency is fixed by the IC and the resonant load is changing over time (temperature) by the presence of the Positive Temperature Coefficient (PTC) resistor. At start the PTC is cold, so that its value is low and it dumps the resonant load. In this condition the electrode current is defined

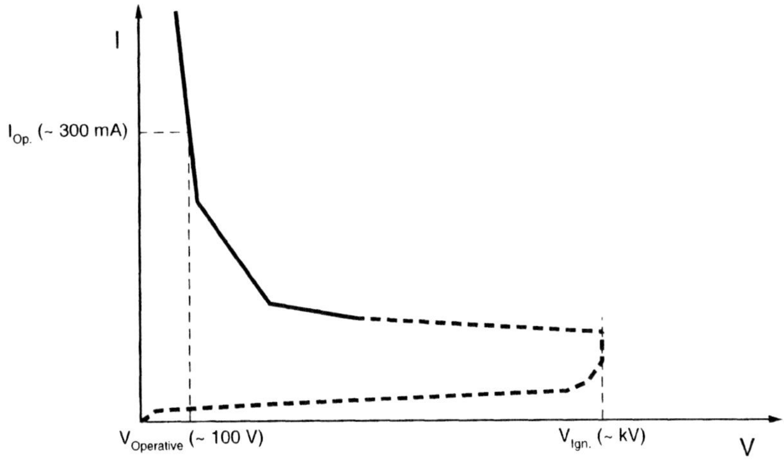

Fig. 8.25 Electrical lamp characteristic

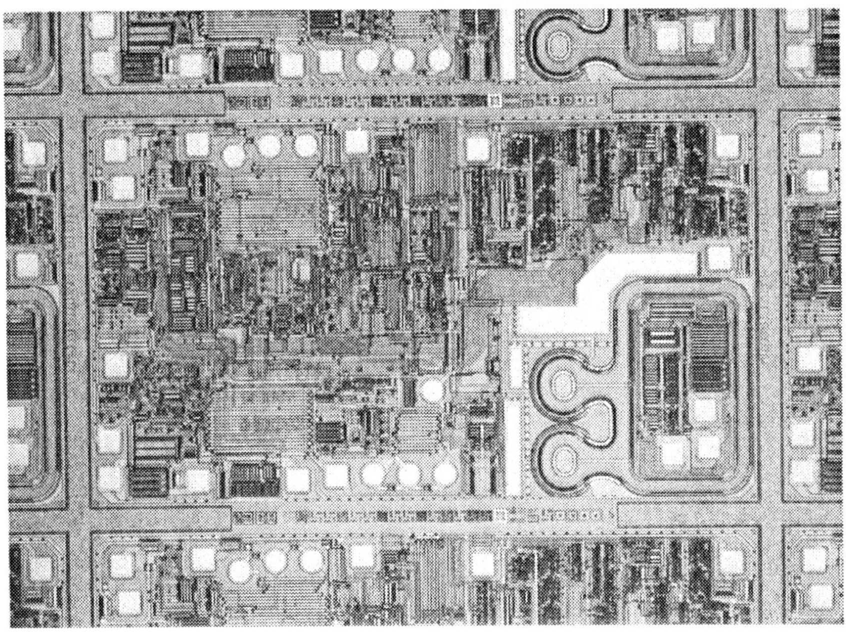

Fig. 8.26 Microphotograph of custom HVIC for lighting applications (SGS-THOMSON)

by the switching frequency, the inductor and the PTC value. This is the preheat phase, whose time is defined by the heating time of the PTC, that increases its value with temperature. In this condition the load becomes resonant and if the switching frequency is close to the resonance value of the LC network, the voltage across the lamp increases rapidly, the gas discharges and the lamp turns on. After turn on the lamp behaves as a low impedance, dumping the resonance and setting the steady state burning conditions. The circuit described is very simple and part of the control features is left to the external discrete components like the PTC resistor. By extra integration it is possible to simplify and reduce the external components count, providing also a better and more accurate control. More sophisticated lamp ballast control IC integrate, beyond the HV level shifter drivers, the feedback current control, the controlled preheat timer, the voltage or current controlled oscillators for frequency shift in the different phases and the protections against anomalous working conditions such as lamp failures. The integration of frequency shift stages allows to avoid the external PTC that is an expensive component and provides a non accurate preheat function. IC solutions that integrate, together with the half bridge drivers and ballast control, the Power Factor Controller function on a single chip are also available. Fig. 8.26 shows the microphotograph of a custom HVIC dedicated to lighting applications.

8.8 Examples of HVICs for Motor Control

Different types of motors for off-line applications are available. The three phase induction motor is a simple design, rugged, maintenance free and is used in large volume home appliances (100 W – 1 KW) requiring cost effective solutions. Induction motor control requires a three phase voltage supply with controllable magnitude and frequency. This is obtained from a DC voltage with an inverter including an interface for the gate drive [14]. The three phases DC/AC inverter is implemented by three half bridge legs for a total of six switches (Power MOS or IGBTs). The two switches of each bridge leg are controlled with opposite phases. For the speed control of these motors, a frequency variation of the inverter output voltage is required. The voltage/frequency ratio must be maintained constant, so motor control requires complex control circuitry for the generation of the balanced three phase sine wave outputs. Usually the generation of the three phase PWM signal may be controlled by a dedicated IC, which is driven by a separate microcontroller, and interfaced with the output drivers and power devices. The microcontroller unit calculates the necessary data to control the PWM circuits. HVICs can be used in such applications just as an interface from the dedicated IC to the power devices, integrating in addition to the drivers and the level shifters the protection circuits for the power devices. A fully integrated monochip solution of the controller sections and HV drivers could also be possible depending on the complexity of the system. A different solution is to replace analog PWM functions with software. In this case the "sine waves" are digitized, with data being permanently stored in the microcontroller memory.

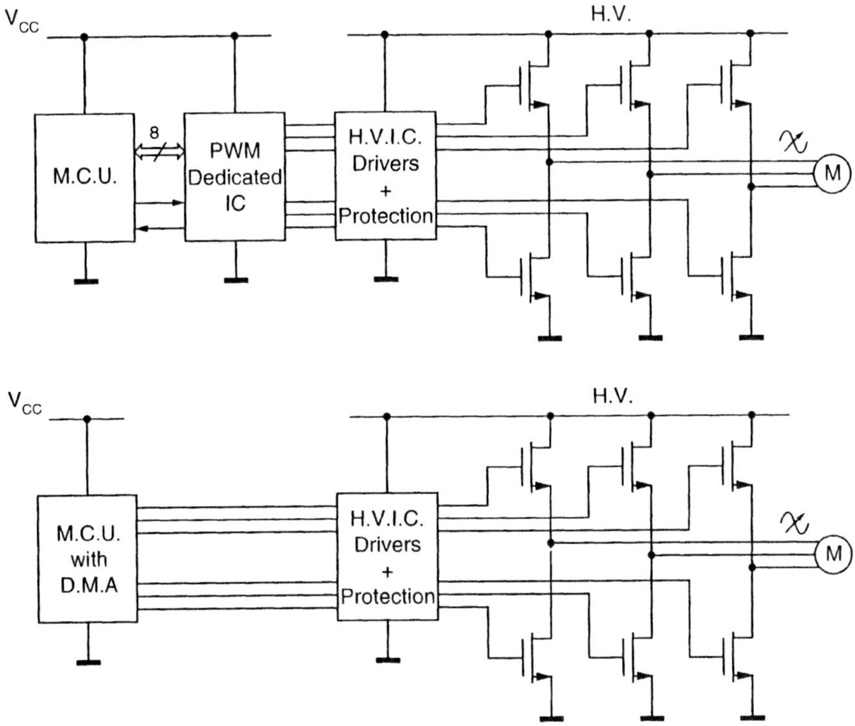

Fig. 8.27 Three phase PWM Induction Motor. Application circuit

The ON and OFF states of the six power switches, necessary to synthetize one period of the three-phase sinewave, are stored in the ROM (or EPROM) internal memory of the microcontroller unit and are sequentially transferred to the six bits of a parallel output port. This transfer from memory to output port is performed by a Direct Memory Access (DMA) device. The CPU only operates when the last byte of an elementary table appears. Consequently the CPU is only lightly loaded and is therefore available for other motor and environmental controls. This solution reduces the need of dedicated hardware (replaced by software). This solution allows the direct interface of the microcontroller with the HV driver. Figure 8.27 shows the block diagram of the two solutions.

8.9 Conclusions

The chapter was intended to provide a view on the growing field of HVICs for off-line applications. An application segment that was historically dominated by electromagnetic components and power discrete devices, today begins to be covered also by ICs (both low voltage and high voltage). The reason is due both to the more sophisticated requirements in terms of performances needed for these applications (for instance Power Factor Correction regulamentation), and to the present availability of high voltage ICs technology in the range of 600 V–700 V, together with dense CMOS and Bipolar low voltage control, that allow the economical implementation of the integrated solution. Examples have been given in the field of: Power Factor Correction, Power Supplies, Lighting, Motor control. This trend will probably continue to develop greatly when it will be possible to have economical Dielectric Isolated technologies, that will allow more flexibility for integration.

References

1. Andreini A, Contiero C, Galbiati P (1986) A New Integrated Silicon Gate Technology, combining Bipolar linear, CMOS logic, and DMOS power, Transactions on Electron Devices, vol ED-33, n 126
2. Rummenik V (1985) Power Devices are in the Chip, IEEE Spectrum, pp 42-48
3. Appels JA, Collet MG, Hart PAH, Vaes HMJ and Verhoven JFCM (1980) Thin Layer High Voltage Devices, Philips Journal of Research, vol 35, No 1
4. Shoofs F (1994) High Voltage ICs for Mains Applications, Proceedings of the Workshop Advances in Analog Circuit Design, Eindhoven University of Technology, Editors: van de Plassche RJ, Huijsing JH, Sansen W
5. Characteristics of the Electricity Supply Voltage for Low Voltage Consumers, Report of the UNIPEDE, Revue Generale de l'Electricité, N° 9, September 1981
6. EN 60555 European Standards
7. Albach M (1989) An AC-DC Converter with Low Mains Current Distortion and Minimized Conducted Emissions, EPE, Aachen
8. Designer's Guide to Power Products, Application Manual, SGS-THOMSON
9. Linear and Switching Voltage Regulators, Application Manual, SGS-THOMSON
10. Alimentatori stabilizzati a commutazione, Edizioni Philips/Elcoma, 1° ed Maggio 1980
11. Cini C, Diazzi C (1989) 250 V Integrated Full Bridge with Driving Stages and Logic Inputs, in Multipower BCD Technology, Proceedings of High Voltage and Smart Power ICs Symposium, vol 89-15, pp 446–465, Los Angeles
12. MOSPOWER Design Catalog, Siliconix, January 1983
13. Wegener A, Amato A (1989) A High Voltage Interface IC for Half-Bridge Circuits, Proceedings of High Voltage and Smart Power ICs Symposium, vol 89-15, pp 457–468, Los Angeles
14. Maurice B (1992) Simplified Digital Control for Three Phase Induction Motor Drive, EPE Journal, vol 2, n° 3

CHAPTER 9

Automotive Electronics

M. Morelli · G. Maggioni · V. Poletto

9.1 Introduction

The electronics in a car must satisfy increasingly different needs and its flexibility is the main reason for the widespread use in the automotive applications, enhancing the possibility to find the best solutions, satisfying various requirements which may often be in contrast with each other and must coexist in a decidedly hostile environment.

It is possible to identify some application areas with different characteristics and much are analysed and dealt with different instruments and methodologies.

Two devices are presented as examples, illustrating how the most suitable technology was chosen for each one, a technology which allowed the realisation of functions otherwise not possible neither technically nor economically: the multifunction voltage regulator and the injector driver.

For each of these we describe the functions and the used technologies, highlighting the reasons for the choice and the principal electric circuits that characterise them.

9.2 Multifunction Voltage Regulator

9.2.1 Need and Characteristics

The voltage regulator is one of the most important components in any electronic control unit, as its failure may be cause of total malfunction of the whole unit itself.

Sensors, microcontrollers and actuators show increasing complexity and more and more sophisticated features, hence they need to rely on a power supply with very good characteristics. The most important of them are the following:

- to keep a constant output voltage despite even rapid variations in supply voltage or in output current.
- to supply the load with rated service voltage and current even when the input voltage is extremely close to the rated load voltage (the difference between the two voltages may be as low as 0.5 V).

- to keep the rated voltage and current at the load, even in the event of a temporary interruption in supply, i.e. an input voltage lower than the output voltage, including negative values.
- to maintain operation as per the above point for as long as possible. This depends on the amount of energy stored in a capacitive element and the speed of its discharge.

Voltage regulators typically involve conflicting requirements including frequency compensation while at the same time maintaining accuracy of the system. The parameters normally characterizing voltage regulators include:

- Error. Defined as a variation in the ouput voltage in relation to a reference input voltage, due to numerous factors affecting mass production. It must be measured under steady conditions using a definite input voltage and output current.
- Regulation. Defined as a variation in the output voltage due to variations of input voltage and output current within a definite range. It must be measured in steady conditions.
- Settling time. The time taken to restore the output voltage to the correct value following a rapid variation in the input voltage or output current. It must be measured with a definite variation speed and amplitude.
- Peak error. Defined as the maximum deviation of the output voltage from its normal operating value, in the presence of a rapid transient which must be measured under the same conditions as in the above point.

Fig. 9.1 L 9170 Voltage Regulator Chip Photomicrograph

The current market demand is for voltage regulators with extremely good error and regulation characteristics.

In fact, on the one hand, a system should be designed to prevent a partial failure from jeopardizing the overall operation of the whole system, whereas a certain amount of degradation is normally acceptable. This requires dividing the voltage regulating function into various sections, so as to prevent a failure in one from involving the others.

On the other hand, an increase in the sophistication of the system, demands for a consequent increase in precision. This is the case of microprocessor systems involving digital/analog and analog/digital conversions, where the integrity of digital data widely depends on the ratio of analog data (ratiometric operation).

At the very last, therefore, the regulated voltages must be practically identical, thereby also yielding the demand for superior error and regulation characteristics.

The device described here includes four 5 V voltage regulators. They have output current rated at 1 mA, 500 mA and the last two 30 mA each (see Fig. 9.1).

9.2.2
Technology

The device has been developed using the SGS-Thomson BCD60I technology. The 60 V mixed Bipolar - Cmos - Dmos process has been quite useful to enhance many key features, such as: the stand-by quiescent current, the frequency stability vs load current and capacitance variations, logic functions like power latch, inrush current, reset, thermal shut-down in short circuit condition.

In addition this process provides very efficient lateral PNP power transistors for the low drop outputs. Their efficiency is measured in terms of silicon area, forward and reverse operating voltages, freedom of parasitic effects that may appear at very high speed voltage transients, or near the saturation region.

The following are the main technology features involved in the chip design:

- Drain extension resistors = 10 KΩ/ square
- Gate oxide capacitors = 380 pF/mm^2
- Lateral PNP BVceo (direct) = 60 V
- Lateral PNP BVeco (reverse) = 25 V
- Lateral PNP saturation resistance = $3\,\Omega \cdot mm^2$

9.2.3
The Power Section*

The power section includes the pass transistors and the circuitry driving their bases. Figure 9.2 shows the simplified schematic for the main output (V_{OUT}). The remaining three regulators have a similar configuration.

* SGS-THOMSON Microelectronics proprietary patent.

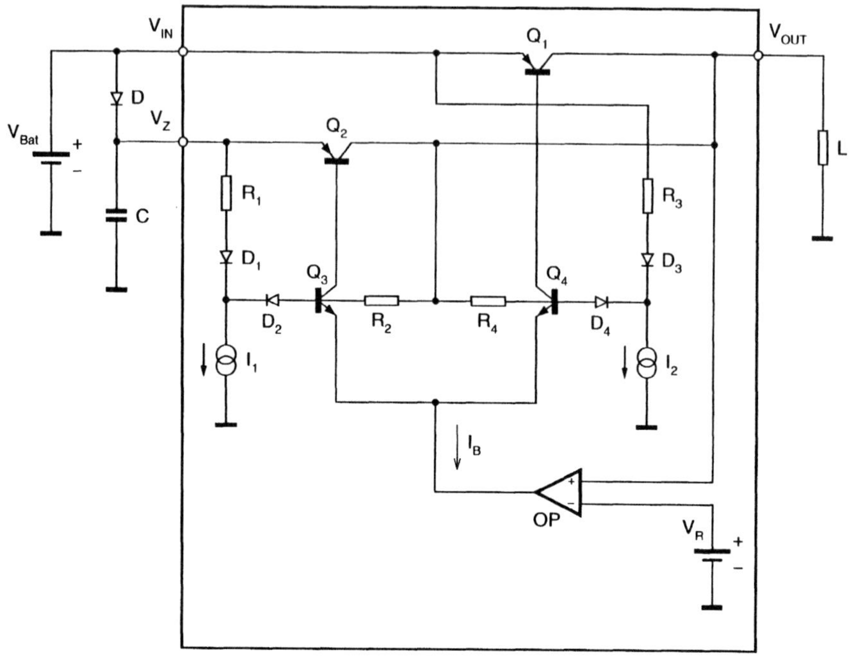

Fig. 9.2 Power Section

With a normal battery voltage the capacitor $C = 220\,\mu F$ (typical) is charged by the external diode D to a voltage around 12 V. The diodes D_2 and D_4 are reverse biased by the network R_1, D_1 and R_3, D_3 respectively. Since R_2 and R_4 have no current other than the Q_3 and Q_4 base currents, these transistors are both conducting, and I_B is split between them according to Q_4 and Q_3 area ratio. Being $\text{Area}(Q_4) = 2 \cdot \text{Area}(Q_3)$, the Q_1 base current is twice as much the Q_2 base current and we may reasonably expect their collector currents have approximately the same ratio. This is done because of the power transistor area ratio: $\text{Area}(Q_1) = 2 \cdot \text{Area}(Q_2)$, this choice will be explained later on. Anyway the power transistors Q_1 and Q_2 are both in a conducting state, contributing at the same time to the output current delivered to the load. This way the current density becomes equally distributed between the power transistors, for a higher reliability and a longer life.

In a situation where the battery voltage at V_{IN} input slowly drops to 5.5 V (Cold Cranking), the capacitor is discharged by the Q_2 emitter current and its voltage at V_Z input follows V_{IN} lowered by the diode D drop $V_D = 0.7$ V. The diode D_2 becomes forward biased when $V_{IN} = V_{OUT} - V_{D2} + V_{D1} + I_1 \cdot R_1 + V_D$ (see the network involving the mentioned parameters). Being $V_{OUT} = V_R = 5$ V, $V_{D1} = V_{D2} = V_D = 0.7$ V, $I_1 \cdot R_1 = 1$ V, this V_{IN} threshold is 6.7 V, below which Q_3 turns off and the share of I_B current flowing from the Q_2 base switches to the Q_1 base across Q_4. At this point the output current is entirely supported by Q_1, and V_{IN} may decrease to its minimum value when Q_1 saturates, without deteriorating the output performances.

In a condition where the battery voltage at V_{IN} input quickly drops or reverses, the diode D switches off and the capacitor C supplies the Q_2 current at the V_Z input with a negative voltage ramp. If V_{IN} drops below the threshold $V_{IN} = V_{OUT} - V_{D4} + V_{D3} + R_3 \cdot I_2 = 6\,V$ (considering $V_{OUT} = 5\,V$, $V_{D4} = V_{D3}$, $R_3 \cdot I_2 = 1\,V$), I_2 directly biases D_4 turning Q_4 and Q_1 off and switching the full amount of I_B through Q_3 to the Q_2 base.

Hence Q_2 has to stand the full amount of the output current, entirely coming from the capacitor C. Since no current is necessary to the V_{IN} input, its voltage may become even negative (down to $-30\,V$) while the device keeps on operating regularly. The capacitor C acts as an energy reservoir and it is discharged by the output current (500 mA) plus the device bias current (30 mA). Under these conditions V_Z may ramp down from 12 V to 6 V in $(12\,V - 6\,V) \cdot 220\,\mu F / 530\,mA = 2.5\,ms$. If the battery voltage at V_{IN} pin resumes its normal value, higher than 5.5 V, before the 2.5 ms elapses, the device keeps on working virtually with no discontinuity.

Since Q_1 has to deliver the maximum output current (500 mA) with the lowest V_{IN} to V_{OUT} voltage drop (0.5 V), while Q_2 may work in more relaxed conditions (500 mA, 1 V), the Q_1 and Q_2 area ratio must be two.

The block OP includes all the circuitry for the output voltage regulation and the frequency compensation. It is described in the following section.

9.2.4 The Voltage Regulator Loop and its Frequency Compensation

Due to recent demand for limiting electro-magnetic interferences, regulated voltage lines must include capacitors with extremely low equivalent series resistance (ESR). For technical reasons such capacitors are low valued. In addition, regulated voltage lines are also connected to higher valued capacitors, for sustaining loads requiring high instantaneous currents, but the ESR of which may become quite high, especially at low temperatures (−40 °C). The required precision for the regulated output voltage is typically 5 mV for a load current ranging form 0 to 500 mA. This means that the open loop gain at low frequencies must be the following:

$$Gm = \frac{\Delta I_{OUT}}{V_{IN}} = \frac{500\,mA}{5\,mV} = 100\,A/V$$

Notice that a low drop voltage regulator with a PNP type power transistor shows an open loop behaviour like a transconductance amplifier. In fact the output signal is the collector current of the power transistor, while the feedback signal is the voltage error.

Being 100 nF the max. load capacitor available with low ESR, the gain-bandwidth product at low frequency must be:

$$GBl.f. = \frac{Gm}{2\pi C} = \frac{100\,A/V}{2\pi \cdot 100\,nF} = 159\,MHz$$

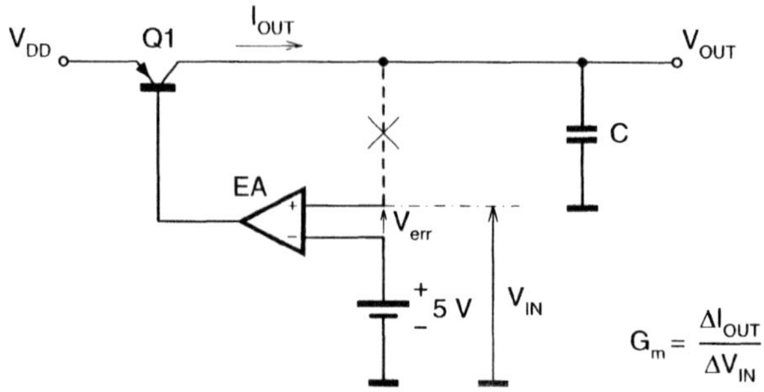

Fig. 9.3 Low frequency block diagram

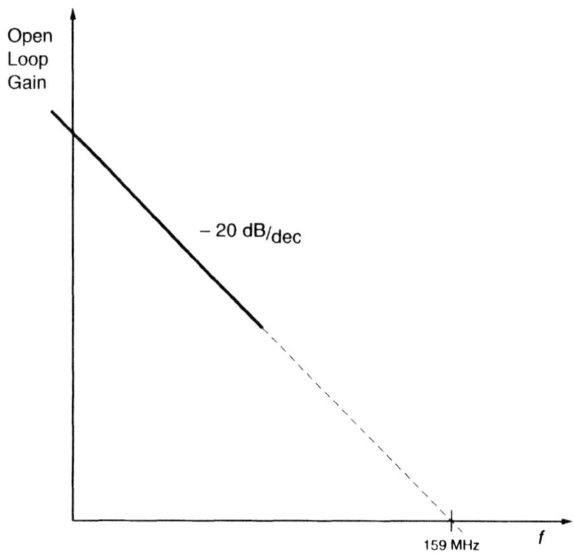

Fig. 9.4 Low frequency open loop gain Bode plot

Figure 9.3 shows the block diagram of the low drop voltage regulator at low frequency.

The Bode Diagram of the open loop transfer function for the low drop voltage regulator at low frequency is shown in Fig. 9.4.

In case the error amplifier EA plus transistor Q_1 of Fig. 9.3 had a constant G_m with respect to frequency, the open loop gain of Fig. 9.4 would actually cross the zero dB level at 159 MHz. In that case the frequency stability would be seriously compromised, since many parasitic poles, not shown in Fig. 9.4, located at frequencies lower than 159 MHz, would deteriorate the phase margin.

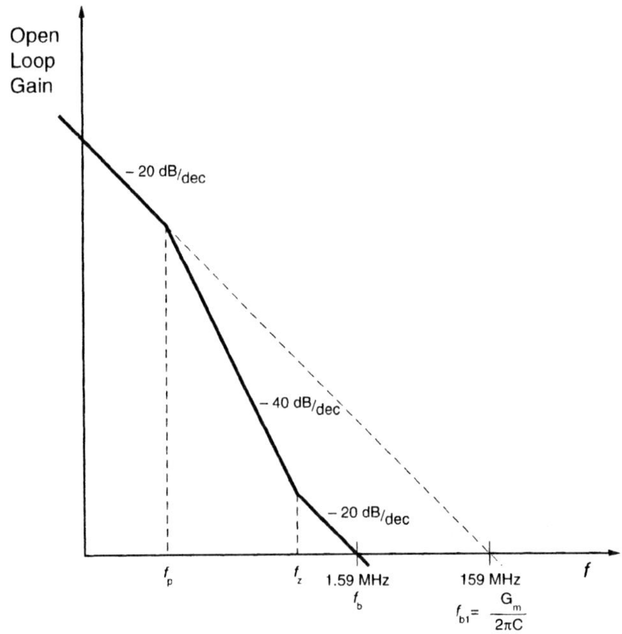

Fig. 9.5 Complete open loop gain Bode plot

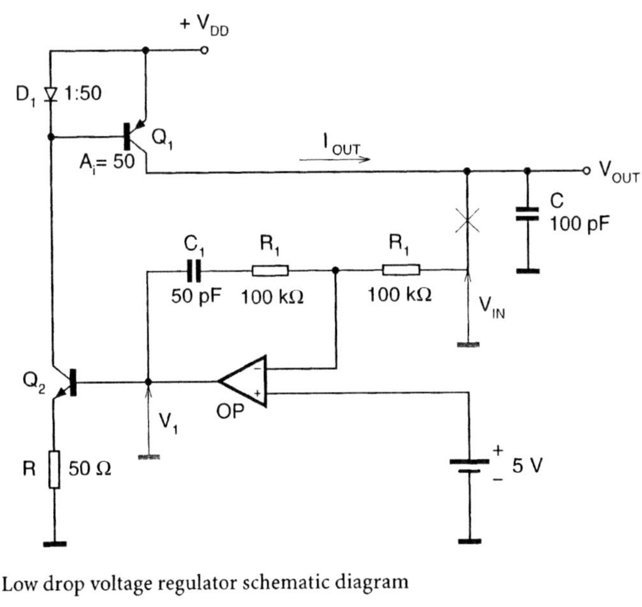

Fig. 9.6 Low drop voltage regulator schematic diagram

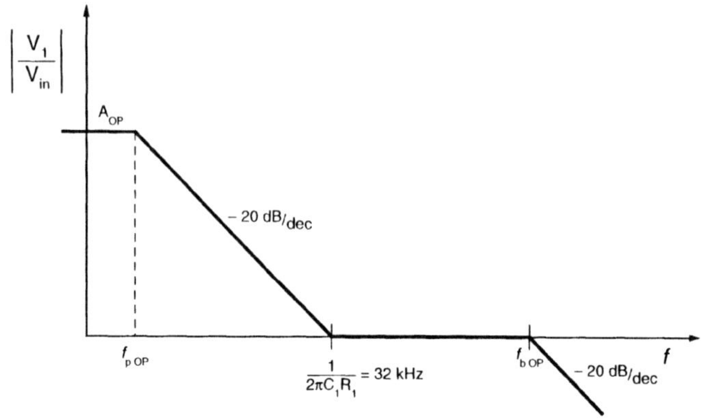

Fig. 9.7 Error Amplifier Bode Plot

Fig. 9.8 Output section Bode Plot

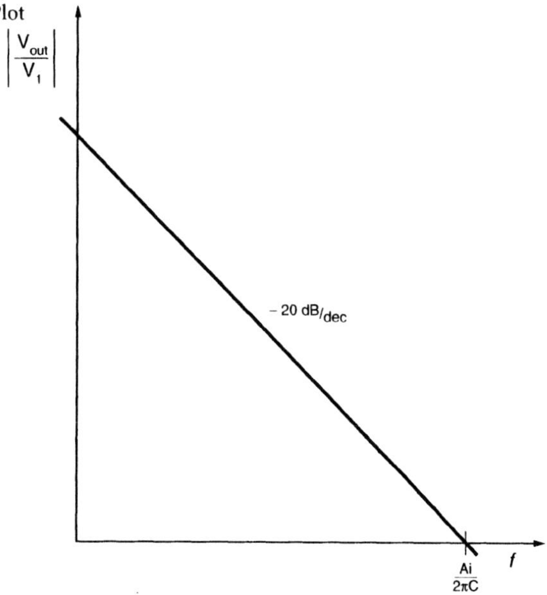

For this reason the zero dB crossing must be set at a considerably lower frequency, for example 1.59 MHz (two decades lower than the low frequency gain-bandwidth product).

In Fig. 9.5 the open loop gain for the low drop voltage regulator described so far is reported.

As it can be noticed, the low frequency line at −20 dB/dec slope must be connected to the high frequency line at −20 dB/dec slope through an intermediate line at −40 dB/dec slope.

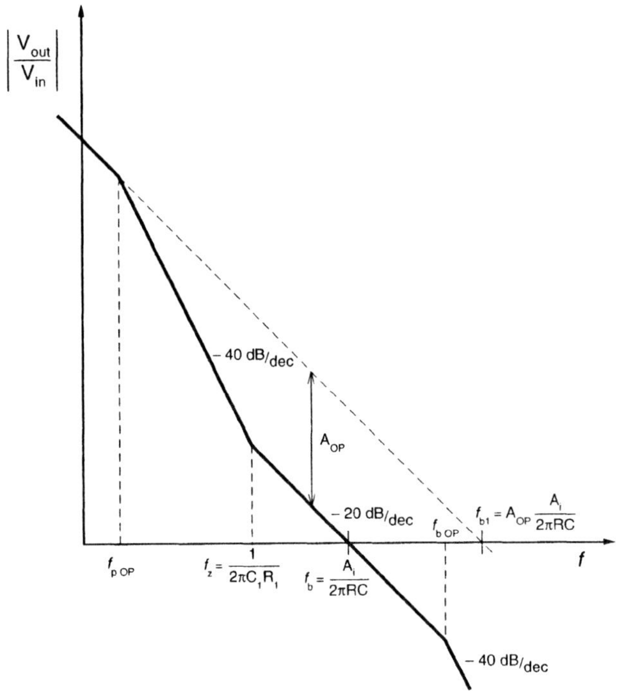

Fig. 9.9 Low drop voltage regulator open loop Bode Plot

Figure 9.6 shows the circuit that approximates the transfer function of Fig. 9.5*. The real Bode plot related to Fig. 9.6 can be calculated by cascading the transfer functions V_1/V_{IN} (see Fig. 9.7) and V_{OUT}/V_1 (see Fig. 9.8).

Considering Fig. 9.6, Aop is the low frequency open loop gain of the operational amplifier OP, f_{pOP} is the low frequency open loop pole of OP, f_{bOP} is the 6 dB frequency of the OP open loop gain. Figure 9.9, that shows the Bode plot for the circuit in Fig. 9.6, is obtained by multiplying the transfer functions of Figs. 9.7 and 9.8.

Comparing Figs. 9.9 and 9.5 it can be seen, from the low frequency gain-bandwidth product f_{b1}, that the low frequency transconductance can be calculated as follows:

$$A_{OP} \cdot \frac{Ai}{2\pi RC} = \frac{Gm}{2\pi C} \qquad \text{hence:} \qquad Gm = \frac{A_{OP} \cdot Ai}{R}$$

Using the values R = 50 Ω and Ai = 50 as shown in Fig. 9.6, the load regulation requirements (asking for a Gm > 100 A/V as previously explained) are satisfied

* SGS-THOMSON Microelectronics proprietary patent.

for:

$$Gm = \frac{A_{OP} \cdot Ai}{Gm} = \frac{A_{OP} \cdot 50}{50\,\Omega} > 100\ \text{A/V} \qquad \text{hence:} \quad A_{OP} > 100$$

This means that any operational amplifier with a low frequency gain higher than 40 dB does the job. Since Aop is usually at least 60 dB, the typical load regulation for a 500 mA max. load current voltage regulator, can be calculated as follows:

$$\Delta V = \frac{1}{Gm}\,\Delta I = \frac{R}{A_{OP} \cdot Ai}\,\Delta I = \frac{50\,\Omega}{1000 \cdot 50}\,500\ \text{mA} = 0.5\ \text{mV}$$

The plot of Fig. 9.9 can be described as a function of a complex frequency (f) with the following equation:

$$\frac{V_{OUT}(f)}{V_{IN}(f)} = \frac{f + \frac{1}{2\pi C_1 R_1}}{f \cdot (f + f_{pOP})\ (f + f_{bOP})} \cdot \frac{Ai}{2\pi RC} \cdot f_{bOP}$$

From this equation the root locus can be drawn as shown in Fig. 9.10 where the constant values are the following:

$$fz = \frac{1}{2\pi C_1 R_1} = 32\ \text{KHz} \qquad f_{pOP} \cong 1\ \text{Hz} \qquad f_{bOP} = 1.6\ \text{MHz}$$

The multiplying factor (frequency constant) $\frac{Ai}{2\pi RC} f_{bOP}$ should be the parameter of the root locus increasing from zero to infinite, but in this case C is an important parameter that can practically have many different values.

For this reason the root locus can be more conveniently plotted with a decreasing C and assuming constant values for $Ai = 50$ and $R = 50\,\Omega$. Fig. 9.10 shows some important points (P_1 to P_4) in which the closed loop poles lie:

P_1: two complex poles at 32 KHz (plus a third one at around 1.6 MHz)
$C = 2\ \mu F$, $f_b = 128$ KHz

P_2: two real poles at 64 KHz (plus a third one at around 1.6 MHz)
$C = 1\ \mu F$, $f_b = 128$ KHz

P_3: two real poles at 800KHz (plus a third one at around 32 KHz)
$C = 300$ nF, $f_b = 380$ KHz

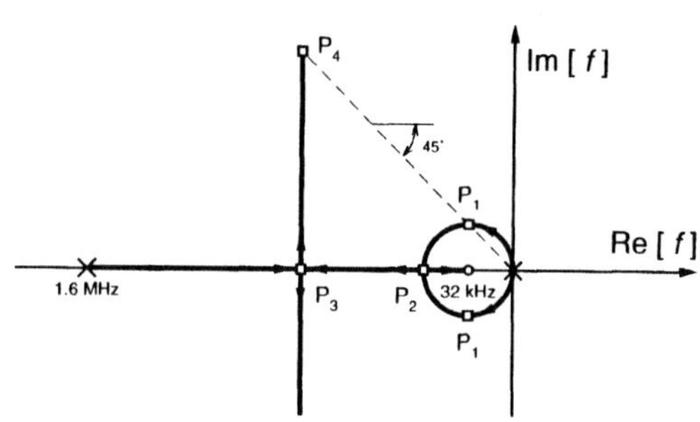

Fig. 9.10 Low drop voltage regulator root locus

P_4: two complex poles at 800 KHz (plus a third one at around 32 KHz)
$C = 170$ nF, $f_b = 700$ KHz

It is evident that a good closed loop frequency stability is assured when closed loop poles vary between positions P_1 and P_4, corresponding to capacitor values in the range:

$$170 \text{ nF} < C < 2 \text{ μF}$$

Also Fig. 9.9 shows that the load capacitor C has both upper and lower limits. In fact shifting vertically the Bode plot with respect to the zero dB level, it is noticeable that fb may range from fz to f_{bOP} always keeping a good phase margin and having around 45° phase margin in the extreme points. Following this way, the load capacitor range may be calculated as follows:

$$f_z < f_b < f_{bOP}$$

$$f_z < \frac{Ai}{2\pi RC} < f_{bOP}$$

$$\frac{Ai}{2\pi R f_z} > C > \frac{Ai}{2\pi R f_{bOP}}$$

$$\frac{50}{2\pi \cdot 50\ \Omega \cdot 32\ \text{KHz}} > C > \frac{50}{2\pi \cdot 50\ \Omega \cdot 1.6\ \text{MHz}}$$

$$5\ \mu\text{F} > C > 100\ \text{nF}$$

Practically, capacitors with values higher than 5 μF are also possible, since they have an ESR that make them behave as resistors more than capacitors.

The resistors $R = 50\ \Omega$ in Fig. 9.6 is important not only for the frequency stability, but also because it sets the max output current Iout. In fact the max voltage applied to resistor R is around $V_r(\text{max}) = 3$ V, since the OP has a 5 V voltage supply. Hence the max output current $I_{out}(\text{max})$ can be calculated as follows:

$$I_{OUT}(\text{max}) = \frac{V_R(\text{max})}{R} \cdot Ai = \frac{3\ \text{V}}{50\ \Omega} \cdot 50 = 3\ \text{A}$$

Though various factors are omitted in the frequency stability calculation described earlier, this in no way affects the accuracy of the evaluation of the frequencies f_Z, f_b and f_{bOP} and their distances. Hence the stability margin can be set with a good degree of confidence even in a wide range of external conditions, mainly related to the load.

This compensation technique gives the voltage regulator some superior characteristics like:

- Compactness and straightforward design. The schematic in Fig. 9.6 schematic is more or less complete and not just a block diagram. It may even be produced in discrete form using a small number of components.
- Precision. The output voltage matches the input voltage within less than 10 mV including all regulation errors coming from load, supply, temperature and production spreads.

- Versatility. The high degree of stability of the regulator enables it to be employed under critical conditions, like low temperatures or in the presence of electromagnetic interferences. Electrolytic capacitors for frequency stability are no longer required, while small capacitors with a very low ESR may be needed.
- Speed. The open loop transfer function may be easily evaluated with a high degree of precision. Even the first pole f_{bOP}, occurring beyond the cut-off frequency f_b, is precisely located. This is usually a parasitic pole, widely spreading over the frequency.

9.2.5 The RESET Circuit

Cars are receptacles of an enormous amount of disturbances much of which originate in the electrical equipment of the car itself. They are caused, for example, by on-off operation of various types of inductive loads (ignition coils, injectors, relays) or resistive loads (lights). They are also not immune to additional external disturbances, such as those caused by radio and television transmitters and repeaters.

As cars become more and more sophisticated from the point of view of their electronic equipment, such an unfavourable environment cannot be ignored by the designer. In case of microprocessor systems, in particular a strict control of the supply voltage is required.

This is tapped from the battery voltage by means of a voltage regulator which, regardless of its performance, cannot possibly guarantee a correct operation under all working conditions. The purpose behind such control of the supply

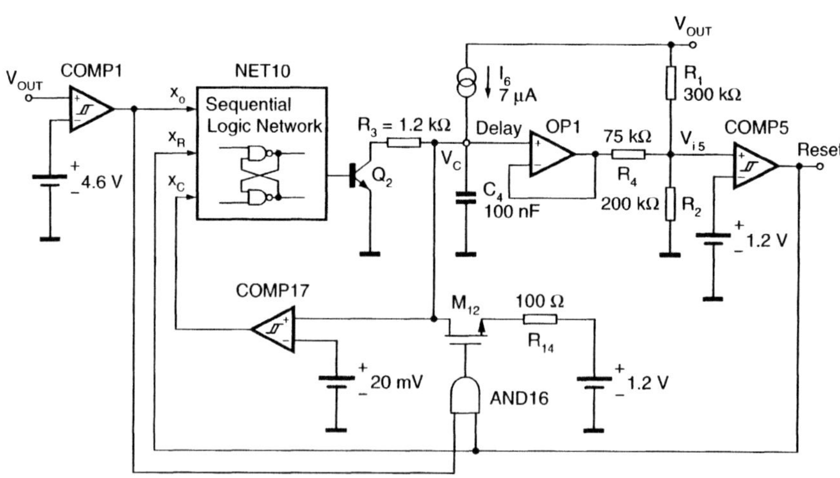

Fig. 9.11 Reset circuit Block diagram

voltage is to disable the operation of the microprocessor when it cannot operate reliably. The information for disabling the microprocessor is supplied by the RESET circuit.

Figure 9.11 shows the block diagram of the RESET circuit*. The output voltage V_{OUT} of the voltage regulator is monitored by COMP1 comparing it to a 4.6 V threshold voltage.

As long as V_{OUT} remains above this threshold, signal X_0 remains high, thus keeping switch Q_2 open (via network NET10) and switch M_{12} closed (via AND16). The capacitor C_4 is then charged to 1.2 V through R_{14}. The input V_{i5} of comparator COMP5 is higher than the threshold 1.2 V, hence the Reset output is high (inactive).

When the V_{OUT} voltage falls below the 4.6 V threshold, signal X_0 switches to zero closing switch Q_2 (via NET10) and opening switch M_{12}, thus discharging capacitor C_4 via resistor R_3.

Resitors R_1, R_2 and R_4 perform a weighted addition of V_{OUT} and V_C (voltage of capacitor C_4) as follows:

$$V_{i5} = \frac{\dfrac{V_{OUT}}{R_1} + \dfrac{V_C}{R_4}}{\dfrac{1}{R_1} + \dfrac{1}{R_2} + \dfrac{1}{R_3}} = \frac{\dfrac{V_{OUT}}{300\ \text{K}\Omega} + \dfrac{V_C}{75\ \text{K}\Omega}}{\dfrac{1}{300\ \text{K}\Omega} + \dfrac{1}{75\ \text{K}\Omega} + \dfrac{1}{200\ \text{K}\Omega}}$$

$$V_{i5} = \frac{V_{OUT}}{6.5} + \frac{V_C}{1.625}$$

When capacitor C_4 discharges through R_3, its voltage V_C is the following:

$$V_C = 1.2\ \text{V} \cdot e^{-\frac{t}{R3C4}} = 1.2\ \text{V} \cdot e^{-\frac{t}{120\ \mu\text{s}}}$$

hence:

$$V_{i5} = \frac{V_{OUT}}{6.5} + \frac{1.2\ \text{V}}{1.65}\, e^{-\frac{t}{120\ \mu\text{s}}}$$

The reset output of comparator COMP5 switches low (active) whenever V_{i5} becomes lower than the 1.2 V threshold. When this happens, the V_{OUT} reset threshold (V_{out} res) changes with time as follows:

$$1.2\ \text{V} = \frac{V_{out}\ \text{res}(t)}{6.5} + \frac{1.2\ \text{V}}{1.625}\, e^{-\frac{t}{120\ \mu\text{s}}}$$

$$V_{out}\ \text{res}(t) = 6.5 \cdot 1.2\ \text{V} \cdot \left(1 - \frac{1}{1.625}\, e^{-\frac{t}{120\ \mu\text{s}}}\right) = 7.8\ \text{V} \cdot \left(1 - \frac{e^{-\frac{t}{120\ \mu\text{s}}}}{1.625}\right)$$

* SGS-THOMSON Microelectronics proprietary patent.

The above expression means that at t = 0 (the instant that the falling V_{OUT} crosses the 4.6 V threshold), the Reset signal is activated if the falling V_{OUT} also crosses the following threshold:

$$V_{out}\,\text{res}(0) = 7.8\text{ V} \cdot \left(1 - \frac{1}{1.625}\right) = 3\text{ V}$$

In other words, the Reset signal is immediately activated when V_{OUT} drops below 3 V starting from normal values (above 4.6 V).

At t = 30 µs (that is 30 µs after V_{OUT} has fallen below 4.6 V), the V_{OUT} reset threshold is:

$$V_{out}\,\text{res}\,(30\,\mu\text{s}) = 7.8\text{ V} \cdot \left(1 - \frac{e^{-\frac{30}{120}}}{1.625}\right) = 4\text{ V}$$

In other words, the Reset signal is activated with a 30 µs delay if V_{OUT} drops to 4 V.

At t = 50 µs, the V_{OUT} reset threshold is:

$$V_{out}\,\text{res}\,(50\text{ ms}) = 7.8\text{ V} \cdot \left(1 - \frac{e^{-\frac{50}{120}}}{1.625}\right) = 4.6\text{ V}$$

In other words the Reset signal is activated with a 50 µs delay if V_{OUT} drops just below 4.6 V. The Reset signal thus depends on both the amount and the duration of the drop in supply voltage. In such a way, a substantial drop in supply voltage for a relatively short time is sufficient to enable the reset signal. On the other hand a small drop in supply voltage can be tolerated for a relatively longer time before the Reset signal is issued.

When the Reset signal is released, the sequential logic network NET10 is informed through its input X_R and keeps switch Q_2 closed as long as the capacitor voltage V_C remains above 20 mV. This way even if the supply voltage drops rapidly enough to cause a reset, and than raises back up over the 4.6 V threshold, the capacitor keeps on discharging down to 20 mV, as detected by comparator COMP17, which informs NET10 by switching signal X_C to the low value. At this point NET10, if V_{OUT} is higher than 4.6 V, opens the switch Q_2 and the capacitor C_4 begins charging via the current source $I_6 = 7\,\mu A$, from a known starting voltage close to zero.

When the capacitor voltage V_C reaches a certain value, the Reset signal switches to the high value (i.e. it is disabled). Consequently the capacitor is charged over a practically constant time period, thus eliminating any uncertainty as to the delay with which the Reset signal is disabled.

On the other hand, the supply voltage V_{OUT} may fall below the 4.6 V threshold, thus switching X_0 from 1 to 0, but without switching the Reset signal, so that X_R remains high, then V_{OUT} may rise immediately above the 4.6 V threshold, switching X_0 from 0 to 1. In this case AND16 drives the switch M_{12} to open first, then to close. Since $R_{14} = 100\ \Omega$ is quite small, it charges C_4 quickly so that in a short time it becomes ready to be discharged again by the following drop in V_{OUT} even if it closely follows the previous one, but still assuring a correct V_{OUT} voltage drop time measurement.

9.3 Injector Driver

9.3.1 Fuel Injection System Description

The fuel electronic injection systems for internal combustion engines have by now become a need for all the car manufacturers.

Even if these systems have been developed to satisfy the requirements of better performances from competition engines or high level cars, for a few years, they will have to be designed in accordance with the international laws, limiting the fuel consumption and the atmospheric pollution.

The most economical way found by engine designers to respect these laws, was to maintain as far as possible under control the ratio between the air in the intake manifold and the fuel injected on all ranges of engine operation in terms of revolution speed and load.

The precision of the injected fuel amount is hardly reachable with the classical carburettor systems and for this reason the electronically controlled injection systems have by now become a necessity. They can be divided in two categories:

SINGLE POINT and MULTI POINT.

In the former, one single injector feeds all cylinders while the fuel distribution works as in the old carburettor systems.

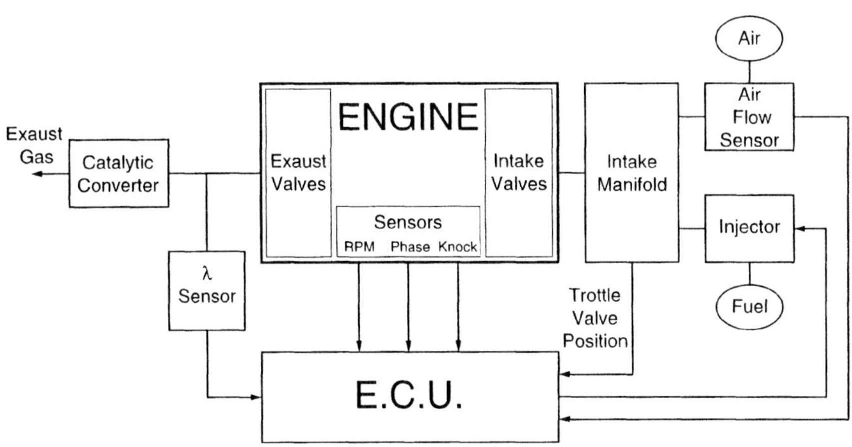

Fig. 9.12 Single point injection system

In the second one, there are as many injectors as the cylinders, and the fuel is injected into the single intake manifold just before the intake valve [6].

Of course the multipoint system is the most precise, even if the single point one presents some advantages, as the cost decidedly lower and the demand of few modifications on an engine designed with carburettor, since the single injector is placed in the same position as the carburettor just after the air-filter.

In this section we will describe a driver for a single point injector which takes into account all the actual requirements in terms of precision and economy of the solution.

The components of a single point system are described in the block diagram of Fig. 9.12.

This injection system detects the air mass coming from the throttle valve and then calculates and injects through the injector the exact quantity of fuel required to obtain the best combustion conditions.

The ratio between the two quantities of air and fuel must be kept as near as possible to 14.7:1 (stochiometric ratio).

This means 14 kg of air every kg of fuel or 10 000 litres of air every litre of fuel.

The ratio between the actual air to fuel concentration and its theoretical (or stochiometric) value is called λ factor:

$$\lambda = \frac{\text{Air/Fuel (actual)}}{\text{Air/Fuel (stochiometric)}}$$

Figure 9.13 shows how the precision of the mixture affects the exhaust gas composition.

If the pressure in the fuel distributor circuit is kept constant, the quantity of the injected fuel is only related to the opening time of the injector.

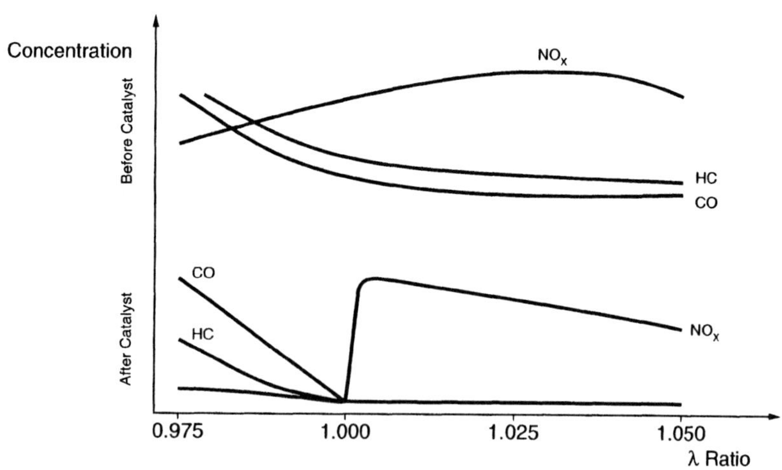

Fig. 9.13 Trend of NOx CO HC concentration

The system has a closed loop control made by the lambda oxygen sensors mounted before and after the catalytic converter. This allows the trimming of the mixture composition by the electronic control unit to achieve the appropriate λ ratio.

9.3.2 Injector Driver Tasks

The main task of an injector driver is to control the electromagnetic valve that introduces the fuel in precise way without any other error than the mechanical errors of the injector itself (see Fig. 9.14).

The injector in a single point system can be modelled by a simple equivalent circuit consisting in a series of a resistor and an inductance. The resistor value is in a range from 1 Ω to 2 Ω and the inductance from 2 mH to 4 mH.

9.3.3 Current Control

The device described here includes a double level switching current regulation to reduce the total power dissipation of the system.

Figure 9.15 shows the current waveform in the injector coil during an operation cycle.

In the first phase (t_0-t_1), after the rising edge of the input command V_{IN}, the whole battery voltage is applied to the injector and the current rises with a

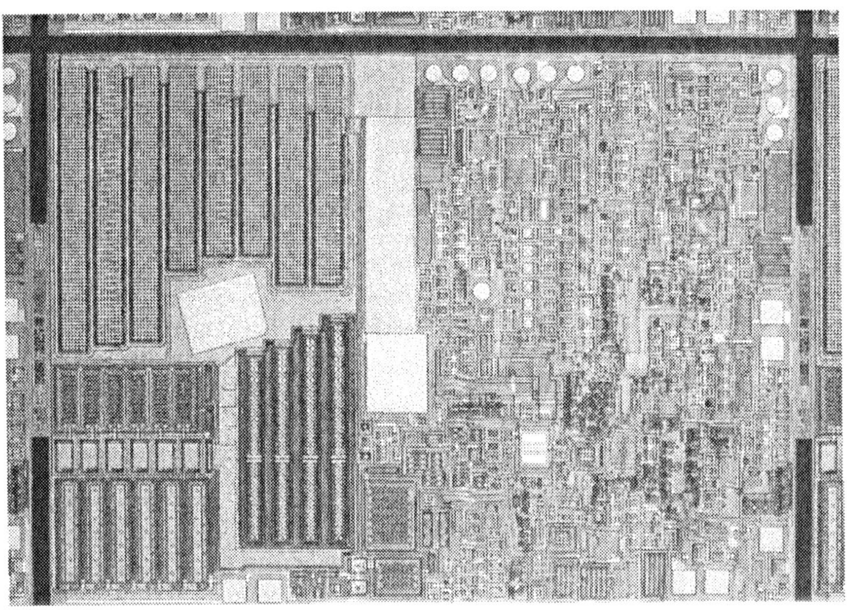

Fig. 9.14 L 9140 Injector Driver Chip Photomicrograph

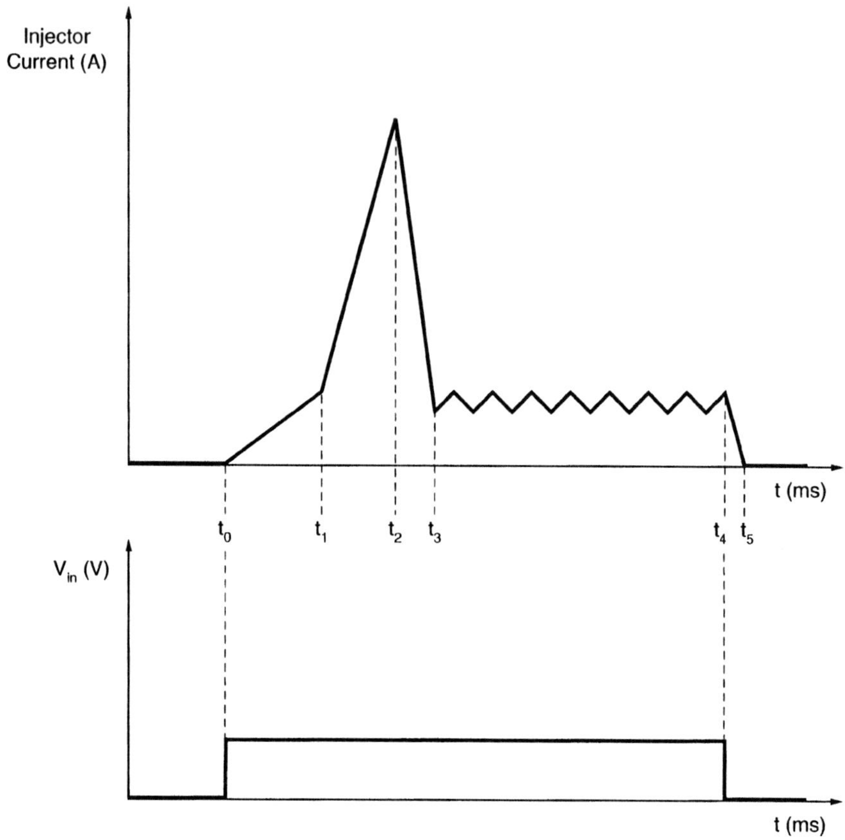

Fig. 9.15 Injector current waveform

rate depending on the time constant of the circuit according to the following low:

$$I(t) = \frac{V_{Bat}}{R}\left(1 - e^{-\frac{R}{L}t}\right)$$

There is a period $t_1 - t_2$ where the slope of the current wave form changes, due to the magnetic field modification caused by the movement of the nozzle needle of the injector in the opening phase. This slope change corresponds to the current level needed to win the mechanical inertia and the force of the opposing springs, therefore it is the current needed to open the injector.

As it is shown in the Fig. 9.15, the peak current is set at a much higher level, to be sure of the injector opening regardless of its mechanical characteristics. Furthermore it can be reduced at a lower one when the injector has to be maintained open (phase $t_2 - t_4$) in order to minimize the power dissipation. On the other hand, the level of the hold current could be fixed with linear or switching

Fig. 9.16 Driver and injector equivalent scheme

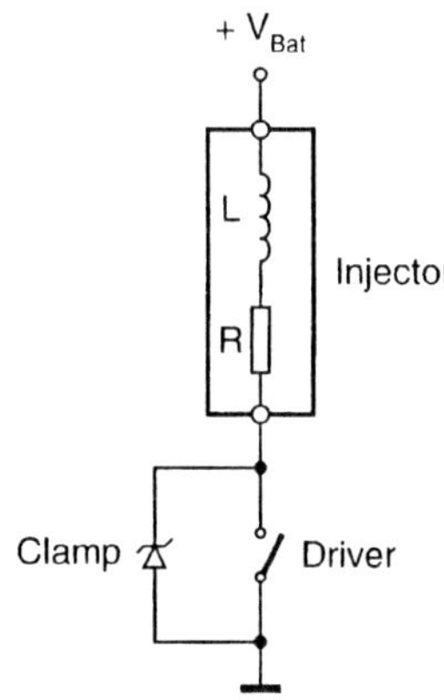

mode, but to further reduce the power dissipation the switching solution has been chosen.

The time duration of the current reduction from the peak value to the hold value (t_2-t_3) and that from the hold to the off condition (t_4-t_5) are also important because they affect the injected fuel quantity; in fact, during the phase t_2-t_3 it is necessary to rapidly reach the hold condition. The same is true during the phase t_4-t_5 because any delay corresponds to a greater injected fuel quantity, larger than the one determined by the control signal V_{IN}.

In Fig. 9.16 the equivalent schematic for the injector, the driver and the recirculation path is shown.

After the inductance charging phase let us analyse the recirculation one.

The formula related to the current decay is the following:

$$I(t) = I_0 e^{-\frac{t}{\tau}} - \frac{V_{REC}}{R}\left(1 - e^{-\frac{t}{\tau}}\right)$$

Where:

$$\tau = \frac{L}{R} \qquad V_{REC} = V_{CL} - V_{Bat} \qquad I_0 = \frac{V}{R}$$

and V_{CL} = Voltage of the Clamp Zener.

It is evident that, in order to obtain short and precise recirculation times, it is important that the value of the hold current be accurate and the recirculation voltage as much high as possible, compatible with the used driving circuit technology.

9.3.4
Voltage Control

The most utilised clamping structures are shown in Figs. 9.17 and 9.18.

In Fig. 9.17 the current flowing in the solenoid winding has to circulate in the zener Z_1, when the power switch PW is off; the voltage across the load is $V_{REC} = V_{Z1} - V_{Bat}$.

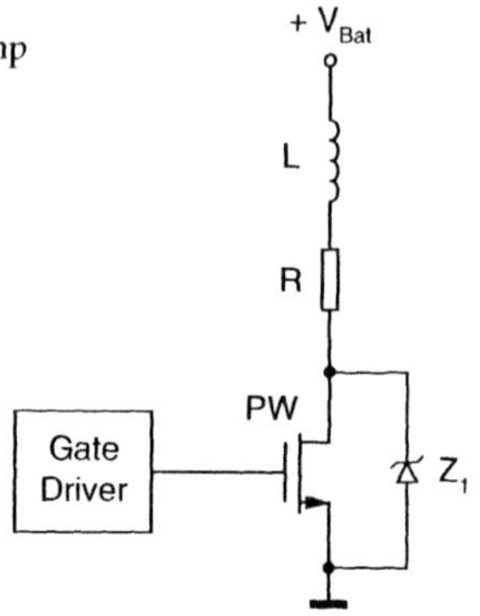

Fig. 9.17 Standard voltage clamp between drain and source

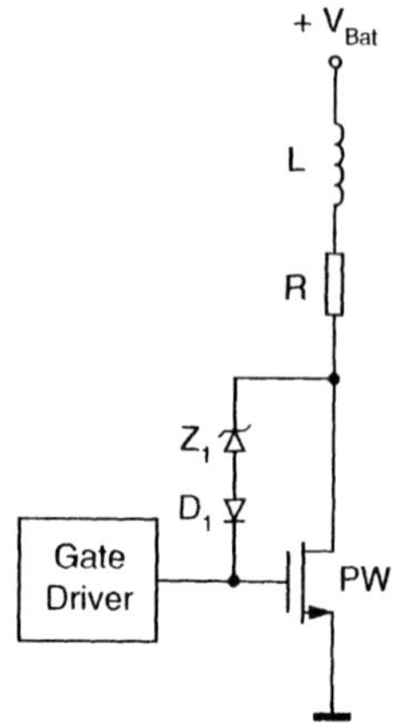

Fig. 9.18 Standard voltage clamp between drain and gate

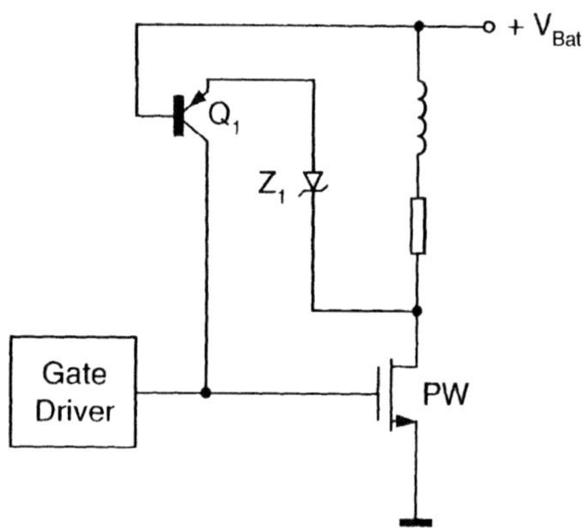

Fig. 9.19 High voltage clamp

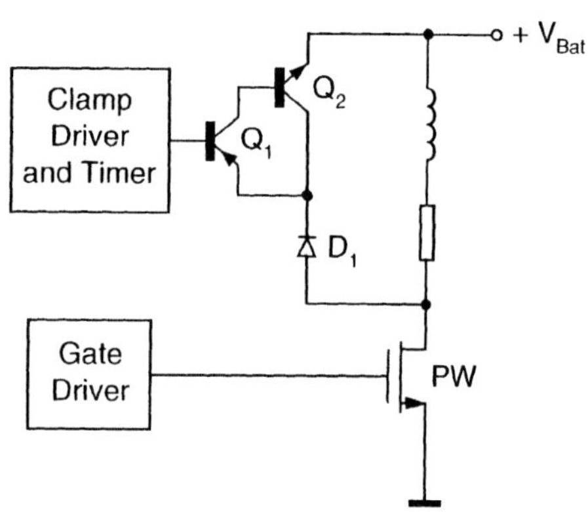

Fig. 9.20 Low voltage clamp

In Fig. 9.18 the discharge of the energy stored in the coil is obtained forcing the current through the same power switch PW whose gate is biased by Z_1 and D_1. In this configuration the voltage across the load is:

$$V_{REC} = V_{Z1} + V_{D1} + V_{GS(PW)} - V_{Bat}$$

In both these solutions it is evident that the recirculation voltage, and therefore the time, depends on the battery voltage and this could generate an error in the quantity of injected fuel.

A solution that overcomes this problem is shown in Fig. 9.19.

In that case the recirculation voltage is

$$V_{REC} = V_{Z1} + V_{BE\,(Q1)}$$

that is independent of the battery voltage.

These remarks are valid for the time intervals t_2-t_3 and t_4-t_5 in Fig. 9.15; during the hold phase (t_3-t_4), when the current is switching, to avoid a too high ripple, it is necessary to utilise a structure with a low recirculation voltage.

Figure 9.20 illustrates a possible solution:

The structure made up by Q_1, Q_2, D_1 guarantees on the load, at typical hold current values of one or two amperes, a maximum voltage of:

$$V_{D1} + V_{CE1(SAT)} + V_{BE2} \cong 2.5\ \text{V}.$$

9.3.5
Technology

The Single Point Injection System needs power switches with good characteristics (currents of 5-8 amperes and breakdown voltages close to 100 volts), precision analog circuits for the current regulation, digital circuits for the timings and interfaces towards the microprocessor of the engine control unit (E.C.U.).

The choice of the integration is of basic importance to minimize the global cost of the function in comparison with a discrete component solution where it is necessary to use about twenty components, among which an expensive power sensing resistor and at least two comparator circuits.

Another reason for an integrated solution lies on the fact that in this way it is possible to save the CPU time required to manage all the timings, because in this case the microprocessor has to supply only the signal defining the time of the injection, leaving to the integrated circuit the control of the injector.

SGS-THOMSON, with the purpose to comply with all the requirements of an Injection system, has developped a device using its BCD100 technology that allows the integration on the same chip of vertical DMOS with BVdss = 100V, bipolar components and CMOS logic circuits [7].

The main features of this technology are listed in the following tables.

High Voltage

Components	Parameter	Value	Unit
Vertical Power DMOS	BV_{dss}	100	V
	V_{th}	3.6	V
	$R_{ON} \cdot$ Area	1.3	$\Omega \cdot mm^2$
H. V. PChannel MOS	BV_{dss}	110	V
	V_{th}	1.6	V
	$R_{ON} \cdot$ Area	10	$\Omega \cdot mm^2$
H. V. NPN	BV_{cbo}	100	V
	BV_{ceo}	30	V
	H_{fe}	220	
H. V. PNP	BV_{ceo}	100	V
	H_{fe}	250	

Low Voltage

Components	Parameter	Value	Unit
Lateral DMOS	BV_{dss}	20	V
CMOS	Absolute max. rating	10	V
	Density	650	TR/mm^2
NPN Pwell Base	BV_{ceo}	30	V
	H_{fe}	220	

9.3.6 Block Diagram

Figure 9.21 shows the simplified schematic of the device.

C_{IN} represents the input comparator of the device and it is directly driven by the microprocessor.

A logic level one causes the power DMOS turn on and therefore the injector opening. C_1, C_2, C_3 are the comparators at whose inputs the voltage generated on the sensing resistor by the load current is applied.

With reference to Fig. 9.21, the comparator C_1 defines the peak current level and the switching time from peak to hold condition, while the comparator C_3 defines the time when the output power is turned on again to start the hold phase.

The comparator C_2 defines the high level of the current during the hold phase.

It most be pointed out that the switching control is realised with a current threshold (C_2) and a fixed turn off time, driven by the timer block that fixes a "time off"of 20 μs.

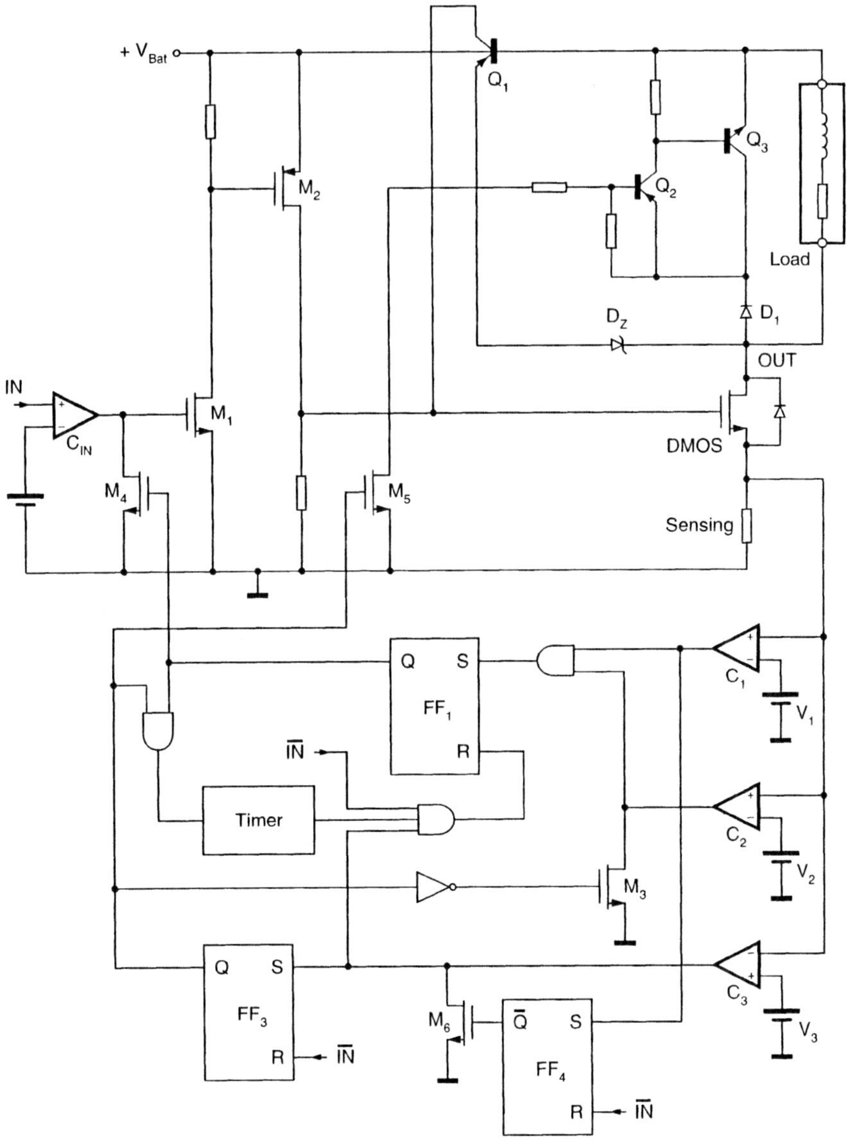

Fig. 9.21 Block diagram of the driver

9.3.7
Voltages References

As it has just been pointed out, to limit the error on the injected fuel quantity, it is necessary to control the current levels of the device with high precision.

A classical structure for the measurement on a power driver, without using external sensing resistors, makes use of the power metal layer as a shunt resistor connected to a comparator with a suitable threshold.

The circuit is shown in Fig. 9.22.

The load current is measured as a voltage drop on the sensing resistor R_S and when the voltage exceeds Vth, the comparator C_1 acts on the driver circuits reducing the load current or turning off the power transistor.

The solution with the internal sensing resistor is the most suitable even if its high thermal coefficient (4100 ppm/°C for Al-Si metal resistor) must be carefully considered as it can cause an unacceptable error in the temperature range of automotive devices that should work from −40 °C up to 150 °C of junction temperature.

Furthermore it is mandatory to be sure that the injector current reaches its peak threshold value, no matter how low the supply voltage might be. In fact, with reference to Fig. 9.16, the injector current cannot be any higher than V_{Bat}/R, hence if the peak current threshold were set at a higher value, the injector current would never reach it. If so the system would get stuck in a situation where the injector is always open, waiting for the peak current to come and never reaching the hold state. A way to insure that the injector properly works in any condition, is to set a peak current threshold as a function of the supply voltage V_{Bat}.

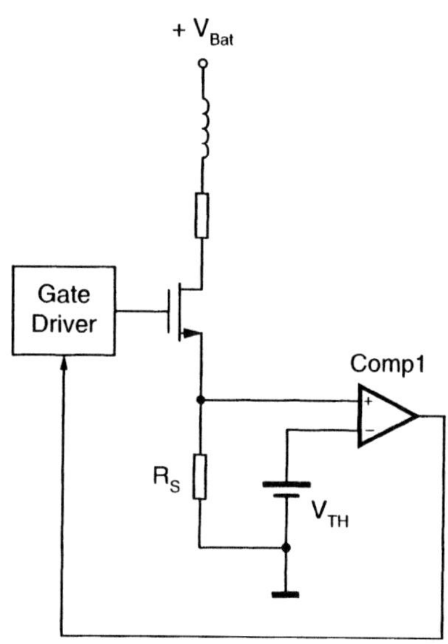

Fig. 9.22 Current sensing

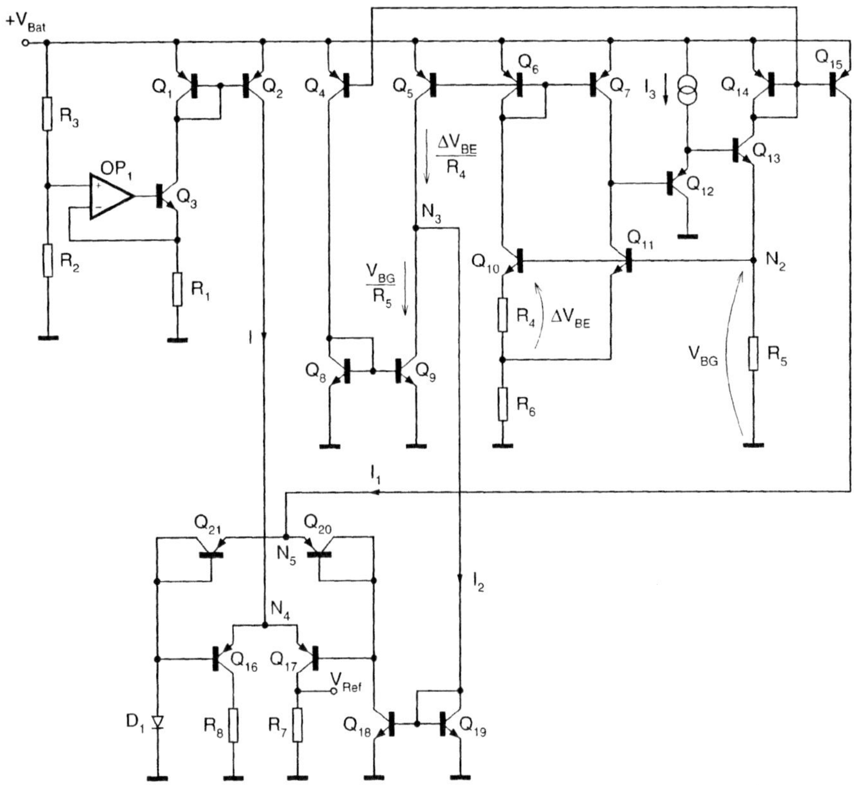

Fig. 9.23 Voltage reference

All this has been realised through a voltage reference having a thermal coefficient equal to that of the sensing resistor plus a linear dependence on the voltage supply.

The circuit configuration relevant to this solution is shown in Fig. 9.23.

The current I is a linear function of the supply voltage, the contribution of I_1 has a zero thermal drift, I_2 is the difference between a zero and a positive thermal drift current.

Thus: $I = \dfrac{K_1 \cdot V_{Bat}}{R_1}$, where $K_1 = \dfrac{R_2}{R_2 + R_3}$

$I_2 = \dfrac{\Delta V_{BE}}{R_4} - \dfrac{V_{BG}}{R_5}$ which depends linearly on temperature [8].

In fact, the difference between the base-emitter voltages of the transistors Q_{10}, Q_{11} is equal to:

$$\Delta V_{BE} = \eta V_T \ln\left(\frac{I_{S10}}{I_{C10}} \cdot \frac{I_{C11}}{I_{S11}}\right)$$

where V_T is the voltage equivalent of the temperature defined by the relation $V_T = KT/q$ (K = Boltzmann's constant, T = absolute temperature, q = electron charge), I_{C10}, I_{C11} are the collector currents of the transistors Q_{10}, Q_{11} and I_{S10}, I_{S11} are the saturation currents of the transistors Q_{10}, Q_{11}.

If

$$I_{C10} = I_{C11}$$

$$I_{S10} = AI_{S11}$$

where A is the ratio between the emitter areas of the two transistors Q_{10}, Q_{11} we obtain:

$$\Delta V_{BE} = \eta V_T \ln A = \eta \frac{kT}{q} \ln A$$

where: η = technological parameter independent of temperature.

The above equation can also be written:

$$\Delta V_{BE} = \eta \frac{kT_o}{q} \ln A + \eta \frac{k(T - T_o)}{q} \ln A$$

where: T_o = reference temperature.

Hence:

$$\Delta V_{BE} = \left(1 + \frac{T - T_o}{T_o}\right) \eta \frac{kT_o}{q} \ln A = (1 + \alpha \Delta T)\, \Delta V_{BEO}$$

The above equation highlights the law of variation of the voltage ΔV_{BE} as a function of the temperature, or rather that of its variation ΔT with respect to the reference temperature T_o, being ΔV_{BEo} the value of the voltage ΔV_{BE} calculated at the reference temperature T_o and α the temperature coefficient given by the expression:

$$\alpha = 1/T_o \; [1/°K]$$

With reference to the circuit formed by the transistors Q_{20}, Q_{21}, Q_{16}, Q_{17}, on the basis of Kirchoff's law it is possible to write:

$$V_{EB16} - V_{EB17} = V_{EB21} - V_{EB20} \tag{9.1}$$

Since:

$$V_{EB} = \eta V_T \ln \frac{I_C}{I_S} = \eta \frac{kT}{q} \ln \frac{I_C}{I_S} \tag{9.2}$$

substituting in Eq. 9.1, assuming that all transistors Q_{20}, Q_{21}, Q_{16}, Q_{17} have the same area, we get:

$$I_{C16}/I_{C17} = I_{C21}/I_{C20} \tag{9.3}$$

If we now apply Kirchhoff's law of currents to the intermediate node N_4 between the emitters of the transistors Q_{16}, Q_{17}, assuming that the base currents of the transistors Q_{16}, Q_{17}, are negligible, we obtain:

$$I_{C16} + I_{C17} = I \tag{9.4}$$

Similarly, if we apply Kirchoff's law to the node N_5 between the emitters of the transistors Q_{20}, Q_{21}, and assuming that the base currents of the transistors Q_{20}, Q_{21} are negligible, we have:

$$I_{C21} + I_{C20} = I_1 \tag{9.5}$$

Finally, substituting the values of I_{C16} and of I_{C21} obtained by Eqs. 9.4 and 9.5, in Eq. 9.3 we obtain:

$$I/I_{C17} = I_1/I_{C20} \tag{9.6}$$

Ignoring the base current of the transistor Q_{17} we have $I_{C20} = I_2$ and therefore:

$$I_{C17} = ({}^{I_2}/{}_{I_1})I \tag{9.7}$$

The value of the output voltage V_{REF} taken across the resistance R_7 is:

$$V_{REF} = R_7 I_{C17} = R_7(I_2/I_1)I \tag{9.8}$$

As pointed out before, the current I has been generated as a linear function of the supply voltage according to the expression $I = (K_1 V_{Bat}/R_1)$. The current I_1 has been generated by a voltage V_{BG} independent of temperature and it has the expression $I_1 = V_{BG}/R_5$, while the current I_2 has been evaluated from the expression:

$$I_2 = (\Delta V_{BE}/R_4 - V_{BG}/R_5) = V/R_4 \tag{9.9}$$

where it was assumed that $R_4 = K_2 R_5$, with $V = \Delta V_{BE} - K_2 V_{BG}$ includes temperature dependence.

Therefore it can be written:

$$V_{REF} = (R_7/R_4)\,(R_5/R_1)\,(K_1 V_{CC}/V_{BG})\,V \tag{9.10}$$

As it can be seen the voltage V_{REF} depends on the supply voltage V_{Bat} and follows a law of variation with temperature that is defined by the voltage V.

The bandgap voltage is constant and independent of the value of both supply voltage and temperature.

9.3.8 Diagnostic

The complexity achieved by the electronic injection systems has, in fact, changed the trouble-shooting procedures in the repair shops.

The components of a system are in great numbers and to facilitate the trouble-shooting it is common to use diagnostic techniques that involve the transfer of information between a tester and the microprocessor of the electronic control unit where the failures of the system are memorised.

All the devices of the E. C. U. have some circuits that detect the various failures and transmit them to the processor.

As for the case of the driver for the injection, the failures that have to be reported are those related to the power output, the short circuit of output to the ground or to the battery and open load.

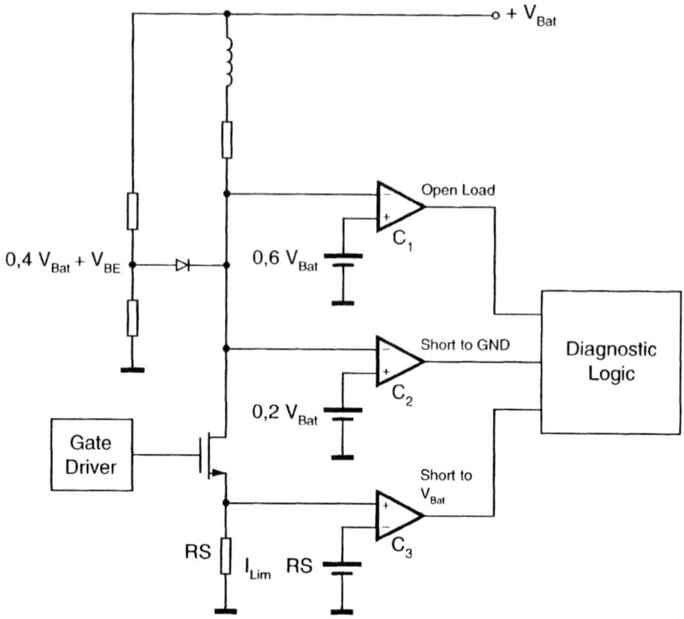

Fig. 9.24 Diagnostic simplified schematic

Figure 9 .24 describes the concept schematic of the diagnostic part of the device.

Some comparators, connected to the output pin, check the status of the load and their outputs, properly filtered and latched, are sent to the microprocessor.

The diagnosis of short circuit to ground and open load is performed while the output power transistor is off. In this condition the voltage of the output pin, if no failure occurs, has to be close to the battery voltage V_{Bat}, while when one of the two failures is present, the voltage of the pin is low.

In order to distinguish these two conditions, the output pin is biased at a known voltage (in the example in Fig. 9.24 this value is 0.4 V_{Bat}) so that this voltage is forced on the output if the open load occurs.

The two comparators C_1 and C_2 used for the control, have threshold voltages which are linear dependent on the battery voltage value V_{Bat} and respectively higher and lower tracking the battery voltage.

If the open load condition happens, the output will reach a voltage of 0.4 Vb and the comparator C_1 will have a high output voltage value while the comparator C_2 a low one.

In short circuit to ground condition both the comparators will show low voltage value.

These two conditions can be easily decoded by a logic circuit.

As far as the short circuit to the battery is concerned, the failure is detected as over current, reading the voltage drop on a sensing resistor on the injector current path (comparator C_3).

9.3.9
Protections

The device is able to protect itself when operates beyond the operating range of voltage and temperature.

When voltage transients occur, the injector drivers are supposed to be normally functioning, as they drive essential components for the engine behaviour.

The driver under examination keeps the voltage constant across the load, but this means that the voltage on the output increases with the battery voltage.

With reference to the Fig. 9.25, the voltage on the pin V_{OUT} will be:

$$V_{OUT} = V_{BAT} + \Sigma V_{ZN} + V_{BE\,(Q2)}$$

when Q_1 is off.

If, as usual, the voltage across the injector during the recirculation phase has to be 50–60 V and the battery voltage can reach up 50 V, then the voltage on the pin V_{OUT} could be higher than 100 V, that is the maximum value allowed by the technology.

The solution consists in reducing the recirculation voltage during overvoltages, turning on the switch Q_1 that results in a short circuit condition of some zener diodes so that V_{OUT} is always lower than the breakdown voltage BV_{dss} of the power element.

As far as the thermal protection is concerned, this function is referred to the microprocessor that considers the signal of high temperature from the device as a diagnostic output.

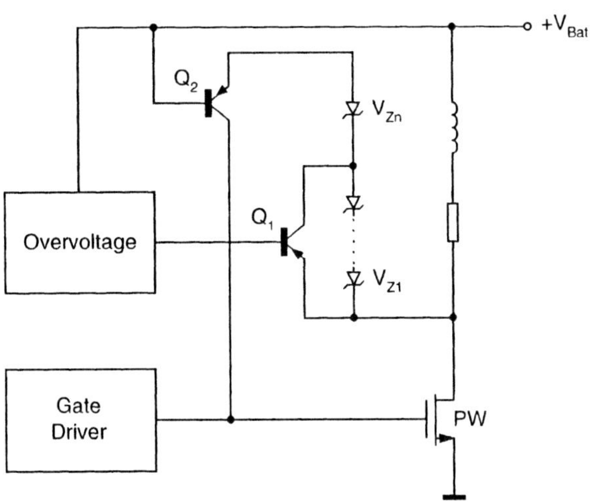

Fig. 9.25 High voltage clamp and overvoltage protection

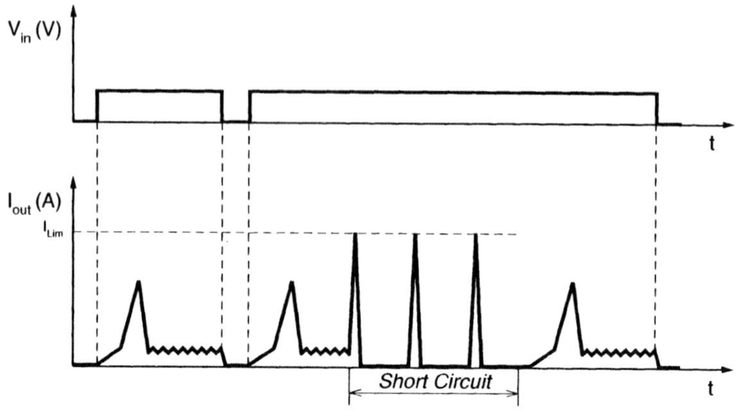

Fig. 9.26 Short circuit behaviour

This kind of strategy is due to the load that has to be driven and directly involves the safety of the motor vehicle. In fact, turning off the injector driver, due to an overtemperature detected by the device, the car should stop, while the microprocessor, having received the information of high temperature, can elaborate strategies in order to reduce the power dissipation.

The overcurrent protection is included in the short circuit detection to the battery voltage; in this case every time that the fault is detected, the power is turned off for 300 μs and then turned on again until the falling edge of the input signal occurs. This allows to control if there is a short circuit condition every 300 μs and to turn on the power when the short circuit is removed (see Fig. 9.26).

Bibliography

1. Glenn F (1984) Versatile Linear Regulator Eases Power Supply Design, EDN Electrical Design News, vol 29, No 10, Boston, Mass, US, pp 161-176
2. Dejoux C (1983) Superviseur d'Alimentation, Toute l'Electronique no 481, pages 33-35
3. Miller ER (1981) Reset Circuit Reacts to Fast Line Hits, Electronics International vol 53, no 2, New York US, p 120
4. Steven Hageman C (1988) Spice Techniques Facilitate Analysis of Feedback Circuits EDN Electrical design News, vol 33, no.20, Newton, Mass, US, pp 73-182, XPOOO252732
5. Millman & Halkias (1971) Integrated Electronics, McGraw-Hill, Tokyo, Japan, pp 501–536
6. Bosch R GMBH edition (1986) Automotive Handbook 2
7. Murari B, Mixed power IC Technology and Small System Integration 19th European Solid state circuit conference
8. Paul Gray R, Robert Meyer G, Analysis and Design of Analog Integrated Circuits

CHAPTER 10

Audio Amplifiers

E. Botti · F. Stefani

10.1 Introduction

In the late '70s MOSFETs acquired an important position when the first commercial audio amplifier, with such output transistors appeared on the market. MOSFETs outputs attracted attention for two significant reasons. Firstly they are high impedance devices that require negligible current at input. Secondly, they are square law devices, while an exponential relationship characterises the voltage-to-current transfer function of bipolar transistors (BJTs). Another feature pleasant to the circuit designer, is the absence of the second breakdown phenomenon, which greatly simplifies protection circuits and enhances reliability. On the other side, they are problematic because of high gate capacitance, intrinsic distortion, and increasing internal impedance (R_{ON}) with rising temperature. The world of monolithic amplifiers, even if aware of the amazing possibilities opened by such components, had to wait until the late '80s for a major advancement to occur. Then, the availability of the mixed bipolar-MOS process allowed the realization of the first fully integrated device [1, 2] characterized by a high voltage DMOS output stage.

The exploitation of such mixed processes yields substantial advantages:

i) the design of high voltage output stages which do not suffer from the thermal discharge that affects their bipolar counterpart;
ii) input stage that can still be bipolar for low noise performance;
iii) design of power circuitry to overcome such traditional drawbacks as high-frequency instability and short-circuit protection trade-offs;
iv) CMOS logic on board to set operation modes and to interface with other devices in the same application (e.g. serial digital bus).

In this chapter the output stage will be described in detail because of its innovative contribution to monolithic linear power amplifiers. The availability on he same process of CMOS logic deserves a comment. If we break the Power ICs market down into two major segments, linear ICs (or pure analogue) and mixed analogue/digital ICs (referred to as mixed-signal ICs), then Smart Power ICs are simply the mixed-signal part of the Power ICs market. Audio Power ICs, traditional analogue devices, can now be moved to the newer segment of Smart Power products. From the standpoint of this chapter, the high speed, high power DMOS transistor, made available by the already-mentioned family of mixed processes, will be considered as the key element. We will first consider its use as a linear element in a class-B amplifier as well as in a higher efficiency topology. Then we will review its implementation as a power switch, specifically in a new class of amplification developed for the automobile environment, which combines the benefits of class-B performance together with improved efficiency.

The chapter will also include the outline of fault protection circuits suited to the DMOS structure and some considerations on packaging and related thermal issues.

10.2 Review of the Characteristics of Output Stages

We will now review some basic characteristics of an audio power output stage. The reader is assumed to be familiar with audio fundamentals. An output stage has to deliver power to the load (loudspeaker) in response to the input signal. In other words, it has to realise a power gain that usually consists in a current gain, being the voltage gain implemented by previous stages. The following is an outline of the general requirements of audio output stages:

a) low harmonic distortion;
b) high voltage dynamics;
c) low output resistance;
d) driving capability of high reactance loads (both inductive and capacitive);
e) high reliability;
f) protection against overloads and short-circuits.

The degree of conversion efficiency from the DC power supply (e.g. the automobile battery) into the useful power that moves the loudspeaker coil, depends on the polarisation, referred to as the output stage class. For the moment, we will only consider class-B stages whose maximum theoretical efficiency is $\frac{\pi}{4} \simeq 0{,}785$ [3]. It is worth mentioning that the full compliance with the above requirements eventually depends on the circuit designers experience: the intrinsic behaviour and ruggedness of DMOS transistors are certainly of help when compared with BJTs, but the result does not come by magic [4, 5]. A second list is necessary in order to focus on the active element chosen for the stage. It has to satisfy the following minimum set of requirements:

a) high current gain;
b) easy handling of high output currents;
c) wide safe operating area;
d) reduced influence of the second breakdown;
e) high gain-bandwidth product;
f) low on-resistance when operating at limiting conditions (clipping).

Other specific requirements can be added depending on the application, but there is no doubt that a DMOS device is appreciated for any of the above items and specifically for its extended Safe Operating Area (SOA). A detailed analysis of this important characteristic can be found in literature [6] and will also be recalled in Sect. 10.4. For our current purpose it will suffice here to give a summary description. The BJT has a transconductance that increases with temperature (roughly, 8 % per °C). If the geometrical layout of the component is not thoroughly equalised, the current tends to focalise into a so-called "hot spot". This spot tends to drain more and more current, because of the above dependence, in a sort of regenerative effect finally leading to the transistor destruction [7]. This will happen even operating inside a rated constant power dissipation curve and the phenomenon is known as second breakdown.

In a DMOS transistor, the transconductance reduces over temperature, allowing a sort of negative feedback on each elementary component (cell) of the transistor layout. In turn, this leads to a more evenly distributed temperature profile and the above phenomenon is not a matter of concern to the designer. In an overall comparison between a BJT and a DMOS as the active element for output stages, one should not limit oneself to the basic performance list, instead one should keep an eye on the technology trends as well [8]. The integrated DMOS has today reached the same compactness in size previously held by BJTs, making its implementation an efficient one in audio amplifiers. In the future, it is expected that power DMOSs will exploit the lithography reduction trend. With submicron rules and extremely compact dimensions, the only concern will be related to their ability to dissipate the generated heat. In fact, the thermal resistance junction-to-case ($R_{th,j-c}$) increases when the area is shrunk. Today, we are not so far from such a physical limit. To better understand the latter statement, a table reporting some R_{th} data follows. The same R_{ON} value is assumed, necessary to give 35 W (square wave output) on a 4 Ω load in car-radio applications ($V_{supply} = 14$ V) evolving through different processes.

The area improvement for the power stage is not so good in passing from 2.5 µm to 1.2 µm, but the table does not show that the signal portion for a similar device would be one half in 1.2 µm. Very impressive though, is the shrinking to 0.5 µm also considering another 30 % gain in the signal part. On the other hand, the ability of BJTs to deliver current is set by their emitter area: current density can be improved by changing the parameters of the diffusion process but there is no such benefit coming to a finer lithography. BJTs will not exploit the trend in the above example and this fact will compromise their competitiveness. Anyway, the final economical impact has to be judged on a case-by-case basis and it is the opinion of the authors that the bipolar solution still has a good perspective, at least for low-voltage, low-power, but high-volume car-radio applications. Needless to say, all of the above considerations are to be intended as a simplified overview of the DMOS versus BJT possibilities.

Table 1

Minimum Lithography	Power Transistors Area (sqmm.)	R_{th} (deg.C/W)
2.5 µm	16.0	0.7
1.2 µm	15.0	0.8
0.5 µm	5.2	1.3

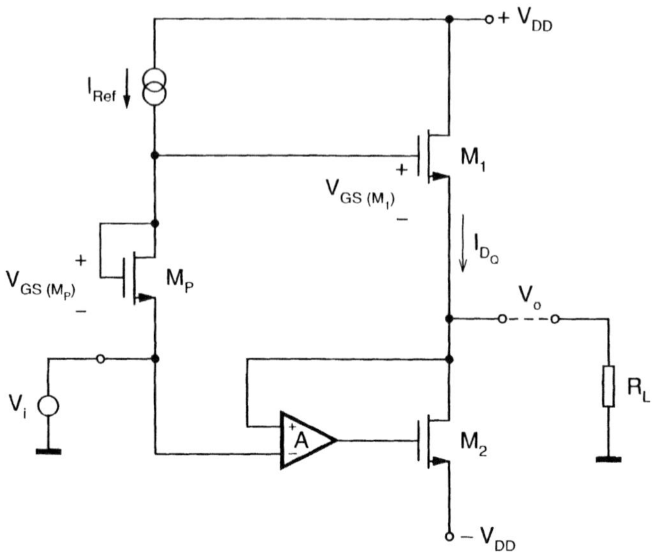

Fig. 10.1 Circuit Schematic for Polarisation Analysis

10.3 DMOS Output Stage

In this section we will analyse in detail a possible circuit structure for an integrated DMOS power stage [9]. Some fully complementary solutions are possible, but the size of a power N-channel is by far a small one when compared with P-channels. In such cases and when the process does not allow complementary power transistors, it is not possible to borrow solutions from the consolidated segment of CMOS. For this reason an original solution has been defined as shown in Fig. 10.1.

As usual when designing audio amplifiers, the main problems to solve are the accurate control of the quiescent current and reduction of non-linearities introduced by the stage, with special attention to the crossover distortion.

10.3.1 Bias Considerations

Referring to Fig. 10.1, we will analyse the static behaviour first. The MOS M_P acts as a polarising voltage source fixed by the current generator I_{Ref}. The feedback around block A, for the moment considered ideal, forces $V_o = V_i$. Then, also:

$$V_{GS,M_1} = V_{GS,M_P} \tag{10.1}$$

Both M_P and M_1 are in saturation region; for each of them the following relation holds true:

$$I_D = \frac{k' \cdot W}{2 \cdot L} \cdot (V_{GS} - V_{th})^2 \tag{10.2}$$

where k′ is a constant depending on the technological parameters of the process and V_{th} is the threshold voltage, while W/L is the channel ratio. Remembering Eq. 10.1 and noticing that $I_{D,Mp} = I_{Ref}$ by definition, the quiescent current I_{Dq} through M_1 and M_2 becomes:

$$I_{Dq} = \frac{(W/L)M_1}{(W/L)M_P} \cdot I_{Ref} \tag{10.3}$$

Some considerations apply to this result. The ratio W/L for the power transistor is usually defined by the maximum current density in the elementary cell and the minimisation of its resistance in the ohmic zone, R_{ON}. The value of I_{Dq} is then fixed by choosing the appropriate values for I_{Ref} and M_P. In this way I_{Dq} is sufficiently immune from the process spreads and from temperature variations. The error introduced by an offset between the inputs of A is negligible, it must be compared with V_{GS}.

10.3.2 Dynamic Behaviour

Now we will consider the dynamic behaviour. For $V_o < 0$, M_2 conducts, and the feedback due to A forces $V_o = V_i$. For $V_o > 0$ we have:

$$V_o = V_i + V_{GS,Mp} - V_{GS,M1} \tag{10.4}$$

the last value being in turn a function of the output voltage.

Being $V_o = I_d * R_L$, then we obtain from Eq. 10.2:

$$V_{GS,M1} = \left(\frac{2 \cdot L \cdot V_o}{k' \cdot W \cdot R_L}\right)^{1/2} + V_{th} \tag{10.5}$$

The transfer function of the stage is then intrinsically nonlinear. It is possible anyway to estimate the maximum harmonic distortion which is, by computer simulations, below 5% under any circumstances. The use of global feedback around the amplifier is good enough to reduce it to acceptable values. The major contribution to the distortion is then caused by the inevitable delays of transistors turn-on after their non-conduction period. Such delays are due to parasitic capacitors, particularly high when dealing with power transistors, and worsening the performance of the stage at high frequencies. It is easy to notice that such a condition does not happen during the lower half waveform: the feedback due to A ensures that $V_{GS,M1}$ be close to its quiescent value for all the conduction time of M_2. On the other hand, when $V_o > 0$, it is:

$$V_o - V_i = V_{GS,Mp} - V_{GS,M1} \tag{10.6}$$

Fig. 10.2 Clamp Circuit for M_2 Gate

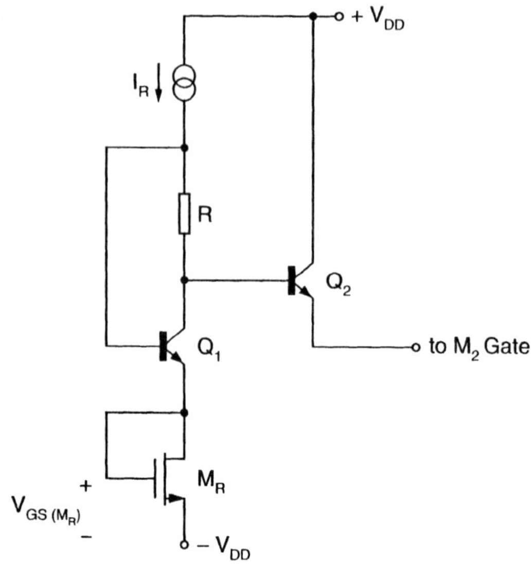

Since M_1 has to deliver current to the load, it will be $V_{GS, M1} > V_{GS, Mp}$; the amplifier A will then be completely unbalanced and will tend to force $V_{GS, M2}$ to zero. This dangerous condition can be avoided by fixing the minimum potential voltage of $V_{GS, M2}$ to a value slightly lower than its quiescent value. This can be done with the simple circuit in Fig. 10.2.

If I_R and M_R are so chosen as to have, in quiescent condition, $V_{GS, MR} = V_{GS, M2}$, and assuming $V_{BE, Q1} = V_{BE, Q2}$, it is evident that the gate of M_2 cannot go lower than:

$$V_{GS\,min} = V_{GS, MR} - R \cdot I_R \tag{10.7}$$

allowing to keep M_2 always in slight conduction through the appropriate choice of R.

10.3.3
The Output Dynamics: Bootstrap and Level Shifting

We will us refer to Fig. 10.3.

In order to obtain an output signal voltage as close as possible to the positive rail ($+V_{DD}$), it should be verified that:

$$V_A \cong V_{DD} + V_{GS, M1} \tag{10.8}$$

This is made possible if we exploit the bootstrap technique [5]. It will be explained with the help of Fig. 10.4, where the current source I_{Ref} is replaced by the current mirror Q_{G1} and Q_{G2}.

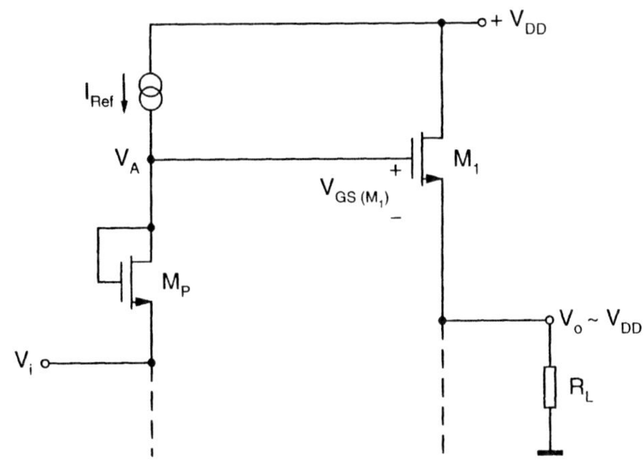

Fig. 10.3 Dynamic Requirements for $V_o = V_{DD}$

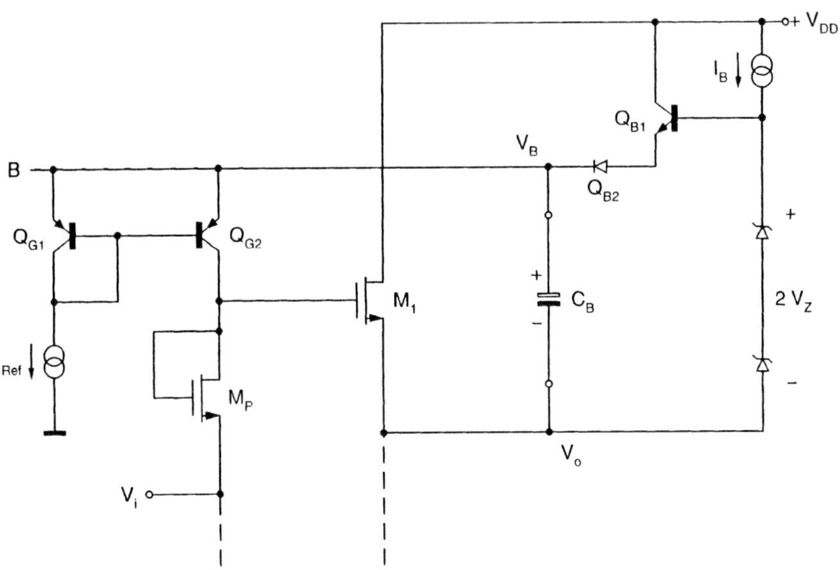

Fig. 10.4 Bootstrap Circuit

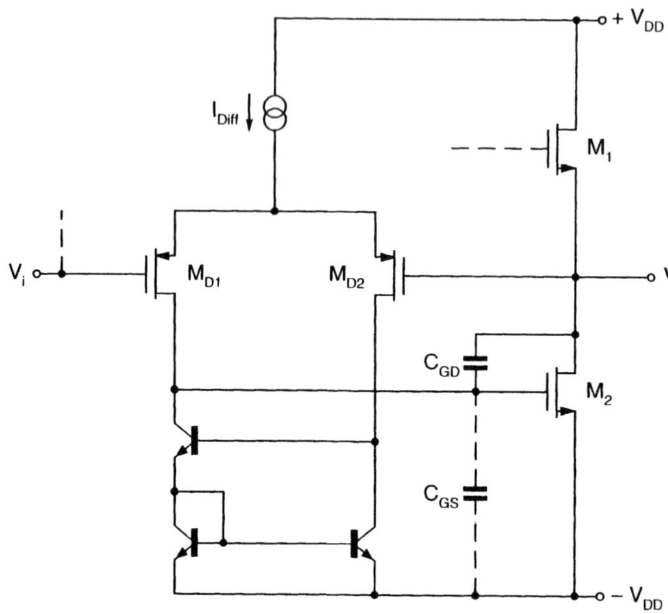

Fig. 10.5 Circuit Schematic for the Sink Portion of the Output Stage

With no signal applied, the capacitor C_B (external to the device) is charged and the voltage across its terminals is:

$$V_{CB} = 2\,(V_Z - V_{BE}) \tag{10.9}$$

and from the positive supply a current is drained equal to 2 I_{Ref} flowing through Q_{B1} and the diode Q_{B2}. When the signal is present, the voltage potential of the line marked "B" is tracking the output voltage V_o:

$$V_B = V_o + V_{CB} \tag{10.10}$$

When $V_B > V_{DD} - V_{BE}$, Q_{B1} and Q_{B2} are cut off and the necessary current 2 I_{ref} is drained from C_B. This means that the capacitor must be chosen in order that, for the minimum frequency to be reproduced, its minimum voltage be:

$$V_{CB\,min} = V_{GS,\,M1\,max} + V_{CEsat,\,QG2} \tag{10.11}$$

In this way we are always sure that M_1 is properly driven. When V_B becomes lower than $V_{DD} - 2\,V_{BE}$, the capacitor is recharged to its quiescent value by $I_B \cdot H_{fe,\,QB1}$. The diode Q_{B2} is a base-collector diode to avoid the breakdown of the emitter-base junction of Q_{B1} during operation. To analyse the situation $V_o < 0$, we will have an insight of amplifier A, until now considered ideal. The reference is Fig. 10.5. The amplifier consists of a simple differential stage, with two p-MOS transistors and a current mirror as the active load. The resulting structure resembles that of a CMOS two-stage operational amplifier, with unity

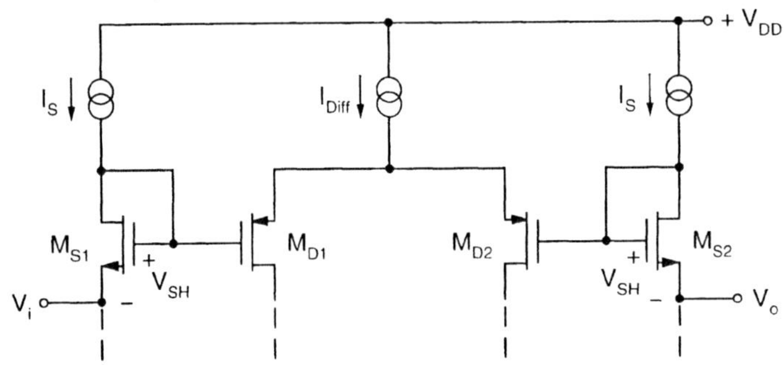

Fig. 10.6 Level Shifter to Optimize the Output Dynamic

gain feedback. Its cut-off frequency can be roughly estimated by noticing that C_{gd}, the parasitic gate-drain capacitance of M_2, acts as the compensating capacitor in the system. Also, transistor M_1 is never cut off and provides a parallel path for the signal to the output: it is a high frequency path because of its source-follower configuration. The combination of the two paths can be made sufficient for the overall stability. Coming to the dynamics of the amplifier toward the negative supply rail ($-V_{DD}$), in order to obtain $V_i \simeq V_o \simeq -V_{DD}$ without M_{D1} and M_{D2} entering their ohmic region, it must be:

$$V_{gs,\,M2} < V_{gs,\,MD1} + V_{th} \tag{10.12}$$

If $V_{th} = 1.5$ V, which is the value for the process under discussion, the gate-source voltage of M_2 is not sufficiently high to drive a low impedance load down to $-V_{DD}$. Therefore a level shifting solution has been adopted with two transistors in series with the gates of M_{D1} and M_{D2} (Fig. 10.6). Eq. 10.12 can now be written as:

$$V_{gs,\,M2} < V_{gs,\,MD1} + V_{th} + V_{SH} \tag{10.13}$$

where $V_{SH} = V_{GS,\,MS1}$ (or $V_{GS,\,MS2}$), leaving adequate swing to the output voltage.

10.4 Efficiency

The standard class-B power amplifier, widely used in the audio field, can become hard to manage when the output power is very high, due to the high power dissipation. In fact, the dissipated power versus output power in case of sinusoidal wave (theoretical) is given by:

$$P_d = (V_{DDt} \cdot V_m)/(\pi \cdot R_L) - V_m^2/(2 \cdot R_L) \tag{10.14}$$

where V_{DDt} is the total power supply, V_m is the peak output signal and R_L is the load resistance. The peak of P_d is not at the full swing, but at $V_m = \frac{V_{DDt}}{\pi}$, and the following relation holds true:

$$P_{d\,max} = V_{DDt}^2/(\pi^2 \times R_L) \cdot 2 \tag{10.15}$$

As an example, we will take $V_{DDt} = 80$ V and $R_L = 4\ \Omega$. In this case we have an approximately 160 W power amplifier, and $P_{d\,max} = 81$ W. The main problem is that P_d decreases only by a small amount if the output power decreases from the worst case; for example, if $P_{out} = 16$ W (then 1/10 of the maximum output power) we still have $P_d = 56$ W. So high a power dissipation amount can become a big problem if both the power devices of the class-B output stage are assembled in only one package, as it often is for the monolithic integrated circuits; hence the necessity arises to explore topologies that can give some advantages. The power DMOS technology is recommended for the basic reasons mentioned in the introduction. One thing is evident and holds true regardless of the topology: the very low and practically negligible amount of drive current. Taking as an example for the bipolar structures a 50 V power amplifier that can deliver around 60 W on a 4 Ω load, the power dissipation due to the base current only can be as high as:

$$P_{d\,base} = (I_{out\,max}/H_{fe\,min}) \cdot V_{DDt} = (4/30) \cdot 50 = 6.6\ \text{W} \tag{10.16}$$

which is not deliverable to the load, unless adopting such output stage structures that usually reduce the maximum output swing.

10.4.1 Class-G

The first topology that we can take into consideration to decrease the power dissipation while maintaining the same output power is the so called class-G amplifier [10–12]. The basic concept is to supply the power amplifier with two voltages: V_{DD1} and V_{DD2}, with $V_{DD1} < V_{DD2}$. The input stage can be supplied with the higher voltage, while the output stage is connected:

a) to the lower voltage (V_{DD1}) if the output peak signal is lower than V_{DD1};
b) to the higher voltage (V_{DD2}) if the output peak signal is larger than V_{DD1}.

The monolithic power amplifier TDA7294 is arranged in such a way to be able to work in class-G condition with the appropriate external components. In fact, in this device two pairs (positive and negative) of power supplies are foreseen: one for the signal part and the other for the power stage. In this way, it is easy to implement the schematic shown in Fig. 10.7 (simplified). As long as the output voltage V_{out} is:

$$V_{out} < V_{DD\,fin} - (V_Z - V_{BE}) \tag{10.17}$$

or

$$V_{out} > -V_{DD\,fin} + (V_Z - V_{BE}) \tag{10.18}$$

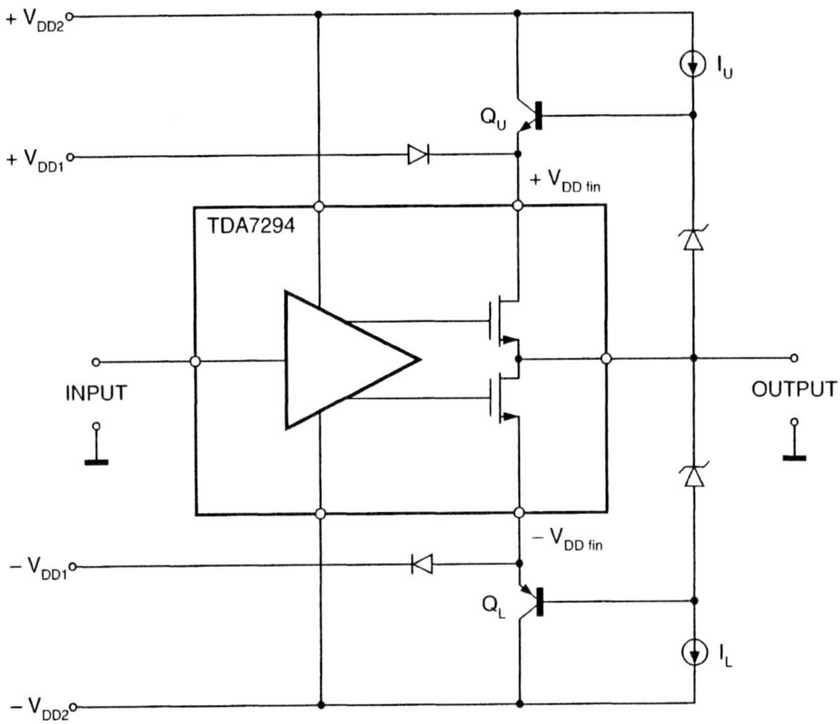

Fig. 10.7 TDA7294 Class-G Application

the power transistors Q_U or Q_L are switched off, and the power stage is supplied through two diodes,from +VDD1 and –VDD1 respectively. When:

$$V_{out} > +V_{DD\,fin} - (V_Z - V_{BE}) \tag{10.19}$$

or

$$V_{out} < -V_{DD\,fin} + (V_Z - V_{BE}) \tag{10.20}$$

then Q_U or Q_L are turned on and their emitter voltage follows the output voltage. In this condition, the final stage is supplied by $+V_{DD2}$ or $-V_{DD2}$. The schematic waveforms of $+V_{DD\,fin}$, $-V_{DD\,fin}$ and V_{out} are shown in Fig. 10.8.

A typical curve of the power dissipation versus output power for a class-G power amplifier, assuming $R_L = 4\,\Omega$, VDD1 = 20 V and $V_{DD2} = 40$ V, is shown in Fig. 10.9, as compared with a standard class-B amplifier supplied at a fixed +/– 40 V.

The shape of the power dissipation versus output power can be qualitatively explained by considering the fact that, up to about 30 W, the power dissipation of this type of class-G is equivalent to a class-B supplied by +/– 20 V, that is 1/4 the power dissipation of a class-B with a double power supply. From 30 W to 200 W,

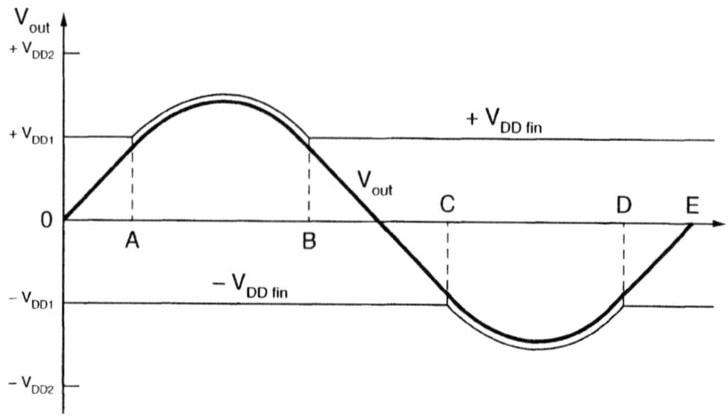

Fig. 10.8 Class-G Waveforms

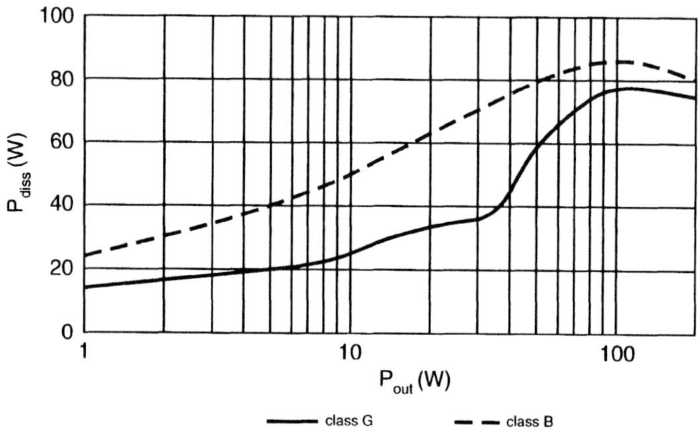

Fig. 10.9 Class-G vs Class-B Power Dissipation (V_{DD2} = +/− 40 V, VDD1 = +/− 20 V, RL=4 Ω)

P_{diss} becomes in any case lower than the equivalent class-B because a portion of the waveform is supplied by V_{DD1} (from 0 to A, from B to C and from D to E in Fig. 10.8). It is important to underline that the maximum advantage of this system is around P_{out} = 20 W. For the dimensioning of the heatsink, this region is the most important. In fact, for music signals, it is common to consider that the maximum average output power is around 1/10 of the maximum output power [13]. The disadvantages of this amplifier structure are essentially three:

1) in the "switching region", around (in this case) 20 V, output peak voltage, a "crossover like" distortion is almost unavoidable, and, in any case, the standard class-B amplifier can offer better distortion performance. Indeed, it is not difficult to reach acceptable results;

2) the amplifier is more complicated, but the "system cost" can be globally less expensive thanks to the improved efficiency, then it is possible to save cost, size and weight in the power supply and the heatsink;
3) the power supply itself is more complicated than the equivalent for the class-B: multiple voltages (4) are required.

10.4.2 Class-SB

The last point can be overcome by using another approach that we call class-SB: instead of switching between two different power supplies (e.g. from +/- 20 V to +/- 40 V), it is possible to switch the amplifier structures from single-ended to bridge. The results, in terms of output power and efficiency, are perfectly equal to the class-G structure. To the authors knowledge there is not any description of such a mode of operation and for this reason a new simple mnemonic name is proposed after the single-ended to/from bridge switching operation.

The principle of operation is shown in Fig. 10.10, and the possible output waveforms in Fig. 10.13.

If

$$V_{Ref^-} < V_{in} < V_{Ref^+} \tag{10.21}$$

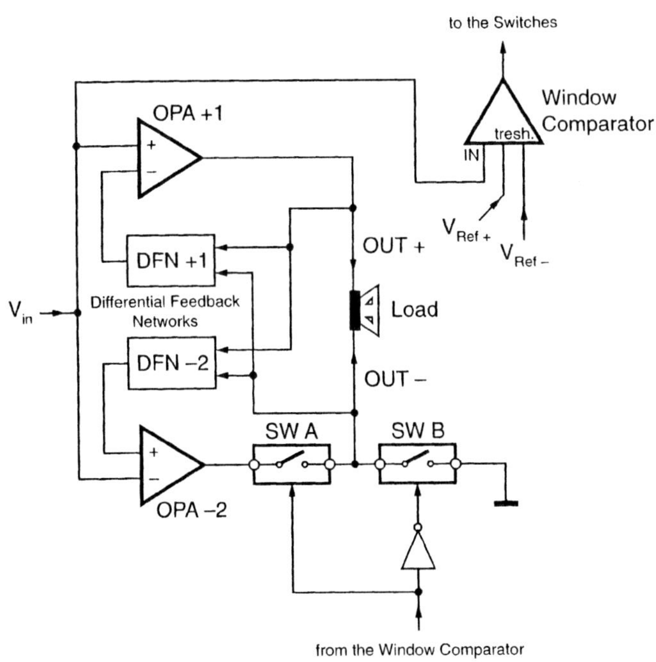

Fig. 10.10 Class-SB – Principle of Operation

the power switch SWA is open and the power switch SWB is closed. In this case the system is equivalent to a single amplifier supplied at +/– V_{DD}, exactly equal to the class-G amplifier in the portion 0-A of Fig. 10.8.
If

$$V_{in} < V_{Ref}^{-} \quad \text{or} \quad V_{in} > V_{Ref}^{+} \tag{10.22}$$

SWA is closed and SWB is open. It is evident that in this case the amplifier becomes a bridge system supplied at $\pm V_{DD}$ that, from the dissipation point of view, is equivalent to a single-ended amplifier supplied at $\pm 2 \cdot V_{DD}$, that we have in the class-G in the portion A-B in Fig. 10.8. Then, the curve of the power dissipation versus output power of a ± 20 V class-SB power amplifier is what is shown in Fig. 10.9, equal to the class-G solution. Needless to say, the power switches have to be DMOS transistors because of the well-known current lost to the substrate if BJTs were used. Such a current would affect the overall efficiency, making the present approach almost useless. An interesting field of application of this type of class is in the car-radio field. In this case the power supply is only around 14 V and there is not a large power dissipation for each amplifier; anyway, in modern car-radios, the output power can reach 4 times 20 W (bridge mode), and recently they are contained in a single package. From the power dissipation point of view, four times 20 W corresponds to a 80 W amplifier and the overall power dissipation can reach about 40 W. Moreover in the small box of the car-radio the space assigned to the heatsink is limited. An efficiency improvement is highly desirable. A first approach that can be done is illustrated in Fig. 10.11.

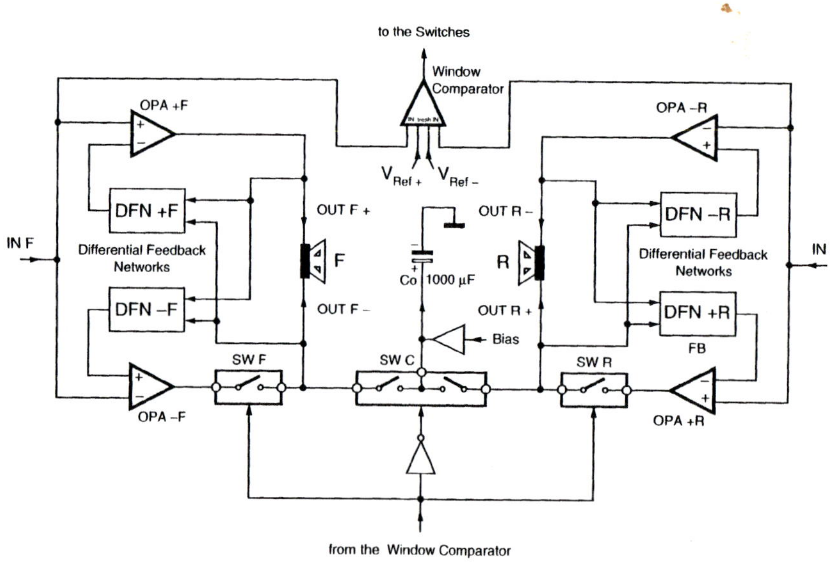

Fig. 10.11 Class-SB – Single Supply

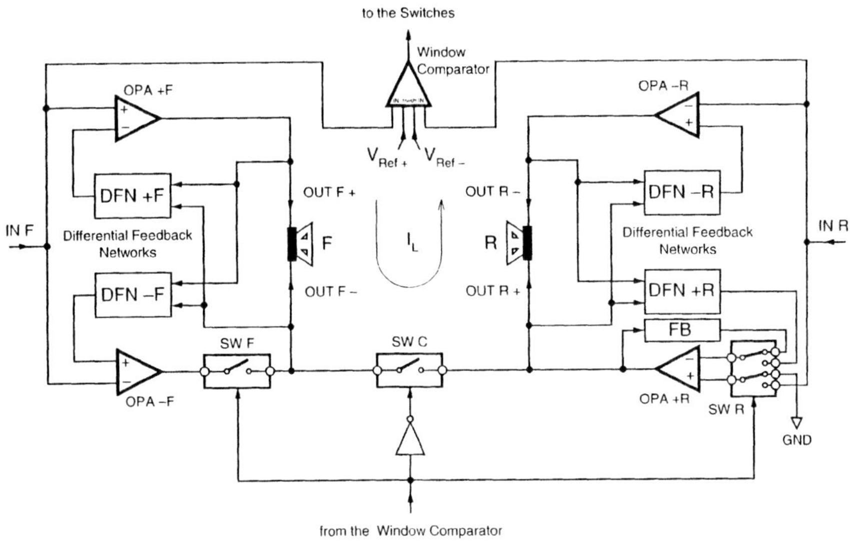

Fig. 10.12 Class-SB – Single Supply without C_0

It is easy to note that this is the system shown in Fig. 10.10 multiplied by two, in which the SWB, substituted by the double switch SWC, is not connected to the ground of the double power supply but to a *virtual ground* made by C_0 biased to half V_{DD}. A second approach is in Fig. 10.12, in which (compared to Fig. 10.11) C_0 and the biasing circuitry are omitted and the SW_C is a single switch, moreover OPA + R is a buffer biased to half V_{DD} for low output power, while it works as a standard amplifier for high output power.

The two channels are the *front* and *rear* amplifiers of the *left* or *right* stereo channel, then the input signals are equal or closely correlated. For small output power (up to about 3 W), SWF is open, SWC closed and OPA + R works as a buffer. If we suppose that the input signals are equal as well as the loudspeaker impedance (4 ohm), we can see that all the current I_L flowing into the loudspeaker F, also drives R. The dissipation is comparable to that of a single bridge driving 8 ohm. For $V_{in} > V_{Ref}^+$ or $< V_{Ref}^-$, SWF is closed, SWC open and OPA + R works as an amplifier. The system becomes a standard double bridge amplifier. The output waveforms are shown in Fig. 10.13a or 10.13b (depending on the type of the feedback network). From 0 to A, B to C and D to E the power dissipation is like a single bridge on 8 Ω. From A to B and C to D the behaviour is the same as in the standard dual bridge on 4 Ω.

As a result, in a quad-power amplifier for car radios power dissipation versus output power is shown in Fig. 10.14. The advantage in terms of reduced power dissipation is evident.

If the loads are not equal and/or the input signals are not at the same amplitude, the difference between the currents through the loads are sinked/sourced

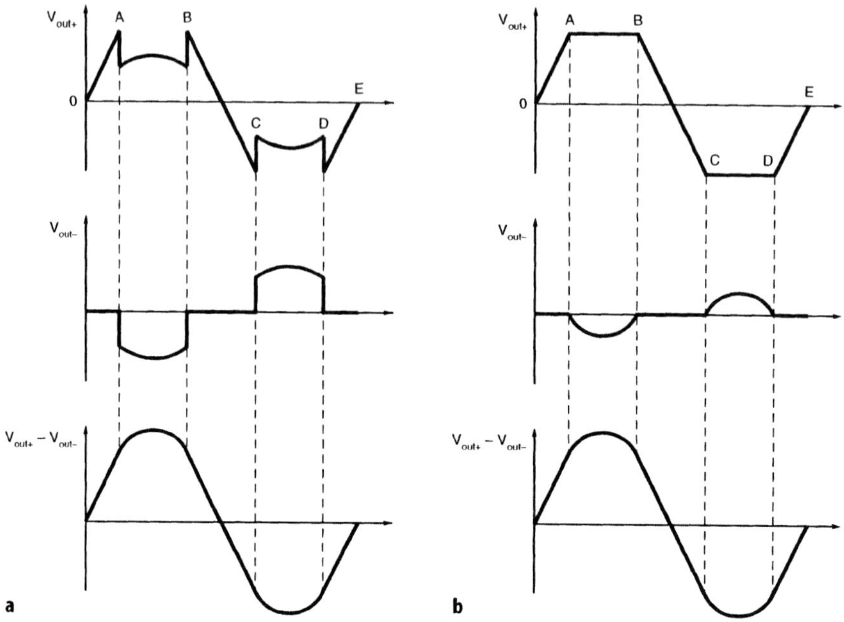

Fig. 10.13 Class-SB – Output Waveforms

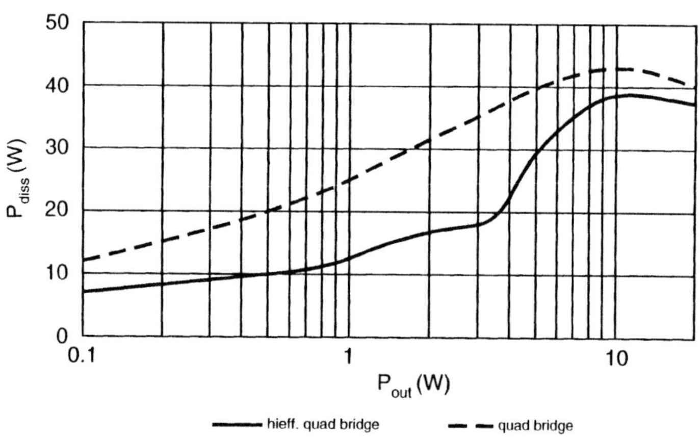

Fig. 10.14 Class-SB – Power Dissipation vs Output Power

by OPA + R working as a buffer. In this case, the efficiency decreases, but it is possible to demonstrate that also the power dissipation is reduced. Overall, the power dissipation in the various cases is aligned to the ideal case, particularly when music signals are of concern. Again, this would be impossible without almost ideal switches like DMOS transistors.

10.5 Protection Circuitry

A very important characteristic of the power DMOS transistor is the behaviour of the V_{GS} vs I_d curve as to the temperature. It is possible to discriminate two regions: A) and B) in Fig. 10.15.

$$\left.\frac{\delta I_D}{\delta T}\right|_{V_{GS}=CONST.} > 0 \tag{10.23}$$

This holds true for low V_{GS} values. In this region, the drain current increases as the temperature increases.

$$\left.\frac{\delta I_D}{\delta T}\right|_{V_{GS}=CONST.} < 0 \tag{10.24}$$

This is for medium to high V_{GS} values, in other words, for high levels of dissipated power. From the protection point of view, this is the most important region: when the temperature increases, keeping V_{GS} constant, I_D decreases. In bipolar transistors, the trend is always positive: the hottest region of a power transistor tends to increase the collector current I_C, then its power dissipation increases and, in turn, its temperature increases as well.

This regenerative loop is the cause of the second breakdown phenomenon that reduces the SOA of the device from the theoretical contour at P_{diss} = con-

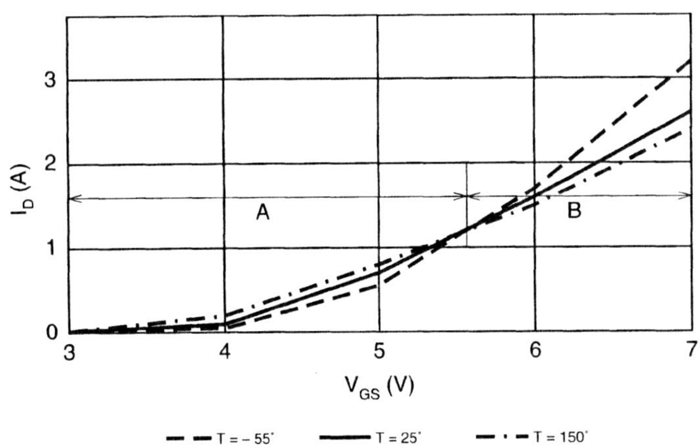

Fig. 10.15 MOS I_D–V_{GS} Transfer Characteristics vs Temperature

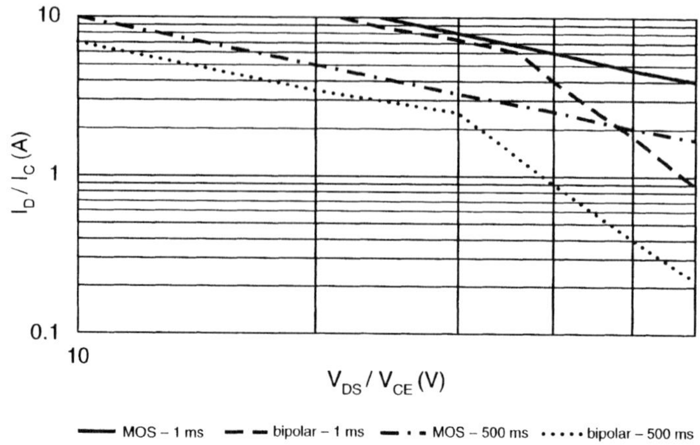

Fig. 10.16 MOS vs BJT Power Transistor: SOA Comparison

stant. Due to the negative slope of the I_D/T characteristic in the B) region of Fig. 10.15, the hottest portion of the power transistor tends to decrease the power dissipation, acting as a thermostat, stabilizing the temperature as well as the whole transistor structure. This effect maximizes the dissipating area and minimizes the thermal resistance as well, with respect to the bipolar solution and avoids the second breakdown problem. As an example, the SOA of a 5000 sqmil N-type DMOS transistor is shown in Fig. 10.16, in comparison with the SOA of an equivalent bipolar transistor.

If we assume that the area of a DMOS power transistor is at the same temperature anywhere, and we put a temperature sensor inside the transistor area, then we can have the exact information about its temperature stress, and we are allowed to take the right decision to avoid the overcoming of dangerous levels. With a standard mixed process the temperature sensor can be, for example, a minimum size NPN bipolar transistor, that can be placed in a small isolated pocket inside the power DMOS. In Fig. 10.17 the layout of the interesting area is shown.

The fact that there is a small but not negligible distance between the dissipating area and the temperature sensor, gives a delay in the response of the sensor itself, due to the low pass filtering effect of the heat transportation inside the silicon [14, 15]. This means that such a type of protection works only after some hundreds of microseconds from the beginning of the power pulse, the precise value depending on the distance between the sensor and the heating area. To avoid that very high power pulses can destroy the DMOS before the sensor intervention, a circuit is also needed that senses such pulses and acts as a power limiter. The simplified circuit schematic of the protections of a power audio amplifier is shown in Fig. 10.18.

The quasi-constant power limiter circuits are indicated in the blocks named as "upper control limit circuitry" and "lower current limit circuitry", while in the

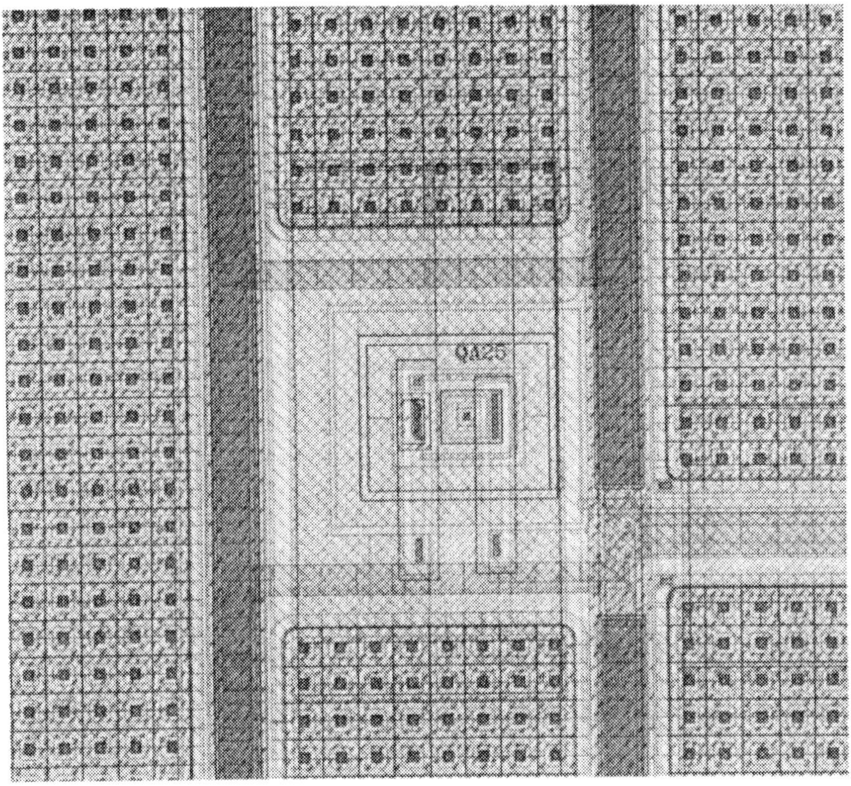

Fig. 10.17 Layout View of the Sensing Bipolar Transistor

blocks named "upper thermal sensor" and "lower thermal sensor" one can find the overtemperature sensors. The combination of the two protections gives a fully protected power DMOS transistor that can work within a very large SOA compared to what is obtainable from the bipolar counterpart. Regarding Fig. 10.18, we can see that the transistors Q_{1U}, Q_{2U}. and Q_{3U}, form a comparator that senses, through the resistors R_{SU}, the output current. When:

$$I_{C,Q3U} > I_U \tag{10.25}$$

then Q_{PU} is switched off. The current generator Iu defines the current limit at $V_{DS} < 15$ V; the network formed by Z_{1U}, Z_{2U}, R_{1U}, R_{2U}, R_{3U} gives the current limitation shape for $V_{DS} > 15$ V. For the lower power transistor, the comparator is formed by Q_{1L}, Q_{2L}, Q_{3L}, Q_{4L} and Q_{5L}. The temperature sensors are Q_{tU} and Q_{tL}, biased by Zener reference generators. As an example, to protect the power DMOS transistors in a 100 W monolithic audio power amplifier, the constant power limiter was set to 180 W, that is the value that a 5000 sqmil DMOS power transistor can withstand for 1 ms. The current pulse during a short-circuit is shown in Fig. 10.19.

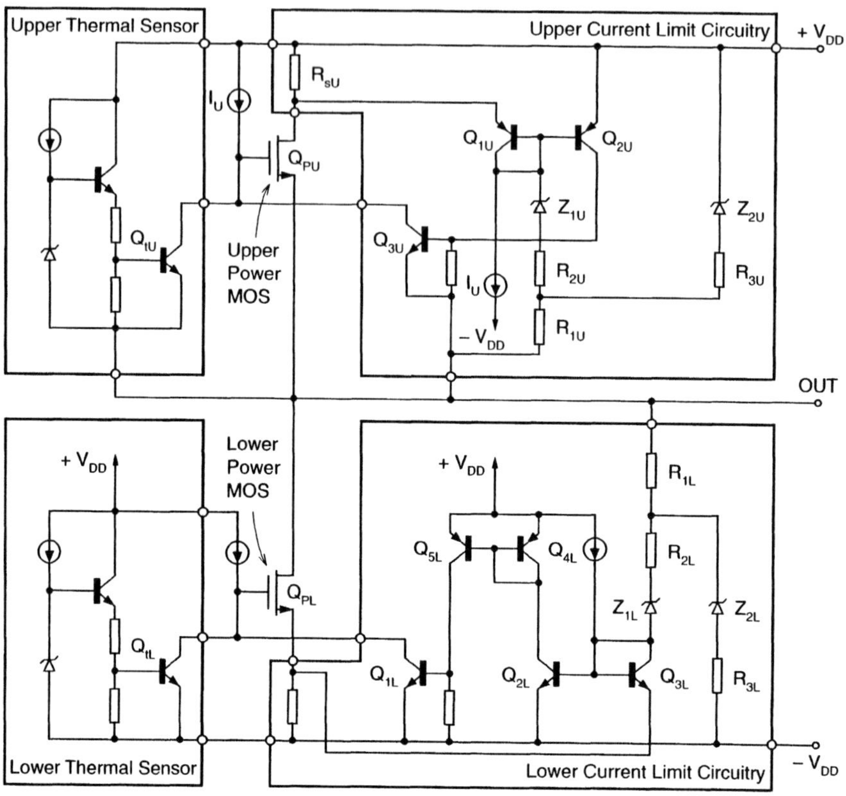

Fig. 10.18 Simplified Short-Circuit Protection Schematic

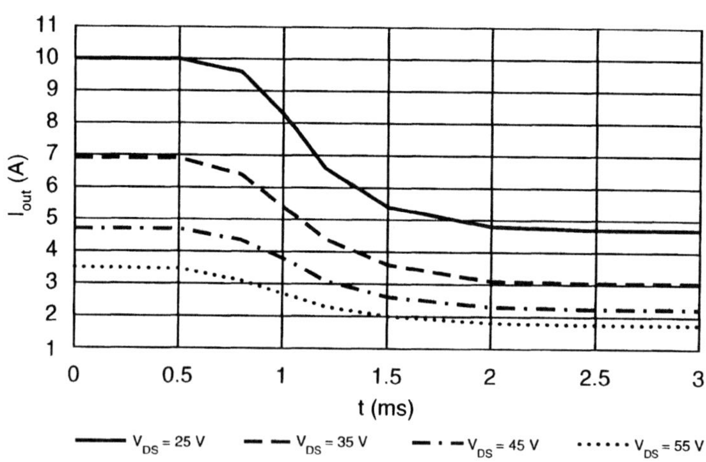

Fig. 10.19 Short-Circuit Current Limiting vs Time at Different V_{DS} Values

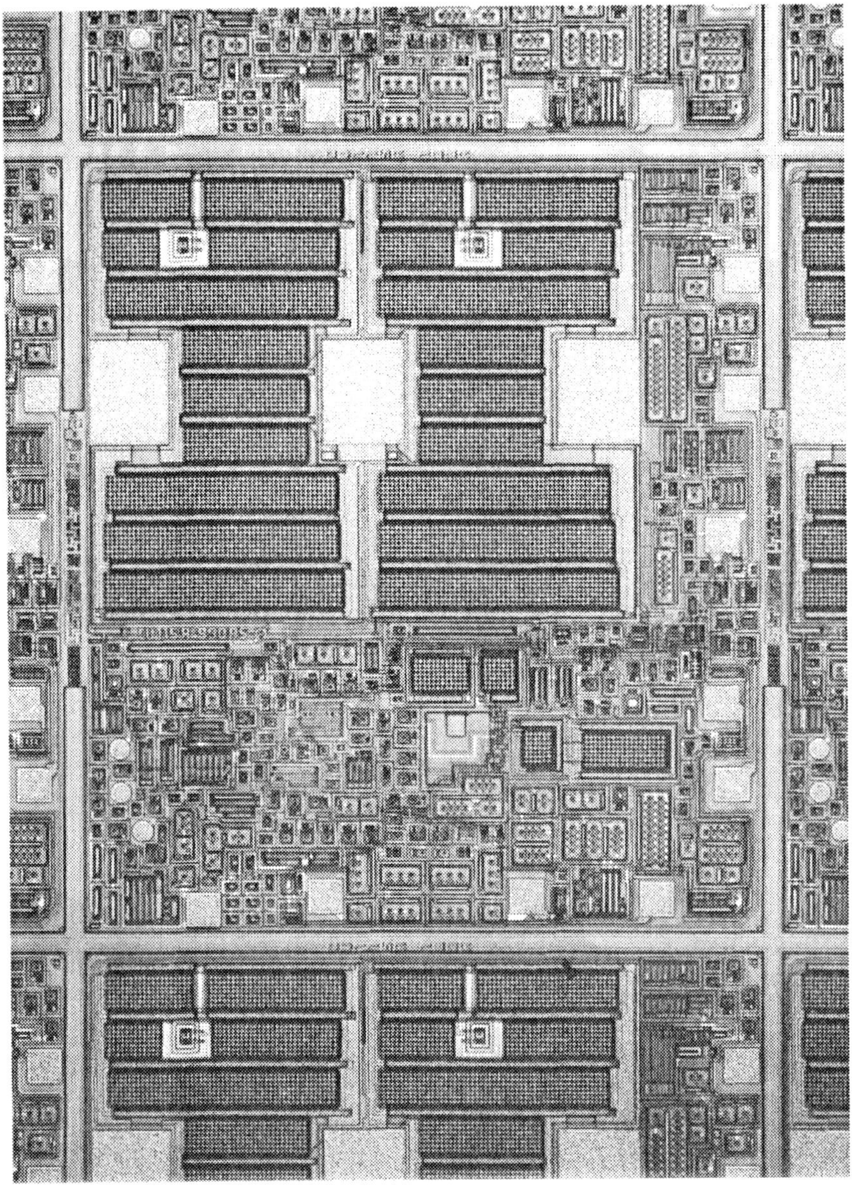

Fig. 10.20 TDA7294 Chip Photomicrograph

Until 0.5 ms, there is a current limitation that limits the power dissipation, after that the temperature sensor takes over and further reduces the output current. The final value of the short-circuit current (and then the steady-state power dissipation) depends on the junction-to-ambient thermal resistance and the ambient temperature. This overload protection allows the maximum output power without any intervention of the protection itself during normal operation together with the absolute safety because it acts only if the temperature of the power transistor increases to dangerous levels, taking into account the environmental conditions. A photo of the chip that uses the protection described is in Fig. 10.20: it is possible to note the temperature sensors placed inside the powers.

In some cases, instead of limiting the power dissipation in overload conditions, as described above, it is useful to:

1) switch-off the power amplifier;
2) sense if a short-circuit is present again and:
   - if not, switch on the power amplifier,
   - if still present, remain in off condition.

In this way there are two advantages:

a) the overloaded amplifier is less stressed,
b) in case of a bridge amplifier, in the load there cannot flow a D.C. current due to a short-circuit of one channel to $+V_{DD}$ or $-V_{DD}$ as shown in Fig. 10.21.

Supposing we need to protect a bridge amplifier as shown in Fig. 10.21. Two types of short-circuit are sketched: across the load and to $-V_{DD}$. The protection circuit is in Fig. 10.22.

The four-input gate OR receives positive levels if one or more output power MOS (M_1, M_2, M_3 or M_4) are overloaded. These signals come from circuits similar to what is drawn in Fig. 10.18, in which the signal present at the collector of Q_{3U} or Q_{1L}, instead of acting directly at the gate of the power MOS, is inverted and applied to the OR inputs. In this case, the flip-flop S/R is set, Q_P is turned on and the power amplifiers are switched off. The two-input gate AND1, thanks to the window comparators, receives positive signals if the two outputs OUT1 or OUT2 in Fig. 10.21 are placed near the ground potential ($+/-V_{Ref}$). This means that there are not (or no more) short-circuits to $+V_{DD}$ or $-V_{DD}$. In this case, the flip-flop, if it was set, will be reset, and the amplifier will be turned on. If the flip-flop was not set, Q is LOW and the AND2 gate, for the signals coming from AND1 is like an open switch. The time constants:

$$T_{set} = R \cdot C_{set} \tag{10.26}$$

and

$$T_{reset} = R \cdot C_{reset} \tag{10.27}$$

with $T_{set} < T_{reset}$, makes the system less sensitive to disturbing spikes, and in case of short-circuit across the load, it also defines the duty-cycle T_{ON}/T_{OFF}. In fact, in this case, the output voltage can remain inside the window $\pm V_{Ref}$ and after the settling, it will be the resetting after a time constant T_{reset} (the AND1

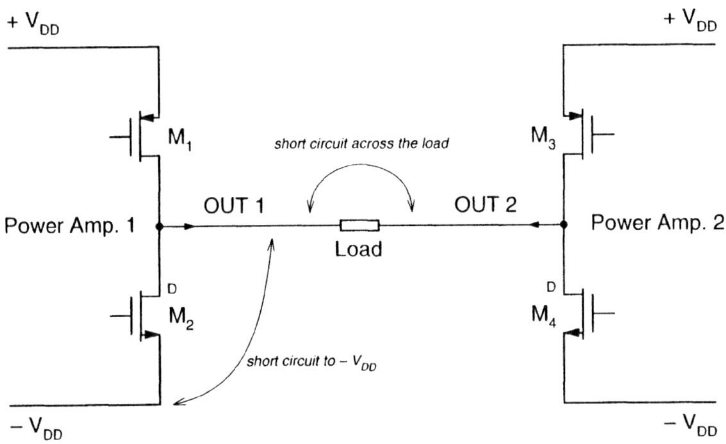

Fig. 10.21 Examples of Short-Circuit Conditions in a Bridge Amplifier

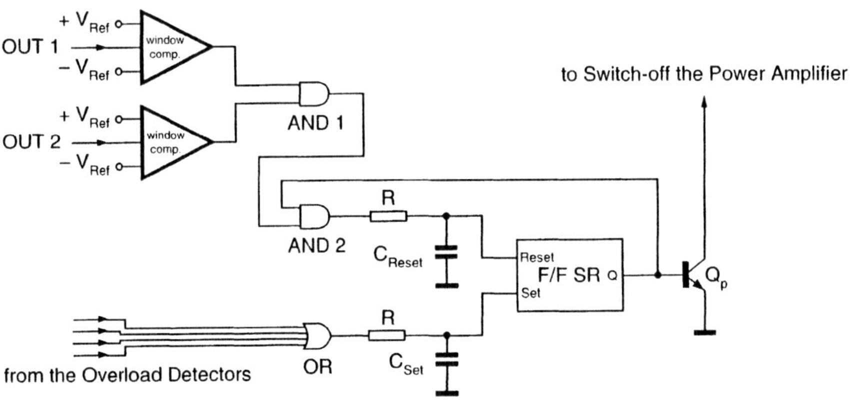

Fig. 10.22 Logic-Type Protection Circuitry

output is always at 1 level). After that the flip-flop can be reset again after a time constant T_{set} if the power transistors are overloaded again.

10.6 Control of Audio Functions Through a Serial Bus

The possibility of using a digital serial bus to control audio amplifier functions is described with reference to Fig. 10.23.

This represents the block diagram of TDA7292, a triple audio amplifier that can be connected either as a stereo plus sub-woofer configuration or as a dual-

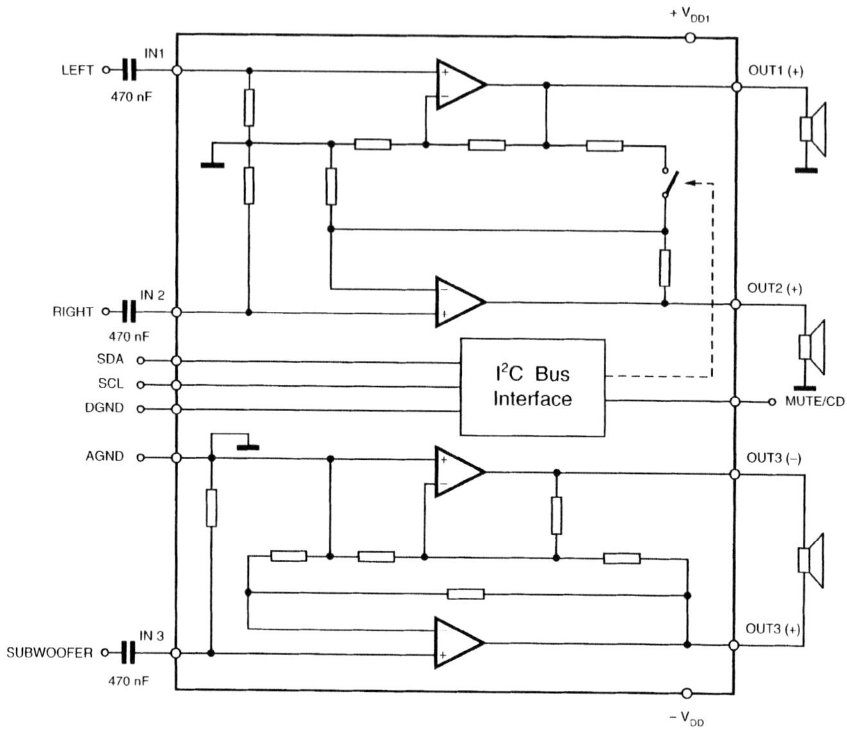

Fig. 10.23 TDA7292 Block Diagram

bridge stereo configuration. Its field of application is for TV-sets mainly. The Inter-Integrated Circuit Bus (I²C-Bus) CMOS-compatible interface sets the different operating modes and feeds back to the controller information about the device status. It is not the aim of the present chapter to give an operative description of the I²C-Bus protocol [16], rather a review will be proposed underlining the possibilities of controlling some audio functions.

10.6.1 I²C-Bus Interface Protocol

The I²C-Bus protocol is used in order to allow different integrated circuits to communicate among each other in a way that avoids every possibility of losing data, confusion, or blockage of information over a serial line. It is based on two serial wires that have the following functions: serial data (SDA) wire and serial clock (SCL) wire. Each device present on the bus is recognized by a unique address. The I²C-Bus interface has the following main features:

a) the device connected to the bus is software-addressable by a unique address and operates in a slave condition whereas the master, usually the microprocessor, operates both in a master-receiver and in a master-transmitter mode;

b) the serial, 8-bit oriented bidirectional data transfer can reach the speed of 400 Kbit/s in the fast mode;
c) both SDA and SCL lines are connected to the positive supply via external pull-up resistors;
d) the data on the SDA line must be stable during the HIGH period of the clock. The state of the data line can change only when the clock signal on the SCL line is in its LOW state; an exception to this case are the start and stop events;
e) both SDA and SCL are bi-directional lines. When the bus is free, both lines are at a logic level HIGH, set by the power supply level.

10.6.2 Device Data Transfer

Every byte put on the SDA line is 8-bits long. For the device under consideration, the number of bytes that can be transmitted in each transfer between the master and the slave is two. Each byte must be followed by an acknowledge (ACK) bit. The most significant bit (MSB) is put first. After the data byte transfer the transmitter releases the SDA line. The receiver can pull down the SDA line during the ACK clock pulse, so that the receiver confirms to have correctly received the preceding byte. In case the first byte is correctly received but the subsequent one is not, the master must abort the transfer. This is indicated by the slave not generating the ACK pulse. The slave leaves the SDA line HIGH so that the master can generate a STOP condition. The same happens if the address is not acknowledged by the slave.

In case a master is involved in transfer as a receiver, it must signal the data end to the slave transmitter by generating a STOP condition. The slave must release the data line to allow the master to generate the STOP condition.

10.6.3 Data Formats

The data format is as follows: after a START condition, a slave address is sent. This address is 7 bits long and the eighth bit (LSB) indicates the data direction; a LOW indicates a transmission (WRITE) while a HIGH indicates a request for data (READ). A data transfer is always terminated by a STOP condition generated by the master. It is not possible to change the bus direction without sending another start and a new address byte.

10.6.4 TDA 7292

As a good example of these must recent developments in the audio field, we will briefly examine the application of TDA 7292. This device can be operated in either of the two configurations:

a) 2 bridge amplifiers
b) 2 single-ended + 1 bridge amplifiers.

The I^2C-Bus interface in the device will perform both control (WRITE) and diagnostic (READ) functions. Naturally, the diagnostic functions are to be intended only as a warning. In fact, if it were necessary to wait for the microcontroller decision in case of a thermal or electrical overload, the device could be destroyed. This is quite a different concept from the thermal protection or shorted load protection described in 10.5. In that case, the device is pushed towards its own physical limits in order to guarantee the maximum output swing under extreme conditions: after that, the protection must intervene immediately. The device protects itself at the expense of a discontinuity in the listening performance. Here we are speaking of monitoring the performance itself to avoid listening interruptions as much as possible. A timely warning allows the microcontroller some preventive corrections, e.g. reduction of bass volume, or switching from bridge to single-ended mode to reduce dissipation. Any other known technique implemented in the audio field can be applied as well to the same target. On the side of shorted load, obviously nothing can be done to save even some minimum performance: in such a case the diagnostics meaning is to inform the repair technician of the type of fault that has occurred. Another diagnostic condition is experienced when the output voltage on any channel has reached the power supply level (output clipping). On top of the availability on the I^2C-Bus, the information is also present on the MUTE/CD pin. This pin deserves a short comment. It can work as an external input for the mute function (hardware MUTE) or as an output for the clipping-detector (CD) signal to the audio processor. The device will start at power-on assuming the mute function mode, switching to the CD function only after a microcontroller command. For what concerns the reading of output voltage clipping, the I^2C-Bus is mainly useful to monitor the lower audio frequencies. This is a direct consequence of the maximum clock speed, which does not allow a reliable reading of the higher signal frequencies. Anyway, this should be enough to reduce the bass as described in 10.6.4. For real-time clipping detection, the MUTE/CD pin should be used with a pull-up resistor, monitoring the low output voltage potential any time the audio output is clipped (hardware CD) [17].

Table 2 I^2C-Bus Output Byte to Slave. Assignment Rules Bit Meaning

	Bit	0	1
LSB	0	Stand-by IN	Stand-by OUT
	1	Mute Channels 1 & 2	Play Channels 1 & 2
	2	Mute Channel 3	Play Channel 3
	3	External Mute IN	CD Out Pin
	4	CD All Channels	CD Channel 3 Only
	5	2 Single-Ended + 1 Bridge	2 Bridge Channels
	6	Unused	Unused
MSB	7	Unused	Unused

Table 3 I^2C-Bus Output Byte from Slave. Assignment Rules Bit Meaning

	Bit	0	1
LSB	0	Device in Stand-by Mode	Device in Mute or Play Mode
	1	Channels 1 & 2 Muted	Channels 1 & 2 Ready to Play
1	2	Channel 3 Muted	Channel 3 Ready to Play
	3	Ext. Mute Pin Input Enabled	Output CD Pin Enabled
	4	No Clipping Occurred on Channel 3 or All, Depending on Setting.	Clipping Occurred on Channel 3 Only, or on One/All of Them Depending on Setting
	5	Thermal Warning Settled, Shutdown not Reached	Thermal Warning Settled, Shutdown Temperature (150 °C)
	6	DC Output Shift, in Mute Condition, $V_{SHIFT} < \|1\|$ V on All Channels	DC Output Shift, in Mute Condition, $V_{SHIFT} > \|1\|$ V on One or More Channels
MSB	7	No Overload or Short Circuit	Short-Circuit to GND or V_{DD} or Overload on One or More Channels

As a final consideration it is worth mentioning that all the warning and clip-detection conditions set on the I^2C-Bus interface are reset only after having been read by the microcontroller. This allows the monitoring of the above conditions or warnings over a very large time lapse. Obviously, this also means implementing an adequate software management of the actions to be taken in response to the warnings in order to prevent unwanted oscillations of the system; however this goes beyond the purpose of the present chapter. The Tables 2 and 3, taken from the TDA7292 Data Sheet, illustrate the read and write bytes used to implement the above items.

Moreover, it is also possible to exclude the I^2C-Bus interface and operate the mute and stand-by functions in a standard (hardware) way. For this reason, negative voltages on the SDA and SCL pins will be recognized as input commands. A voltage potential $-1\,V < V_{SDA} < 0\,V$, will force the device in stand-by condition, while a potential $-V_{DD} < V_{SDA} < -3.5\,V$ will leave the device ready for Mute or Play condition.

A voltage potential $-1\,V < V_{SCL} < 0\,V$, will force the device in Dual Single-Ended + Bridge mode, and a potential $-V_{DD} < V_{SCL} < -3.5\,V$ will set the device in the Dual Bridge mode. The MUTE pin will not give out any clipping information and will work in quite a standard way: for $0\,V < V_{MUTE} < 1.5\,V$, the device is muted, for $3.5\,V < V_{MUTE} < 5\,V$ the device is in Play condition.

10.7 References

1. Brasca G and Botti E (1992) A 100V/100W Monolithic Power Audio Amplifier in Mixed Bipolar-MOS Technology, IEEE Trans on Consumer Electronics, Vol CE-38, No 3
2. Brasca G, Botti E and Cini C (1992) Designing a High-Power Operational Amplifier in BiCMOS Technology, Analog & Mixed Signal Conf, San Francisco, CA

3. Antognetti P (Editor) (1986) Power Integrated Circuits: Physics, Design, and Applications, McGraw-Hill, New-York, Ch 9
4. Brown I (1989) Feedback and Fets in Audio Power Amplifiers, Electronics & Wireless World, p 123–126
5. Brown I (1990) Audio Power, Fets and Feeddback, Electronics & Wireless World, p 343–349
6. Philips Electronics Applications Bulletin (1973), Vol 32, No 4, p 137–153
7. Villa F (1986) Improved Second Breakdown of Integrated Bipolar Power Devices, IEEE Trans on Electron Devices, Vol ED-33, No 12
8. Murari B (1993) Mixed Signal Power IC Technologies and Small System Integration, Presented at the 19th ESSCIRC, Sevilla, Spain
9. Brasca G (1991) Electronic Engineering Degree Thesis, Polytechnic of Milan, Milan, Italy
10. Sampei T, Ohashi S, Ohta Y and Inone S (1978) Highest Efficiency and Super Quality Audio Amplifier Using MOS Power FETs in Class G Operation, IEEE Trans on Consumer Electronics, Vol CE-24, No 3
11. Krauss H, Bostian C and Raab F (1980) Solid State Radio Engineering, John Wiley & Sons, New York, p 468–473
12. Raab F (1986) Average Efficiency of Class-G Power Amplifiers, IEEE Trans. on Consumer Electronics, Vol CE-32, No 2
13. Brambilla D, Brasca G and Fassina A (1992) Dissipation and Efficiency in Audio Power Amplifiers: an Analytical Approach, SAE '92 Intern Congress, Detroit, MI, Techn. Paper No 920387
14. Antognetti P, Bisio R, Curatelli F and Palara S (1980) Three-Dimensional Transient Thermal Simulations: Application to Delayed Short-Circuit Protection in Power ICs, IEEE Journal of Solid-state Circuits, Vol SC-15, No 3
15. Widlar RJ and Yamatake M (1987) Dynamic Safe Area Protection for Power Transistors Employs Peak Temperature Limiting, IEEE Journal of Solid-State Circuits, Vol SC-22, No 1
16. Button M (1994) Busman's Guide to I^2C, Electronics & Wireless World, January 1994, p 24–28
17. Botti E, Fassina A, Stefani F and Tavazzani C (1994) Detection of Total Harmonic Distortion Due to Output Clipping in Audio Power Amplifiers, SAE '94 Intern Congress, Detroit, MI, Techn Paper No 940085

CHAPTER 11

High Complexity Smart Power Devices and Future Developments

G. Pedrazzini · S. Storti

11.1 Introduction

Today an electronic system is often designed using a programmable IC, in order to perform logic and control functions, calculations or specific algorithms, and other ICs to control, by mean of the drive output transducers or actuators, all the power flow. This solution always presents a complete separation between power and logic blocks realized with different ICs and, sometimes, by different boards. The system designer is free to use the right partition choosing the microprocessor and the power stage that can satisfy the application and at the same time optimizing the final cost. Besides, there is the same problem linked to the possible interference between the power environment and the small signal and logic part. These last troubles could be solved using a correct layout board strategy to build up the whole system. Nevertheless, there are some systems in which the space is the most important factor because all the electronic parts must be inserted in a small volume, for example, in a battery package or in the case of a small motor. This kind of application is now growing in number and the management system is becoming more and more sophisticated. The only way to solve this problem is to reduce the number of the components and the connections needed for the application and to integrate small signal, logic and power part into a single chip. Another direction is to use more complicated algorithms and more memory in order to realize a sort of self-trimming and self-adaptive system to avoid precise but expensive extra components and extra calibration operations. In order to follow this tendency the new generation of mixed processes was created, in which it was possible to integrate power DMOS transistor, small signal components, dense CMOS logic and also non volatile memories. With this in mind the chapter will focus on the possible future trends exploiting the possibility of using these kinds of processes and it will describe some typical specific field applications.

11.2 System Partitioning

An electronic system can be partitioned in some different blocks accordingly to the function, the complexity required and the level of the power involved. A typical structure for these systems could be the following one:

1) Input interface
2) Control and elaboration circuit
3) Power supply
4) Power driver

11.2.1 Input Interface

The input interface has the primary function of adapting the signal, coming from the external world, to the other internal blocks of the system. With external world we mean sensors, switches but also, and not last, all the electrical systems connected by transmission lines or by measurement points on power lines (e.g. a voltage measurement across a motor or across a floating shunt resistor to sense a current). The most relevant circuitry is generally analog, and characterized mainly by the precision and the voltage at which it must work. The most important circuits used in the input interface could be the following ones:

a) Circuits for signal conditioning, such as low noise and low offset preamplifiers, digitally programmable gain amplifiers and so on, with the task to amplify low level signals (for example, coming from a photo diode).
   These blocks are using sophisticated circuit solutions in BiCmos technology.
b) Circuits to protect and directly interface the logic part of the system, working at low voltage, with the power and transmission lines working at a voltage significantly higher.
   Some examples are line drivers/receivers, contact monitors and sense amplifiers. Those circuits must use elementary transistors with the same breakdown characteristics needed in the power part.

11.2.2 Control and Elaboration Circuit

The intelligent part of electronics is generally designed employing logic gates and typically this part applies a CMOS approach to reduce the power consumption mainly during the idle state. This state is by far more present in each new project and it is particularly important for the battery operated applications. The speed of this part is in the range of 1 to 10 MHz. If there are not any problems linked to the silicon area, it is better to use a static or pseudo-static logic methodology in order to reduce the influences due to the huge current that the power part is switching. In the design of the coming future, the intelligent part will be a complete microprocessor because, as we will see later in the battery charger example, the algorithm of the controls requires a flexible flow, provided only by a software programmable machine, instead of a fixed control methodology, provided by a classical wired state machine. The most used microprocessors for these applications implement an 8-bit CPU, memory and some other peripheral blocks. Overall the memory blocks use a standard approach but the RAM part must be a static [1] for two main reasons: the first one concerns the data retention during the idle (stand-by) mode when, usually, all the internal clocks are stopped; the second one is that it gives better immunity to the interference linked to the power application.One example of the peripheral block is a 8/16 bit timer to implement a counter function, or just to create a PWM signal for a motor drive. There is always present a watch-dog part that generally is a logic count-down counter, periodically set by the program, that it will reset the CPU if it reaches the zero value. The same function could be implemented also with an analog circuit: in this case a capacitor, charged by a switch and discharged by a constant current (or vice versa) and a comparator replace the digital counter. The analog solution is considered safer just because it will work independently of the CPU. Serial or parallel ports are required because the CPU must send or receive information using these pertinent blocks that can run independently of the master CPU itself avoiding extra programming routine and waste of time. The most used port is the serial one to minimize the number of connections and the communication protocol is specific for each application field. The output of the logic serial port is often connected to a line driver/receiver present in the input interface block to adapt the logic levels to those needed for the transmission line. The major features of the parallel I/O ports is the possibility of changing its I/O configuration by software in order to program different bits, of the same port, in a different output/input mode. The general I/O modes are: input mode, open drain output, push pull output, analog input (to interface with the internal analog parts like the A/D) or one of the previous configurations with interrupt capture. It is important to underline that, in order to test these devices, at least one pin, to force the test mode, and a serial interface to introduce the test pattern are needed. In this way the serial interface could be used in two different ways following the current working mode: test or user mode.

Another very important internal unit is the A/D converter used to sense the analog signals, for example a process control value, coming from the external world. It converts this analog information in a stream of numbers that can be understood and elaborated from the internal CPU. The most used A/D has an 8-bit resolution but the new applications can require a 12/16 bit resolution.

All the peripherals we mentioned, must be inserted in an interrupt chain: in this way, the core of the CPU can oversee all the blocks without reading each line by software. This possibility is important to realize fast or real time control, in fact, the program could be written with its normal flow chart without worrying about checking the critical lines and therefore without losing time in the polling routine.

11.2.3 Power Supply

This is a general purpose part, already described in this book (Chapter 7). Its main function is to supply the energy, with the correct voltage value, from an unregulated supply line, with a relative high voltage, to the input interface and to the signal/elaboration blocks usually working at a lower voltage level. The task of the regulator is also to protect all the supplied block against the line voltage transients (e.g. the load dump in automotive electronics, R.F. interferences, voltage spikes and glitches present on any power line). The choice of the kind of regulator must be strongly dependent on the system supply. Often, a traditional linear regulator is enough; in a lot of battery powered applications and in automotive electronics a low drop regulator (with pnp output stage) is required in order to use at maximum the battery voltage. Sometimes high efficiency is the main point: in these cases the space and the weight of the heatsink must be decreased, the efficiency must be increased and size of the battery (pocket, palmtop application) decreases. In all of these fields it is mandatory to use switching regulators. In particular, in battery powered system, it is common to implement a step up or step up/down regulation to supply logic parts at the standard voltage (e.g. 5.0 V or 3.3 V) starting, for example, from 2 cells batteries that give typically 3.6 V maximum (when in charge) to 1.6 V minimum. The future trend is to distribute, through the system, only one unregulated supply to energize the actuators and to generate locally all the other necessary supplies.

11.2.4
Power Driver

The last but the hugest area is the power circuitry. The main purpose of this part is to adapt the power levels (voltage and current) needed to drive the actuators and output transducers. This could be divided into the power driving one (see Chap. 5) and the real power device to switch the current from the mains supply to the load (see Chaps. 7 and 9).

Usually,the driving part must protect itself and the load from dangerous conditions taking care to measure the main electrical and physical quantities, like the supply voltage or the current in the load and the temperature of the IC, immediately switching off the final power transistors and/or giving a diagnostic message/command to the main control unit.

11.3
Possible Implementations

If we have a look at the practical implementation of the typical system, just described, there are two ways of proceeding: the first one is to use a specialized technology for each block, for example BiCMOS technology for the input interface, a pure CMOS for the signal processor and a power technology for the power supply and power driver blocks. An important point is that the power block, both to improve power efficiency and cost effectiveness, must work at the maximum voltage present in the system; on the contrary, the signal processing part, to reach the needed complexity at the reasonable cost, must work at a voltage in the range of 3.3 V (new submicron tech.) to 5 V (standard tech). In this case a minimum of 2 or 3 ICs are required to build up the complete system.

The second possible way is to integrate all the control and power circuits in only one part using mixed processes, capable of combining high power devices, like power DMOS, with high density CMOS logic and now non-volatile memories in only one die as well. In this last direction the technologies already described in the other parts of the book were developed.

11.4 Main Applications

Today, circuits integrating a software programmable controller with power actuators can be mainly used in the following fields:

a) Industrial Power Actuator and PLC
b) Switch Mode Power Supply
c) Camera Application Control
d) Battery Charger
e) Automotive

and we will have a look at each application and we will underline the two last applications (d) and (e).

11.4.1 Industrial Power Actuators and PLC

The main feature of these ICs (PLC = programmable logic controller) is the capability of interfacing the internal CPU core with a external master unit already present in the overall system integrating into the same chip also the power transistors to drive a high number of relays. In this direction, this kind of chip contains a high speed serial interface that will use standard line protocol communication like RS232 or RS422 and it has a large number of I/O port with high current capability. The complexity of the control part could be used like a remote or post-processing unit: for example, to control the status of each load and realizing a sort of real time local feedback without overloading the transmission line and the master control unit.

11.4.2 Switch Mode Power Supply

The SMPS is a typical application that normally doesn't need, in general, a real micro processor nevertheless an integrated programmable CPU can control the current and the frequency of the output switch according to the connected load, to optimize the efficiency and the usage of the chip. This methodology can fit many applications just changing same control internal register that decide the actual field.

11.4.3 Camera Control

The camera control system includes all the camera electronics beginning with a preamplifier and an external photo diode used to measure the scene light, to the final power driver that energize the shutter motor. The signal coming from a selected photo diode is integrated and then converted into a digital form to be ready for the internal high speed 8 bit microprocessor. In fact, the correct exposure time is calculated, by the CPU, depending on the brightness value of the scene and the subject distance. Others motor actuators or servos, present in the camera, can be energized by the integrated power driver as well. The main constraints of this application are space occupation and the power consumption. The space problem is solved by integrating all the power drivers and logic control into a single chip. The problem of power consumption is due to the fact that the overall system is supplied by a small battery and it is solved using the property of CMOS logic in that it doesn't consume any current in a static condition.

11.4.4 Battery Charger

This application is a battery charger device and for this device we will focus on the hardware problem, the system solutions and the partitioning with a final view of the interferences that could be arise using a high logic complexity part, like a micro processor, near to a huge power transistor. The need to have sophisticated battery chargers is mainly due to the need to reduce environmental pollution that has become compulsory in the last years. The pollution is mainly due to the toxic lead in the lead-acid batteries and Cadmium in the Nickel Cadmium ones, that is even more toxic than lead. There are, luckily, an alternative, in the rechargeable cells that is the new NiMH batteries similar as features to the older NiCd type. From an other point of view the NiCd type is better than the NiMH one in terms of fast charging, overcharging and life but NiMH has a better discharge profile shown in the Fig. 11.1.

A sophisticated programmable charging device can help to solve the problems because it allows to have an hardware independent to the kind of cells optimizing, in this way, the up time of the powered system, minimizing the charging time, maximizing the stored available energy in any condition and the battery life in term of charge/discharge cycles.

The problem we faced was how to recharge a battery and which was the right algorithm to program. For this purpose, looking at the typical charging curve of the two battery types, NiCd and NiMH (Nickel metal hydride), there is a critical point to detect (Fig. 11.2).

This point is immediately before the bump in the voltage curve that can occur after the cell has received too much charge. This overcharging could give a loss of the stored energy and a shortening of the battery life. But, this curve is not the only one that we can use, in fact, a thermal sensor near the battery pack, gives

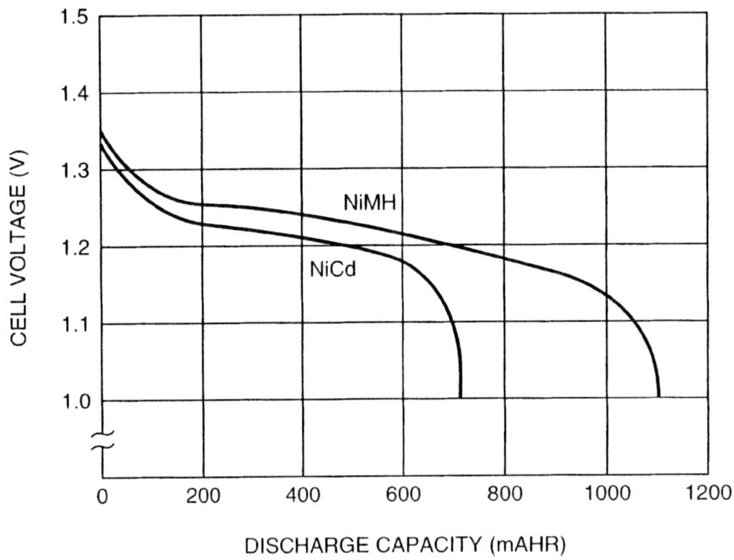

Fig. 11.1 Typical discharging curves

another information in order to detect the overcharging characterized by an increase in the slope of the temperature curve versus time during charging (Fig. 11.2). With this methodology it is possible to recharge a NiCd type in *15/45-min* and the NiMH in about 1.5 hours avoiding boiling in the battery. For this purpose, a sophisticated algorithm is implemented in the battery charger chip and a typical program flow-chart is drawn in Fig. 11.3.

The life and the performances of the battery are also strictly dependent on a lot of parameters: the discharge current, the previous number of recharging cycles (that gives, virtually, the history of the battery), the temperature and the length of the rest period between discharges.

All these parameters can be taken into account only using a complex algorithm for charging and using a current generator that could drive 3 C (C is the total capacity of the battery under charge). For example, using a battery with 1 Ah of capacity we can force at maximum 3 A. The last problem we have to discuss is the discharge of the battery during use. In fact, if the user or the system does not discharge the battery completely between charges, it is possible that a strange phenomenon could happen. This problem arises if the use of the battery is minimum and the cells are not completely discharged. In this condition the cells will not get the overall charge and we can see that they maintain a sort of *memory of their charge*. The solution, in this case, is to discharge completely the battery before all the recharging phases.

With this introduction a sophisticated programmable controller, at first sight, could seem to be too complicated if it is used only to charge (may be also

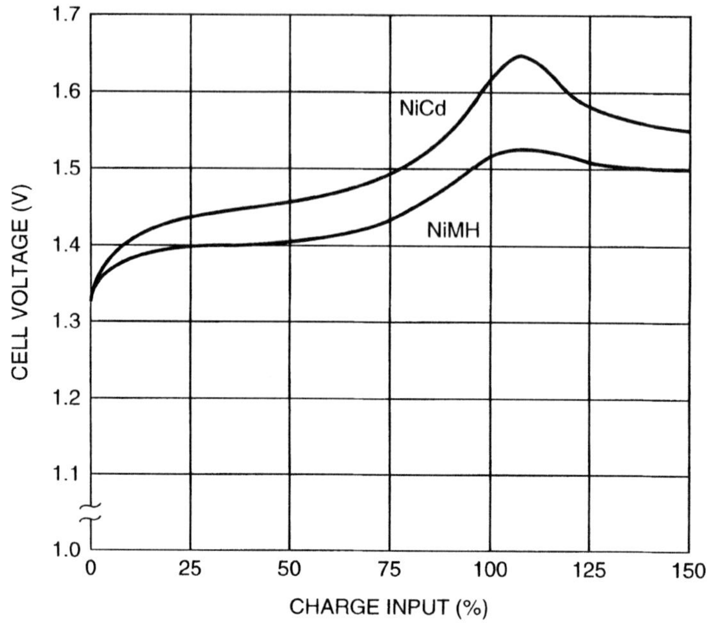

a

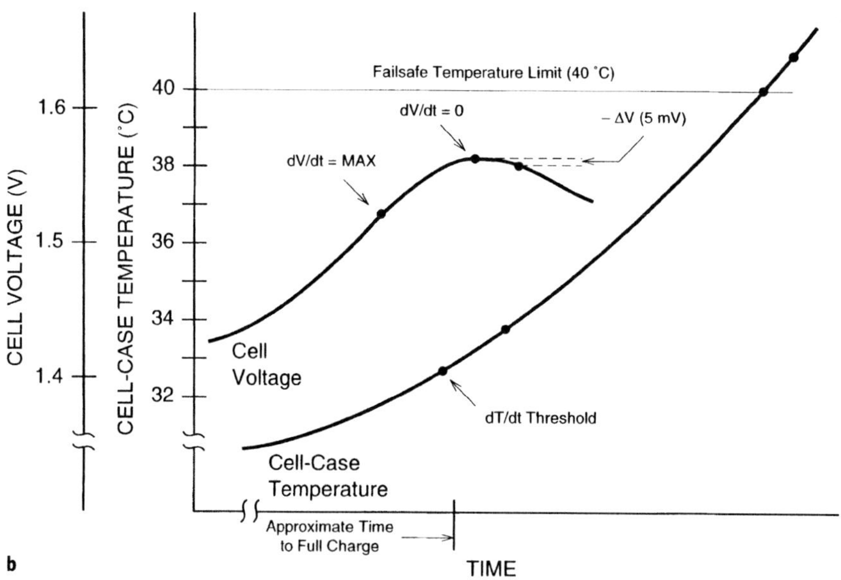

b

Fig. 11.2 Typical charging curves, **a** Bump point in the charging curve, **b**

Fig. 11.3 Simple charging flowchart

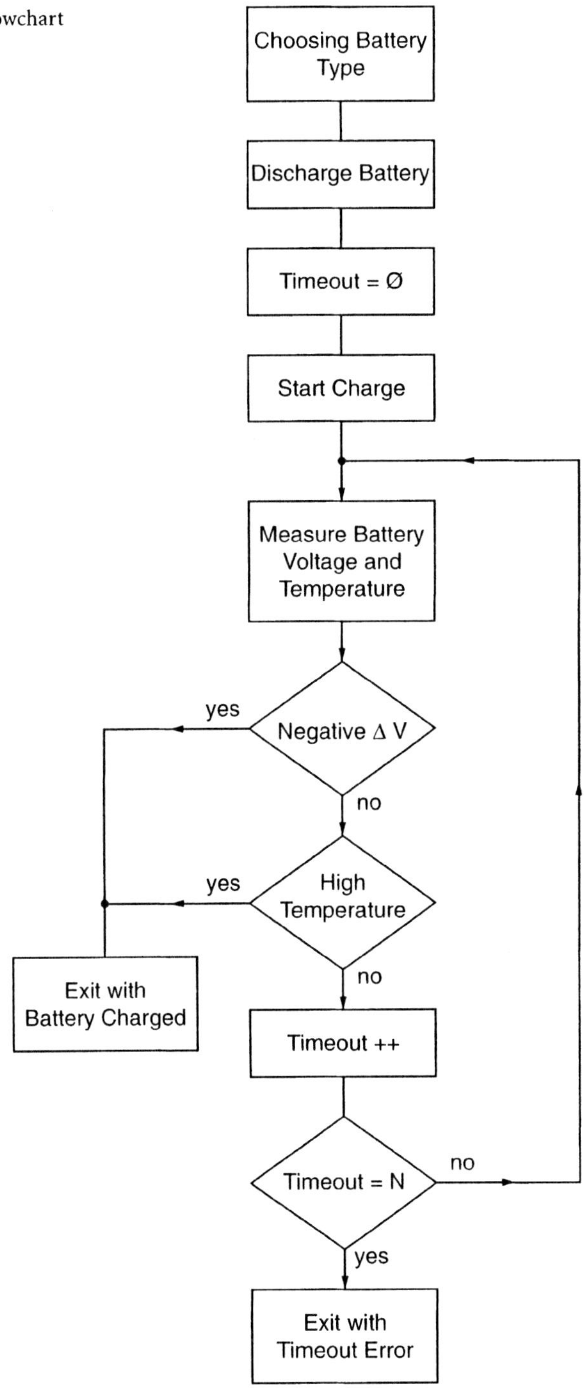

discharge) a simple battery; however, today this is the best solution if it is required to solve, *at the same time*, all of the following problems:

a) to provide fast or ultra-fast battery charging;
b) to maintain the battery in charge;
c) to use completely all the battery energy;
d) to obtain the maximum possible life of the battery during charge/discharge cycles;
e) to be compatible with any type of battery;
f) to be as small as possible;

The chip, we will describe, has the purpose to implement a complete battery charger, in terms of the charging control and in terms of power management, to be inserted in a small package mainly for car application or cellular phones.

The chip is realized in a 1.2 um mixed technology that it is able to integrate the standard 5 V CMOS, bipolar, vertical DMOS that withstands up to 60 V and EPROM cells. The chip was designed to be fully purposeful for the case but it could be used for some other switching applications.

The main parts of the device are (Fig. 11.4):

1) 8-bit CPU
2) 64-byte RAM
3) digital watchdog
4) 8-bit timer with pre-scaler
5) safe oscillator
6) I/O ports
7) A/D converter

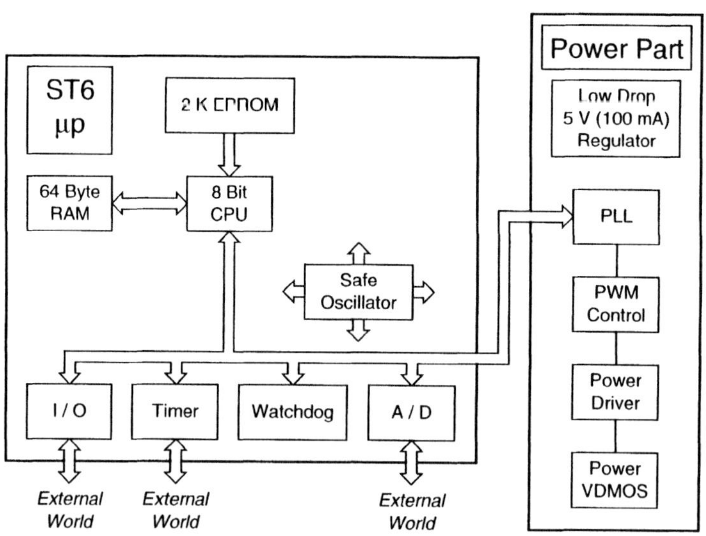

Fig. 11.4 Battery Charger internal block diagram

8) EPROM matrix
9) voltage regulator
10) PLL block
11) complete current mode step down voltage converter

11.4.4.1
The Digital Control Part

The CPU block runs at 8 MHz and it is designed with a serial structure that means that it works on one bit in each clock cycle and it is one bit wide due to the ALU control part design. The instructions set is the standard one without multiplication and division. The instructions are, typically, performed in 3 CPU cycles and one CPU cycle takes 13 clocks times. The reason for this number is due to the address of the program that is 12 bit wide and it must be incremented using the ALU (1 bit at a time) and we add a extra clock just to deglitch some signal to be sure about their values. Transfer with the internal peripherals is parallel in order to increase the total speed. Following this philosophy the serial core of the CPU (ALU block) could be smaller, only one bit wide instead of 8 with the final result that can be transferred in parallel to the addressed peripheral. The surrounding peripherals are *custom* blocks and their *modularity* give an extra advantage to the *system approach*. The internal RAM is not huge, only 64 byte, but for the application we will fit, it is the right number. All the peripheral blocks are mapped like a standard RAM register and for this feature it is useful to have the same instruction for the RAM register and for the peripheral register used to configure it. The digital watchdog is mandatory in this application and may be used in all the industrial application. It is implemented in a standard way, count-down counter, with the possibility of disabling its original function and using it like an extra counter. The timer is a key feature to monitorize the charge time just to have a time out signal if the external sensor does not give the right information or the battery does not work well. All the blocks previously described have a total area of 4 mm^2 and are the standard microprocessor named ST6210 [2]. The following part we have to analyze is the oscillator block: this drives, with a frequency of 8 MHz, all the internal clock signals and it is able to stop the main clock under software control. In this way, it is possible to enter in a sort of low consumption mode. All the internal operations, of the other blocks of the microprocessor, are synchronous with the main clock and for this reason it is mandatory to have a good clock signal in terms of stability, duty cycle, rise and fall time and *spurious glitches*. With this constraints the oscillator contains a part designed to filter out all the possible spurious commutations of the clock signal and is also able to generate a second extra clock signal, with a lower frequency (300 KHz), that is automatically used when the main one has troubles or is missing. This feature is very important in this application and in all the others that require a constant active control. In order to start the synchronous operation, only when the clock source is stable and the oscillation start up transient is finished a delay, created by a counter, is inserted into the circuitry. With this circuitry the external reset time width is increased until the internal clock has a good waveform. The oscillator can work using a normal external quartz or using a simple external resistor with an integrated capacitor that can create a cheaper but less accurate clock source. The oscillator block contains also a reset control circuit. This part monitorizes the most important internal signals in terms of time between transitions. In this way, if one of these do not respect some precise timing limit, a reset will be generated in order to restart the whole system. With this approach each operation is under control as concerns signal generation to avoid the possibility of looping in a wrong and unknown routine.

The I/O ports are divided in analog and digital. The analog I/O ports have the purpose of interfacing, with the internal A/D, the external analog signal from a temperature sensor and from the battery voltage. There are also two high current (10 mA) output analog I/O ports used to supply external leds. These leds display the information about the status of the battery and the actual running operation. The digital ports are pure CMOS I/O and they can be fully programmed bit per bit. These pins could also realize, if needed, the connection with the internal A/D but usually they are configured in digital way and they could be linked with an internal interrupt signal used for real time operations. The digital ports already described have an extra important use during test mode. In fact, during this time the same pins will begin only input or only output in order to send to the CPU the test vectors coming from the tester in a first phase and to read back the result of the internal elaboration. Because of the one-bit-wide internal structure and to save pins, these operations are serial therefore causing a longer testing time.

In order to be a real general purpose microcontroller, it is mandatory to insert also an analog-to-digital converter to interface the external analog measure with the internal program flowchart. For this issue, the chip contains an 8-bit A/D converter that works using a successive approximation algorithm.

The reference voltage used by the comparator of the SAR block is a voltage divider. The choice of the voltage divider instead of another topology was made because it assures both a good noise immunity and a high linearity of the conversion. This resistor string is realized using 256 resistors with the same value and the selection of the value, to compare with the external analog one, is realized by a binary tree of switches. These use a CMOS transfer gate that is driven by the successive approximation logic register (Fig. 11.5).

The conversion time takes 30 time clocks for any analog value. The area of the A/D converter is around 1 mm^2 mainly due to the resistor string and it is adjusted to the current consumption. The EPROM block has a size of 2 Kbyte, for the program and data of the micro, in an area of 3.9 mm^2. The EPROM cell needs a 12.5 V in order to be programmed and for this reason the complete EPROM blocks are divided in 3 fundamental part: the 5 V logic part used to interface the matrix with the micro and to create all the control signals, a 14 V CMOS part for the decoder block that addresses each cell in order to be read or written and the matrix in which the data will be stored. All EPROM verification and programming operations are performed only in test mode forcing a sort of programming that can scan each cell to be read or to be written. In order to decrease the pin number and to separate the normal test time and the programming time, the pin of the test mode could detect 3 different situations. In the first mode, user mode,

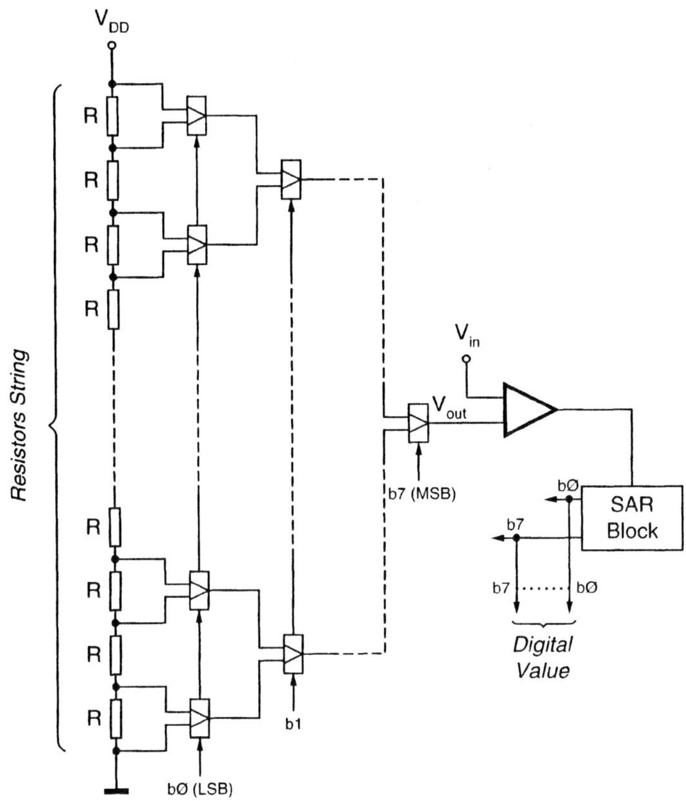

Fig. 11.5 A/D structure

the pin has a low level; if the pin has a 5 V level a normal test will be performed and if this pin is pulled up over 12 V, the programming phase is activated always using the test approach.

11.4.4.2
The Analog and Power Parts

The analog part of the chip is composed of the regulator (Fig. 12.6), the step-down converter and the PLL block designed to *reduce the interferences* between the analog and the digital commutations.

The regulator block has the main function of supplying all the internal IC blocks and due to his current capability it is able to supply also external circuitry.

The regulator is composed of two main blocks: the bandgap voltage reference and the power amplifier. The bandgap circuit uses a standard topology in which two transistors, with different area, are forced to sink the same current to compensate the negative drift of the V_{be} with the positive one of the DV_{be} created by

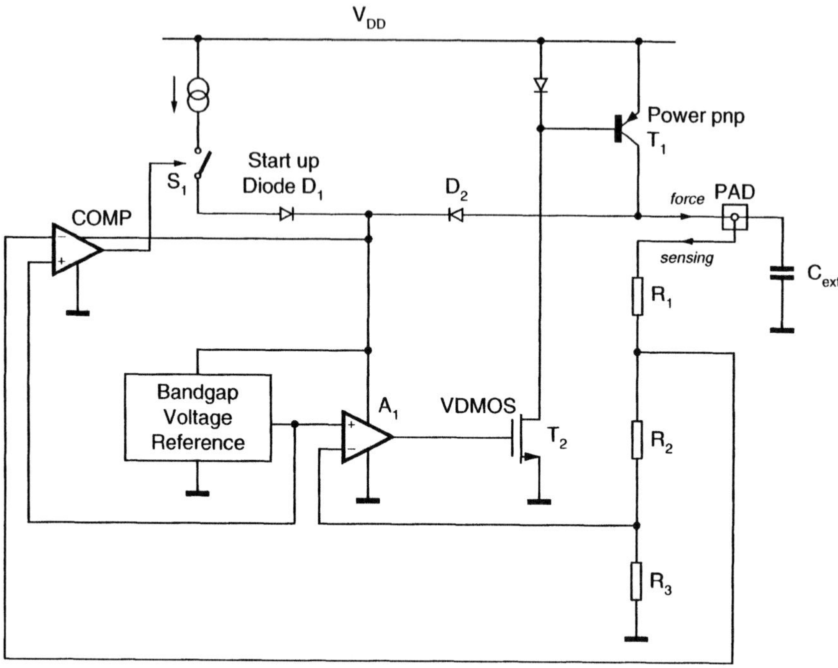

Fig. 11.6 Regulator block diagram

the same two transistors [3]. The trimming of the bandgap reference value, to obtain the final precision of 2%, is realized during the testing phase of the overall device and it is implemented by a sort of metal fuse approach that includes some resistors in the bandgap loop to reach the target value. All the trimming could also be realized using some EPROM cells but, the choice of design was different because the all EPROM cells can be deleted by the user to change the program losing the initial trimming. Another good solution is a trimming using the EEPROM cells available in the process, but EEPROM is not present in the chip. The power amplifier [4] uses the output of bandgap as a reference voltage and produces a 5 V output with a current capability of 100 mA. Because of its structure, using a pnp (T_1) for the current output stage, it is intrinsically instable and it needs an external capacitor to create the dominant pole (C_{ext} = 10 uF). Due to the automotive application, with high voltage spikes and a noisy voltage supply, it is mandatory to use a regulator structure able to withstand these transient and also the load dump overvoltage. For this reason, a high voltage pnp (T_1) for the output stage is required. In order to increase the power supply rejection ratio, the bandgap and the power amplifier are powered by the output through D_2. The output voltage value is monitored by a comparator block (COMP) and when it is about 4.7 V the supply voltage for the bandgap and the amplifier (A_1) are switched from the battery voltage to the output of the overall

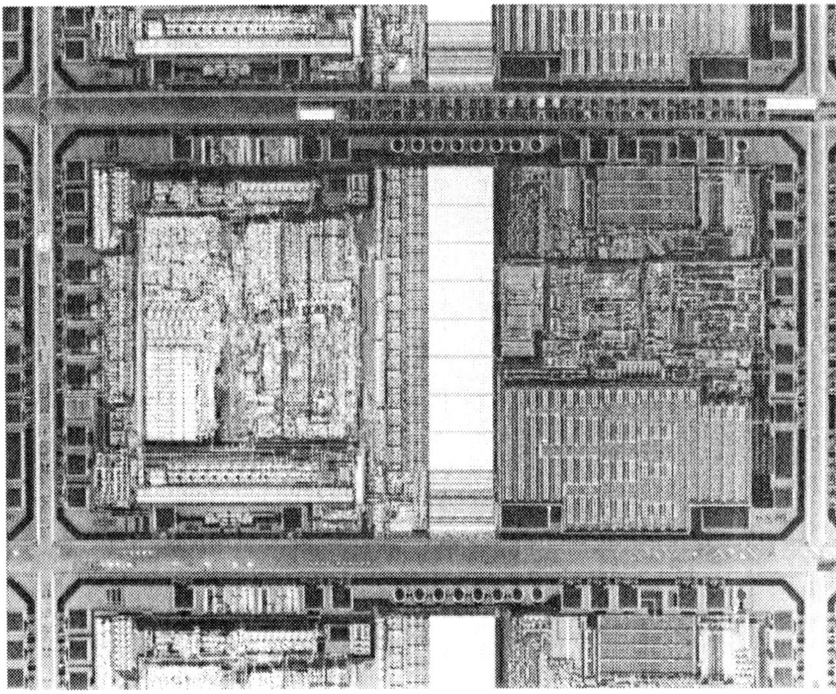

Fig. 11.7 Battery Charger device

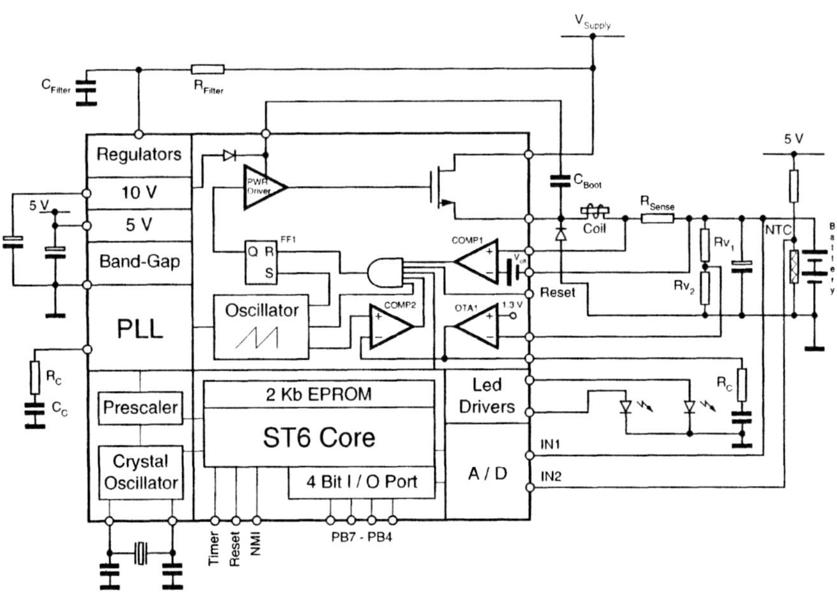

Fig. 11.8 Battery Charger application

regulator by the switch S_1. In this way, it is the regulator itself that supplies its internal blocks increasing the supply line rejection. The total area of the regulator is around $1mm^2$ due to the large area of the power pnp (T_1).

The PLL block performs a sort of sawthooth generator with a maximum frequency of about 250 KHz for the PWM control synchronized with the CPU clock. In fact, the main problem is to create a ramp voltage with a fixed frequency but also with a fixed swing range. The first problem could be solved using a simple pre-scaler with the CPU clock as input but the swing limitation is not so trivial having a digital waveform as an input. For these reasons, a PLL approach has been chosen as it creates right signal wave and can guarantee that all the signal commutations are *synchronous*, minimizing the interference problems between the blocks. The other block realizes a real current mode pwm control able to force a precise current into the load (battery). The current is set by an external resistor choosen according to the battery type and size. The control is a standard one; it needs an external RC compensation network and it will work also in voltage mode when the battery pack is removed to clamp the output value at a non dangerous voltage level. The frequency of this switch mode power supply and some other control feature are *mapped in an internal register* of the CPU. This allows to create different applications using the same chip with different program parameters representing the PWM configuration. The pwm signal is generated in the standard way with an error amplifier and a comparator to compare the error with the sawthooth signal. To complete the pwm converter a 3 A, 60 V DMOS transistor with a RDS_{on} of 0.160 ohm is integrated. The gate of the DMOS, N-channel, is driven with an external bootstrap capacitor. The area of the overall pwm control is about 2.1 mm^2 and the power DMOS is in a 3 mm^2.

The final chip area is 26 mm^2 and it is shown in the Fig. 11.7.

The package used for the IC is not a power one, due to the high efficiency of the SMPS; in fact, a SO28 package has been used without any heatsink or other package innovation. For the type of memory inside the chip, EPROM, the package needs a window but the final production will be an OTP or a ROM version to reduce the package cost. The final application is drawn in Fig. 11.8.

11.4.5 Automotive

As already seen (Chap. 9), Automotive Electronics is a good example of a system that must take into account a big number of trade-offs between cost, complexity, possibility to work in extreme ambient and electric conditions assuring, at the same time, the highest condition of reliability and of robustness of the electronic devices.

In order to highlight some difference in the applications in this field, we can put on one side motor, power train and battery charging control; on the other side, instrumentation and body electronics. Body electronics is intended all the electronics extended in the body of the car to control the electric appliances such as motors, lamps, heating resistors.

The first is the field of the most traditional system partitioning.

If we take as an example a motor control unit, we can find signal analog circuits to interface sensors (E.G.: the lambda sensor to monitor exhaust gas), power driver circuits (E.G.: injector drivers-Chap. 9), and a microcontroller to carry out the logic processing. This kind of partitioning is not foreseen to change in the near future, because a higher and higher computational power and memory size is required from the microcontroller to optimize each working point of the engine also possibly complying with the wishes and the needs of the driver moment by moment. The difficult task of the microcontroller is in other words to optimize engine efficiency, pollution, and driveability, taking in to account a lot of input parameters. (For instance, exhaust gas, throttle position, temperatures ...).

The best partitioning is therefore to use specialized technologies for each circuit block.

For instrumentation and body electronics, on the contrary is not so important to reach the maximum of performances (at a reasonable cost). It is important to reach the minimum overall system cost (at a given level of performances). If we take, for example, the mechanic actuators, servomotors and switches placed in a front door of a car, such as mirror positioning motors, door lock actuators and window lift motors, there are mainly three solutions competitive at cost level to reach a similar standard of features and performances.

- To use one wire for each sensor, each actuator and each switch and a minimum amount of electronics, placed on a central part of the car.
- To use some standard multiplex system [5], putting in each door one peripheral station of the system that will control all the local functions using only one bus exchanging coded information with some other local or central unit.
- To use some purposeful multiplex, similar to the previous in concept, but in which both the peripheral stations and the transmission protocol are minimized tailoring them on each particular application.

The first approach is the most traditional one and the most used today, but the increase of the number of the overall components to control (already more than 50), and the increase of assembling cost will make economical multiplexed power supply and control systems. These systems are composed of a single common power line and a multiplexed signal network for the control. In this way, it is no longer necessary to have one wire for every load, because all the information is carried serially by a common transmission line for all the loads.

For example, a multiplexed control system can be made with a central unit near the dashboard, for the user interface and some peripheral units near the loads, connected by a serial data transmission line (Figs. 11.9, 11.10).

The multiplexed systems not only make possible an overall reduction of the weight and dimensions of cable harness, now critical in some places (E.G.: junction between the vehicle body and the doors), but also it makes possible to have, without extra wires and cost, bi-directional signals between peripheral units and the central unit. This is useful for the transmission of the status of each actuator, fault detection, and data transmission.

Fig. 11.9 Example of a multiplexed system

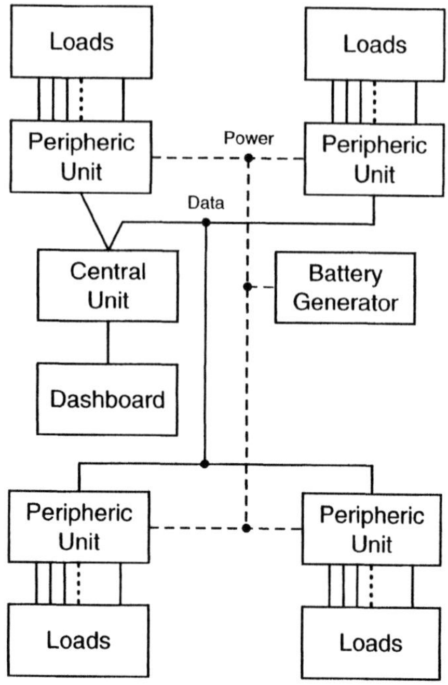

Fig. 11.10 Example of block diagram of a peripheral unit

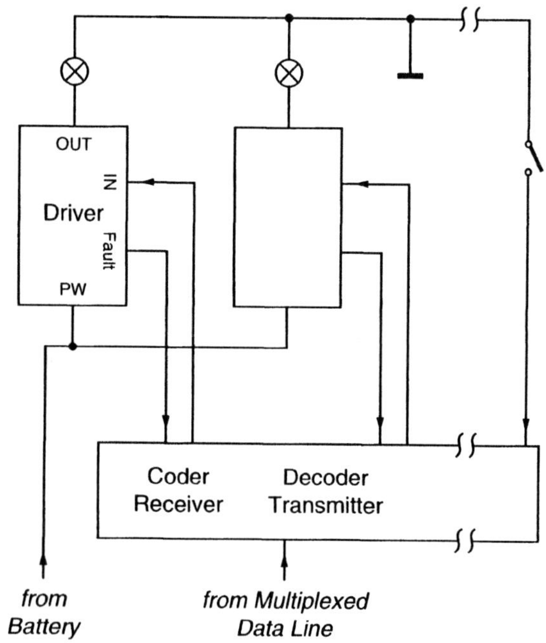

Today the key point from system engineering point of view is data transmission, whereas for semiconductors technology electronic power switches are also an important detail [6, 7, 8, 9, 10].

The standard multiplex solutions are generally given with a traditional partitioning system. For example, in a door a box is fitted containing, as input interfaces, one line receiver/transmitter and some contact monitor circuits, as output actuators, some ICs of the kind described in the previous chapters (mainly 3, 6, 9) and as signal control block some logic that can be both wired or software programmable according to the features and so the intelligence really needed from each unit. To note that, with this topology, the quantity of cable needed for a car is greatly reduced, but not the number of connections and the overall number of pieces to install (there is one box more...). The volume of the electronic circuitry is not relevantly important, because the box containing the peripheral unit into the door is not so strictly limited in dimension. The way to reduce the costs is to standardize and optimize as much as possible each part: provided that protocols and physical implementation of transmission line are standard (examples J1850 – VAN – CAN), standard can be line transceivers and logic parts that handle the protocol. Contact monitor circuits, voltage regulators and power actuators can be standard as well, and so microcontrollers. As a conclusion, this kind of multiplex is generally structured with circuits made with specialized technologies.

The other way to realize a multiplex system is to integrate the electronics in the same case of an already existing part, for example, the door lock or the window lift mechanism, the fairing of an external mirror, an electric connector or the back side of a keyboard. In this way, by minimizing the number of overall parts and connections, it is possible to realize an economical solution. The volume available for each peripheral unit, near or into the electro mechanical actuator or sensor, is very small and so it is really important to minimize the count of ICs necessary to carry out the function.

In this field, the possibility of mixed technologies to integrate as much as possible on a single chip, is really convenient.

Of course, in this second implementation, it is important to reduce the cost also tailoring the protocol for the serial connection of the system and the intelligence present in any single unit, to the minimum necessary. For this reason small subsets of standard protocols or pertinent protocols are used.

11.4.5.1
Door Lock Unit Example

An example of a system solution can be a door lock unit, to be mounted with the motor and the mechanism itself. This unit is one peripheral interface. Each peripheral interface has a protocol handler, a transmission handler, its own specific I/O block and their relevant analog blocks. Each of these chips monitors the words on the bus and decides, by checking the address, if any peripheral activity is to be carried out. Each chip also monitors information (status or error) from its peripheral and prepares it for transmission via the bus to the

Fig. 11.11 Door lock unit using one power and one logic IC

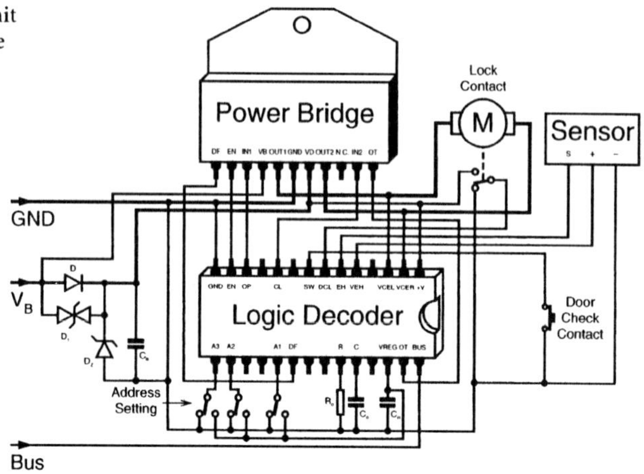

Central Unit that communicates with each peripheral unit using the appropriate interface via a single wire bus.

For one realization the starting specification was, to drive one single door lock motor, with a stall current of 7 ÷ 10 A, and to monitor the position of the locking mechanism (locked, unlocked), of the door (open,closed) and of another sensor (key turned in an extra position). Due to the relatively high current level for the motor,it was decided to divide the unit in two ICs, one power driver, H bridge, in pure bipolar technology and an interface circuit using Bipolar, CMOS DMOS mixed technology 60 V rated (Fig. 11.11). These two devices, with some other discrete components (such as a resistor and a capacitor to set the clock oscillator time constant plus a capacitor, a diode and a zener to take care of supply polarity inversion and transient over the maximum ratings), can be used to build a unit small enough to be placed in the case of the motor.

The H bridge is a very simple IC: it contains only the power transistors, the interfaces to drive them from a logic signal referred to ground, a sensor to monitor the chip temperature and another to detect an extreme electrothermal overload condition [11].

The interface circuit has three basic functions:

1) To interface the serial bus.
2) To interface the power bridge both for driving and for protection against overcurrent, overtemperature and overvoltages.
3) To interface all the sensors present in the application.

The distribution of functionality amongst the devices was mainly decided on the basis of the maximum current capability and the minimum number of devices needed.

For this reason, interface circuits specific to each application have been developed in mixed technology with the capability of supporting overvoltages of up to 60 V, thus allowing direct interface with the automotive electrical environment.

The use of programmable and/or high density logic circuits was excluded in this case because the application and transmission protocol were fixed and sufficiently simple.

The main features of the protocol are the following:

- The Bit is coded in PWM, that is there are a rising edge and a falling edge for each bit transmitted. The first is the synchronizing one, after that a "long" pulse means "1" a short "0", like old "Morse" code. A bit is long several clock cycles (16) to allow an appropriate filtering.
- The word is 12 Bit long, comprehending address, data and checksum, and is preceded by a start field, different from bit symbols.
  With a nominal frequency of 32 KHz, the data flow is 2 K Baud.
  The system is designed to synchronize itself on the start bit and to re-syncronize every rising edge, in this way it is possible to allow a tolerance of 30% in clock frequency between C. U. and the peripherals.

Each door lock peripheral unit is addressed differently by programmable input pins, the remaining part of the address (substantially dependent from the "kind" of unit) is fixed.

The data field of the transmitted message indicates the status of the switch inputs and of the actuator (all ok, overload or open circuit).

The data received from the bus are only put in the parallel form by the interface and directly used to control the power bridge.

However, in the case of a motor stop, the interface has also the significant task of controlling the timing sequence. This is due to the absence of a diode between the outputs of the power bridge and the positive supply (silicon area needed too large) while one exists between outputs and ground (substrate collector junction on NPN power transistors on the low side of the bridge). As a result, there is no re-circulation path back to power supply to allow the energy that was loaded in the motor inductance to be discharged during switch off. To overcome this limitation, the switch-off of the power devices should follow a well defined timing sequence, controlled by the interface circuit: before a completely switching off of the bridge, both the low-side transistors must be on for quite a long time (100 msec). During this time, the motor is dynamically braked and also the recirculation of all the currents is assured, taking care that at least one of the low side transistors is always on. This is an example of the way in which it is possible to save power components, and therefore silicon area, slightly complicating the control sequence.

Another important feature of the interface is the sleep mode: if the motor is completely off, and there are no registered variations on the bus and on the switch position, the interface, after a delay, goes in a sleeping state, reducing its power consumption. In the case of change of lock position or when a door is opened, the interface must be able to wake up all the multiplex system, if it was in a stand-by state, forcing a low level for a relatively long time (~ 25 msec) on the bus, which allows the central unit to carry out its reset and wake up routines.

The power bridge is mounted in a multiwatt package, the interface circuit in S. O. package.

Like this unit, other units for window lift, mirror positioning and keyboard monitoring were developed; is to be noted that the latter is designed with only one IC because power is not involved.

11.4.5.2
The Monolithic Door Lock Unit

From this design, made few years ago, to the designs running now, something has changed, mainly:

1) From the technological point view: the 1.2 μ mixed process is available.
2) From the mechanical point of view: the current needed for the motor has decreased from 7 A to 2 A.
3) From the system point of view: more flexibility is required.

For all these reasons, now it is preferred to integrate, together the power section, a programmable microcontroller core and some specific analog and logic circuits (Fig. 11.12).

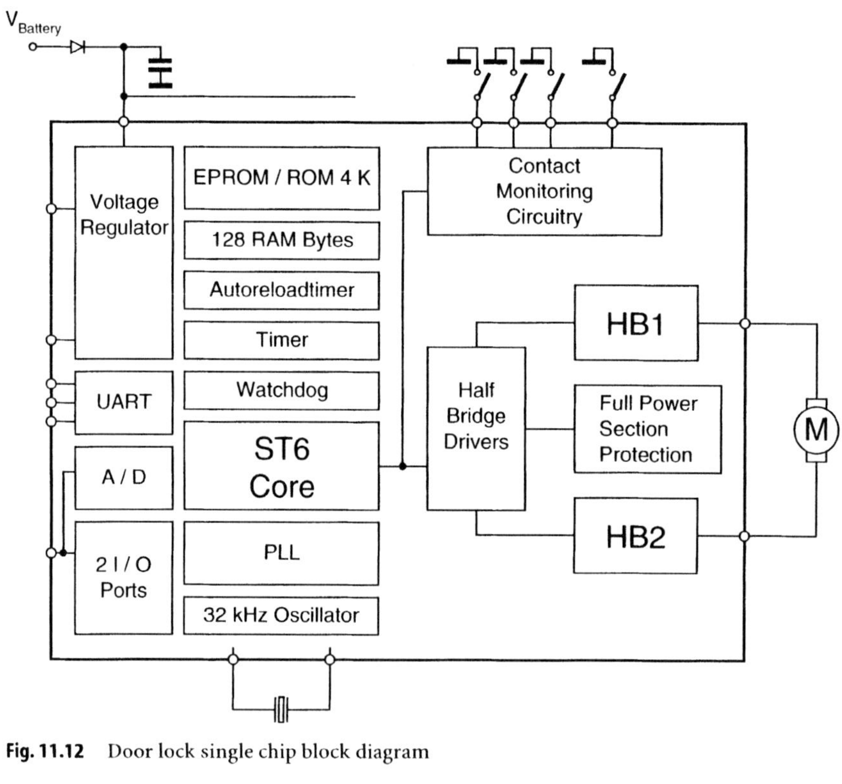

Fig. 11.12 Door lock single chip block diagram

Compared to the previous design, the count of ICs for the application is reduced from 2 to 1 and the flexibility has greatly improved integrating a software programmable machine as well.

Now we will proceed with the description of the main blocks.

Power section: it is made with an H bridge, to drive the bi-directional DC motor of the locking mechanism. Each switch is a power DMOS transistor characterized by a Rds_{on} of 300 mΩ and a BV_{dss} of 60 V. The nominal design current is 2 A, the maximum current is limited by an overload detection circuit, with the threshold settled at 4.5 A. The current is sensed by an integrated shunt resistor. The circuit is fully protected against short circuits to ground, to the power supply and across the load. The circuit is protected also against overtemperature and overvoltage.

The operation of the power part is controlled by the microprocessor via two specific 8-bit registers. In these registers, it is possible to write/read the commands for the bridge, and the status of the protections. It is also possible, when the motor is not powered, to monitor the continuity of the load circuit. After the intervention of a hardware protection, the power part switches immediately off and gives a diagnostic feedback to the microcontroller, and can be switched on only by an appropriate signal from the microcontroller itself.

Microcontroller section: this section is composed of a small, general purpose 8-bit microprocessor, for a more detailed description you can see Paragraph 11.4.4.1.

The microcontroller section comprises the following macrocells:

1) Microcontroller core (arithmetic logic unit).
2) Memories: RAM 128 bytes wide and ROM 4 K bytes wide.
3) Timers: general purpose one, autoreload timer to generate PWM signals, digital watchdog timer.
4) Standard 5 V I/O port (8 I/O lines).
5) 8-bit successive approximations A/D converter.

Protocol Handling: The serial transmission/reception on the line is managed by a specific peripheral unit of the microcontroller called UART (Universal Asynchronous Receiver Transmitter). The use of this logical standard peripheral unit is a compromise between a completely wired protocol handler and the choice to manage the transmission/reception completely by software. In other terms, a good degree of flexibility is allowed in the choice of the protocol without overloading the CPU of the microcontroller too much.

11.4.5.3
Interface and Power Supply Blocks

This kind of circuits, is the most original for this kind of applications and for this reason are here described in more detail. The following examples are used in both the realizations with minimum modifications.

The line driver: It is a circuit characteristic of this kind of application, in particular it must withstand both the maximum positive voltage, and also voltages below

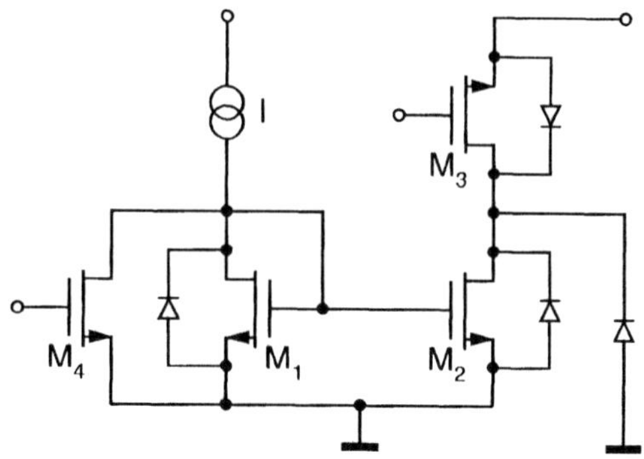

Fig. 11.13 Simplified schematic diagram of a line driver. The transmission line is connected to the pad, source of M_3

to the substrate potential. For this reason, it is not possible to use the standard 5 V I/O ports of the microcontroller, but components that must exploit at maximum the features of mixed processes, even if they are not power transistors.

In general, the external transmission line, when is in the recessive state, logic high, can be pulled up by a single resistor connected only to one unit.

When a unit transmits a logic low state (dominant), it sinks a current higher than the current sourced by the pull up resistor.

The transmitter can be conveniently made (Fig. 11.13) with a high voltage DMOS current mirror M_1 and M_2 driven by a normal N-channel transistor M_4 connected to the logic. The reference current I can be both constant or proportional to the main power supply if the external bus pull-up resistor is connected respectively to a reference voltage or to the power supply.

Using a limited current to drive the transmission line to zero, the advantages are that the circuit is intrinsically protected against the shorts at the supply and that the falling edges have a controlled slope, avoiding to radiate to much energy.

If the drain of M_2 is directly connected to the pad when the transmission line is pulled down by another transmitting unit and if the ground level of this unit is lower than the substrate (source of M_2) it is possible a current flow through both the source drain diode of M_2 and through the EPI (drain) substrate diode. In particular, this second current can have parasitic effects on the circuit.

Moreover, if for some error, the substrate is connected to the power supply and the transmission line output to ground (possible simply exchanging the wires of the bus) the transmitter is destroyed.

To avoid these effects the simplest solution is to put a diode in series with the output. A most sophisticated way to do the same, without loss in a diode drop imposing the low level, is to put another DMOS transistor M_3, in series with M_2, with its drain connected to the drain of M_2, thus the D-S diode of M_3 is forward biased in normal operation, reverse if the potential of the line is lower than the ground (substrate) potential of the transmitter.

If the gate of M_3 is driven to a positive supply in normal operation but kept to its source potential (by R_2) when this is below ground, M_3 itself acts like an ideal diode that has no forward drop (Fig. 11.14).

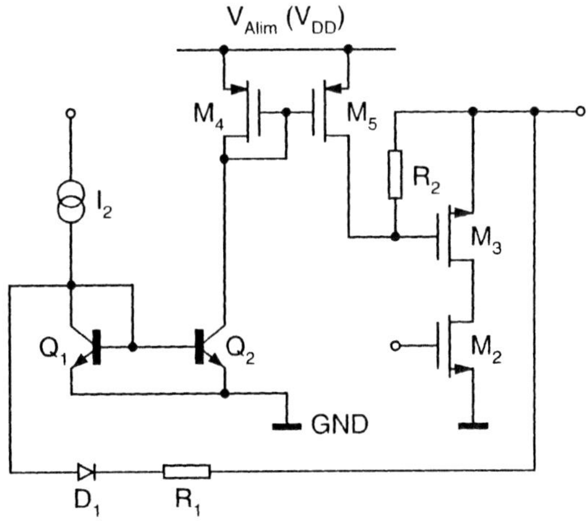

Fig. 11.14 Simplified schematic diagram of a line driver with the M3 DMOS, to avoid reverse conduction, and its control circuit in evidence

Under normal conditions the current I_2, mirrored by Q_1, Q_2, M_4 and M_5 generates a voltage drop on R_2. This guarantees that M_3 transistor is conducting, which in turn means that the voltage drop between its drain and its source is not longer the V_f of the associated diode, but it is equal to the drop across the Rds_{on} on resistance of the M_3 DMOS itself.

However, if the external line goes below the substrate potential, D_1 and R_1 absorb the current I_2 causing M_5 to switch off. The current path from the substrate to the output is avoided because R_2 discharges the gate of M_3, which switches itself off. It should be noted that M_2 resistance is in a non-biased epitaxial pocket in order to guarantee the functionality of this block under all possible conditions.

The contact monitor circuit: Another circuit that does not require high power, but that can require components that must withstand the same voltage as the power supply, is the contact monitor circuit (Fig. 11.15).

The status of the output (out signal) can be established by comparing the currents on the FCI node, one which is proportional to the power supply voltage (Valim) the other that is proportional to the voltage on the input node.

For a first approximation, we can neglect the voltage drops on the Q_1 and Q_4 mirrors. If we assume that OUT is a logic high value, a current of

$$I_{FI} = \frac{\text{Valim}}{R_2} \tag{11.1}$$

is sourced by the Q4-Q3 mirror and M1- M4 mirror (supposed equal emitter area and channel ratio W/L).

At the same time a current of

$$I_s = \frac{V_{\text{input}}}{R_1} \tag{11.2}$$

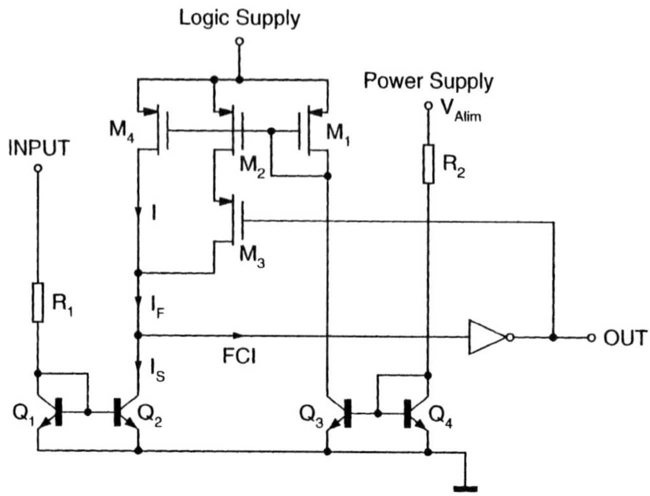

Fig. 11.15 Example of contact monitor circuit. Input terminal to be connected to supply by the contact to sense

is sunk from the $Q_1 - Q_2$ mirror. Since $R_2 = 3\,R_1$ the threshold value (obtained when $I_F = I_s$ I. E. forced current = sunk current) is

$$\text{Vinput, treshlow} = \frac{1}{3}\,\text{Valim} \tag{11.3}$$

If the input drops below this value, the potential of the point FCI increases, and the output of the inverter goes at a logic low value, switching on also M_3 transistors. The M_4 and M_2 current generators are therefore connected in parallel. As a result, the current forced on FCI node by the logic supply becomes

$$I_{fh} = 2\,\frac{\text{Valim}}{R_2} \tag{11.4}$$

and so the input threshold voltage is

$$\text{Vinput, tresh high} = \frac{2}{3}\,\text{Valim} \tag{11.5}$$

In this way, a comparator with hysteresis has been realized, with thresholds at $\frac{1}{3}$ and $\frac{2}{3}$ of the power supply voltage. This kind of comparator is useful to detect the status of a switch that can connect or not the Input point to the power supply itself.

It is to be noted that, excluding resistors, all the components can be operated at a logic voltage, independent of the supply V_{alim}. With a limited number of high voltage components (input PNP or P channel mirrors) it is possible to make the complementary version to detect the status of a switch closing versus ground.

Using some high voltage MOS switch in series with R_1 and R_2 a low consumption idle state can be added.

A high current state (some mA), to verify the quality of the contacts and to clean them can be imposed both, by dimensioning properly R_1 resistor and $Q_1 - Q_2$ mirror ratio, or by putting some resistor or current generator in parallel to R_1.

11.4.5.4
Mirror Driver

An interesting application of the mixed processes is the control of the external mirrors in a car.

The simpler controllers must drive, accordingly to the simple command sent on a serial line, two or three motors (for mirror orientation and folding), and some other load (e.g.: defrosting resistors).

Such circuits can easily be integrated into one IC with traditional mixed technologies putting together power transistor and a wired logic part.

The most advanced controllers add an extra feature that is the memorization of some different positions depending on different drivers. Each different driver can then set his preferred mirror position by just pushing a button.

The electronics contained in the mirror must be able, in this case, to set the required position in a short time, with a good precision and without oscillations. In other words, there must be present a complete feedback loop with position sensors (potentiometers), controller, and power actuators (Fig. 11.16).

To make a flexible control loop, that can be adapted to different kinds of mirrors (different inertia, friction, power of motors), it is useful to convert the data in a digital form and to use a microcontroller, the output of this block converted in PWM, can drive the power bridge. The control algorithm can be stored in an EPROM or in a ROM, and the mirror characteristics as well. The settling points for each different driver, can be memorized on a RAM or on an EEPROM.

In this application, if a software programmable controller is integrated a lot of extra functions can be implemented easily, like for example:

- coordination with seat positioning via serial line
- extra set points for automatic positioning during rear gear
- scan on a programmed angle when turn left (right) signal is switched on and a lot of other similar functions.

So this circuit can be taken as an example of the usefulness to integrate into the same chip the power part, like the bridge motor drivers, the control part, like the microcontroller,and all the kinds of memory like ROM, RAM EPROM and possibly also EEPROM.

11.5
Trade- Off's and Limits

The first trade off to consider is the cost and the benefits, in terms of space saving and/or added functions and programmability, of the one chip solution versus the traditional solutions. The biggest cost, for the integrated system, is due to the surface of the chip. If we make the future projection of this figure, the

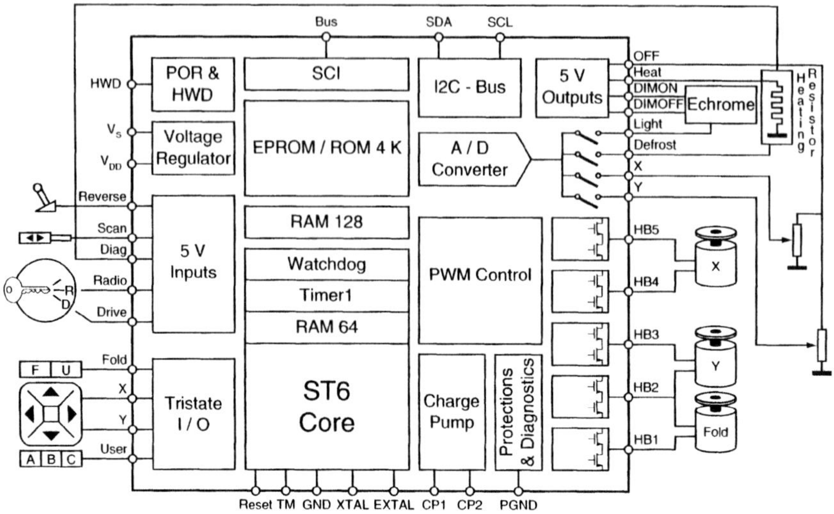

Fig. 11.16 Single chip mirror driver block diagram

density of the power and analog parts is slowly improving with the lithography (Chap. 1) as usual for this kind of processes, but the logic-cmos part can gain the full advantage of the fine lithography exactly in the same manner as VLSI circuits. For this reason, there can be foreseen a big development of the logic part towards more powerful controllers while the analog part and the overall level of power manageable by complex circuits increase only relatively slowly. The second limit is given by the package: traditional high pin count packages cannot dissipate more than a few watts and traditional power packages have no more than 25 pins. Hopefully, progress in surface mounting techniques and packages (powerSO and PQFP – see Chap. 13) allows to have a sufficient pin count with high power dissipation capabilities. Nevertheless, the package pin count limits the applicability of the concept to the fully integrable systems in which the number of pins is limited to the necessary input, output and supplies with a minimum of service pins (clock, test) and excludes the systems in which a very high connectivity is needed (E. G.: parallel processors or processor that are using external parallel busses). The third issue to take into consideration are the parasitic effects (Chap. 1). It must be considered, on one side, that CMOS logic circuitry is largely less susceptible than the bipolar analog one because of the noise margin concept intrinsic in the logic design and because the epitaxial pockets, victims of the most dangerous parasitics, in the CMOS structures are directly strapped to the positive supply and being the body of the p channel MOS, are not connected to any active node; on the contrary, in bipolar structures the epitaxial pockets are the base of PNP or the collectors of NPN or drain of N-channel DMOS transistors. For this reason, a current sunk from the epitaxial pockets, that is the most common effect of parasitics, induces some error in bipolar-DMOS based circuits and has no effect on CMOS, if properly designed to avoid extreme latch-up problems. On the other hand, the decrease of the lithography and the increase of the cut-off frequencies of the components, make the circuits more susceptible, decreasing the intrinsic filtering effect, particularly important in the logic. The increase of the switching frequency of the power part, and consequently of the rise and fall edges slopes increase the possibility of interference. It must be noted that decreasing the lithography, the parasitic due to the capacitors versus the substrate becomes more and more important versus those due to the direct current injection from junctions more negative than substrate potential. Making more complicated and dense chips, the usual precautions (Chap. 6) are becoming more difficult to be implemented. Also the system techniques to generate a "blanking" signal to make some circuit not operating when the power section is switching or the techniques to preview and control the interference transistor by transistor can be no longer extensively used when the complexity increases. Furthermore, they cannot be used by principle in the software programmable parts, because the circuit designer does not know exactly in which way the part is used. This can be the biggest concern when integrating programmable logic with power, but the most recent designs and experiences, have demonstrated that the most susceptible circuits are still the analog ones specific to the protection of the power section: that is, the problem and the solutions are well- known and above all, the logic part is the most immune and therefore not limited, even in complexity, by the parasitics. Worth noting also that, in some cases, the dielectric isolation technologies, can be the ultimate solution to the problem of parasitics (Chap. 4).

Bibliography

1. Weste N, Eshraghian K (1985) Principles of CMOS VLSI Design, Addison-Wesley Publishing Company
2. ST62 General Purpose DATABOOK – SGS-Thomson Microelectronics
3. Brokaw AP (1974) A Simple Three Terminal IC Band Gap Reference, IEEE, J Solid State Circuits, VOL SC-6
4. Storti S, Menniti P (1984) Low Drop Regulator with Overvoltage Protection and Reset Function for Automotive Environment, IEEE J Solid State Circuits, Vol SC-1, No.3
5. Jurgen RK (1986) Coming from Detroit Networks on Wheels, IEEE Spectrum, Vol 23, pp 53-59
6. Wrathall RS (1985) The Design of a High Power Solid State Automotive Switch in CMOS – VDMOS Technology, in Pro. IEEE PESC
7. Einzincer J et al. (1986) Monolithic I.C. Power Switch For Automotive Applications, in ISSCC Digital Technology Papers
8. Cini C et al. (1986) High Side Monolithic Switch in Multipower BCD Technology, in Proc 3rd Int Macroelectronic Conference (Munich, Germany)
9. Storti S, Consiglieri F, Paparo M (1988) A30A – 30V DMOS Motor Controller and Driver, IEEE Journal of Solid State Circuits, Vol 23
10. Contiero C et al. (1987) Design of High Side Driver in Multipower BCD and VIPOWER Technologies, in IEDM Techn Dig (Washington, DC)
11. Treek G Van, Ferrari R, Storti S (1991) Hochstrom – Motortreiber ICs, Elektronic 14

CHAPTER 12

Modeling, Design and Simulation of Power Electronic Devices and Circuits

W. Fichtner

12.1 Introduction

The continuing advances of semiconductor device technology have made it economically feasible to place several thousands to millions of devices on a single piece of silicon. Current developments in semiconductor manufacturing show no sign of a slowdown in this trend towards higher complexity. Triggered by these staggering advances in semiconductor technology, the field of power integrated devices and circuits has seen major new developments during the last few years. Power integrated circuits embrace a wide spectrum of different applications ranging from more or less conventional analog circuits that are operated at non-standard voltage levels, to smart power circuits, and to circuits composed of parallel arrangements of similar devices such as power MOSFETs and insulated-gate bipolar transistors (IGBTs).

Caused mainly by the advent of merged bipolar and MOS (BIMOS) technologies, analog integrated circuits with voltage levels of 20 V or more have emerged for telecommunications and automotive applications. Similar improvements can be witnessed for single power devices such as power MOSFETs, IGBTs or other types of MOS controlled power devices, all of which have become very-large scale integrated circuits with cell densities exceeding one million. In the other chapters of this book the reader can find many examples for this trend.

The design of a power integrated circuit encompasses a large number of individual steps, ranging from the initial specification to architectural considerations and the layout and verification phase. Table 1 gives a schematic representation of the major components in the design of a power integrated circuit.

Table 1 Phase model of Power IC development

Development phase	Task	Result
Study	Definition of desired properties	Specification sheet(s)
Specification	Development of concept and testing strategy	Chip and test specification
System design	Development of modules and testing environment	Functional structure Testing frame
Implementation	Construction of entire chip layout	Verified Layout
Integration and Test	Wafer fabrication and Packaging	Prototype

The wide variety of different power electronics systems that contain semiconductor devices does not allow a simple categorization of the various hierarchy levels in Table 1. For example, an analog integrated circuit acting as a LCD driver will require different system and architecture modeling if compared to an IGBT switch in a step-down converter or a power supply. It will therefore come as no surprise that the development of more or less strict system design guidelines has found no equivalent effort in power electronics. If we restrict ourselves to the chip level, however, concepts begin to converge.

12.2 Layout Tools for Power Device and Circuit Design

Similar to the VLSI world, the design of power devices and circuits relies heavily on the modern CAD tools that are available today. With the advent of the affordable scientific workstations, the design of power devices and circuits is not different today to the more conventional VLSI applications.

The diverse group of tools in a modern CAD system relies on a single component, the master technology file, for technology encapsulation. The information in this file includes technology characteristics and process parameters as well as information that governs behavior of graphics devices used by the design tools. Because all design tools use a single source of technology information, consistency is guaranteed.

The technology database permits designers to specify completely the target process on both the mask and the symbolic level. This has several advantages in both the active layout generation phase and the layout verification phase (rule checking).

To illustrate the power of this approach, the following material illustrates how a rather complex MOS-controlled thyristor chip with an area of more than 1 cm^2 can be assembled from a single cell [1]. At the start, the layout of one unit MCT cell is optimized based on the process available (For an equivalent CAD effort,

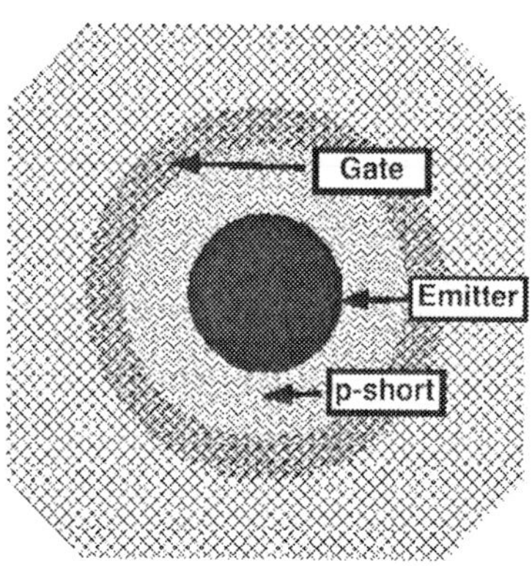

Fig. 12.1 Layout structure of a circular single-cell MCT layout structure

see [2]). Figure 12.1 shows this layout. It was generated by hand in less than one hour by an experienced designer, verified against the technology database by a design rule check, and subsequently put into the design database.

The full power of a modern layout tool becomes obvious once one starts to generate larger cell ensembles. This can be simply achieved by writing "code" in a high-level language that accesses the basic building blocks and arranges them according to the software specification. The possibilities are manifold, and can be tuned to any situation that might arise.

The following program creates a circular MCT assembly consisting of a central gate and three "turns" of evenly spaced single MCT cells (Fig. 12.2). The number of turns as well as the spacing between the MCT cells are parameters to the procedure. The language is Mainsail and VLSItools VIP (Procedural Layout Language) modules are used for the actual generation of the CIF code [3].[1]

```
begin "circurlar_mct"
# module for generating a circular MCT assembly, consisting
# itself of circular MCT cells.
# 'max_turns' = number of turns
# 'd' = spacing of the cells
sourcefile "(usevip)";
STARTOF (mctrnd);
DEFSTART ("circ_mct", [max_turns,d], [<define local variables]);
 #draw the central gate cell
 DRAW("il","vcell:vcell",
       PARAMETERS("[ly]single_mct_cell",ENDARGS));
 FOR k:= 1 UPTO max_turns DO BEGIN
       n := <Calculate number of cells per turn)
       FOR i:= 0 UPTO n DO BEGIN
       x,y := <Calculate position coordinates of the cell
       DRAW(nxy(x,y),"i","vcell:vcell",
       PARAMETERS("[ly]singel_mct_cell",ENDARGS));
       END;
 END;
ENDDEF;
ENDOF;
```

The same approach has been utilized to generate real MCT chips. The above code should be understood to be only one of the key sections for drawing the complete MCT assemblies. The pad for the connection of this assembly and the layers covering the entire area of this structure are added in another section of the module. Figure 12.3 shows a plot of the final layout.

The advantages of such an approach are manifold. Apart from being correct-by-construction, compiled designs can be easily changed to other cell configurations. Suppose the technology would have changed during the design phase and some design rules have to be adapted accordingly. With a minimum of CPU time, the complete layout can be regenerated – again correct-by-construction.

[1] Mainsail is a trademark of Stanford University, and VIP is a trademark of Vlsi Technology, Incorporated.

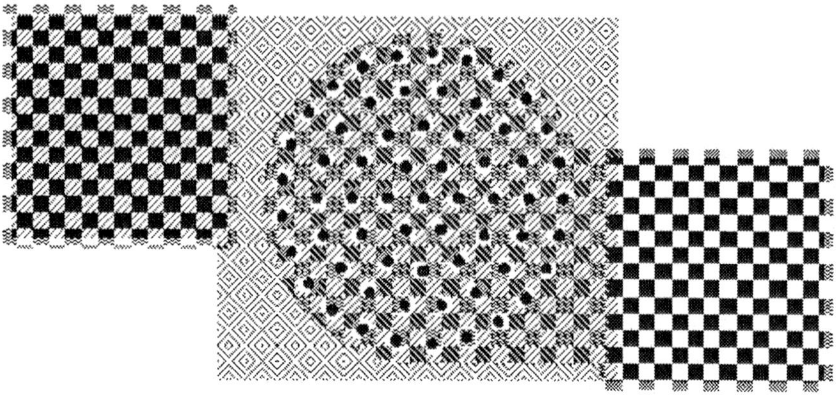

Fig. 12.2 Circular MCT ensemble generated from single cells

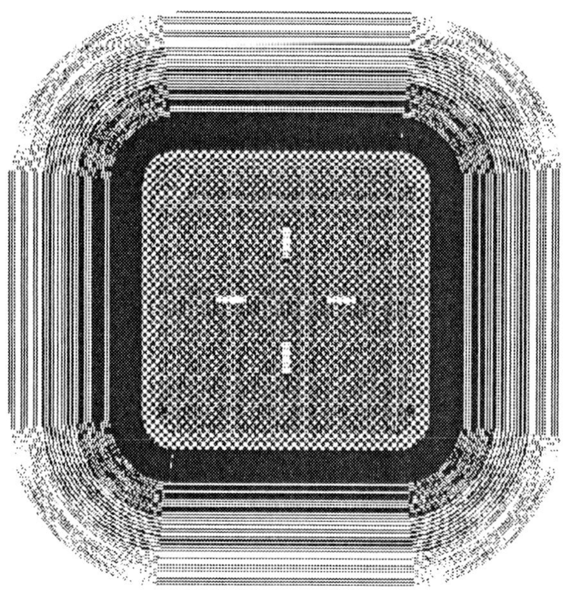

Fig. 12.3 Complete MCT chip designed through automatic compilation [1]

12.3 Simulation Tools for Power Devices and ICs

Simulation plays an important role in the design of power integrated circuits and devices. Using simulation, the designer can determine both the functionality and the performance of a design before committing to the expensive and time-consuming step of fabrication. This is true not only for the more classical integrated circuits, but also – and probably even more so – for power electronics integrated devices and circuits. It is important to understand that simulation is more than a mere convenience. Through ways that are otherwise impractical or even impossible, a designer can explore function and feasibility of her/his circuit. Influences of manufacturing and the ambient can be analyzed without the need to actually implement the design; testing prodedures can be developed; and the design can simply be debugged. For example, it becomes possible to study influences of certain fabrication steps without actually fabricating a circuit or device.

In Fig. 12.4 the simulation and design hierarchy as it applies to power ICs is illustrated. For these circuits, three levels of the simulation hierarchy are of particular importance: circuit, device and process simulation. To use a circuit simulator, the designer enters a design into the computer, usually in the form of a list of components (the *netlist*) that are connected to one or more nodes. The simulator bases its predictions on models describing the operation of these components. Depending on the hierarchy level under consideration, component models vary widely in complexity. The behavior of a component can in general be expressed in terms of an interconnection of ideal elements. Usually, only a small repertoire of ideal elements needs to be supported. *Resistive Elements* relate currents to voltages. They include two-terminal elements (i.e. resistors and independent voltage and current sources), and four-terminal controlled sources. *Energy Storage Elements* relate the state of the storage element (capacitor, inductor) to one variable (e.g. voltage for a capacitor). These equations might be linear or nonlinear. Circuit simulation has been recently applied to system related aspects such as snubber [4] and distortion-sensitivity [5] analyses.

Success and failure of a power circuit simulation hangs and falls with the quality of the models used for the semiconductor devices. While excellent models have been developed over the years for the more classical power bipolar and MOS devices [5–10], models for "real" generic power devices are still in their infancy. Examples are models for the IGBT [11] and the SITH [12].

One reason for the dissatisfying state of CAD models for power switches can be found in their complex physical modes of operation. It will therefore come as no surprise that power device researchers have taken a more basic approach through numerical process and device simulation.

In process simulation, one tries to simulate all fabrication steps of importance – implantation, diffusion, oxidation, etc. – with a "best effort" physical model, usually at the level of differential equations that are solved numerically in one or multiple dimensions. Over the years, several software packages have been developed that enable excellent process characterization and forecasting (see [13] and references therein for work performed up to 1988. References [14] and [15] are detailed user's guides for software tools used today).

As a representative example obtained with a modern process simulation tool, Fig. 12.5 shows a surface plot of the net impurity concentration near the surface of a single IGBT cell. This result was obtained through a full one- and two-dimensional numerical simulation of the complete process flow for the fabrication of a modern IGBT process, obained through [14, 15]. Based on the

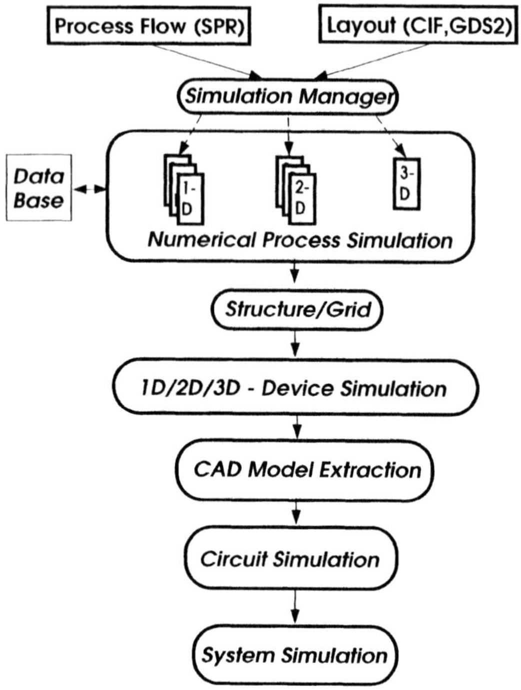

Fig. 12.4 The CAD simulation hierarchy

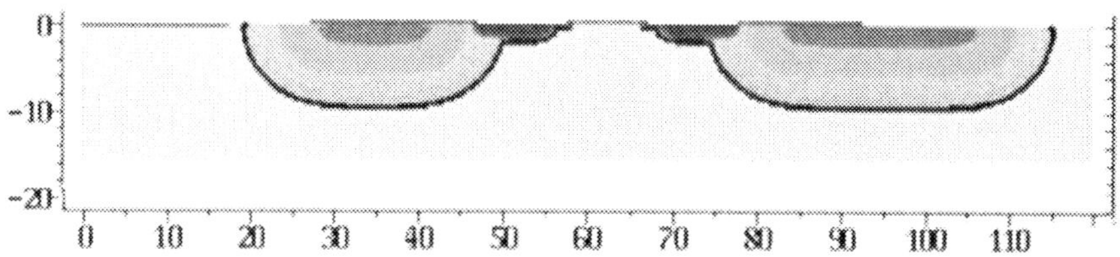

Fig. 12.5 Doping profile of a single IGBT cell

results of such a simulation, process optimization becomes considerably easier and more straightforward, and narrow margins in critical fabrication steps can be usually avoided.

In Fig. 12.6 and 12.7, zoomed views of the same result are given in order to better illustrate the doping profile distributions near the active gate area of an IGBT. From results such as these, it becomes evident that process optimization can be achieved rather easily if the results of such simulation have been calibrated against experimental results [16].

Once process simulation results are available, however, it becomes necessary to transfer the relevant doping and geometry data of the structure into a format suitable for device simulation. While this transfer is obvious for one-dimen-

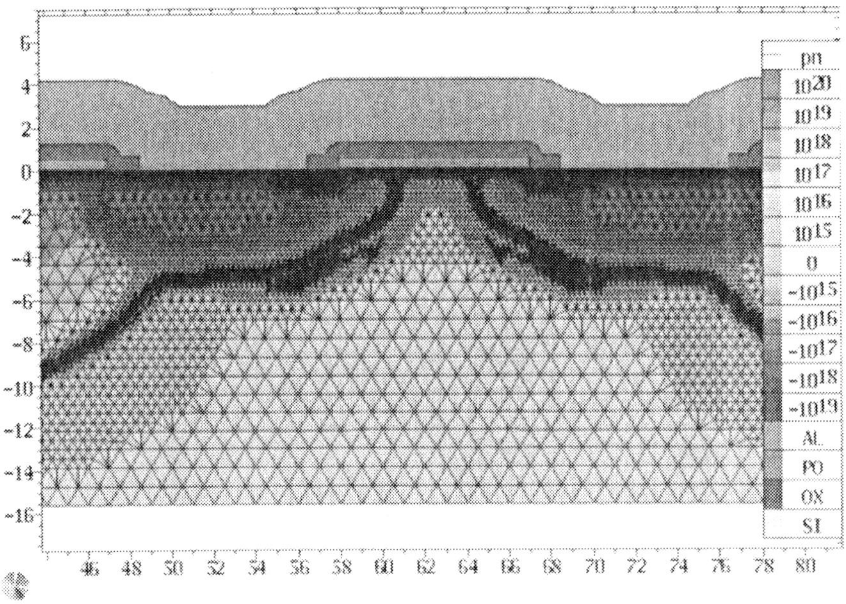

Fig. 12.6 Two-dimensional doping profile in the surface region of a modern IGBT

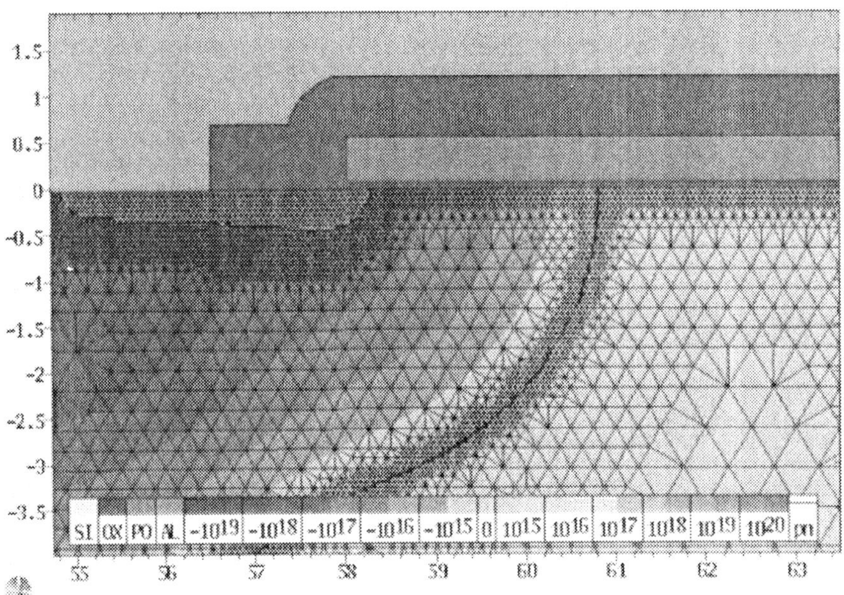

Fig. 12.7 Zoomed view of the surface region of a modern IGBT cell

sional simulations, it is not at all the case for two and three dimensions. The particular structure of the semiconductor device equations requires physically based finite-element discretizations that obey certain restrictions in terms of element shapes and interior angles. In Fig. 12.8, we show how the process simulation result can be incorporated into a full wafer IGBT structure. Using modern grid generation techniques, a mesh like the one in Fig. 12.9 can be generated in a matter of a few seconds on a scientific workstation.

Using the regridded results from process modeling in the form of geometries and doping distributions, accurate device investigations can be performed, again on a numerical basis in one or more dimensions. Device simulation programs such as PISCES [17, 18], BAMBI [19], TONNADDE [20], GENSIM [21], SECOND [21] or DESSIS [22] are now in widerspread use for power devices. Older results obtained in the 1980's can be found in [23, 24] and [25]. Over the past few years, device simulation possibilities have been extended considerably.

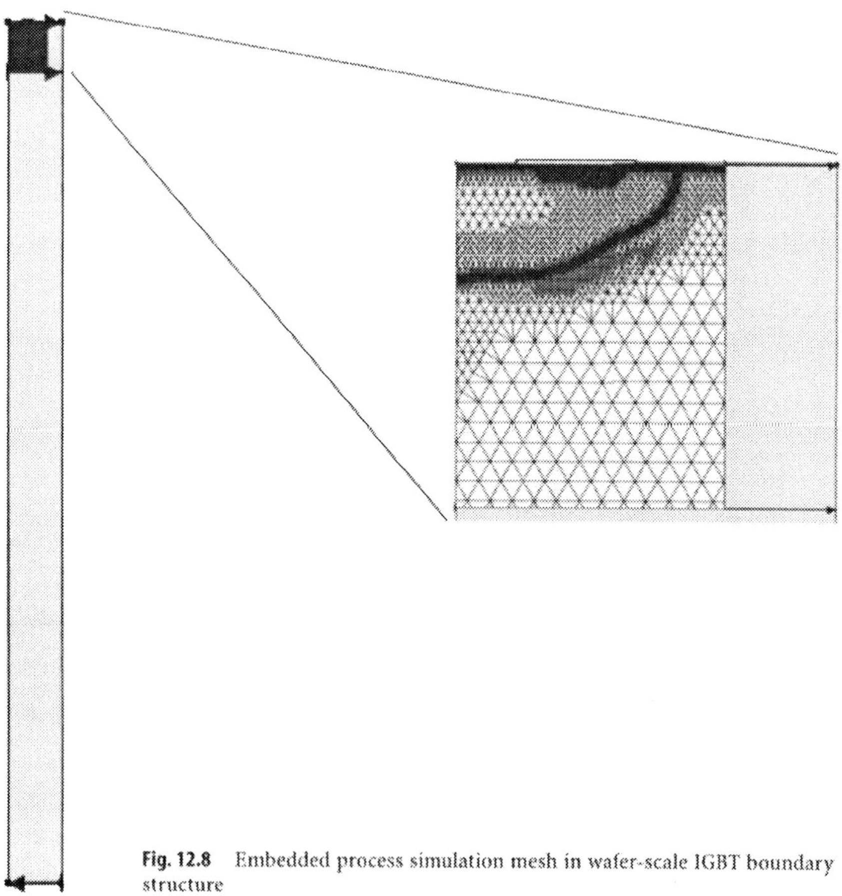

Fig. 12.8 Embedded process simulation mesh in wafer-scale IGBT boundary structure

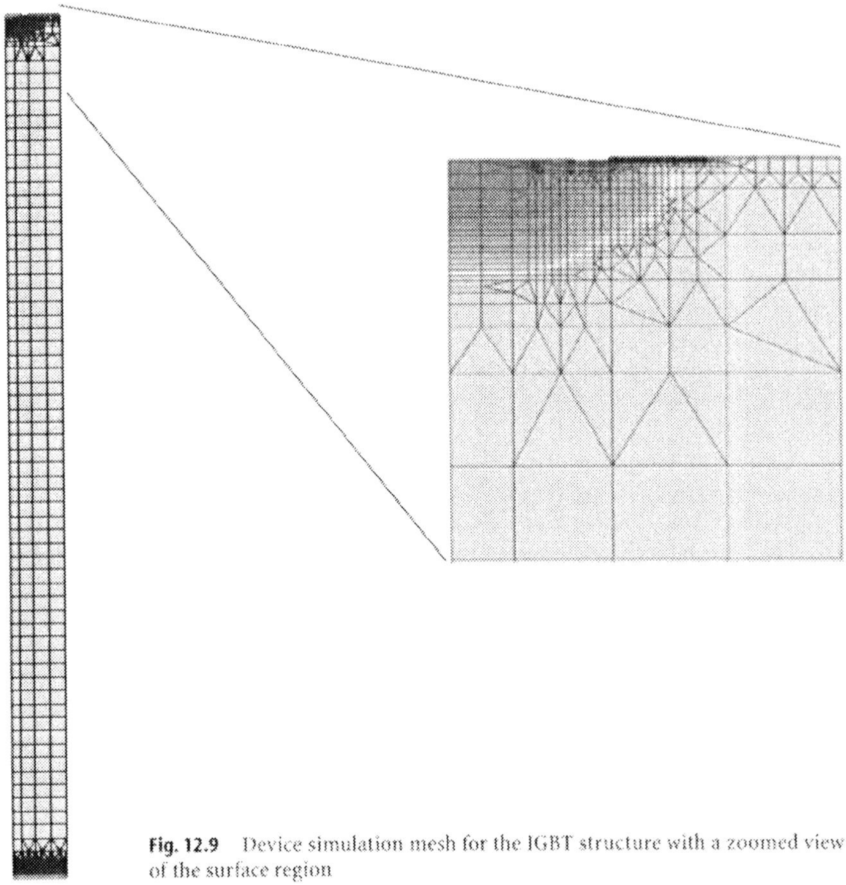

Fig. 12.9 Device simulation mesh for the IGBT structure with a zoomed view of the surface region

The capabilities of modern CAD programs are manifold allowing both steady-state and transient conditions with external circuitries such as snubbers and gate drive circuitries. Compared to the more conventional device simulations performed in the VLSI environment, the inclusion of drive and protection circuits in one simulation together with the power semiconductor switch becomes particularly important. In [26], a more detailed review of mixed-mode device-circuit simulation is given.

In Fig. 12.10, a realistic example for a mixed-mode device circuit configuration is shown as must be simulated and optimized in typical smart-power applications. The circuit depicts a simplified half-leg configuration of an inverter with several semiconductor devices and lumped circuit elements. Depending on the voltage and power level, the semiconductor switch in this circuit could be a power MOSFET, an IGBT, or even a GTO (Gate Turn-off Thyristor). Similarly, depending on the case under consideration, the switch might have to be repre-

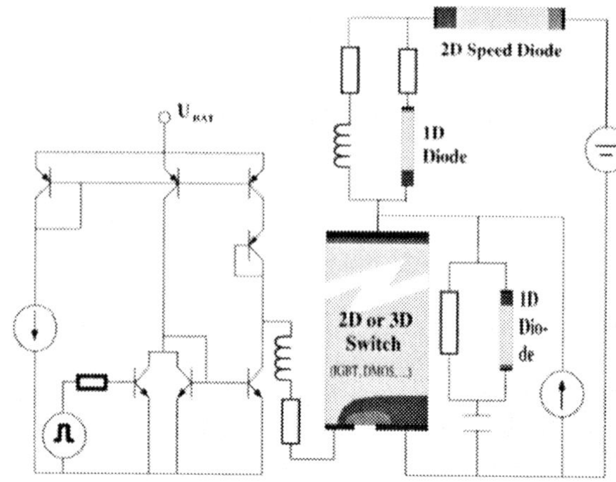

Fig. 12.10 Simplified IGBT inverter circuit with finite element and lumped circuit models

sented by a finite-element model in one, two or three dimensions. In most cases, two-dimensional devices are used today due to the restrictions of the available computing power.

Apart from the active power switch itself, finite element representations are needed for certain components of the protection and drive elements in order to achieve realistic results. In Fig. 12.10, a two-dimensional model for a free-wheeling diode indicates the usual case as it occurs in inverter design. One-dimensional models are used for the diodes that are used in conventional snubber elements as transient protection devices. The books by Dettmer [27] and Lendenmann [28] provide extensive examples on how modern CAD techniques can be applied to power device design.

For the gate-drive circuitry, it suffices to use lumped circuit elements for the MOS and bipolar transistors, diodes, and passive devices such as inductors, capacitors and resistors. A vast literature exists for lumped circuit modeling, and we refer to [29] and [30].

Figure 12.11 depicts a typical result as it is obtained by simulating the circuit in Fig. 12.10. While only the voltage waveforms are shown here, a multitude of results can be extracted from such simulations. By comparing the waveforms for the input voltage, Vin, with the voltage seen by the IGBT gate, Vgate, it becomes obvious how important it becomes to incorporate the gate drive circuit. By analyzing the oscillatory behavior of the anode output voltage, Va, one notices the influence of the parasitic resistive and inductive elements of the circuit.

Technology-based simulation shows its real power when it comes to cases of device and circuit failure. With their inherent instabilities during operation, high power devices such as GTOs, IGBTs and MCTs are prone to transient failure mechanisms such as dynamic avalanche and filamentation. In [28, 31] and [32], MCT failure mechanisms were investigated by two-dimensional device simulation.

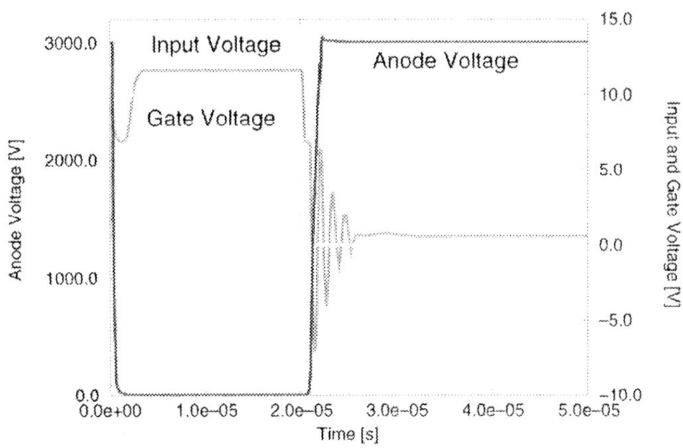

Fig. 12.11 Calculated Result of the switch-off behavior of the circuit in Fig. 12.10

Among the many interesting failure mechanisms in semiconductor devices, effects related to electrical overstress (EOS) and electrostatic discharge (ESD) are of foremost importance, especially for devices and circuits operating at lower voltage and power levels. In the case of smart-power ICs, protection devices have to be designed and incorporated on the chip. For the best summary of current developments in this field, please look at [33] and other proceedings of this annual conference.

With the steady advandces in feature size reduction and device scaling, ESD protection becomes more difficult because the performance of the protective device has to be improved while decreasing the real estate available. While an improvement can be achieved by changes to the layout or the process, considerable know-how and experience are needed in order to be successful.

To simulate and extract important ESD related parameters, TCAD software packages can be used successfully today. To obtain the most actual geometry as well as accurate doping profiles of the investigated ESD devices, it is absolutely necessary to simulate all relevant process steps. After discretization, the resulting geometry and doping profile are taken as input for thermo-electric device and circuit simulations. The CPU time needed to complete a ESD device simulation varies between a quarter hour and several days depending on the number of used grid points (3,000 – 10,000) and on the complexity of the problem.

To study the switching behavior of our ESD devices and to find out how the failure occurs, transient thermo-electrical mixed-mode device simulations using networks of discrete devices to generate square-like and HBM (Human Body Model), CDM (Charge Device Model) or MM (Machine Model) like pulses must be used [34].

Figure 12.12 shows the cross section of a bipolar protection element as it is used in smart power applications. The collector of this device is connected to the

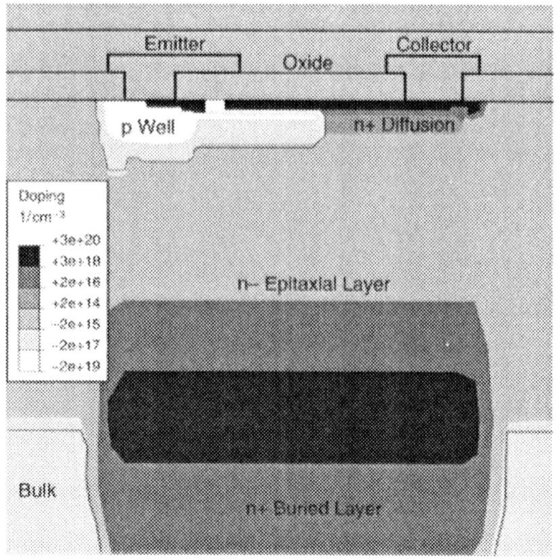

Fig. 12.12 Two-dimensional doping profile of a bipolar protection device as used in a smart power technology

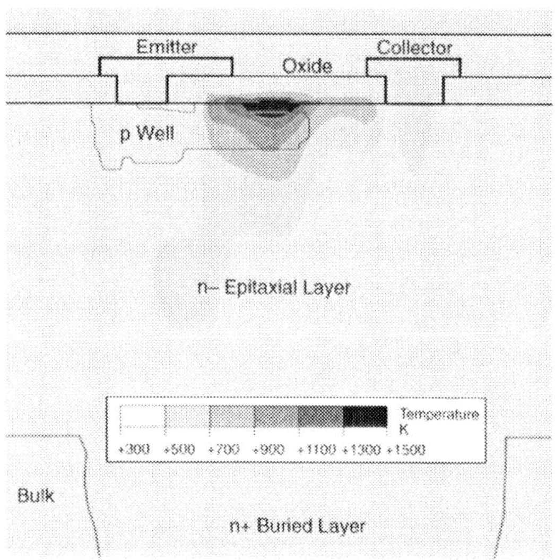

Fig. 12.13 Lattice temperature distribution during an ESD event.

outside world. ESD signals are applied at this contact in order to study the failure mechanisms.

During an ESD event, a considerable amount of energy is dissipated in the semiconductor structure of the protection element. As long as the temperature rise associated with the energy dissipation stays below critical values such as the melting points of the materials in the structure (e.g. Al, Si, etc.), a protection element can recover from even very bad situations. It is therefore of utmost interest to understand the current flow in a protection device under these stressed conditions. As experimental methods cannot provide any information of the device interior, device simulations are the only possibility to obtain information.

In Fig. 12.13, the lattice heat distribution is shown in the interior of the bipolar protection transistor after the ESD pulse has hit the collector contact. The maximum device temperature is close to 1500 K, and the current transport in the device has already become intrinsic in the hot spot.

12.4 CAD for Package and System Applications

Over the past few years, there has been increased interest into the electrical, thermal and mechanical behavior of complete power systems on a common substrate. With the advances in computing methods, a fulfillment of this wish has come within reach [36, 37].

In Fig. 12.14, the problem to be analyzed is depicted in a simplified manner. A power converter (like the one used above) is assembled together with its control circuitry on a common substrate. During operation, the power device heats up, and part of this heat is being transferred into the package and onto the control chip. As a first result of this transfer, the control chip changes its electrical behavior, which in turn might strongly influence the power device itself. For more information, the reader is referred to [36].

A second effect of the heat transfer is the influence upon the package itself. Through the difference in the material parameters of chip, solder and package,

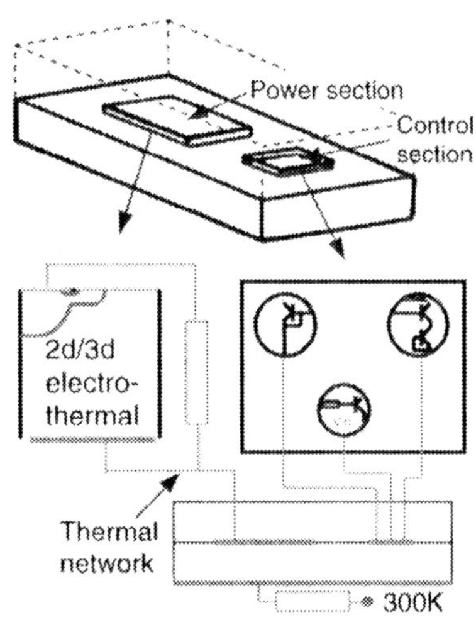

Fig. 12.14 Schematic view of the assembly of a smart power chip and its control element on a common substrate [36]

Fig. 12.15 Two-dimensional cross section of the substrate showing the heat flow in the package

Fig. 12.16 Van Mieses stress distribution in the power package

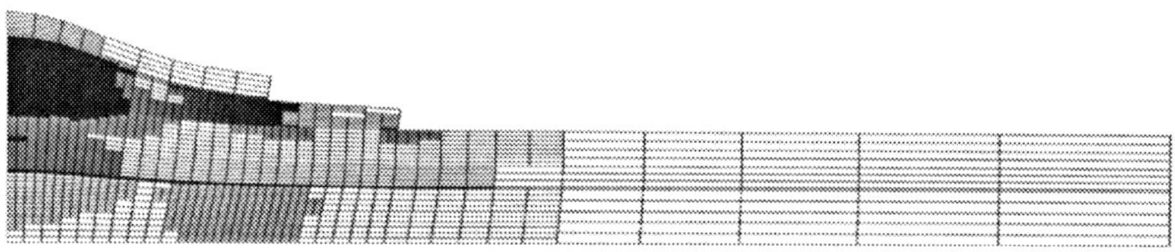

Fig. 12.17 Magnified view of the deformation of the package-solder-chip assembly

the heat flow through the package leads to a mechanical deformation of the complete assembly.

Using modern finite element techniques, such effects can already be calculated today [37]. In Figs. 12.15 through 12.17, simulation results from a two-dimensional analysis of the heat flow, stress distribution and final deformation of the package are shown as a representative example. Apart from the computational intricacies, such types of simulations depend strongly on the availability of material parameters in the system [38].

12.5 Conclusions

In this paper, we have tried to show through examples how modern CAD tools can become a significant factor in the overall design cycle of power integrated circuits. A rather optimistic picture can be painted for the possibilities of modern layout tools. Originating from VLSI requirements, these tools offer unique advantages through automated compilation and verification, and fast turn-around.

On the side of power IC simulation, a more critical outlook must be given. While numerical process and device simulation tools have become high quality, indispensable aids for process and device development, circuit simulation and CAD models need additional efforts before they will reach the same level of usability and quality. It should not be forgotten that in the final analysis – CAD tool or not – all things become a matter of economics. Increasing productivity is important because design time is the critical issue in the fabrication of all products. Expert designers and technologists are a rare commodity and expensive. The better the CAD tools, the more work can be done in the same amount of time with considerably lower risks involved. If one looks at these rather convincing arguments, it would be unfair not to mention that modern CAD tools have also negative aspects. As a first point, one should never forget that modern last generation CAD tools are extremely complicated software products. Their use requires educated users willing to take up a challenge that should not be underestimated. Secondly, modern CAD tools require modern computer hardware platforms. While today's workstations can be purchased at reasonable prices, they are not free, and they have to be maintained. Through their business, power electronics companies are traditionally more conservative than corresponding VLSI units. Nevertheless, they will have to follow the VLSI bandwagon – not only in technology, but also in CAD.

12.6 References

1. Fichtner W, Bürgler J, Dettmer H, Lendenmann H, Müller S and Westermann M (1991) Computer Aided Design Tools for Power ICs, in Proceedings ISPSD, pp 11–15
2. Elias NJ (1987) A Case Study in Silicon Compilation Software Engineering HVDEV-High Voltage Device Layout Generator, in Proceedings of the 24th Annual ACM/IEEE Design Automation Conference
3. VLSI Technology CAD Manuals, 1991
4. Undeland TM, Petterteig A, Hauknes G, Ådnanes AK and Garberg S (1988) Diode and Thyristor turn-off Snubbers Simulation by KREAN and an easy to use design Algorithm, in Proceedings IAS, pp 657–654
5. Kuroe Y, Haneda H and Maruhashi T (1986) A new Computer-Aided Method of Distortion Sensitivity Analysis and its Elimination Scheme for Power Electronics, IEEE Transactions on Power Electronics, vol PE-1, pp 200–209
6. Getreu I (1976) Modeling the Bipolar Transistor, Tektronix
7. Xu CH and Schröder D (1988) Modeling and simulation of power MOSFET's and Power Diodes, in Proceedings of the 18th Annual IEEE Power Electronics Specialists Conference
8. Thajur DK, Khanna VK, Kumar A, Swami NK, Mahajan SK, Good SC and Khokle WS (1987) Characterization and Transient Analysis of Monolithic Power Darlington Transistors for AC Drives Using CAD as a Tool, in Proceedings IAS, pp 504–511

9. Xu CH and Schröder D (1989) A Power Bipolar Junction Transistor Model Describing the Static and the Dynamic Behavior, in Proceedings of the 19th Annual IEEE Power Electronics Specialists Conference
10. Tsividis Y (1987) Operation and Modeling of the MOS Transistor, McGraw-Hill
11. Hefner AR (1988) Analytical Modeling of Device-Circuit Interactions for the Power isolated Gate Bipolar Transistor (IGBT), in Proceedings IAS, pp 606–614
12. Metzner D and Schröder D (1990) A SITH-Model for CAE in Power Electronics, in IPEC Technical Digest
13. Fichtner W (1988) Process Simulation, in VLSI Technology (S.M. Sze, ed.), McGraw-Hill, pp 422–465
14. Strecker N (1994) DIOS-ISE User's Guide, ISE Integrated Systems Engineering AG, Zurich, Switzerland
15. Feudel T (1994) TESIM-ISE User's Guide, ISE Integrated Systems Engineering AG, Zurich, Switzerland
16. Dallmann G, Feudel T, Syhre H, Lendenmann H and Fichtner W (1994) Two-Dimensional Dopant Profile Characterization for MCT and IGBT Structures, ISPSD 1994 Technical Digest, pp 305–308
17. Pinto MR, Rafferty CS and Dutton RW (1984) PISCES II: Poisson and Continuity Equation Solver, tech rep, Stanford Electronics Laboratory, Dept of Electrical Engineering, Stanford Univ, Palo Alto
18. Pinto MR, Rafferty CS, Yeager HR and Dutton RW (1985) PISCES IIB: Supplementary Report, tech rep, Stanford Electronics Laboratory, Dept of Electrical Engineering, Stanford Univ, Palo Alto
19. Franz AF, Franz GA, Selberherr S, Ringhofer S and Markowich P (1983) Finite boxes – a Generalization of the Finite-Difference Method Suitable for semiconductor device simulation, IEEE Trans Electr Dev, vol ED-30, pp 1070–1082
20. Nakagawa A and Nakamura S (1990) Application of General Purpose Power BIMOS Simulator TONNADE-II to double gate lateral bipolar MOSFET design, in ISPSD Technical Digest, pp 42–47
21. Bürgler J, Conti P, Heiser G, Paschedag S and Fichtner W (1989) Three-dimensional Simulation of Complex Semiconductor Device Structures, in Proceedings of the 1989 International Symposium on VLSI Technology, Systems and Applications, IEEE
22. DESSIS User's Guide, ISE Integrated Systems Engineering AG, Zurich, Switzerland, 1994
23. Masada E, Tamura M and Nakajima T (1988) Simulation of Switching Processes in turn-off Thyristors, in IEEE PESC Technical Digest, pp 91–98
24. VanDell WR and Dyle WK (1988) SEMINET: further advances in Integrated Device/Circuit Simulation for Power Electronics, in Proceedings of the 18th Annual IEEE Power Electronics Specialists Conference
25. Danielsson BE (1985) Studies of turn-off effects in Power Semiconductor Devices, Solid State Electronics, pp 375–391
26. Fichtner W, Feudel T, Kells K, Lilja K, Litsios J, Müller S and Strecker N (1993) TCAD in Power Device Design and Engineering, Proceedings IEDM 93, IEEE Press
27. Dettmer H (1994) MOS Controlled Power Devices for High-Voltage Application, Hartung-Gorre
28. Lendenmann H (1994) Operation and Performance Limits of MOS controlled Power Devices, Hartung-Gorre
29. Antognetti P and Massobrio G (1988) Semiconductor Device Modeling with SPICE, McGraw-Hill
30. Meta-Software, HSPICE User's Manual Elements and Models, Meta-Software Inc, Campbell, CA, 1992
31. Stoisiek M, Oppermann K-G and Wachutka G (1989) Turn-off Behavior of GTO's with Small Emitter Elements, AEÜ, vol 43, pp 320–327
32. Lendenmann H (1994) Turn-off Failure Mechanisms in Large (2.2 KV, 20 A) MCT Devices, ISPSD 1994 Technical Digest, pp 207–212
33. Muncie CH (1985) Proceedings of the 1985 EOS/ESD Symposium

34. Stricker A (1995) Layout Optimization of an nMOSFET Protection Element by Experiment and Simulation, 1995 EOS/ESD Symposium Technical Digest
35. Yasukawa A and Sakamoto T (1988) Simulation for Designing Semiconductor Packages with High Reliability under Thermal Cycling, in ISPSD 1988 Technical Digest, pp 36–41
36. Litsios J, Schmidhüsen B and Fichtner W, Large Scale Thermal Mixed-mode Device and Circuit Simulation, SISDEP 1995 Technical Digest
37. Korvink J (1995) SOLIDIS 1.0 Reference Manual, ISE Integrated Systems Engineering AG, Zurich
38. Baltes H and Korvink J (1995) SISDEP 1995 Technical Digest

CHAPTER 13

Packaging

C. Cognetti · E. Kobayashi

13.1 Introduction

The most evident progress in electronics is *system miniaturization*. It took place in the early '80s, with the commercialization of very compact consumer sets; but since the last few years it is so common that new designs in almost all market segments have the same imperative: "Small, Light and Cheap". This is especially the case of portables, like PCs and phones.

Basically, miniaturization is obtained from the improvement of both semiconductor and Printed Circuit Board (PCB). As a consequence, reduction in the component count is achieved for the same function, with substantial improvement in miniaturization.

But this concept would have severe limitations without new PCB technologies.

Also PCB is in permanent progress toward higher density of interconnections in smaller volumes. This is recognized as major advantage offered by the Surface Mount Technology (SMT) which was born 15 years ago and is now arrived to maturity, with very fine line at 100 micron pitch and reliable, cost effective multilayers.

But, even if in so close relationship, semiconductor and PCB are also very different.

The need exists of some "bridge" between the sophisticated, fragile, submicron world of silicon and the robust, submillimeter world of printed board.

This "bridge" is the semiconductor package, assisted by a family of interconnection technologies.

In the last five years, package and interconnections merited as much effort as the silicon itself with the development of new structures, materials and processes in large number.

In order to quantify this progress, three parameters can be used as an example:

1) *the lead pitch,* which moved from 2.54 mm typical of Dual-in line packages (DIP) to 0,3 mm used at present in the most aggressive Plastic Quad Flat Packages (PQFP);
2) *the pin count,* which was not higher than 40–60 in DIPs and is now close to 500 in the Ball Grid Array (BGA) family;
3) *the efficiency of interconnection,* intended as ratio between the active chip and the package footprint which was below 1–2 % in DIPs and is now close to 50 % with memory packages like Thin Small Outline Packages (TSOP).

In this chapter, packaging for Smart Power ICs will be presented and discussed.

13.2 Concept of Smart Power IC Package

Smart Power IC is the integration of components capable of delivering a significant amount of power to a given load, with circuits providing several logic functions, like control, detection, sensing and diagnostics; it is used in applications (automotive, industrial, consumer, computer peripherals) very sensitive to cost and reliability.

The four main functions provided by Smart Power IC package are:

1) *electrical interconnection* between the silicon and the external circuit (PCB), by means of a *large number of leads* (up to 100) requested by the complexity of the "Smart" section of the chip;
2) *mechanical and chemical protection,* for very low failure rate which in automotive systems will be fractions of ppm;
3) *device cooling,* by means of thermally conductive elements (slug) able to transfer the heat generated in operation (1–40 W) from the silicon directly to the ambient or to external heatsinks;
4) *electrical isolation,* (up to 4000 V) especially in case of technologies based on vertical power MOS transistors. As the transistor drain is connected to package slug, the need exists of using isolated packages for cost and safety reasons.

Driving forces and design criteria of Smart Power IC packages are summarized in Table 13.1.

13.3 Smart Power IC Package Types

Packages for Smart Power IC's can be classified according to the level of power dissipation.

Table 13.1 Driving forces and design criteria of Smart Power ICs Packages

Driving force	Design criteria
Power dissipation (1–40 W)	Power enhanced structure
Logic functions	Large pin count (5–100)
Low cost (< 0.01 US $/pin)	Plastic self isolation (2500–4000 V)
Low failure rate (< 1 ppm)	Material selection

Two families are commonly considered:

a) medium power packages, up to about 2 W;
b) high power packages, for more than 2 W up to 30 – 40 W.

13.3.1 Medium Power Packages

A simple way to achieve power dissipation up to 2 W is to transform the available signal packages and modify their leadframe in order to have high thermal conduction patterns from the silicon to the printed board.

This solution is used for all of the most common package types (Fig. 13.1):

a) for Through Hole Technology (THT) in DIP, with pin count up to 28;
b) for Surface Mount Technology (SMT) in SOP with pin count up to 28; in PLCC (Plastic Leaded Chip Carrier) for pin count up to 44 – 68; for PQFP (Plastic Quad Flat Package) for pin count up to 80 – 100.

Modification involves a number of lead (heat transfer leads) which are connected to the die pad in order to minimize the junction to pin thermal resistance θjp and to improve the thermal exchange from chip to substrate.

Fig. 13.1 Plastic packages of different generations: Dual In Line (DIP), Small Outline (SOP), Plastic Leaded Chip Carrier (PLCC) and Plastic Quad Flat Pack (PQFP)

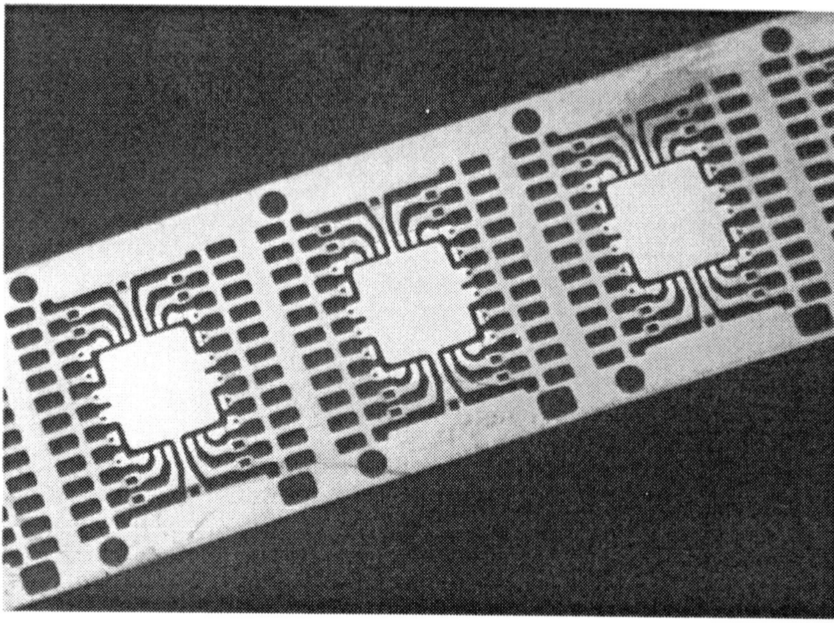

Fig. 13.2 Leadframe of medium power SOP, with heat transfer leads connected to die-pad

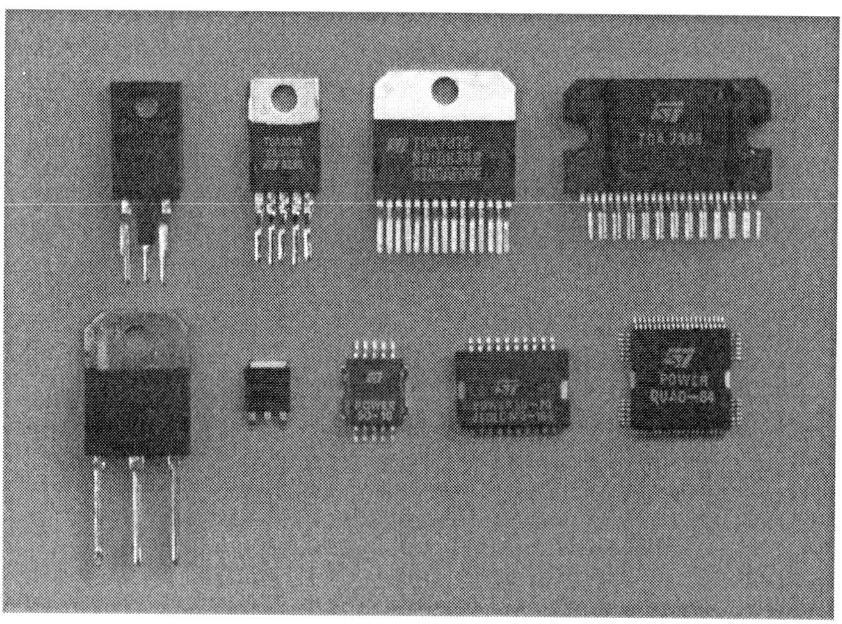

Fig. 13.3 Evolution of plastic power packages, from insertion to surface mount

As the external dimensions of the package are not changed, the undoubtful advantage is obtained of being compatible with existing handling and testing tools.

Modified leadframes are shown in Fig. 13.2. Due to the particular shape of their lead frames, medium power packages are called "batwing" or "crab" packages.

13.3.2 High Power Packages

For thermal dissipation higher than 2 W, external heatsinks are used, whose dimension is choosen according to the dissipated power. In order to have efficient heat transfer from the silicon to the heatsink, massive copper slugs are integrated in the body of high power packages.

Evolution of this family is shown in the roadmap of Fig. 13.3, where three main subgroups can be identified:

a) insertion packages
b) isolated packages
c) surface mount packages

13.3.2.1 *Insertion High Power Packages*

The group is originated from the two most widely used power packages, both with 3 leads: the TO-220 and the SOT-93 (TO-218), shown in Fig. 13.4.

The main difference between these two packages is the size, hence the size of die which they can accomodate. This also defines maximum operating current and power dissipation capability. Their characteristics are summarized in Table 13.2. (see section 12 for definition of θjc junction to case thermal resistance).

These packages are inserted in THT boards and secured to the external heatsink using screws through the fixing hole existing in the upper portion of the slug. In recent years substitution of screws with clips has become more common for cost reasons.

For more than 3 leads, a number of packages have been derived from TO-220 and SOT 93 structure (Fig. 13. 5).

with 5 leads:	Pentawatt, PentaSOT
with 7 leads:	Heptawatt, HeptaSOT
with 8, 11, 15 leads:	Multiwatt
with 15, 21, 25, 27 leads:	Flexiwatt

Table 13.2 Characteristics of TO-220 and SOT-93 (TO-218)

	Max die Size (mm^2)	Max Current (A)	θ_{JC} (C/W)
TO-220	25.50	60.00	0.5 – 1.5
SOT-93 (TO-218)	54.00	80.00	0.3 – 0.7

Fig. 13.4 Plastic power packages for discretes: TO-220 and SOT-93 with their isolated versions

13.3.2.2
Isolated Packages

In many applications a number of devices are secured to one single heatsink. If package slug is connected to the drain of a vertical MOS transistor each individual package must be isolated from the heatsink.

This can be achieved by placing a thin sheet of mica or adhesive mylar between slug and heatsink. This solution is relatively expensive, due to the manual assembly of the isolation layer and is not very consistent for what refers the thermal performance.

In order to minimize above problems, isolation can be conveniently obtained during the molding process, if a thin layer of molding compound is formed on the external surface of the slug. The solution is very competitive, especially if high thermal conductivity compounds are used (Table 13.3).

Fig. 13.5 Power insertion plastic packages for integrated circuits

Table 13.3 Comparison of isolation methods (with equivalent die size and dissipated power)

Package	Isolation	θ_{JC} (C/W)
TO-220	None	1.3
TO-220	100 micron mica	3.3
TO-220	Mylar	4.4
Isowatt-220	None	3.0
Isowatt-218	None	2.1

The isolated version of TO-220 and SOT-93 (TO-218) are called ISOWATT-220 and ISOWATT-218 respectively. Both exist with 3, 5, 7 leads.

ISOWATT-221 is the high voltage version of ISOWATT-220.

ISOWATT packages are shown in Fig. 13.4

13.3.2.3 *Surface Mount Power Packages*

Very recently, the extension of surface mount technology to power applications, requested the development of new package structures, which can associate miniaturization with good thermal characteristics.

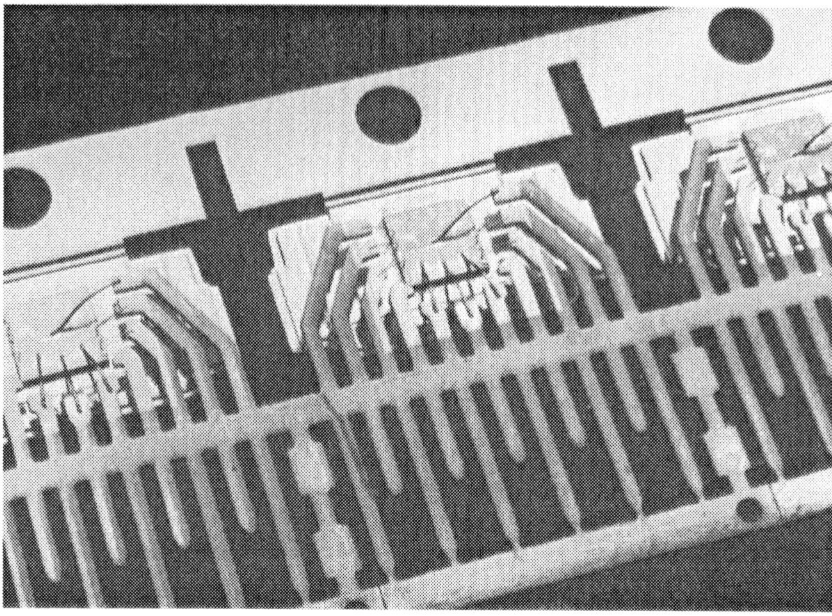

Fig. 13.6 Mixed wire interconnection for high current IC

Fig. 13.7 Recently developed power surface mount IC packages: Power SO-10 (MO-184) and Power SO-20 (MO-166)

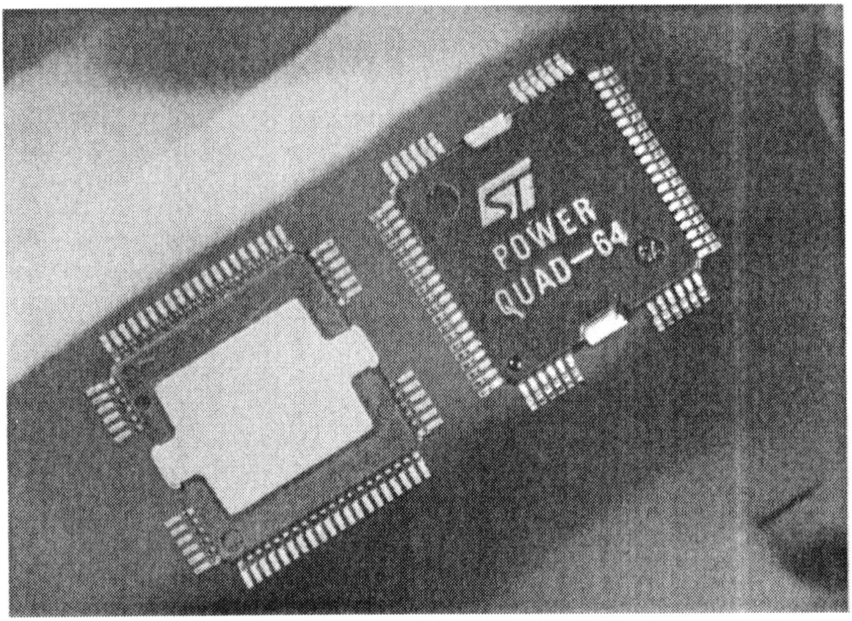

Fig. 13.8 Surface mount Power Quad, for Smart Power IC up to 64 I/O

Table 13.4 Smart Power IC Package Types

	Medium power < 2 W	High power 2–40 W	Isolated
Insertion	Dip	TO-220 SOT-94 Pentawatt PW Heptawatt HW Multiwatt MW Flexiwatt FW	Isowatt-220 Isowatt-221 Isowatt-218
Surface Mounting	SO PLCC PQFP	PowerSO PowerQUAD	

They are ideal packages for Smart Power Technology of last generation, whose complexity needs a relatively high pin count, up to 80–100.

They are purpose designed packages having leads on two sides – like PowerSO-10 and PowerSO-20 (see Fig. 13.7) or on 4 sides – like PowerQUAD – (see Fig. 13.8).

They are used directly on laminate boards, on high conductivity Isolated Metal Substrates (IMS), and in contact with external heatsinks, for dissipation ranking between 2 and 20 W.

Package types described in this paragraph are summarized in Table 13.4.

13.4 Thermal Management

13.4.1 Thermal Resistance

Thermal resistance is the physical quantity which indicates how "easily" the heat flows between two points A and B. If thermal resistance is small, heat is transferred from A to B with little temperature difference between A and B. If thermal resistance is large, transfer of the same quantity of heat requires a much higher temperature difference in the two points.

Thermal resistance is defined as:

$$\theta_{AB} = \frac{\Delta T_{AB}}{P}$$

(Note: θ is often written as R_{th})

where P is the power, or the rate of flow of heat between the two points, while ΔT_{AB} is the temperature difference between the two points.

In electronic devices, the two most important temperatures are the ambient temperature T_a and the temperature T_j reached by the junction when a given thermal power P_d is generated in the silicon.

Thermal resistance junction to ambient θ_{ja} is the ratio between junction temperature increase and dissipated power:

$$\theta_{ja} = \frac{T_j - T_a}{P_d} \tag{13.1}$$

θ_{ja} depends upon a number of factors, including size of the device, thermal conductivity of the materials, size and shape of the heatsink.

Device failure rate if affected by the junction temperature and about doubles for every 10 °C increase of T_j.

The main aim of thermal design is to keep the junction temperature within defined limits, which is usually lower than 150–170 °C for discretes, and 125–150 °C for ICs. At the same time, systems are increasingly being used in ambient temperatures up to 125 °C, for example in "under the hood" car electronics. In this case, $T_j - T_a$ can be as low as 20–25 °C. This causes particular problems where large amounts of power (10–20 W) are to be dissipated and θ_{ja} must be limited to a few C/Watt.

θ_{ja} is controlled by placing a suitably sized heatsink in contact with the power package slug. From a physical viewpoint, a thermal chain exists from the silicon to the ambient through the die attach, the slug, the interface between the slug and the external dissipator and finally the dissipator itself (see Fig. 13.9)

This can be represented as follows:

$$\theta_{ja} = \theta_{jc} + \theta_c + \theta_s \tag{13.2}$$

where θ_{jc} is the thermal resistance from the junction to the case, defined by the conduction properties of the device from the junction to the portion of the case

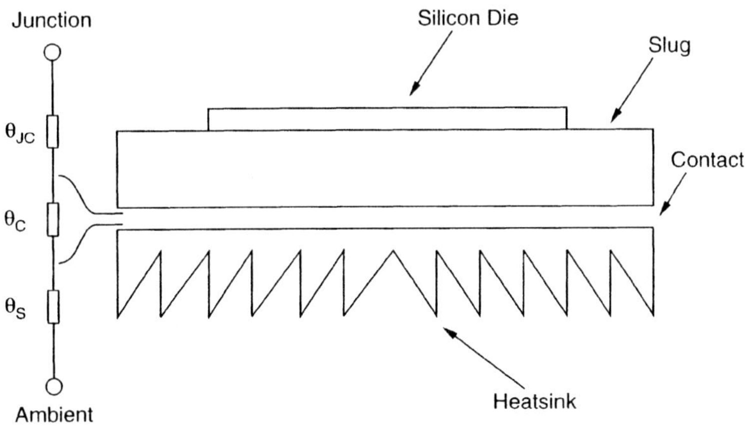

Fig. 13.9 Thermal chain from silicon to external heatsink

where the heatsink is applied; θ_c is a parasitic element describing the loss at the contact between package and heatsink; θ_s is the thermal resistance of the external sink, which indicates its efficiency in dissipating heat through convection and radiation.

13.4.2
Thermal Resistance and Heat Transfer Mechanisms

The flow of heat from the silicon to the ambient is ruled by three important mechanisms: conduction, convection and radiation.

13.4.2.1
Thermal Conduction and θ_{jc}

Conduction is the heat transfer mechanism due to thermal excitation at molecular level, by direct contact. No motion of material is associated to this mechanism. Thermal resistance depends on the thermal conductivity c of the material and its dimensions: length L and cross sectional area A according to the relationship:

$$\theta = \frac{1}{c}\frac{L}{A} \tag{13.3}$$

Equation 13.3 can be applied to the J-C pattern as shown in Fig. 13.10.

In this simple model, the heat generated at the junction flows in a 45° cone through the silicon, the die attach alloy and the copper slug up to its external surface. Table 13.5 shows the thermal conductivity of some materials commonly used in microelectronics.

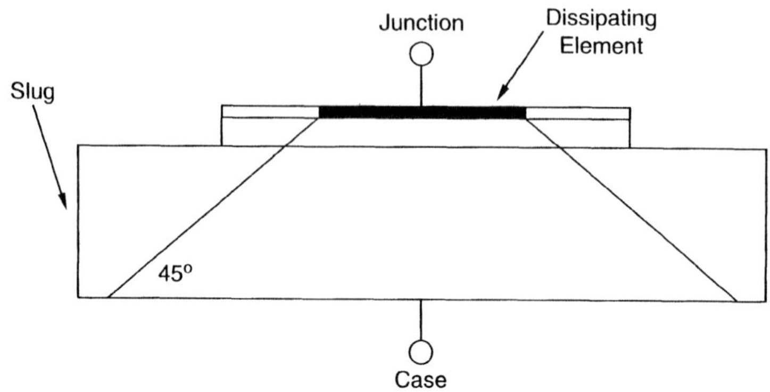

Fig. 13.10 Thermal dissipation pattern in power IC package

Table 13.5 Thermal conductivity of materials commonly used in packaging

Material	Thermal conductivity ($Wm^{-1} K^{-1}$)
Copper	390.00
Silicon	150.00
Glass (Silicon Oxide, SiO_2)	0.60
Silicon Carbide (SiC)	85.00
Aluminium	210.00
Aluminium Nitride (AlN)	150.00
Aluminium Oxide (Al_2O_3)	30.00
Resin (Standard packages)	0.71
Resin (Isolated packages)	2.51
Mica	0.50
Air	0.03

The following points should be noted:

a) θ_{jc} is mainly affected by the chip size and by the size of power element(s) in the chip. The theoretical relationship is given in Fig. 13.11 [1].
b) The thermal conductivity of silicon increases with an increase of temperature. In the temperature range considered in microelectronics a variation of 20–30% is common.
c) The dimensions of the slug are important. In principle, a thinner slug should provide a lower resistance (as there will be a shorter conducting path); in practice, due to the exitance of parasitic effects at the contact with the external heatsink (uneven surface, air, dust, etc.) a thicker and larger slug performs better: indeed it can spread the heat much more than the 45° considered in Fig. 13.10 and reduce the losses at the contact because of the larger area of contact.

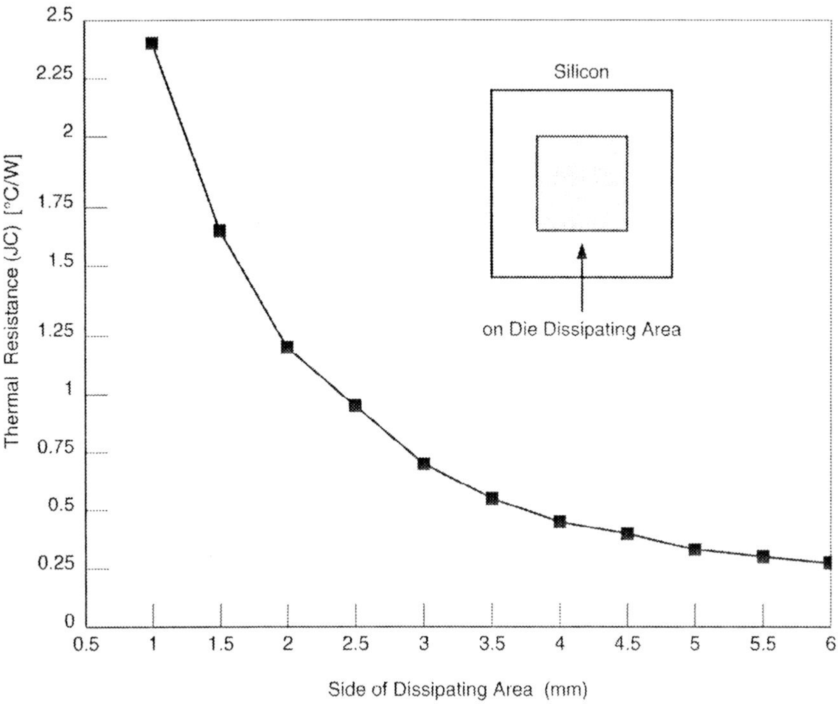

Fig. 13.11 Typical Junction to Case thermal resistance in power IC package (Multiwatt)

13.4.2.2
Convection, Radiation and θ_{ja}

Thermal Convection

Convection is the transfer of heat associated with the movement of material in a fluid (liquid or gas). It is due to the heating of the fluid in contact with the source, which moves and is replaced by cooler fluid upward. The process is ruled by the heat-transfer coefficient h depending on the source temperature T_s and the ambient temperature T_a.

The quantity of heat removed by convection is given by the relationship:

$$Q = h(T_s, T_a) \cdot A \cdot (T_s - T_a) \qquad (13.4)$$

where A is the area of the heat exchange surface. In case of forced cooling, h is a function of the speed of the cooling fluid and is always higher than 1 in Eq. 13.4. The following considerations are important:

a) the size of the heatsink has the greatest effect on the final performance of the cooling systems. As the exchange surface dominates in Eq. 13.4 it is common to increase it with the use of fins and grooves. However, this solution is most effective in forced cooling.

b) The higher the source temperature, the stronger the convection effect. This can compensate for the negative effect of the temperature on the θ_{jc} considered in the previous paragraph, and the final result is a θ_{ja} which decreases by increasing power dissipation.
c) The thickness of the heatsink is important, in order to spread the heat in the sink by conduction, and to have a larger exchange surface.

Thermal Radiation

The transport of electromagnetic energy is associated with radiation, which involves a transmitting surface at temperature T_s and a receiving ambient at temperature T_a.

The heat Q exchanged by radiation from a surface area S is:

$$Q = \sigma.\alpha.\ S.(T_s^4 - T_a^4) \tag{13.5}$$

where σ is the Stefan-Boltzmann constant and α is the emissivity of the material, which varies between 0 and 1.

Black surfaces have emissivity close to 1, indicating the maximum radiation energy theoretically possible at T_s. Polished, white metals (like aluminium) have the lowest emissivity, between 0.1 and 0.3.

Due to the 4th power relationship with the temperature, radiation dominates only when T_s is much larger than T_a (in absolute units). This is generally not the case for applications in electronics; however the effect of radiation cannot be ignored.

As an example of values associated with convection and radiation, if a 1 mm · 10 cm · 10 cm red (natural oxide) copper sheet is used as a heatsink positioned vertically at a T_s of 100 °C, the thermal resistances due to both radiation and convection are around 7 °C/W, giving an overall value of 3.5 °C/W.

13.4.2.3
Contact Thermal Resistance θ_c

Power packages are secured to the heatsink using screws, bolts or rivets fixed through the hole in the upper part of the copper tab (Fig. 13.12a). In order to reduce the cost of mounting, the use of clips is becoming more common, with the pressure applied to the top of the plastic body, at the centre (Fig. 13.12b).

The contact thermal resistance θ_c can be minimized by adding a thin layer of silicone grease, uniformly distributed between slug and dissipator and applying an appropriate force.

In case of screw mounting, this force can be controlled by choosing the torque (in kg · cm) applied to the fixing screw, by means of a dynamometric screwdriver.

Experimental value of θ_c are shown in Table 13.6 with different torque levels, for both conditions, with and without grease.

In practical cases, the recommended value of torque is 6 – 8 Kg · cm.

Fig. 13.12 Methods for mounting power packages onto massive external heatsink: screw and clip

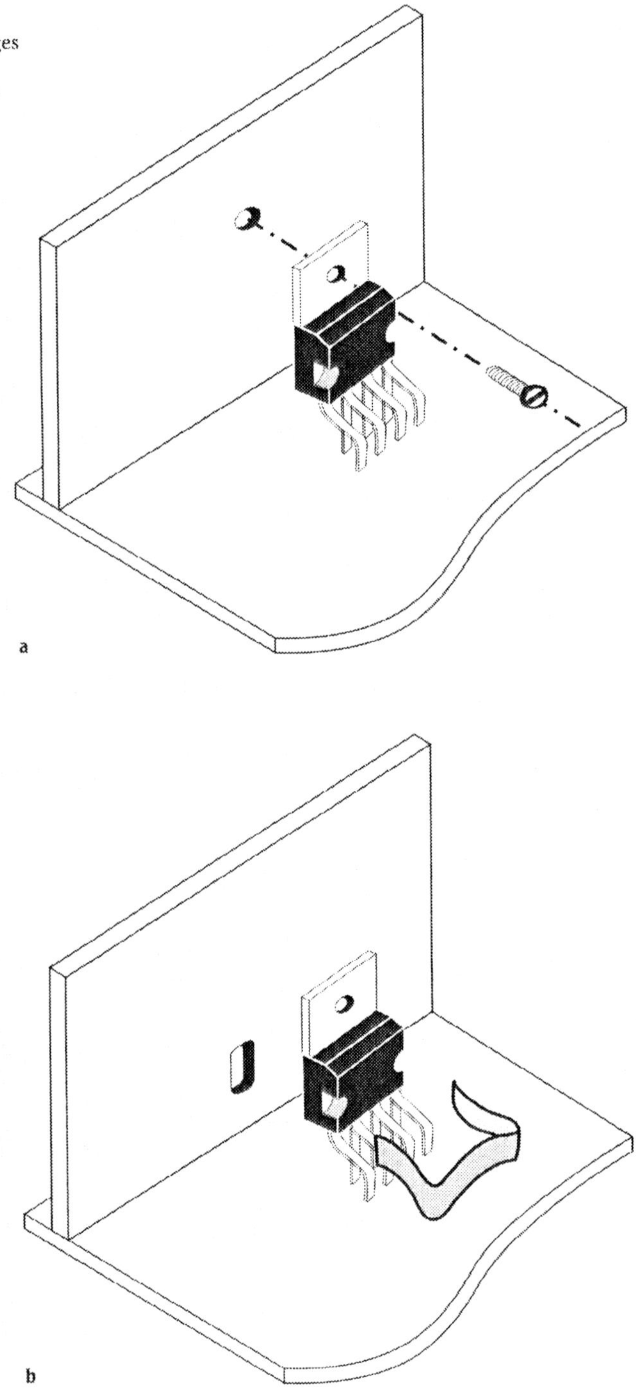

Table 13.6 Contact thermal resistance for TO-220 package (experimental data)

Torque kg · cm	θ_c with grease °C/W	θ_c without grease °C/W
3.00	0.27	0.90
4.00	0.22	0.82
5.00	0.17	0.78
7.00	0.12	0.75
10 – 15	< 0.1	< 0.7

When a spring clip is used, θ_c is also decreased if the force F of the clip is higher; a minimum of 5 kg is recommended.

13.4.3 Thermal Impedance

In section 13.4.1, thermal exchange was studied in steady state conditions, that is once the system had reached thermal equilibrium. However, in many situations the power dissipation of the system is not uniform, for example when the device operates in pulsed conditions. In this situation thermal capacitance of the system must be taken into account [2].

Thermal capacitance exists because the elements in the heat transfer path take a finite amount of time to heat up or "charge" to their equilibrium temperature. In practice this has beneficial effects: if excessively high dissipation peaks occur, it is possible that the maximum junction temperature will not be exceeded, provided that the pulse is not too long, just as an electrical capacitor will prevent voltages exceeding certain limits in the presence of current spikes.

An important parameter is the thermal capacitance C_{th} associated with the various structural elements of the thermal chain from the junction to the ambient. This parameter can be expressed as:

$$C_{th} = c \cdot d \cdot V \tag{13.6}$$

where c is the specific heat, d is the density and V is the volume of the material.

In a simple model, the above elements are the silicon, the slug and the external heat sink. They can be modeled as RC cells, as shown in Fig. 13.13.

The first cell represents the thermal behaviour of the silicon with a small volume and a correspondingly low thermal capacitance (a few mJ/°C). The thermal resistance between the junction and the silicon/slug interface is about 0.2 – 2 °C/W depending on the die size and on the size of the dissipating element on the chip. The time constant τ (= RC) of this first cell is typically in the order of few milliseconds.

The second cell represents the copper slug, with a relatively large thermal capacitance and a time constant of a few seconds.

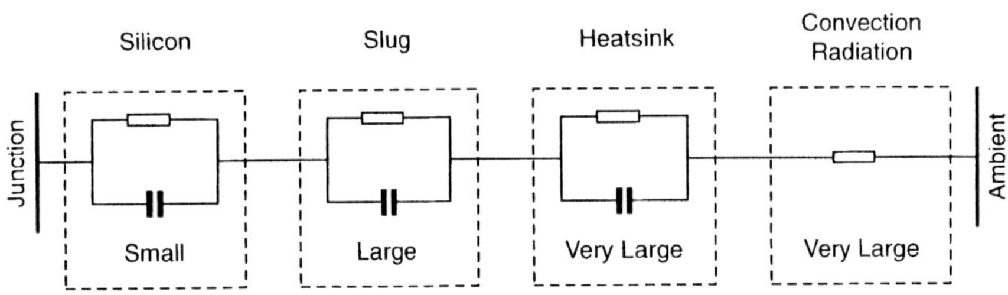

Fig. 13.13 RC equivalent of the thermal chain from junction to ambient

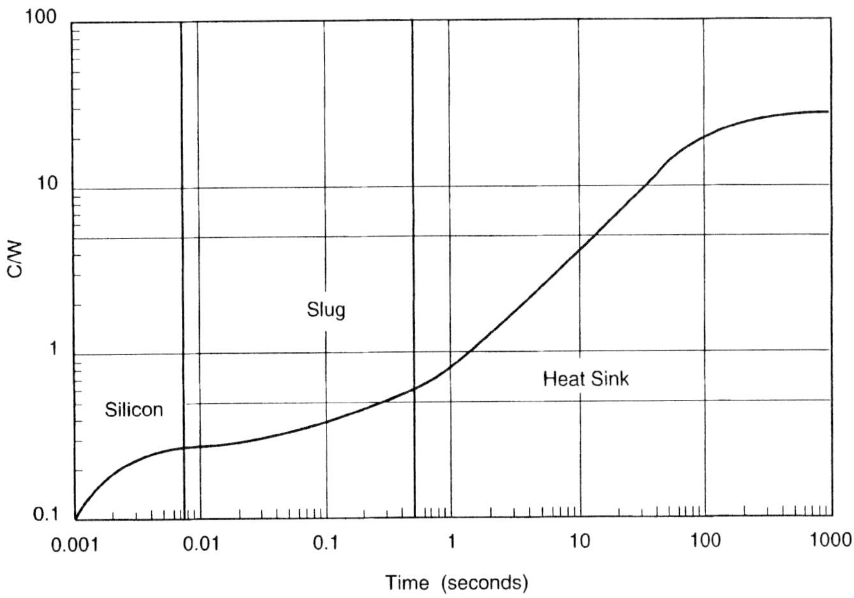

Fig. 13.14 Typical thermal impedance of power IC package (Multiwatt)

When an external heat sink is used the third cell can have a time constant of hundreds of second and equilibrium can take several minutes. Each element is ruled by the theoretical relationship

$$\Delta T = \theta \cdot P_d(1 - e^{t/\tau}) \tag{13.7}$$

and the actual behaviour of the package is a combination of the effects of the three cells.

A large plateau occurs up to 1–2 sec, due to the copper slug, limiting the thermal impedance to 0.2–2.0°C/W (Fig. 13.14). This effect is often used in switching applications, as a large number of power pulses can be dissipated before a large variation in the junction temperature takes place.

13.5
Process and Materials [3, 4]

The operation of plastic packaging assembly is a sequence of five main processes: (Fig. 13.15): die attach, wire bonding, molding, lead plating, lead forming. Some important difference exists between medium power and high power packages, which will be outlined in the next few paragraphs.

Due to their strong impact on device performance and reliability, materials like tin based alloys, encapsulation compounds, and surface plating deposits, have vital importance in plastic packaging technology, and its evolution towards cost effective solutions and low failure rates. In Fig. 13.16, thermomechanical characteristics of most important packaging materials are compared.

13.5.1
Leadframe

The leadframe is the mechanical support and the electrical and thermal interconnection of silicon chip. Leadframe for low-medium dissipation have a uniform thickness of 5–10 mils (125–250 micron); most frequently, copper alloys are choosen (instead of the very popular Fe/Ni Alloy 42), due to the excellent thermal and electrical conductivity [5, 6].

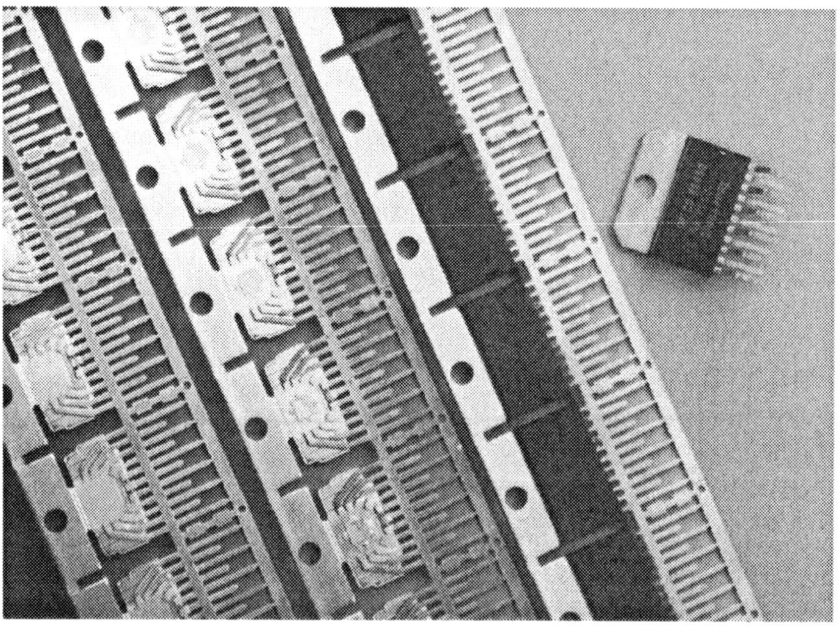

Fig. 13.15 Assembly process, from leadframe to finished product through: die attach, wire bonding, molding, lead plating, lead forming

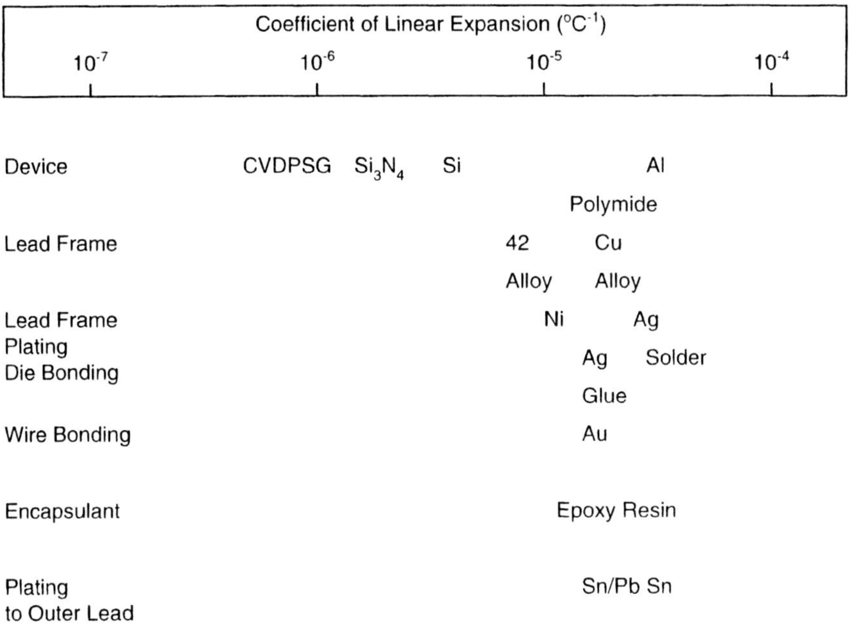

Fig. 13.16 Thermomechanical characteristics of typical packaging materials

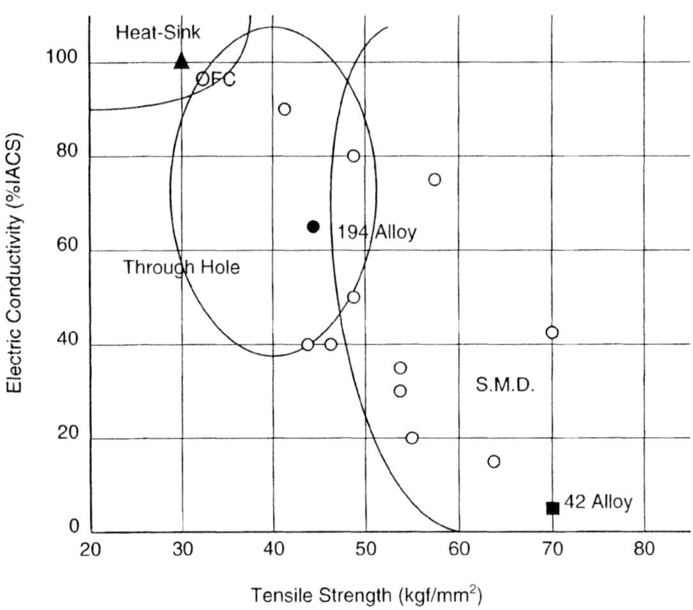

Fig. 13.17 Thermomechanical characteristics of copper alloys used in slug and leadframe fabrication

As Surface Mount Devices (SMD) are progressing toward fine pitch and multipins, lead position and planarity after assembly and testing are very critical issues. Due to this reason, special copper alloys have been developed offering a good compromise in mechanical stability and thermal conductivity.

Leadframes for high power devices have a copper slug of appropriate size, in order to reduce the junction to case thermal resistance down to a minimum value and provide some thermal capacitance in pulsed application. Oxigen Free Copper (OFC) or de-oxidized copper is often choosen, in order to have as high thermal conduction as possible.

In Fig. 13.17, thermomechanical characteristics of different copper alloys are shown for the different applications.

Both medium power and power leadframe, are silver or nickel plated in the die attach area and on lead tips, where bonding wire is welded.

13.5.2 Die Attach

After wafer sawing (die fab) high power die is brazed onto the slug using high melting temperature (> 280 °C) tin based alloys, as PbSn, PbSnAg, SnAgSb.

This process is critical in view of thermal performance and long term reliability; it must produce a uniform, void free joint between silicon back metallization (CrNiAu, CrNiAg, TiNiAu, etc.) and copper slug, in order to avoid hot spots in the active area and mechanical degradation in thermal fatigue.

Characteristics of several solder alloys are given in Table 13.7 [7].

For low-medium power die, packages in SO, PLCC and PQFP, epoxy or polymide glue die attach is commonly used. As it is compatible with the process needed by signal devices assembled in huge quantities in the same lines, this choice is effective in cost, while maintaining acceptable thermal performace. Thermal conductivity of glue is improved by adding a metal filler, commonly Ag.

In Table 13.8, physical characteristics of different glues (basically epoxy glue with and without solvent and polymide glue) are summarized.

13.5.3 Wire Bonding

After die attach, electrical interconnection between die bonding pads and leadframe is obtained, by means of gold or aluminium wire [8]. They are ultrasonically bonded to the chip and to selected areas in the leadframe, which are Ag plated in Au wire bonding or Ni plated in Al wire bonding.

Choice of Au or Al wire is performed after several considerations:

1) Process speed, which is higher for Au bonding, with 6 wires bonded in one second, more than twice the Al speed;
2) reliability in epoxy encapsulation, related to the corrosion resistance, critical with Al wires and totally missing in Au wires. Due to this reason, a minimum Al wire diameter in plastic packages is 5–6 mils (125–150 μ) versus 1.5 mils (38 μ) in ceramic packages;

Table 13.7 Properties of soft solder

	Alloy (wt %)	Temperature Solidus (°C)	Temperature Liquidus (°C)	Tensile Strength (Kg/mm²)	Elongation (%)	Hardness (HB)	Electrical Conductivity (IACS %)	Specific Gravity (g/cm³)
Pb/Sn	Pb/2 Sn	316	322					11.2
	Pn/5 Sn	300	314					11
	Pb/10 Sn	268	301					10.7
	Pb/20 Sn	183	279					10.2
	Sn/38.1 Pb	*(E) 183					8.4	
	Sn/40 Pb	183	190					8.5
Pb/Ag	Pb/2.5 Ag	*(E) 304		3.1	42		8.8	11.4
	Pb/3.5 Ag	304	317	4.2		8	8.7	11.3
	Pb/5 Ag	304	364	3.5	35		7.2	11.3
Pb/In	Pb/5 In	310	314	3	52	6	5.1	11.1
	Pb/10 In	294	300					10.8
	Pb/25 In	255	265					10
Pb/In/Ag	Pb/5 In/2.5 Ag	304	307	3.2		9	6.9	11
	Pb/5 In/5 Ag	290	310					11
Pb/Sn/Ag	Pb/2 Sn/2.5 Ag	299	304	2.5				11.2
	Pb/5 Sn/1.5 Ag	*(E) 296					7.9	11.1
	Pb/5 Sn/2.5 Ag	296	300	4.5		10	8.1	11.1
	Pb/10 Sn/2 Ag	275	300					10.8
Pb/Sb	Pb/2 Sb	300	314	1.9	51		8.1	11.2
	Pb/5 Sb	252	298	5.8	24	11	6.3	11
Sn/Ag	Sn/3.5 Ag	*(E) 221		2	73	40	14	7.4
	Sn/5 Ag	221	250	5.6	30	14	12.6	7.4
	Sn/10 Ag	221	295					7.5
In/Sn	In/45 Sn	117	119					7.3
	In/48 Sn	*(E) 117		1.2	83	5	11.7	7.3
	In/50 Sn	117	125					7.3
In/Ag	In/3 Ag	*(E) 141						7.4
	In/10 Ag	141	230	1.1	61	2.7	22.1	7.5

* E = Eutectic Alloy.

Table 13.8 Physical characteristics of Ag glue

	Epoxy solvent	Epoxy non solvent	Polyimide
H_2O absorb ratio	○	Δ	Δ
Peeling strength	○	Δ	⊙
Die share strength	○	Δ	⊙
Void in die bonding	○	○ ~ Δ	Δ
Thermal conductivity (cal/cm s °C)	$4 \sim 5 \cdot 10^{-3}$	$4 \sim 5 \cdot 10^{-3}$	$1 \sim 2 \cdot 10^{-3}$
Electric resistivity (Ω cm)	$1 \sim 10 \cdot 10^{-4}$	$1 \sim 10 \cdot 10^{-4}$	$1 \cdot 10^{-4}$

○ fair
Δ good
⊙ very good

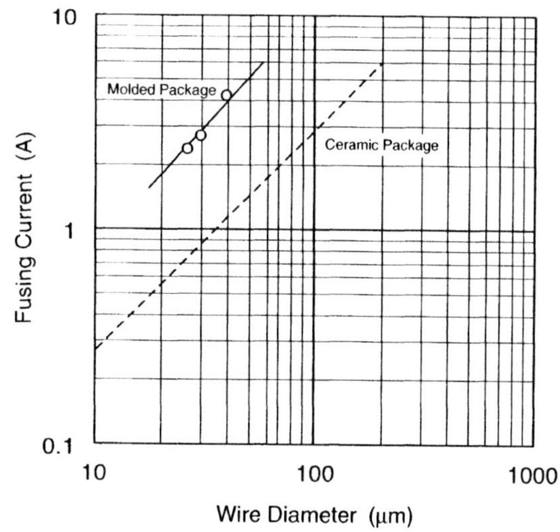

Fig. 13.18 Fuse current for Au and Al wires

3) capability in handling current, which is related to material, diameter and environment. About wire diameter in plastic package, the empirical rule is 1.3 A/mil as max DC current. Fuse current is showns in Fig. 13.18; it can be noticed that in plastic package, due to the intimate contact between wire and encapsulant, the fuse current is about 3 times higher than in cavity ceramic packages [9];
4) bonding design rules, in terms of bonding pad size and loop hight control. Especially in low profile power SMD, limitation of this parameter should be taken into account (Fig. 13.19).

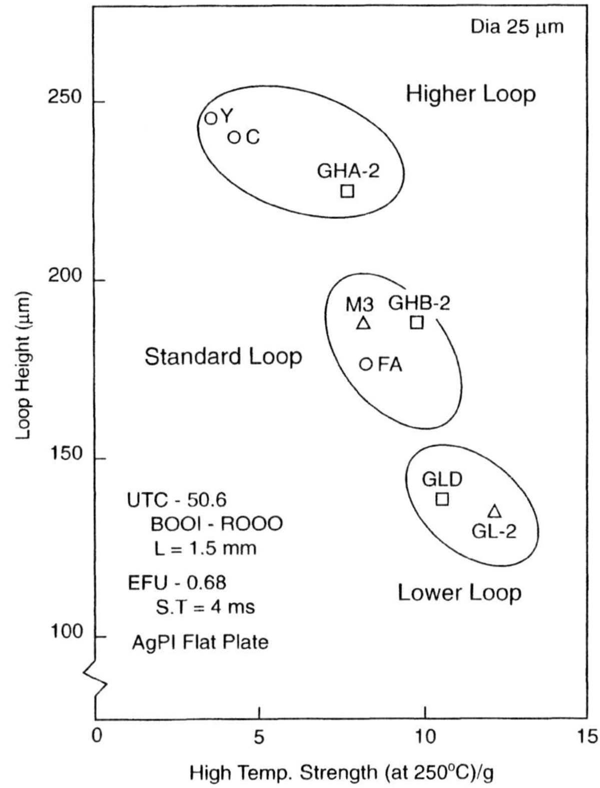

Fig. 13.19 Relationship of the wire loop vs the characteristics of bonding wire

13.5.4 Molding

Molding is the third step of assembly, when the leadframe strips are positioned in molding cavities which are then pressure filled with liquid thermosetting compound. After 40–80 s at about 180 °C, the compound solidifies in the mold (curing) in order to provide hard, reliable and cost effective encapsulation.

The molding compound is a key factor in package fabrication; it is in permanent evolution toward lower mechanical stress, lower impurity content and better cost [10].

The base material is thermosetting epoxy, filled with fused silica to enhance reliability and control thermal expansion coefficient. A different filler like Al_2O_3 or SiC can be used in order to increase thermal conductivity of the molding compound.

One of the most important properties is the natural absorption of moisture from the ambient. During the board mounting of surface mount devices, due to the relatively high soldering temperature (220–240 °C) moisture is suddenly desorbed, with possible cracks and delaminations in the package structure. This event known as "popcorn effect" is a major cause of failure in plastic SMD [11, 12].

13.5.5
Lead Plating and Lead Forming

The last major steps in semiconductor assembly are lead plating and forming.

After molding, external portion of the leadframe is electroplated by a 6 – 15 μ thick solder alloy (Pb Sn); this tin rich alloy is a low melting temperature coating, which provides a "wettable" surface when devices are soldered on the printed board.

Due to this reason, lead finishing has impact on quality and reliability of the solder joint, at system level. Lead solderability is measured by means of especially developed tests, as described in the next Chapter.

References

1. Tiziani R (1989) Thermal Characteristics of Multiwatt Package, SGS Thomson Application note
2. Hopkins T, Cognetti C, Tiziani R (1988) Designing with Thermal Impedance, Semitherm
3. Ishikawa T et al. (1985) High Reliability and Full Automation in Assembly Technology of Semiconductor, Ohyo Gijutu Shuppan Co Ltd Tokyo, pp 427
4. Tanaka Kikinzoku Kogyo, Technology and Production Guide, Tanaka Kikinzoku Kogyo, pp 20
5. Furukawa Electric (1986) Co Ltd Metal Div, High Performance Copper Alloy, Eftec Series
6. Ohyama Y et al. (1987) High Performance Copper Alloy Eftec-64 and 64T for Lead Frame Applications and their Properties, Furukawa Jihou
7. Sumitomo Metal Mining Co Ltd, Alloy Preforms and Clad Materials for Semiconductors Technical Data, Sumitomo Metal Mining Co Ltd
8. Asaka J (1970) Bonding Technology, Nikkan Kougyo Sinbunsha, Tokyo, pp 144
9. Iwasaki Y et al. (1977) On Fusing Current of Gold Wire in Plastic Package, 7th IEEE Reliability and Maintenability Symposium
10. Kakiuti H et al. (1990) Epoxy Resin, Shoukoudo, Tokyo, pp 280
11. Murakami G et al. (1988) Mounting Technology and its Reliability in Surface Mount, Type LSI Packages, Ohyo, Gijutu Shuppan Co Ltd Tokyo, pp 514
12. EIAJ (1992) Standard of Electronic Industries Association of Japan, Environmental and Endurance Test Methods for Semiconductor Devices, ED-4701, Japan

CHAPTER 14

Reliability of Smart Power ICs

L. Lonzi · M. Piccoli · F. Speroni

14.1 Introduction

The severe stress conditions at which Smart Power ICs can be submitted during their life on the field, the complexity of circuits and of their applications, the low level of defectiveness to be guaranteed make the reliability assurance of these ICs a very challenging task. In this Chapter, the present approach for the Smart Power ICs reliability, the related stress tests and the relevant failure mechanisms are discussed, while the last paragraph is dedicated to the new reliability approaches.

Power ICs, ten years ago, were realised only with bipolar technology and limited to the audio amplifier application with a power handling of a few watt; the reliability evaluation and qualification were very simple and mainly based on international standards like MIL STD.

Today, the power ICs scenario is more complex, for the applications, which practically cover all the fields, as widely described in the previous Chapters and for the large number of silicon and package technologies involved. As concerns the silicon, in addition to the standard bipolar processes, the mixed BCD (Bipolar, CMOS, DMOS) technologies are becoming more and more important (see Chap. 1). For the package (see Chap. 13) besides to the old ones typical of power application, such as Versawatt, Pentawatt and Multiwatt, some important packages for Surface Mounting Device (SMD), like SO-Power, have to be considered. The power handling involved in smart power application can range from a few watt up to 200 W.

Also the approach to assure the reliability performance has evolved in these years (see Fig. 14.1) and in particular two important aspects have been considered: field application and environment, stress induced by customer manufacturing process. The new methodology is based on a long experience developed with customers and a deep analysis of all the failure mechanisms encountered during years and years of reliability experiments and extensive failure analysis.

Today, new severe requests are coming from the market: less than one digit PPM of defectiveness during all the life long of an IC in extreme working conditions is not unusual. The ICs suppliers are going to accept these new challenging targets and, to face them, new reliability strategies become necessary. First of all, in the design phase, where a concurrent engineering approach is activated to realise the concept of robust design. Later on, during the reliability evaluation,

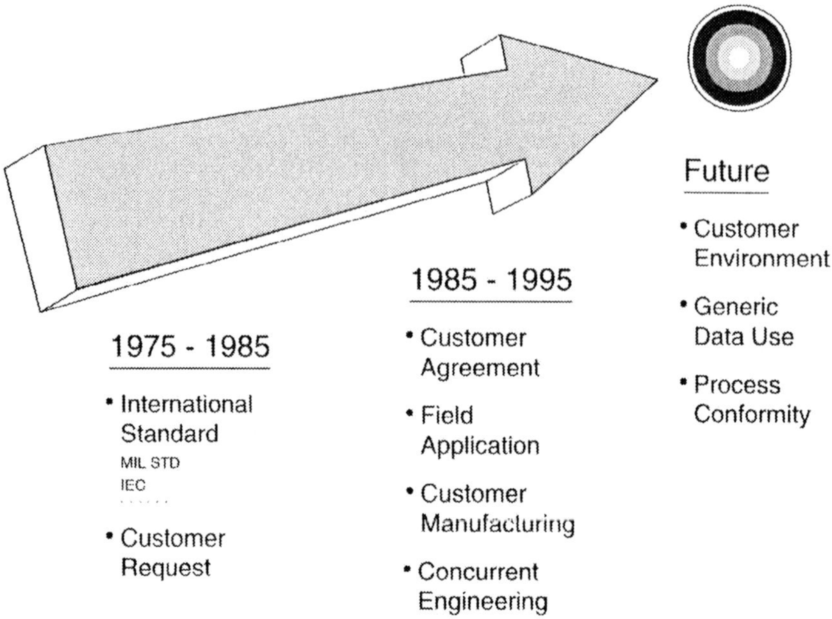

Fig. 14.1 Reliability approach evolution

a deep analysis on the sample before and after the test is performed to single out possible weaknesses and marginalities and the "quasi defect" concept is emphasised.

14.2 Reliability Approach for Smart Power ICs

Discussing the approaches to assure the reliability of Smart Power ICs, five main factors dominating the scenario have to be considered: complexity of the circuits, complexity of ICs applications, severity of customer manufacturing process, low defectiveness requirements both in early and useful life.

Fortunately all these factors are not present at the same time, or have the same weight, for the various market/application segments and different approaches must be used in accordance to the different situations. Just as an example in computer application we have a very short life time and a low defectiveness request in an unrestrained environment. In the telecom field, the main issue is to guarantee long term reliability. In the automotive segment, the environment is very severe and to have a low early life defectiveness it is very important to cover the warranty period.

Different applications are going to use a different approach in system design and manufacturing, consequently also the reliability approach must follow this trend to guarantee customer needs and to cover their new requests. A good approach in reliability is to follow all the phases of the life of the devices and for each phase to develop a specific activity. Only using this methodology it is possible to get a good reliability performance and to assure its continuous improvement.

Practically we can consider six phases:

- Customer system analysis;
- Supplier engineering;
- Supplier manufacturing;
- Customer manufacturing;
- Field application.

We will consider in detail all these steps because only starting from their deep analysis it is possible to build a satisfactory reliability performance.

Customer System Analysis

First of all, the reliability performances must be completely defined, including all the kinds of stresses involving the IC, the life to be guaranteed and the relevant failure rate of the system. After this analysis, it is possible to define the full reliability performance of the IC. Therefore, the supplier is committed to design a suitable IC to meet the system application characteristics.

Before choosing the assembly and silicon process it is necessary to define the real active life of the component during the life of the system. To clarify, we will consider the example of the telecom line card: in this case, it is necessary to guarantee at least 25 years of life. However, by the system use analysis we discover that for 90 % of the life span the system is on stand-by and only for the remaining 10 % it is activated. Other examples are the application in the automotive field where, during 10 years of life, only a few months are in real operative condition, in this case, the number of key-on and key-off of the car is a very important because of the stress generated in these phases.

After this basic system analysis we evaluate, on the basis of the historical data on the consolidated processes, the feasibility to design a circuit with the expected performances using an already existing process. In case we are not able, something new, in terms of silicon or package, has to be considered.

On the basis of the main expected failure modes we can try to estimate "a priori" a reasonable failure rate for the new circuit. It is mandatory to consider in all these analyses the important effect of the stress induced by the customer manufacturing process that can strongly influence the final reliability performance.

Supplier ICs Design

During this phase, a concurrent engineering approach is widely used and the reliability engineer contributes to implement suitable design rules and guidelines to meet application requirements. For Smart Power ICs the main points to be considered are related to:

- Electromigration;
- Current effects in wires and bonding;
- Thermomechanical effects;
- Parasitic effects (surface and bulk);
- Thermal performance.

Suitable tests are designed and implemented to investigate these aspects and to highlight the reliability ruggedness of the IC.

In these phases a specific test matrix is implemented focused on the main expected problems in order to have a fast feed-back on the design performance. It is not important at this stage to test manufacturability and to evaluate defectiveness. The test matrix is realised with highly accelerated short tests and the acceptance criteria are mainly by attribute (go-no go) but sometimes, when necessary, also electrical parameter drifts analysis is performed to detect possible parasitic effects related to layout mistakes.

Supplier Engineering

The compatibility of the design with silicon and assembly processes spreads is evaluated in this phase. For this purpose, a complete reliability test matrix is implemented to verify the full circuit performance in the short and long term.

The acceptance criteria are mainly by variable (electrical parameter drifts analysis [1]) at least for the electrical tests. A deep investigation, by using destructive analysis techniques on good and failed parts, is mandatory to have a complete "perception" of the device reliability.

Sometimes, for advanced device design with new silicon technology or new package, stress limit tests are implemented to verify the extreme performance and the reliability margins of the project. The test matrix includes also customer manufacturing process simulation to avoids problems in the field not evaluated during qualification exercise.

Also in this phase the reliability defectiveness measurement is not possible because of the limited number of samples and manufacturing lots that are usually available.

Supplier Manufacturing

This one is a key phase in the reliability strategy in order to:

- Measure defectiveness;
- Detect anomalous manufacturing trends and take corrective actions;
- Collect generic historical data on the technological ICs families;
- Collect stress limit data.

To assure a good reliability manufacturing control it is necessary to implement two kinds of test matrix: the first one consists of fast and high accelerated tests, called real time controls, which allow to have a feedback on weekly basis on the ongoing manufacturing lines situation. The second one is based on long term tests reproducing qualification tests to verify that the requested reliability performance is assured.

The key point of this control is the sampling plan that has to be periodically updated considering the following factors:

- Qualification exercise results;
- Customer field returns;
- Maturity level of the technological family involved.

A pertinent analysis on the results of these manufacturing monitorings is normally performed.

Customer Manufacturing

Only during the last five years this phase has been correctly considered, in particular, after the extensive use of SMD (Surface Mounting Device). In fact, these techniques have created a lot of problems not clearly understood.

By performing specific reliability exercises in close co-operation between supplier and customer it was possible to put in evidence reliability problems created by contamination or thermomechanical stresses during customer assembly process. In this way, it was possible to highlight and to reproduce potential reliability problems as, for instance, delamination effects and interlevel metal short circuits.

This kind of co-operation made possible to improve the defectiveness in the field of one decade at least. A lot of supplier/user joint activities are in progress to fix a standard temperature profile, representative of the assembly process, to be used in qualification exercise. At present, JEDEC in USA and CECC in Europe have defined two different standards.

Field Application

Customer field returns is a very important source of information for ICs suppliers and a good co-operation with customers to have an accurate classification of the different failures is in progress. Anyhow, sometimes, for the customers it is difficult to have traceability of the failed parts, in particular if they have not a definite network to collect failed cards.

Users operating in the telecom field have started this activity ten years ago and they are in a good situation from this point of view. Automotive customers have very big difficulties in organising this activity because the products are largely spread and sometime the cards are substituted without any kind of feedback to the car producer.

Another important source of information on field failure comes from the results of customer burn-in and field trials which have to be compared with supplier burn-in and manufacturing monitoring data.

Suppliers have organised in these last years important failure analysis laboratories dedicated to the investigation of field returns in short time and with a high level of accuracy.

14.3 Reliability Stress Tests

It is very useful to classify the reliability tests [2] to be performed in accordance to the different kind of evaluation that we want to implement. The most important are: tests for failure mechanisms investigation, tests for ICs qualification, tests for manufacturing control.

14.3.1 Tests for Failure Mechanisms Investigation

These tests are generally performed on dedicated structures to stress a single mechanism; we will discuss them later in this chapter when the different main failure mechanisms will be described in detail. In addition, mathematical models of interpretation will be presented including acceleration models to evaluate useful life during application conditions.

Sometimes it is very important to run in parallel both tests for ICs qualification and tests for failure mechanisms characterisation (on elementary structures) to compare their results. This exercise is particularly important for phenomena like electromigration where the tests on the elementary structures are not able to simulate all the stress of the real application.

14.3.2 Tests for ICs Qualification

The purpose of the qualification exercise is to validate the IC for manufacturing. The test matrix to be used includes different types of reliability tests [3, 4]:

- Electrical;
- Environmental;
- Combined.

14.3.2.1 *Electrical Tests*

These kinds of tests are divided into two main groups:

- Tests to simulate the field application;
- Tests to verify process problems and die package compatibility.

Electrical Test to Simulate the Field Application

To simulate field application it is possible to proceed in accordance with two methodologies. The first one is to use the same sequence of electrical states that you have in real situation. For instance, in the telecom application for a telephone set you can have the following sequence:

Stand-by – Ringing – Conversation – Stand-by – Dialling – Conversation

In this case, there are no doubts about the perfect simulation of the field application, but we have the disadvantage of a low acceleration of the test and consequently a very long test duration. Another disadvantage is that, in case of failures, it is difficult to discriminate which electrical state or stress has caused them.

The second methodology is to consider all the electrical states of the application, including transitions, separately and to accelerate each state as much as possible. In this way we have the following two advantages:

- To increase the acceleration of the test and hence to reduce the test time;
- To separate the different electrical stresses and to have a clear visibility of the failure mechanism.

It is difficult to say which of these two methodologies is better. Normally when we have a consolidated process situation and a very innovative circuit to test, we prefer the first one that can avoid any doubts about correlation with field application. The complexity of the test implementation in this case is not negligible and must be carefully considered.

If, on the contrary, the device is an advanced device and either new technologies or new package are present, the second methodology is preferred. In addition, to investigate the technological aspects specific tests are sometimes implemented.

After this "philosophical" discussion on the electrical tests for field simulation, we can describe some of them like:

- Operating Life Test (OLT);
- Intermittent Life Test (ILT);
- Thermal Fatigue Test (TFT);
- Static Life Test (SLT).

Operating life test is performed in accordance with the application, but in extreme conditions as concerns:

- Absolute maximum voltage supply;
- Maximum current on the load;
- Maximum junction temperature (not exceeding 150 °C for plastic packages).

In case not all these conditions are compatible the priority order is: current load, junction temperature, voltage supply. Sometime field application simulation requires more complex tests like ILT and TFT.

Intermittent life test was developed to simulate the electrical stress that arises when a power equipment is switched on. This was used for the first time for TV set applications and later on also for automotive applications where it was called key-on and key- off test. The electrical configuration that we have is the same as that used in OLT, but the electrical parameter prioritised is the voltage supply that must be as high as possible. This test aims at stressing the silicon versus junction breakdown during electrical commutation. The frequency of commutation and the test duration should be designed to cover the useful life span of the device in the field application. When the t_{on} is lower than

4–5 ms (silicon thermal constant time) the test is stressing only silicon and it is a pure die oriented test. This test is very important because in Smart Power ICs applications the life of the equipment is strongly influenced by the frequency of switching on and off. ILT requires specific boards and equipment because of the difficulty of commuting a high number of devices under test. Normally the scanning technique is used to avoid the overload of voltage suppliers and to minimise inductive effects.

Thermal fatigue test is similar to ILT but in this case the priority is not in maximising the voltage supply but the current load; the electrical configuration is the same as that of OLT and ILT; the focus is on three parameters: ΔT_{case}, t_{on}, t_{off}. Considering that this test is addressed mainly to stress die attach and wire bonding it is easy to understand the importance of ΔT_{case} which ranges from 70 °C to 100 °C. The two times t_{on} and t_{off} should be compatible with the thermal capacitance of the package to have a meaningful level of stress. The used value range from 2–3 min. The number of cycles is between 10–30 Kcy. The parameters checked are the thermal resistance (Rth-jc) of the die attach and the pin to pin continuity; usually a degradation in Rth-jc of 20 % after 20 Kcy is allowed. It is a good method to observe the wire bonding and die attach state after test through a destructive physical analysis to assess if the reliability margins are adequate. This test was developed and used for discrete devices; for ICs the limitation is the thermal resistance measurement using the substrate diode, because this is not a good diode and it is difficult to dissipate power during measurements.

Static life test was originally designed for voltage regulators and in the last ICs applications it is used to simulate stand-by condition. Due to the trend of power reduction in stand by this test is becoming very similar to HTRB test that will be explained later. The importance of this test was in the past related to wire and wire bonding investigation in high current and temperature stress.

Electrical Test to Verify Process Problems and Die Package Compatibility

The electrical tests for field simulation, previously described, supply information to evaluate die package compatibility and to highlight process problems. We will now describe two other very important tests, dedicated to these reliability investigations:

- High Temperature Reverse Bias (HTRB);
- Thermal Humidity Bias (THB).

High temperature reverse bias test comes from discrete bipolar devices which was aimed to stress base collector junction in reverse mode monitoring junction leakage and current gains degradation. In Smart Power ICs, this test is very important because it can emphasise parasitic effects related to: layout defects, silicon contamination, resin related effects.

For layout defects, today, CAD tools are able to help in parasitic MOS detection. The advantage given by a good HTRB test is to consider all the involved aspects, that sometimes are difficult to include in an evaluation based only on CAD tools. A good HTRB test is done following these rules:

- Maximum voltage supply compatible with process and circuit limitations;
- Junction temperature around 150 °C for plastic and 170 °C for hermetic packages; in any case it must be compatible with circuit thermal shut-down protection or limitations;
- Power dissipation as low as possible;
- As many junctions as possible in reverse mode. Particular attention should be paid to avoid overstressing of the base emitter junctions;
- All the states of the IC must be set, no floating states are allowed;
- The electrical key parameters to be checked during the read out steps are those influenced by silicon surface behaviour, as gains, reference voltage and leakage currents.

Silicon contamination is detected measuring parameter drifts during the test. Specific computer tools (database) to perform statistical analysis of this reliability data have been developed [1]. Usually to detect these kinds of problems it is necessary to continue the test for a few weeks due to the constant time of mobile charges within the different die dielectric layers. A faster degradation is an indication of heavy contamination problems. For this reason, this test is fundamental for process changes and transfer qualification silicon oriented.

HTRB test can supply in short time (a few hours or days) very good information on the effects of the moulding process on IC reliability. In fact any anomalous situation in the moulding process induces a different polymerisation structure of the resin, this can cause different behaviour, in terms of polarisation of the resin, when it is submitted to the electrical field. Due to this property the resin can transfer the electrical field from one area to another one in the device through a "sky effect", activating in this way possible parasitic structures.

The fundamental importance of this test is that it can be easily implemented, not expensive and very fast to set. However, some precaution must be taken when dealing with it:

- Electrical testing must be done within no more than 4 hours from the end of test;
- Electrical testing must be done at room temperature;
- Testing equipment must have a high level of accuracy and repeatability; reference samples are mandatory.

When this test is well managed it is able to supply very fast reliability indication. This can be obtained using the initial linear region of the exponential law typical of the surface effects.

Thermal humidity bias (85 °C, 85 %RH) test evidences problems of die package compatibility related to phenomena activated in wet conditions such as electrochemical corrosion. The test is done using the same schematic of HTRB but reducing the voltage supply at the data sheet nominal value or less to reduce power dissipation. Sometimes for power ICs the voltage supply reduction is not enough and consequently some other techniques are used to reduce the junction temperature that makes the test insignificant if it rises above 95 °C. In this case the switching technique is used, setting t_{on} and t_{off} parameters compatible with thermal impedance of the package and the junction temperature requested. Sometimes to reduce junction temperature a suitable heat-sink is used, but the results are not always good.

To increase the test effectiveness and to reduce the stress time an alternative test, called High Accelerated Steam Test (HAST), can be adopted; typical conditions can be, for instance: 130 °C, 85 % RH. Acceleration models, such as those of Peck or Reick Hakim are used to correlate HAST and THB tests (see par. 14.4.3.2); the results and correlation are not always satisfactory and HAST implementation needs caution and skills in test management.

14.3.2.2
Environmental Tests

Two main types of environmental tests are considered: thermomechanical and moisture resistance stress tests.

Thermomechanical Tests

These tests are very important to focus on thermal mismatches among the different layers and elements present in an IC. They are important in particular for Smart Power ICs because of the asymmetric structure of power packages (plastic and metal from two different sides). The tests generally performed are: *Temperature Cycling* (TC) and *Thermal Shock* (TS). These tests are different for technique: air-to-air for TC, liquid-to-liquid for TS and furthermore shocks are 3–4 times faster. The adopted temperature excursion is in accordance with the mentioned standard like: MIL 883, JECEC, CECC. These tests are typical process oriented tests, they are very useful for new product qualification but also for the evaluation of change and transfer of the assembly process. Additional uses are in manufacturing control and for screening with/or in alternative to burn-in.

Moisture Resistance Tests

The key test in this case is *Pressure Pot* (PP). This test is very powerful, fast and easy to implement. It is very effective to point out the critical water entry paths and the consequent phenomena of corrosion, oxidation, delamination, etc. The standard generally adopted is 2 atm absolute in 100 % water vapour saturation. Sometimes other standards such as 3 atm are adopted in manufacturing control to reduce test duration.

14.3.2.3
Combined Tests and Sequences

Combined stress tests are becoming more and more popular especially in the automotive market segment because of their effectiveness and ability to simulate

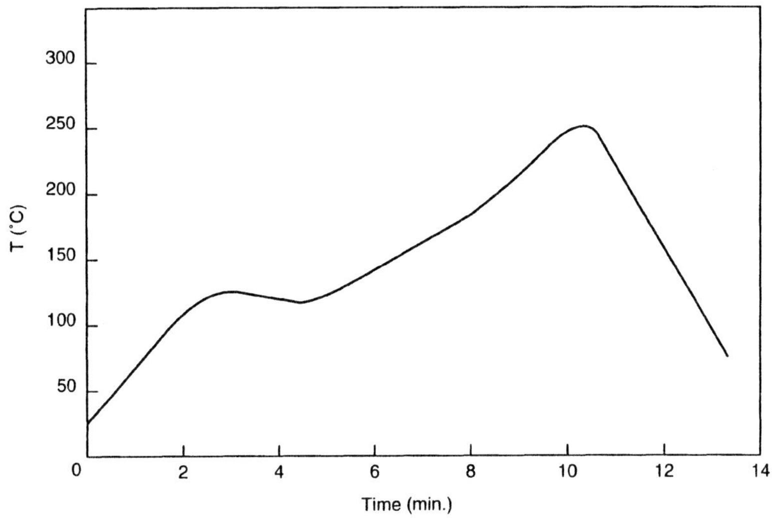

Fig. 14.2 Customer assembly typical temperature profile

field application. The most important one is *Power and Temperature Cycling* (PTC). This test is realised performing an OLT stress in a conditioned environment, cycling between −40°C and 85°C, in the meanwhile the DUT (Device Under Test) is switched on and off every few minutes. This test needs powerful and expensive equipment; however, it is very effective, especially to focus on die attach and wire bonding problems.

Other test sequences, in which different elementary stresses are applied in series, are often used; with this aim several environmental sequences have been defined. A particular case of sequence has been introduced to simulate surface mounting that, as explained in par. 14.2 is an important stress induced by customer manufacturing on SMDs. Taking it into account, during qualification exercise a soldering simulation is introduced as pre-conditioning before the other classical reliability tests; Fig. 14.2 shows a typical temperature profile adopted for soldering simulation.

14.3.3 Tests for Manufacturing Control

Manufacturing control needs to be fast, effective and focused on the main expected problems. The criteria for planning and sampling are based on:

- Customer returns;
- Qualification results;
- Process to be controlled;
- Volume of sales;
- Diffusion and assembly lot dimension and composition.

The reliability tests matrix must be implemented in one week including electrical testing control, to allow fast reaction in manufacturing line at high volume. The main difficulties come from testing because of quality and availability of testing itself. Usually testing equipment is the same as that used in the manufacturing line for mass production and it is difficult to have its availability when reliability control is necessary. Another difficulty is having possibility of manufacturing datalogging for parametric drift due to the capability reasons above explained and for repeatability problems of the testing system itself.

Manufacturing reliability control should move in the future toward process parameters in-line control, after an in-depth investigation of the correlation between reliability and process parameters. The measurement, for instance, of the in-line shear resistance could be equivalent to perform a certain number of temperature cycles, after that the correlation has been investigated in depth. At present, the monitoring of failure mechanisms that could affect the diffusion process is in some cases approached by the Wafer Level Reliability Control [5] methodology. It is based on very fast tests focused on specific problems such as metal integrity, oxide integrity and hot carriers phenomena that are evaluated on dedicated elementary structures.

The reliability tests necessary to perform a good reliability monitor have to be characterised by these following main features:

- To be able to focus on process deviations;
- To be easily implemented;
- To be done at least on weekly basis;
- Not to be expensive;
- To be shared with customers in terms of effectiveness.

After all these considerations, we think that a good reliability test matrix could be:

- PP: 2 or 3 atm.; 96 hrs
- TC: −65 °C (15′)/+150 °C (15′); 100 cy
- HTRB: $T_j = 150\,°C$; 96 hrs
- OLT: $T_j = 150\,°C$; 48 hrs

PP and HTRB tests can highlight any problems related to the moulding compound process and generic ionic contamination. OLT and TC can detect bonding and die attach problems.

A second type of manufacturing control is based on the same long term tests matrix performed for IC qualification and it is typically carried out on quarterly basis. This kind of tests supplies generic reliability database organised according to package lines or to silicon technological families. Today these databases are very important to give to the customer a good confidence level of supplier process reliability performances and to allow new product qualification on the ground of generic data. The major Smart Power ICs users are moving in this direction that offers two big advantages:

- To reduce qualification cycle time;
- To reduce qualification cost.

The use of generic data for qualification is sometimes not so easy in the Smart Power ICs field but, as discussed in par. 14.5.2, new techniques are being developed so as to allow qualification on generic data also in this case.

14.4 Main Failure Mechanisms Review

Many different physical phenomena can cause the failure of integrated circuits during their operating life. This paragraph is dedicated to a review of the most important ones.

These failure mechanisms can lead to a degradation of the electrical performances, parametric failure, or to a complete lack of functionality, catastrophic failure, such as in the case of metal stripe interruption.

Failure mechanisms can be classified as intrinsic or extrinsic and these differences involve different reliability investigations.

Intrinsic Failure Mechanisms

Intrinsic failure mechanisms are induced by the functioning of the elementary structures itself, such as hot carriers for CMOS (parametric) or electromigration phenomena in metal stripes (catastrophic failure).

As the word Intrinsic implies, practically they cannot be avoided; it is only possible to reduce their impact defining conservative operating conditions. It is, in fact, mandatory to define design rules to minimise their impact during the useful life of the IC. In the particular case of wear-out phenomena the constraint placed on as integrated circuit design rules is the only way to guarantee IC reliability in the field.

The definition of design rules, to minimise the impact of intrinsic failure mechanisms, is the result of the failure mechanisms characterisation. It is usually done performing experiments related to empirical models that allow the estimation of the time to failure for the mechanism under investigation. These experiments are carried out on elementary structures that are usually submitted to DC stress. It has to be underlined that the acceptance criteria defined many years ago by the reliability world-wide community are usually too conservative and not compatible with the new advanced technologies.

New approaches for intrinsic failure mechanisms characterisation and for a more correct evaluation of their impact on device reliability are today requested. To answer this question three different complementary approaches can be identified:

- Reliability simulation tools [6, 7];
- AC stress on elementary structures [8];
- An innovative approach to the evaluation of reliability data from stress tests performed on ICs during qualification phase. The stress condition applied to elementary structures during, for example, OLT ought to be identified by circuit simulation and in the same way the IC electrical parameters potentially affected by elementary structure degradation ought to be singled out.

Extrinsic Failure Mechanisms

These are the ones induced by defects or more generally, by weakness due to problems of compatibility among different factors such as IC design and manufacturing.

Extrinsic failure mechanisms have to be investigated using stress tests, that can be performed both at device and elementary structures levels. The aim of these tests is just to examine weakness in the design, both of device and technology, related to manufacturing process. Examples of these kinds of phenomena can be given by the problems induced by mobile ions (Na^+, K^+ mainly) with reference to the physiological content of these species [9] in dielectric layers.

Also in this case some design rules are available which do not give a guarantee against wear-out problems but only define the safe operating region. For these phenomena, just because they are not intrinsic, extrapolation models do not exist, specific studies have to be performed case by case.

Having in mind this general classification, the main failure mechanisms affecting Smart Power ICs will now be analysed with reference to the stress factors activating them:

- Electrical (voltage and/or current);
- Thermomechanical;
- Chemical-physical.

14.4.1
Failure Mechanisms Induced by Electrical Stress

The failure mechanisms induced by electrical stress here following are discussed separating those affecting active structures, induced by parasitic elements and of oxide ageing from those causing the degradation of interconnections.

14.4.1.1
Electrical Stress Affecting: Active Structures, Parasitic Elements and Oxide Ageing

Dealing with this class of stresses we mainly subdivided them as: Surface effects (those causing degradation close to the Si/SiO_2 interface) and Bulk effects (those that take place in the bulk of the silicon far from that interface). Electrical stress causing the oxide ageing problem is also separately discussed.

Surface Effects

Surface effects are all those phenomena affecting reliability because they cause a change in the electrical features (electrical field distribution, charge carriers mobility) of the silicon at the Si/SiO_2 interface. Different phenomena can be classified as surface effects, they range from carrier injection (including hot carriers), to dielectric/resin polarisation, to mobile ions and electrical conduction of dielectric surface.

Smart Power IC applications are characterised by high voltage, up to hundreds of Volts, and high chip temperature due to power dissipation or by the ambient itself (such as in the case of the automotive field). These two stress factors enhance in particular those phenomena related to mobile ions. More emphasis will be given to these phenomena while as concerns hot carriers, for example, because the high temperature condition is a deceleration factor, they will be shortly reviewed.

As regards dielectric mobile ions, first of all we have to specify that we are mainly considering Na^+, K^+ positive ions for which there are experimental evidences [9] of their presence in the dielectric. Their presence, even if it can be heavily reduced by cleaning and by gettering techniques, cannot be completely avoided and furthermore, the scaling down trend of the technologies, causing the increasing of the electrical field, makes this kind of phenomena more and more critical. In the following, we will focus our attention mainly on the reliability effects induced by mobile ions. We also want to remind that different techniques, using dedicated structures, for quantitative measurement of the dielectric mobile ions content, such as TVS (Triangular Voltage Sweep), have been developed (see [9] for a complete discussion).

Referring to the effects induced by mobile ions and also by the conduction of dielectric surface we can consider two different cases. A first one, when the change in the distribution of the mobile ions externally induced influences the functioning of the elementary structures. A second case, when mobile ions active electrical interactions among different ICs elementary structures inducing parasitic elements. In both cases the stress factors are the high temperature that increases the ion mobility and the presence of electrical field that causes their motion.

In the first case, the most critical elementary structures are those so called lateral, such as: lateral PNP, lateral DMOS and in particular those high voltage structures working on the basis of RESURF effect, see Chap. 2.

In general, with reference to MOS structures, today's cleaning and gettering techniques have eliminated the problem of gate oxide contamination. The shift of threshold voltage, V_{th}, in the case of p-type substrate is given by the formula [10]:

$$V_{th} = V_{FB} + 2|\Phi_P| + \frac{1}{C_{OX}} \sqrt{2\varepsilon_s q N_a (2|\Phi_P|)} \tag{14.1}$$

where:

V_{FB} is the flat-band voltage;
Φ_P is the substrate Fermi potential;
C_{OX} is the gate oxide capacitance for unit area;
ε_s is the silicon permittivity;
q is the elctron charge;
N_a is the acceptor density.

Flat-band voltage can be written as:

$$V_{FB} = \Phi_{MS} - \frac{Q_f}{C_{ox}} - \frac{1}{C_{ox}} \int_{\theta}^{x_{ox}} \frac{x}{x_{ox}} \rho(x) dx \qquad (14.2)$$

where:
Φ_{MS} is the work function voltage;
ρ is the oxide charge density;
Q_f is the fixed interface charge density;
x_{ox} is the oxide thickness.

By these expressions it is quite evident that a change in the oxide charge density causes a shift of the flat-band voltage and hence of the threshold voltage. The same analytical expressions are also valid for threshold voltage of parasitic MOS elements that, as also discussed later on, can be switched on if a change in the charge distribution of its relevant gate oxide is caused by mobile ions.

Focusing now our attention on lateral structure, on principle, we can say that a change in the mobile ions distribution can give rise, for example in the case of increasing the ions concentration, to inversion phenomena in p-type silicon and accumulation in n-type silicon layer just below the Si/SiO_2 interface. Depending on where this enhancement of concentration takes place, the degradation of breakdown voltage, an increasing of leakage current or a degradation of current capability (R_{on}) can be caused. In general, these kinds of structures are critical because their functioning is strongly dependent on the electrical characteristics of the silicon just below the Si/SiO_2 interface and the biasing applied induces possible lateral ions motion.

These kinds of effects typically induce extrinsic failure mechanisms. a general model able to predict the degradation behaviours does not exist and it is mandatory to investigate these phenomena by stress test at high temperature and high voltage (like HTRB). The correct interpretation of the drift, if any, is demanded for an in-depth analysis of the physical functioning of the structure and of the effects of the applied stresses. The model has to be defined case by case usually performing different experiments at different temperature and voltage stress. In these evaluation experiments, a fundamental step is the baking at high temperature of failed parts. When the degradation is caused by mobile ions the recovery can be obtained usually after a few hours at 150 °C, because the high temperature causes, by diffusion, the restoring of the original distribution of mobile ions. Good examples of these kinds of problems are described in [11, 12].

The second case concerning mobile ions and dielectric surface conduction has to be referred directly to ICs because it involves the interaction between active structures. This interaction is due to parasitic elements that can modify the device functioning. In this case for parasitics we refer to MOS elements that can be activated due to phenomena leading to unexpected inversion. Practically, a channel can appear because of the change of mobile ions distribution, as above explained. The specific ICs stress test that is intended to evaluate this problem is

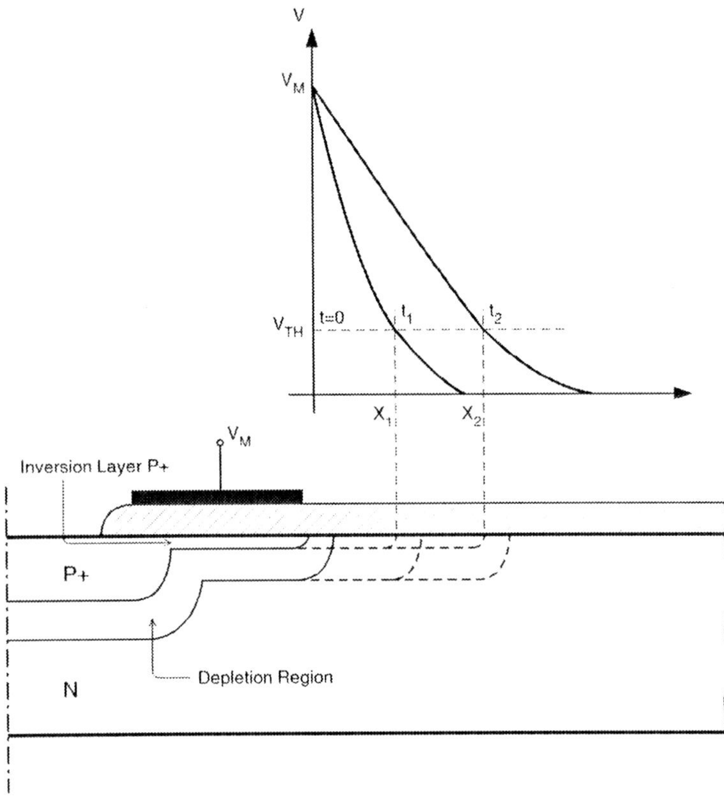

Fig. 14.3 Biasing propagation due to surface conduction of dielectrics

the HTRB test, in which the electrical field and the temperature are maximised. Such as for elementary structures, the recovery of this degradation by high temperature storage of failed parts is a first experimental step for their identification. Different critical situations can be identified. For example, we can consider the case in which two p-doped regions are separated by an n-doped region above which a metal is running. This is potentially a P-ch. MOS structure that could be switched on if the bias on the metal layer is greater than its threshold voltage. This dangerous situation could take place due to a change of the original value of the threshold voltage caused by a re-distribution of the mobile ions.

The same kind of phenomena could also be activated even in absence of a biased metal layer running directly over the n-doped region [11]. A negative bias could be induced over this region by propagation of bias due to conduction phenomena of oxide surface. In this case, the oxide capacitance and the surface film resistance acts as an RC circuit, the behaviour of which can be described by the following expression (see also Fig. 14.3):

$$V = V_M \operatorname{erfc}\left(\frac{x}{2\sqrt{t/RC}}\right) \tag{14.3}$$

where:

V_M is the metal potential;
R is the sheet resistance of oxide surface;
C is the oxide capacitance;
x is the distance from the metal from which the bias propagates.

Another phenomenon could give rise to the same effects, i.e. the propagation of unwanted bias; in fact, it could also be induced by dielectric (including resin of package) polarisation. The high voltage bias, at which the Smart Power ICs can work, makes these kinds of effects very critical. To avoid them different layout design rules are defined. Their experimental verification by HTRB stress tests on device is mandatory for a full qualification of ICs concerning the compatibility between design rules and process manufacturing.

A different important cause of change in the electrical characteristics of the silicon close to the Si/SiO_2 interface can also be due to injection phenomena (Fig. 14.4). In this case, the degradation cannot be recovered after only a few hours of high temperature (150 °C) storage.

Charge injection can cause mainly two different effects: degradation of charge carrier mobility and electrostatic effects causing a change in the

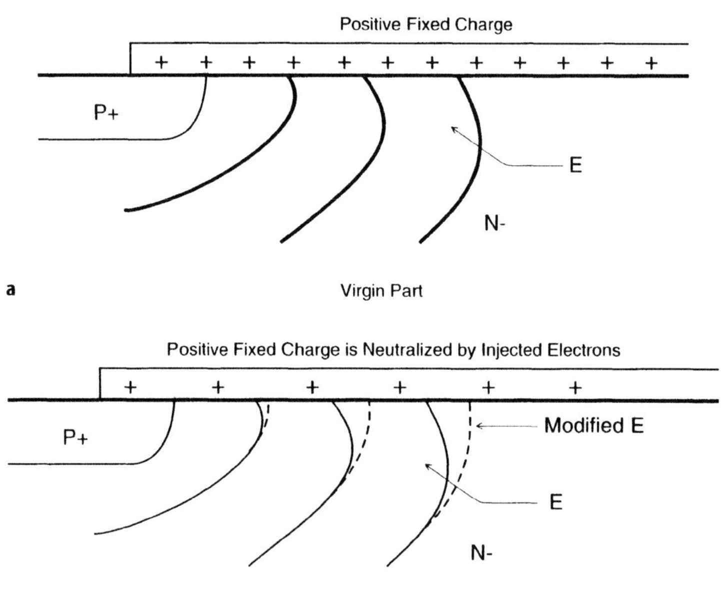

Fig. 14.4 Changing in the field distribution due to injected negative charge

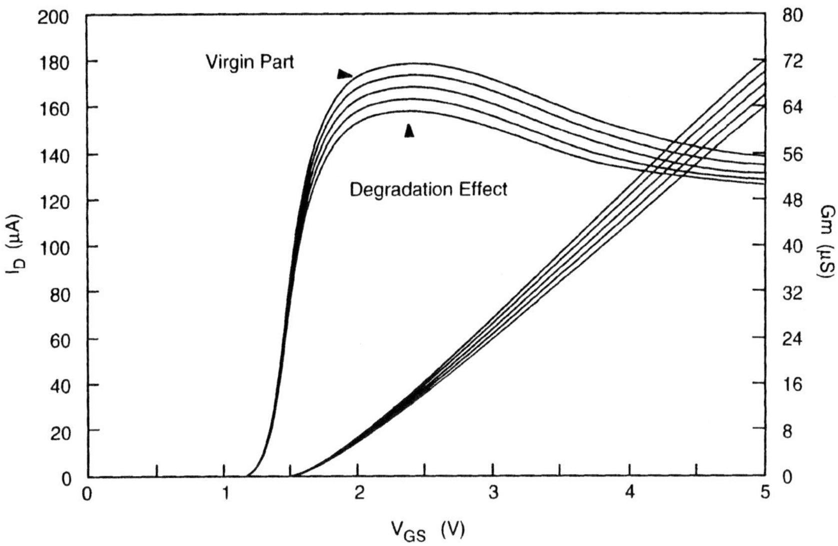

Fig. 14.5 Transconductance degradation due to hot carriers

electrical field distribution. For example, we can refer to the well-known junction breakdown walk-out phenomena just caused by a sort of electrostatic field induced by the injected carriers over the space charge region of the junction.

A particular case of charge injection is represented by the well-known phenomena of hot carriers affecting MOS, mainly N-ch, structure. These carriers are called hot because their energy is higher than that corresponding to thermal equilibrium at room temperature. They acquire this positive delta of energy by the electrical field in which they are moving. The de-acceleration factor due to the increase of the temperature is hence understandable, which causes a decrease of the carriers mean free path in the silicon lattice with a consequent decrease of their positive energy delta. A description of this phenomenon and of the related empirical model is given here following.

The electrical parameter more sensitive to the degradation induced by hot carriers phenomena in N-ch. MOS transistor is usually the transconductance in the linear region, but other parameters such as the current in the linear region of functioning and the threshold voltage can be also affected. Figure 14.5 shows the transconductance degradation for a N-ch. MOS structure of a BCD (Bipolar CMOS DMOS) process.

Hot carriers phenomena are a typical case of intrinsic failure mechanism causing a parametric degradation. It is one of the most studied failure mechanisms, many papers have been published on this subject. It is exhaustively described in the most important original ones [13,14]. In particular following the

E. Takeda, N. Suzuky model [13] we have that the transconductance degradation can be modelled as:

$$\Delta G_m / G_{mo} = A \cdot t^n \tag{14.4}$$

where:
t is the stress time
A is practically the amplitude of the degradation

Factor A has a strong dependence on drain bias, V_{DS}, and a fairly negligible dependence on the gate bias; it can be expressed by the following formula:

$$A \propto (\alpha / V_{DS}) \tag{14.5}$$

Using the two previous experimental relationships it is possible to define the time to failure, τ, as a function of the drain source bias for a certain amount of transconductance degradation (usually 10 % decrease of its initial value).

$$\tau \propto e^{\frac{b}{V_{DS}}} \tag{14.6}$$

where: $b = \alpha / n$

On the basis of this expression it is possible, knowing the relevant parameters, to extrapolate the use condition data from accelerated stress test and hence to define an adequate design rule. The reliability characterisation of this phenomena, adopting Takeda's model, can be done by fitting methodology on the basis of at least three experimental points. The reliability characterisation with this aim is mainly based on two steps.

In the first step at least three different stress tests, at three different drain source biasing, are performed collecting degradation data. These stress tests are performed in the worst stress conditions, i.e. the condition maximising the amount of degradation. The worst stress condition in terms of temperature, considering what has been mentioned above, is the generally assumed to be room temperature. As concerns bias it is usually experimentally verified to be $V_{GS} \cong V_{DS}/2$. The correct ratio has to be stated experimentally: according to the maximum peak of substrate current or directly performing stress tests. In these preliminary stress tests V_{DS} is fixed while V_{GS} varies; the analysis of the worst degradation after the same stress duration gives the worst biasing condition. The results from the three experiments, performed in the defined worst stress condition, are hence three couples of values, ($\tau^{(n)}$, $V_{DS}{}^{(n)}$) with $n = 1, 2, 3$, that represent the time for the 10 % of degradation at $V_{DS}{}^{(n)}$. In Fig. 14.6 a graphical representation of this first experimental step result is given.

The second step is the extrapolation to the use condition of the time to failure: the maximum operating voltage V_{DSmax} is defined. This value represents the design rule as concerns hot carriers phenomena on N-ch. MOS. For drain source biasing less or equal to V_{DSmax} it is guaranteed that the degradation of transconductance will be less or equal to 10 % of its initial value in the useful life. As useful life it was usually defined 10 years; as mentioned above, considering that it is

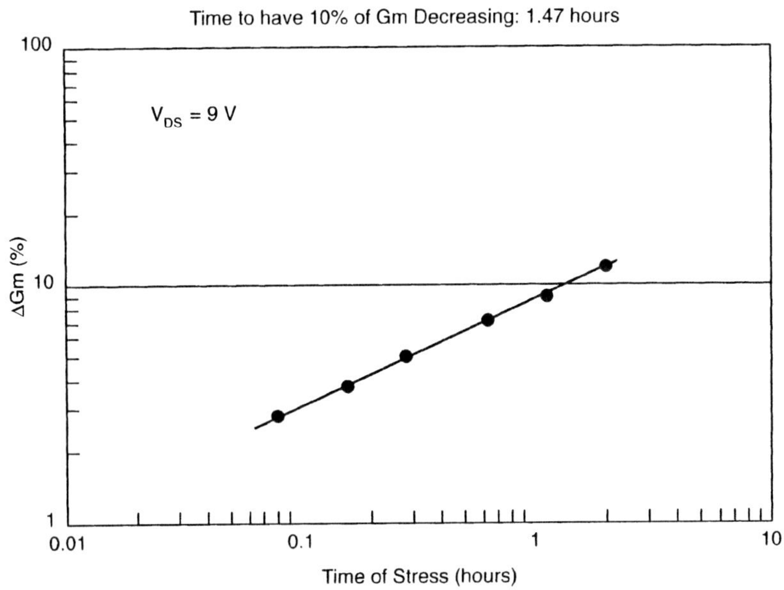

Fig. 14.6 Hot carriers degradation trend in an accelerated stress test

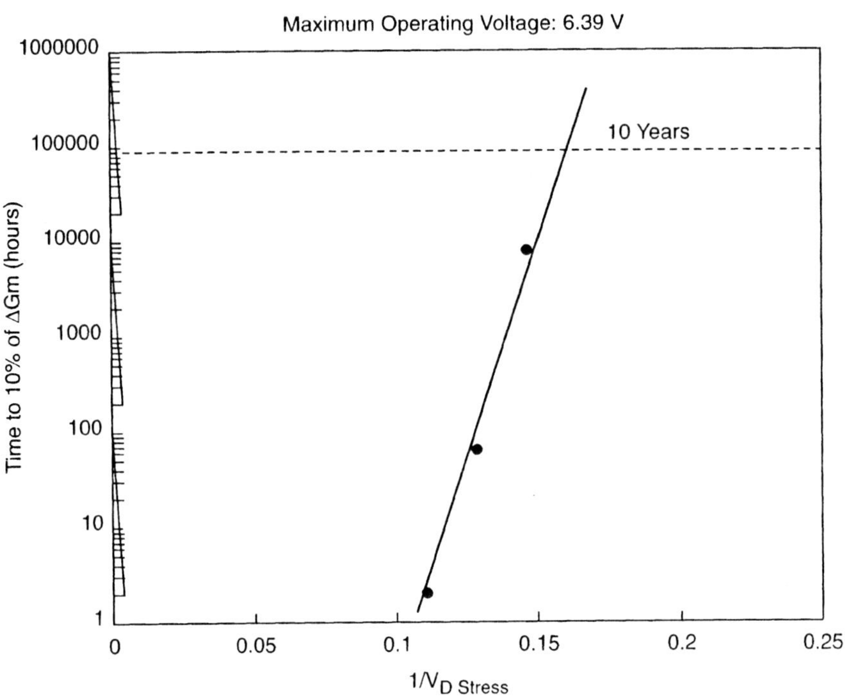

Fig. 14.7 Hot carriers maximum operating voltage extrapolation

referred to DC stress in the worst bias case this is a very conservative rule. In Fig. 14.7 this extrapolation is represented.

Oxide Ageing

The effects of oxide ageing is a typical intrinsic failure mechanism causing catastrophic failure: oxide breakdown. As for hot carriers also in this case empirical model are available and the reliability characterisation of this phenomenon is hence experimentally performed with reference to them. Two main models have been proposed usually called, with reference to electrical field dependence, E [15] and 1/E [16]. The E model will be shortly described. For this model we have that the Acceleration Factor (AF) due to electrical field stress is empirically defined as:

$$AF_{EF} = e^{\left(\frac{E_O - E_S}{E_{EF}}\right)} \qquad (14.7)$$

where:
E_O is the electrical field in the use condition;
E_S is the electrical field in the stress condition;
E_{EF} to be experimentally determined.

For this failure mechanism an acceleration effect induced by temperature can be considered according to the usual Arrhenius law:

$$AF_T = e^{\left[\frac{Ea}{k}\left(\frac{1}{T_S} - \frac{1}{T_o}\right)\right]} \qquad (14.8)$$

where:
T_o is the temperature in the use condition;
T_S is the temperature in the stress condition;
Ea activation energy to be experimentally determined;
k Boltzmann constant.

Also in this case the characterisation is performed in two main steps. In the first one at least three different stress tests are performed at three different electrical fields with the aim to collect three different failure distributions versus stress time. The stress conditions are not a worst case of possible operating condition (such as for hot carriers) because to have failure distribution in a reasonable time the electrical stresses have usually to be very close to the minimum value of the oxide breakdown distribution. These minimum values are typically far from the operating conditions. The failure distributions collected in this way are represented in a Weibull chart.

Elaborating the data represented in the Weibull chart it is possible to determine E_{EF} from Median Time To Failure values and, from the shape parameter of the Weibull distribution, to extrapolate the experimental data to the use condition for any different percentage of failure. In the telecom field, for example, a percentage of failure equal or less than 100 PPM has to be guaranteed after 25 years in operating conditions. It is quite evident that this low level of failure percentage has to be investigated only by extrapolation, otherwise, by experiment, a sample size of at least 10000 parts ought to be necessary for each stress condition.

For the two analysed intrinsic failure mechanisms, hot carriers and oxide ageing, we can now underline an important difference related to the fact that the first one causes parametric degradation whereas the second one catastrophic failure. For intrinsic parametric failure mechanisms practically all the parts degrade in the same way during their use, whereas, in the case of intrinsic catastrophic failure, the evaluation of the percentage of failure vs time becomes fundamental. This is done by representing the failure distribution, collected during stress tests, in adequate statistical models such as Weibull for oxide or Log-Normal for electromigration (see [17]).

Bulk Effects

With this term we refer to various phenomena which take place in the silicon bulk and cause abnormal current consumption. They can be inherent to a single active component itself or involve parasitic interaction between different components. Depending on the level of energy involved and on the degree of localisation of the phenomenon, it can either simply cause reversible anomalous device operation or induce permanent catastrophic degradation in the silicon. The sensitivity to these phenomena is a feature related to the technology and the design of the IC and bulk phenomena generally do not evolve in time as for the surface effects previously discussed. However, a high sensitivity to these phenomena can result in a poor reliability as the device cannot be able to withstand even small electrical overstresses or temperature rises. For these reasons these kinds of problems are often detected during operating life tests in IC qualification exercise.

A systematic analysis of the bulk phenomena that can take place in the advanced mixed technologies used for Smart Power ICs is very complex and beyond the purpose of this paragraph. Among the effects affecting a single component we just want to recall those related to the power transistors. Bipolar Junction Transistors (BJT) are known to be sensitive to second breakdown phenomena; this instability is basically caused by the negative temperature coefficient of base-emitter voltage that decreases roughly at a rate of 2 mV/°C causing an increase of collector current [18]. This leads to a formation of localised regions in the power BJT with current density well over the average value, triggering in this way a positive feedback between temperature and current increase with thermal runaway and final melting of aluminium metallization. To avoid the phenomenon the power capability has to be restricted and a suitable Safe Operating Area (SOA) is defined. Improvements of power structure and layout have been implemented [18] to reduce thermal-induced second breakdown.

The power MOS transistors are less sensitive to the thermal instability above described. In fact the positive temperature coefficient of drain-source on-resistance can compensate the negative one of the threshold voltage. However, second breakdown failures can still arise because the power DMOS has in its structure an intrinsic parasitic bipolar (T1 in Fig. 14.8) that can be activated in dynamic condition during the transients [19], for instance, during the commutation of inductive loads.

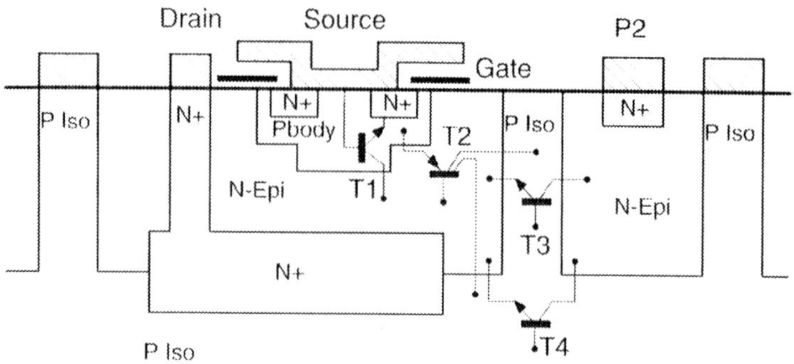

Fig. 14.8 Parasitic bipolar structures present in an integrated vertical DMOS

Regarding the parasitic effects among different chip components, we recall the well-known phenomena of latch up in CMOS (see for instance [20]) and the parasitic interaction between different epitaxial pockets, important for the mixed technologies used in Smart Power ICs. In order to understand the last effect we will consider, for instance, the situation in Fig. 14.8; if, for some reason (e.g. during recirculation phase), the drain of the DMOS is brought negative, the n-epi works as emitter of the parasitic bipolars T3 and T4 injecting stray currents in the substrate which can be collected by another epi-pocket (P2 in the figure) biased at high positive voltage. These phenomena can cause abnormal supply current increase or give rise to spurious interaction between power and piloting sections of the IC. To avoid these effects design guidelines and layout skills have been developed by Smart Power ICs designers.

14.4.1.2
Electrical Stress (High Current) Inducing Degradation on Interconnection

The high current capability required in many Smart Power ICs applications and the consequent stresses induced on interconnection system, is an important aspect to be considered for ICs reliability and the prevention of the related degradation phenomena can have a strong impact on IC design.

Here following we will now focus our attention on the degradation of aluminium alloy die interconnection, analysing first of all electromigration phenomena and their basic theory, then discussing the complex stresses typical of Smart Power ICs due to the remarkable temperature gradients and excursions combined with high current conditions. At the end of the paragraph the degradation of wires will be shortly reviewed as well.

Electromigration of Die Interconnection Metallization

Electromigration is a mass transport process that occurs due to the high current density, in the range of 10^5 A/cm^2, present in the thin film of aluminium alloy used in ICs interconnection. Such large current densities are rarely encountered in bulk metal while they are possible in thin films because of the very effective thermal dissipation through the silicon substrate.

The electromigration phenomenon is a result of the interaction of the flowing electrons and the lattice ions of the metal film. At the above-mentioned current densities there is a meaningful momentum transfer between the electrons and the lattice ions, and the effect of this so called "electron wind" is a preferred diffusion of M^+ ions in the direction of electron flow. The force induced by the electron wind prevails on the force exerted by the electric field that tends to move the metal ions in the opposite direction. Considering both the two contributions, the force applied to a lattice ion [21] can be written as:

$$F = Z^* qE = (Z^*_{El.} + Z^*_{W.})\, qE \qquad (14.9)$$

where Z^* is the effective charge number.

The first term ($Z^*_{El.}$) is the contribution of the electrostatic field whereas the second term ($Z^*_{W.}$) takes into account the effects of the electron wind. As this last effect prevails for aluminium films, Z^* is negative. The previous equation can also be written in terms of current density as:

$$F = Z^* q\rho J \qquad (14.10)$$

where ρ is the resistivity of the material.

Due to the force F a flux of metal ions, J_M, takes place when a current is flowing in the film. Remembering the Einstein relation we can write for J_M:

$$J_M = (ND/kT)\cdot Z^* q\rho J = (ND_o/kT)\cdot Z^* q\rho J\cdot e^{-Ea/kT} \qquad (14.11)$$

where N is the density of mobile ions and $D = D_o e^{-Ea/kT}$ is the diffusion coefficient which depends on the lattice temperature according to Arrhenius law with activation energy Ea. The mass transport phenomena predicted by the Eq. 14.11 result in the generation of voids in the region near the current source compensated by accumulation in the other regions of the film. Furthermore the generation of voids can take place in points where, for any reason, the divergence of mass flux is negative. The growth of these voids and their coalescence eventually can result in metal interruption and hence give rise to an electrical failure.

Discussing more profoundly the onset of electromigration phenomena in thin aluminium films, first of all the polycrystalline structure of the film should be considered. The mobility of aluminium ions in fact depends on where the ion motion takes place: grain bulk, grain boundary, film surface, etc. Activation energies in the previous Eq. 14.11 are different for the different diffusion mechanism; according to [22], for pure aluminium we have:

- Grain bulk: Ea = 1.4 eV
- Grain boundary: Ea = 0.4 – 0.5 eV
- Grain boundary to bulk: Ea = 0.62 eV
- Defect to bulk: Ea > 0.62 eV
- Film surface: Ea = 0.28 eV

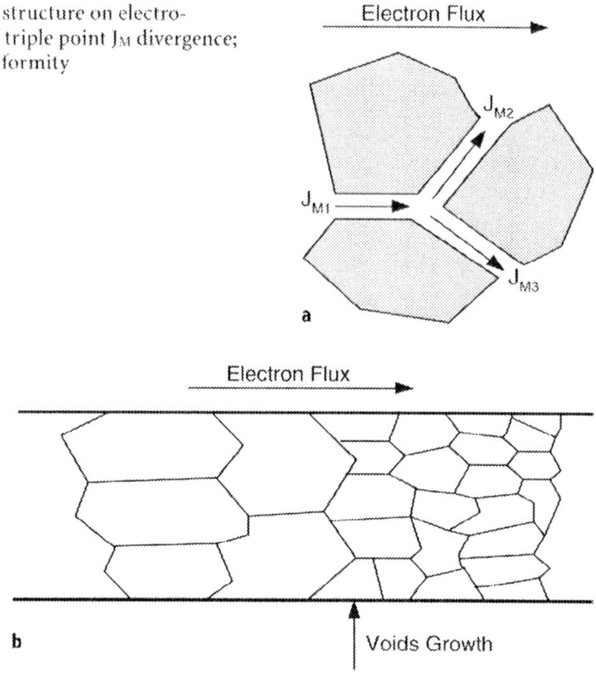

Fig. 14.9 Influence of film structure on electromigration: **a** explanation of triple point J_M divergence; **b** effects of structure disuniformity

In polycrystalline films, grain boundary provides a preferential path for atomic flux and at the grains triple points connection, atomic flux divergence intrinsically takes place. With reference to Fig. 14.9, if J_{M1} is lower than $J_{M2} + J_{M3}$ a void grows in the triple point.

The resistance to electromigration of thin film aluminium interconnection is related mainly to three technological features:

- Film structure;
- Type of Al alloy;
- Type of passivation.

As mentioned, electromigration in aluminium films takes place at grain boundary, hence large grains with homogeneous size strongly improve the film performance as concerns electromigation. On the other hand, non-uniform film structure is very dangerous, as it causes mass flux divergence strongly enhancing electromigration as illustrated in Fig. 14.9.

The presence of impurity in the film composition strongly influences electromigration phenomena. In fact, these impurities tend to segregate at the grain boundary affecting the grain boundary diffusion process. In particular, the addition of copper strongly improves electromigration performances. The addition of silicon is also common but mainly to improve contact reliability, as it prevents the dissolution into the metal of the contact silicon.

Also the passivation plays an important role in preventing electromigration; on one hand, it strongly reduces mass transport phenomena at the film interface; on the other, it acts as mechanical constraint confining the metal and causing a back-flow pressure in the region of material accumulation.

Time to failure for metal line electromigration depends on the stress conditions according to Black's empirical equation which in its general form states:

$$MTF = A \cdot J^{-N} \cdot e^{Ea/kT} \qquad (14.12)$$

where:

MTF is the Median Time to Failure;
T is the metal temperature;
J is the current density;
Ea is the activation energy;
k is the Boltzmann constant;
N is the current density exponent;
A is the constant which depends on technological and geometrical features of the stressed metal.

Black's equation is the basis for the characterisation of intrinsic electromigration phenomena through accelerated life tests. This methodology consists in stressing, under large current density and high temperature conditions, suitable metal elementary structures (generally straight stripe with Kelvin contacts); a set of these tests at different T and J is carried out and the failure distributions are determined. These distributions can be generally assumed of Log-Normal type and their interpolation allows to determine the time to failure to have a certain failure percentage. By the MTFs at different stress conditions it is possible to estimate the parameters Ea, N and A in Black's equation and to predict the time to failures at use conditions. An example of this kind of characterisation is shown in Fig. 14.10. For an Al/Si film the predicted life time for 1% of failures at $J = 2 \cdot 10^5\ A/cm^2$ and $T = 130\,°C$ is roughly 15 years in continuous operating conditions. As mentioned before, even better results are obtained by adding copper to the metal alloy.

Design rules adopted in ICs typically allow maximum current density around $2 \cdot 10^5\ A/cm^2$ in the metal lines. As shown by the previously mentioned example, this value is conservative enough to prevent on principle electromigration phenomena in the majority of ICs applications. In Smart Power ICs however, in addition to high current density, further stress factors contribute to metal degradation and have to be considered. The high power density locally dissipated inside the IC by the power components, bipolar and DMOS, causes high temperature gradient on the die. Temperature profile obtained by infrared emission mapping and thermal simulation results [23] show temperature differences up to 50–60 °C between power transistors and other die positions. Furthermore, in the power ICs, critical layout situations exist where sudden current density variations take place; for instance in the power transistors with interdigitated structure the finger termination can be critical, in fact in that point the metal collects all the finger current and joints to the main metal line. As shown in

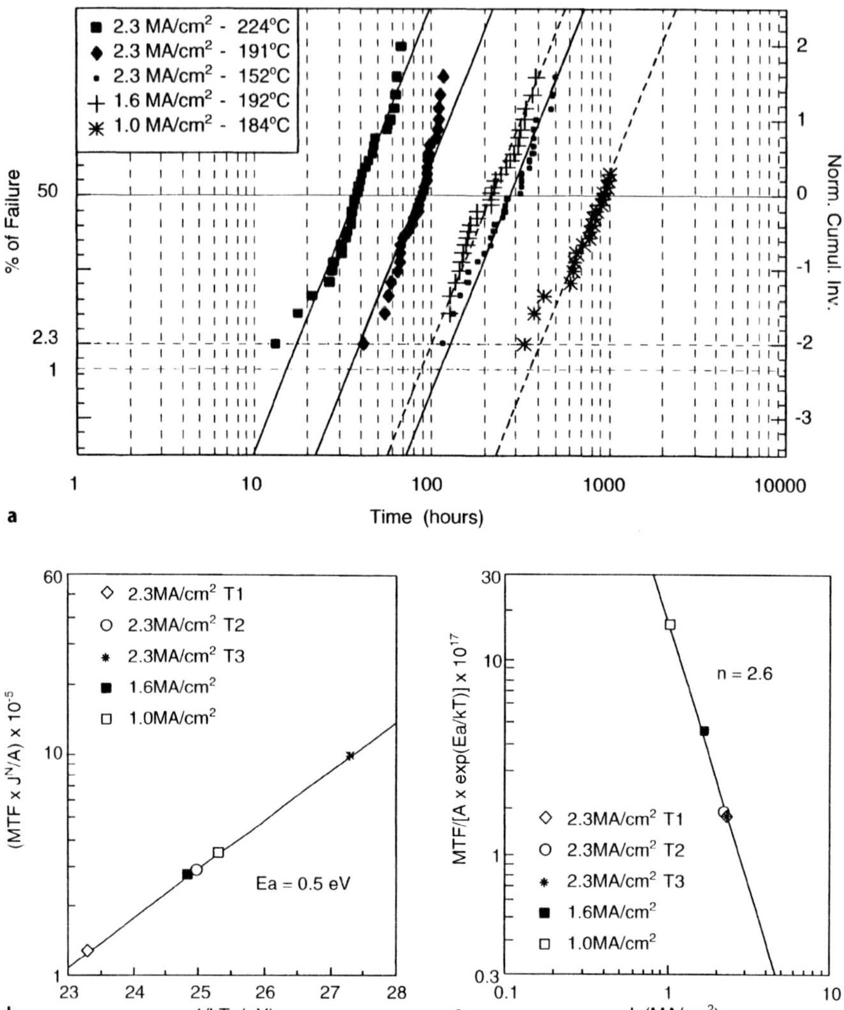

Fig. 14.10 Electromigration characterization in an Al/Si film: **a** Failure distributions for different J,T stress conditions; **b** Activation energy, Ea, estimation; **c** Current density exponent, n, estimation

Fig. 14.11, it is clear that both the current density gradient and the temperature gradient can cause a divergence of mass flux and hence give rise to electromigration voids.

The phenomena discussed above are not considered in the electromigration tests performed on elementary structures during metal film characterisation. Their combined effects can result in heavy metal degradation also in situations that, considering only the mean temperature and current density, are not critical

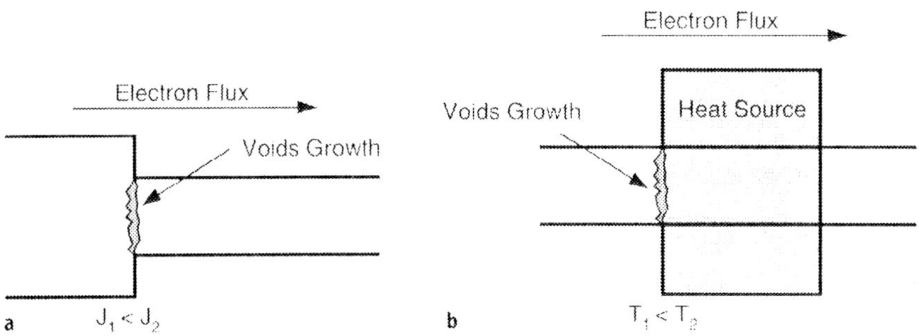

Fig. 14.11 Effects on electromigration of current density gradient **a** and temperature gradient **b**

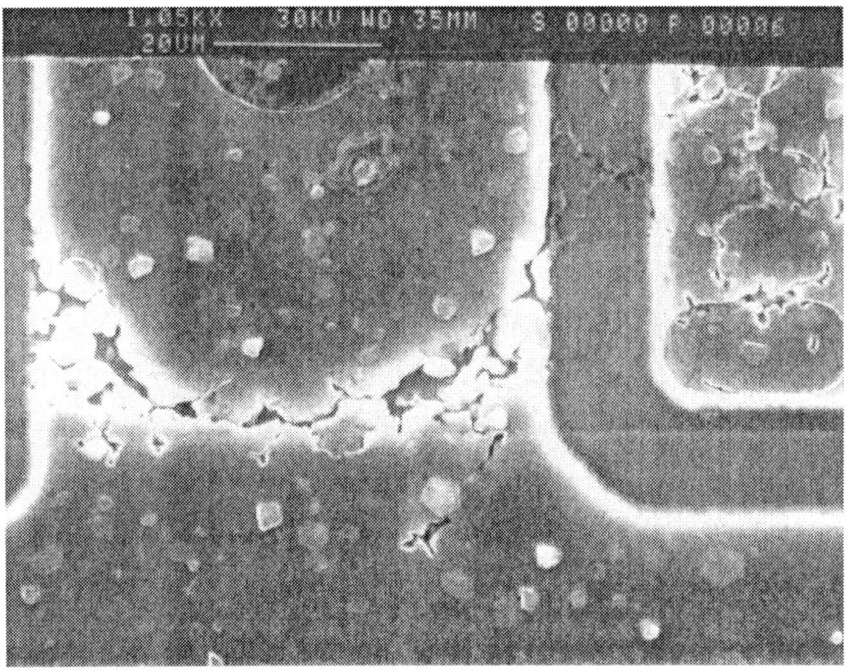

Fig. 14.12 Metal degradation in a bipolar device after high accelerated life test

for electromigration. These phenomena are specific of the particular circuit and application and can be effectively investigated only on the IC in operative working conditions. This is done during the qualification exercise by means of specific long term life tests carried out under high stress dynamic or static conditions in order to enhance the phenomena and to single out possible circuit weaknesses. An example of degradation is shown in Fig. 14.12 for a bipolar IC in which heavy damages were detected after dynamic life test in the metal connections of the NPN BJT output power transistors. The observed degradation can not be explained in terms of current density, which is in the range of 2–3 10^5 A/cm^2 during the test, but can be understood only considering the high gradient of temperature and current density present in the region where the failure occurred.

Another important stress factor to be considered for power ICs metal connections is the remarkable thermal excursion experienced during peaks of power which can take place during device dynamic working, for instance during the switching-off of an inductive load due to the re-circulation energy that has to be dissipated into the output power transistor. In extreme cases, peaks of temperature in the range of 50–80 °C above average junction temperature have been observed. During this temperature cycling, thermomechanical stresses arise for the different Coefficent of Thermal Expansion of aluminium (CTE = 23 PPM/°C) compared with those of the other layers which are lower: 2.5 PPM/°C for Si, 0.5 PPM/°C for SiO_2, 2.8 PPM/°C for Si_3N_4. These kinds of stresses play an important role in degradation of Smart Power ICs Al metal connections.

For the complex degradation phenomena just described, the current density is not always meaningful for the definition of the stress level. A stress parameter more correlated to the observed Al degradation seems to be the power density locally dissipated by the power structures [24]. On the ground of our experience for power density over 10 W/mm^2 noticeable metal degradations begin to appear.

High Current Induced Bond Wire Degradation

During Smart Power ICs operations the bond wires can be exposed to severe stresses due to the high current level. In these conditions, important degradation affecting both aluminium and gold wire can take place causing the increase of wire resistance and in some extreme cases leading to wire burn-out. In Fig. 14.13, for instance, the resistance trend is shown, highlighting an expressive increase, for a gold wire submitted to a conduction test.

Comparing the current induced degradation in the wire with the equivalent phenomena described in the previous paragraph affecting die metallization, it should be considered the worst thermal dissipation condition for the wire which can cause a remarkable temperature increase for Joule heating at current density relatively lower than those necessary to cause electromigration in the metal film. Due to the high temperature reached by the wire submitted to high current stress conditions, structural modifications affect the wire which can become brittle and break; an example of this degradation is shown in Fig. 14.14.

Structural modification and re-cristallization phenomena of the wire can be enhanced by pulse operation due to the remarkable temperature excursion experienced by the wire. The degradation under pulsed current was investigated in [25]. According to this author, on a Al/Si wire of a diameter of 25 μm and length of 3.5 mm a thermal time constant of about 15 ms was evaluated; For pulse width in this range a significant wire heating takes place and Al degradation is observed.

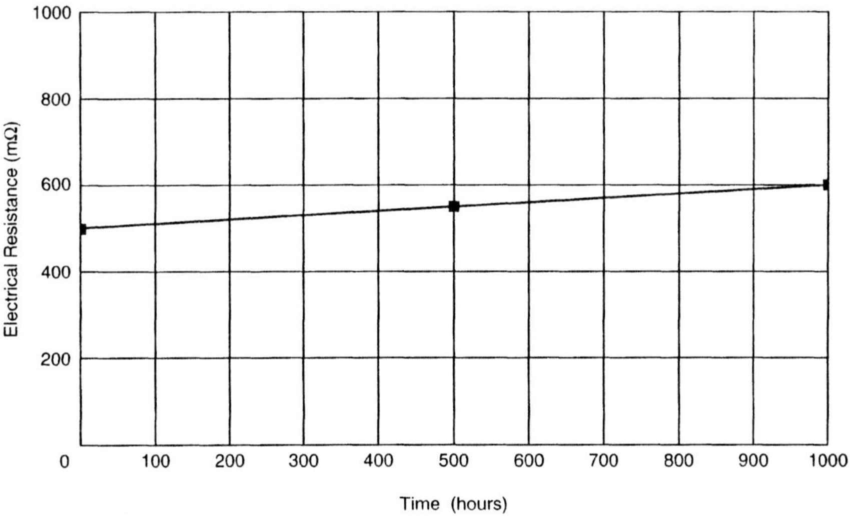

Fig. 14.13 Resistance increase of a gold wire (diameter 25 μm; length 1cm) submitted to conduction test

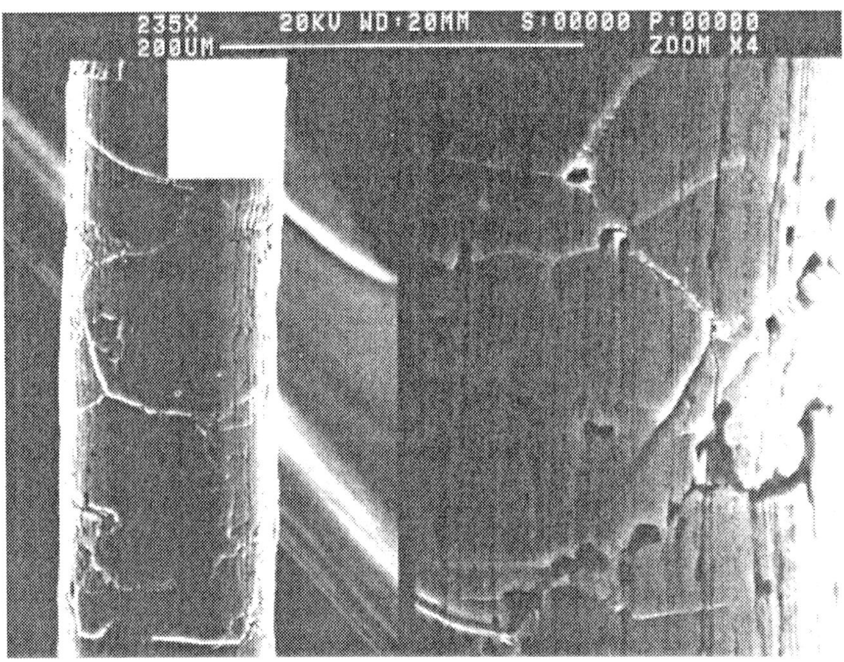

Fig. 14.14 Crystalline structures modification of an aluminium wire after conduction test

Plastic packages are less critical for these phenomena than hermetic due to the more effective heat transfer assured by the moulding resin. The wire length is very important and long wire can result very critical. Wire material has a heavy influence: a gold wire is able to withstand higher currents of an equivalent one made of aluminium.

14.4.2
Failure Mechanisms Induced by Thermomechanical Stress

As regards plastic assembly the mismatch in the Coefficent of Thermal Expansion (CTE) of the different die and package material can result in high thermomechanical stresses, especially when the IC experiences high temperature excursion as can be the case for Smart Power ICs.

For commonly used molding resin, CTE is in the range of 15–20 PPM/°C, for temperature below the so-called glass transition temperature, while above this temperature, which typically varies between 160–170 °C, CTE is higher and near to 70 PPM/°C (see Chap. 13 for more details). These values have to be compared with that of silicon which is remarkably lower (CTE = 2.5 PPM/°C). As a result of this mismatch, high stress levels are induced in the IC when it is cooled down after the moulding operation. For instance, Fig. 14.15 shows the map of the shear stress at die surface as measured by a suitable stress gauge chip based on the piezo-resistivity properties of diffused silicon resistors. This type of stress is very important for metal displacement and bonding degradation phenomena which can take place at the die corners where the shear stress is higher, in particular for large die size.

Another important thermomechanical mismatch to be considered is between the die and the metallic frame on which the die is mounted. In particular for power asymmetric packages, such as Multiwatt or similar, the frame is also the heat-sinker and is generally made of copper alloy to assure a good thermal conductance. The high CTE of this material, typically 16–17 PPM/°C compared

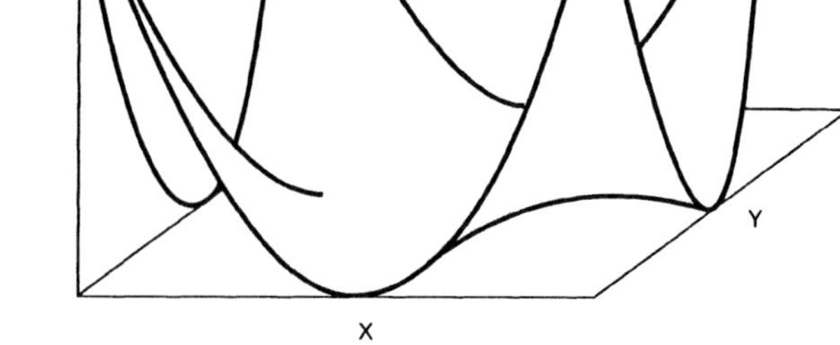

Fig. 14.15 Map of shear stress (τ_{xy}) in a dual in line package

with silicon causes a high stress at the die frame joint and in particular in the die-attach layer. The presence of the heat-sinker and the asymmetric structure of the package cause also important and complex stresses on die surface and on bonding wires. In order to reduce them the frame structure has to be designed to avoid, as far as possible, sliding between resin and frame.

A remarkable excursion can be experienced by the device both during customer manufacturing process and in the field operation. The first type of stress is important in particular for SMDs due to the temperature profile (see Fig. 14.2) during soldering on customer board. The high temperature peak reached in this phase can cause: delamination at die-resin and frame-resin interface; resin cracks for "Pop-Corn" effect if enough water is absorbed; heavy thermomechanical stresses on bonding wires and on die surface. Also in the field operation the ICs can be submitted to remarkable thermal excursion due to the environmental variation itself or due to self-heating of the device for power dissipation. Both these aspects are very important, in particular for Smart Power ICs in automotive application where the environment is very severe and remarkable stresses arise in key-on and key-off phases.

To reproduce these types of stresses during qualification exercise, the devices are submitted to temperature cycling and thermal shock tests (see par. 14.3.2.2). The acceleration factor (AF) vs. temperature excursion range (ΔT) was considered, for temperature cycling, by [26]:

$$AF = \left[\frac{(\Delta T)_2 - (\Delta To)_2}{(\Delta T)_1 - (\Delta To)_1}\right]^Q \tag{14.13}$$

where:

$(\Delta T)x$ is the total temperature excursion range;

$(\Delta T_o)x$ is temperature excursion in elastic range (not contributing to degradation);

Q is to be experimentally determined and depends on the activated failure mechanisms.

If (ΔT_o) is negligible compared with (ΔT) the previous equation becomes:

$$AF = \left[\frac{(\Delta T)_2}{(\Delta T)_1}\right]^Q \tag{14.14}$$

which is commonly used in the calculations.

More exhaustively we will consider the failure modes activated by thermomechanical stresses and focus our attention on:

- Effects on die interconnections and dielectrics;
- Degradation of wire bonding;
- Degradation of die-attach.

Effects on Die Interconnections and Dielectrics

As shown in Fig. 14.15, a remarkable shear stress is present at the die surface, in particular at the edges and at the corners. For large die size, during temperature

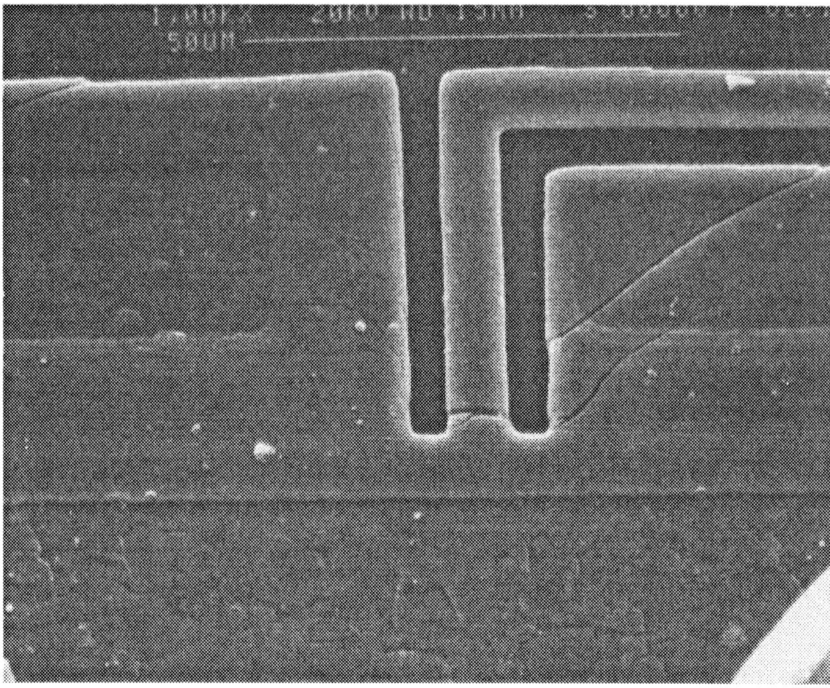

Fig. 14.16 Failure for metal interruption after temperature cycling test

cycling, this type of stress can induce passivation cracks and deformation of aluminium metallization (see for instance [26]) causing electrical failures, either opens or shorts, in critical layout situation. An example is shown in Fig. 14.16. Even if no electrical failures take place for metal displacement, the damages of the passivation can make the device more sensitive to other metal failure modes, like aluminium corrosion (see par. 14.4.3.2), introducing potential reliability risks.

The Smart Power more than the other ICs are critical for these degradations; this is due to the type of packages and to the use of thick metal films, often necessary to have enough current capability, which give rise to a rough die "topography" making the shear action of the resin more effective. On the ground of the failure analysis experience, suitable corrective actions, in terms of metal layout design rules and passivation improvement, have been introduced allowing a strong reduction of the impact of these degradations.

Another important failure mode affecting interconnection and enhanced by thermomechanical stresses is the short circuit between metals of different levels in the overlapping region. Cracks in the intermediate dielectric layers can occur during temperature cycling, especially in structural weakness points as on the metal steps; both the action of the resin and the internal stresses in the die due to the mismatch of the various materials (metal, dielectrics, silicon) can contribute to this phenomenon. Furthermore cracks in the intermediate dielectric layer can be caused by particle or microdamage in some process phase. In any cases, during temperature cycling, the aluminium, due to its high CTE, can fill these cracks causing an interlevel short circuit. The short circuit can also take place subsequently when an electrical stress is applied, due to isolation loss of the intermediate dielectric layer.

The incidence of this failure mode has been strongly reduced in Smart Power ICs by using planarized processes to improve the robustness of the intermediate layers and keeping a tight process control in diffusion and assembly lines to reduce the defectiveness induced by particle and bad handling.

Degration of Wire Bonding

Thermomechanical stresses can induce degradation of the bonding of the wire connecting the die with the lead of the package. The failures can take place both at the bonding on the die contact pad and in the welding between wire and lead. For the first type of failure, it is very important the shear stress induced by the package resin but also, in the case of gold wires, the stress inherent to the Au/Al joint itself, as discussed in par. 14.4.3.1. An examples of degradation of the wire to pad bonding is shown in Fig. 14.17.

Fig. 14.17 Detachment of gold bond ball by aluminium pad after temperature cycling

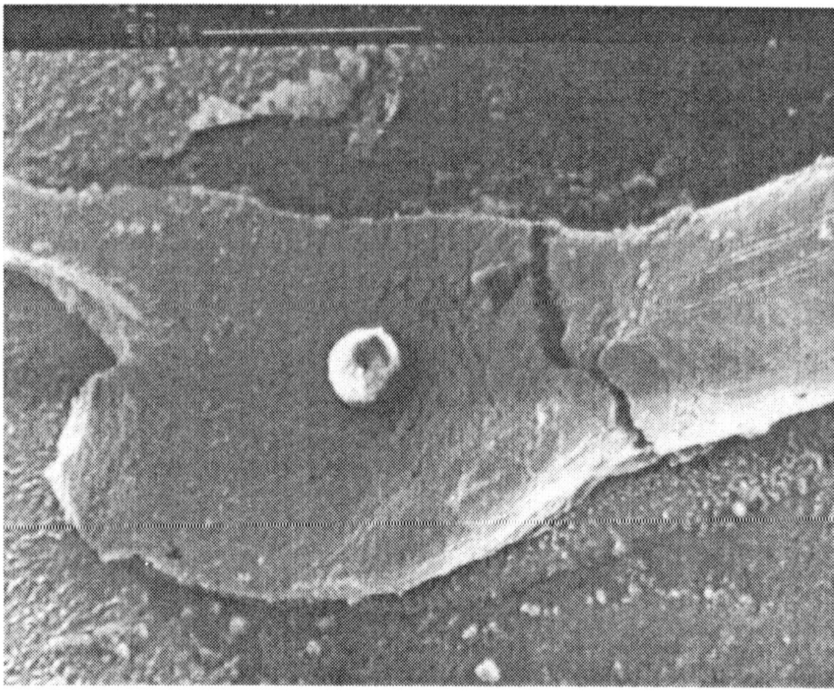

Fig. 14.18 Degradation of a gold wire welding on the frame after temperature cycling

An example of degradation of the wire at the lead bonding is shown in Fig. 14.18. In this case a fracture of the wire near the weld occurs because of the traction and bending action of the resin.

Sometimes during temperature cycling, the degradation of the wire itself has been observed; the wire structure can be modified and damaged by thermal fatigue with cracks that can appear in a similar way as discussed for current induced wire degradation (see par. 14.4.1.2).

Die-Attach Degradation

This degradation can affect ICs with die of large size in asymmetric packages like Multiwatt. As previously mentioned, the mismatch induced by the different CTE of silicon, copper of the frame and Pb/Sn die-attach alloy, besides the action of the resin, can cause remarkable stress at die-frame joint. This, in some cases, can even cause die cracking. More frequently fractures in die attach Pb/Sn layers can occur as shown in Fig. 14.19.

Die attach degradation has to be carefully considered for power devices as it can result in an unacceptable worsening of thermal dissipation and hence in a junction temperature incompatible with the IC functionality and reliability.

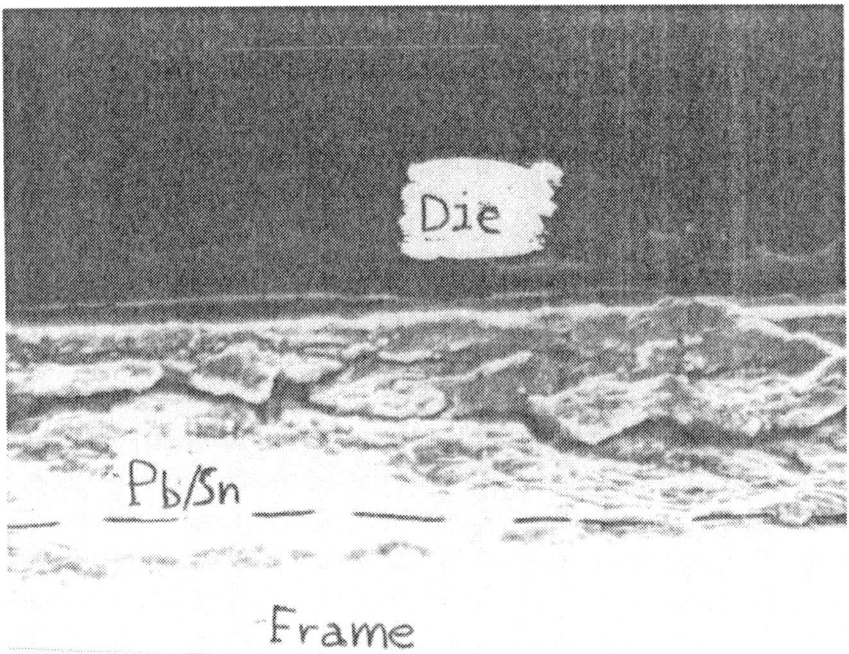

Fig. 14.19 Microcracks in the Pb/Sn die attach after thermomechanical stress

14.4.3
Failure Mechanisms Induced by Chemical-Physical Effects

In this section we will consider those phenomena of chemical-physical nature affecting some IC elements which can evolve causing the final failure of the device. In particular we will analyse effects related to gold-aluminium joint metallurgy and corrosion phenomena affecting aluminium interconnections. Both these two failure mechanisms can be very important for Smart Power ICs reliability.

14.4.3.1
Gold-Aluminium Joint Degradation

The degradation of the bond between gold and aluminium at the joint between wire and pad has been a major problem for quality and reliability of silicon semiconductor devices, and for this reason it has been the object of an intensive and continuous study.

Today the improvement of bonding technology and a systematic line monitoring have meaningfully reduced, in general, the incidence of this problem. However, in the case of Smart Power ICs Au/Al bond degradation is still to be considered an important failure mode. This is due to the high temperature at which these devices can operate, both for the high power level required and for the severity of the field environment typical of some Smart Power ICs applications. In automotive field, for instance, the request to guarantee the IC at junction temperature up to 170 °C also in plastic package is not unusual. In this case Au/Al bond degradation problems are a key reliability issue and a major constraint to reach this target.

As already mentioned, the mechanism of Au/Al intermetallic formation and evolution was studied in detail to understand the cause of the mechanical and electrical degradation which takes place in the bond. First of all, it should be considered that when gold and aluminium are soldered different intermetallic compounds ($AuAl_2$, AuAl, Au_2Al, Au_5Al_2, Au_4Al) can be formed as shown in the phase diagram in Fig. 14.20.

Practically, in the wire bonding, gold and aluminium are joint in solid-state conditions and the time is too short to reach the thermodynamic equilibrium. Due to the insufficient supply of one of the metals and to the different kinetics of nucleation and growth of the various intermetallic compounds, the amount of these phases can be very different and some of them can be missing. Obviously, Au-rich phases will be present at the gold side of joint while Al-rich phases will prevail at the other side. Philofsky [28] studying Au/Al intermetallic formation in the range of temperatures between 200–460 °C found that at the beginning

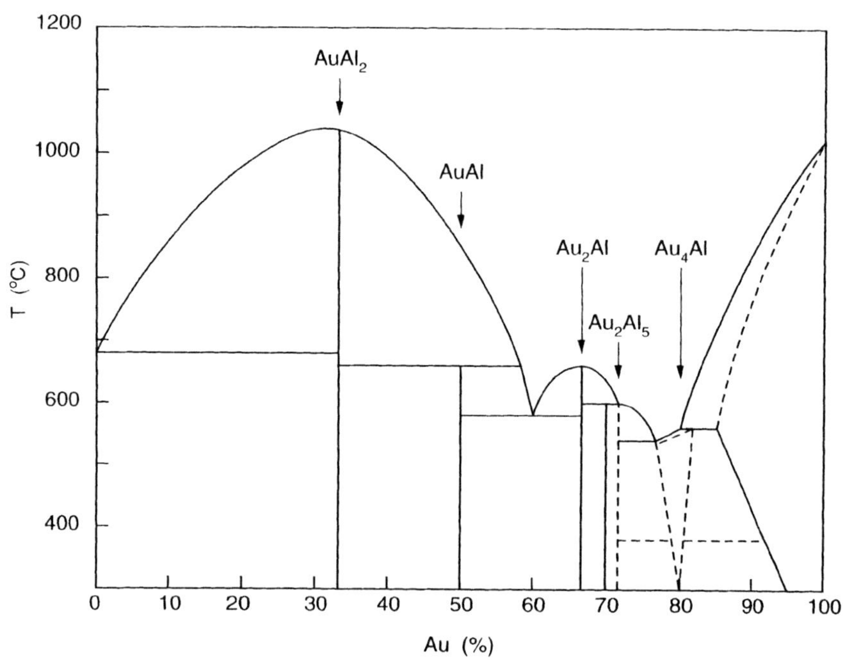

Fig. 14.20 Aluminium – gold phase diagram (after [27])

the predominant phase is Au_5Al_2 while the other phases appear later after annealing. The growth rate of intermetallic formation during the interdiffusion of metals follows a parabolic law:

$$x = (K \cdot t)^{1/2} \tag{14.15}$$

where:
x is the intermetallic layer thickness;
t is the time;
K is the rate constant.

K depends on the temperature following the Arrhenius law:

$$K = K_0 \cdot e^{-Ea/kT} \tag{14.16}$$

with Ea = 0.69 eV.

The above equation is valid under the assumption of a sufficient supply of Au and Al in the bonding area; in this case, the intermetallic growth is dominated by the phase Au_5Al_2. However, if there is a shortage of gold or aluminium the diffusion continues slowly and the intermetallic compounds already formed evolve toward those rich in the metal in excess, hence the growth of $AuAl_2$ or Au_4Al respectively prevails.

The intermetallic compounds exhibit poor electrical conductivity; their resistivity ranges from 7.9 $\mu\Omega \cdot cm$ for $AuAl_2$ to 37.5 $\mu\Omega \cdot cm$ for Au_4Al [29]; while

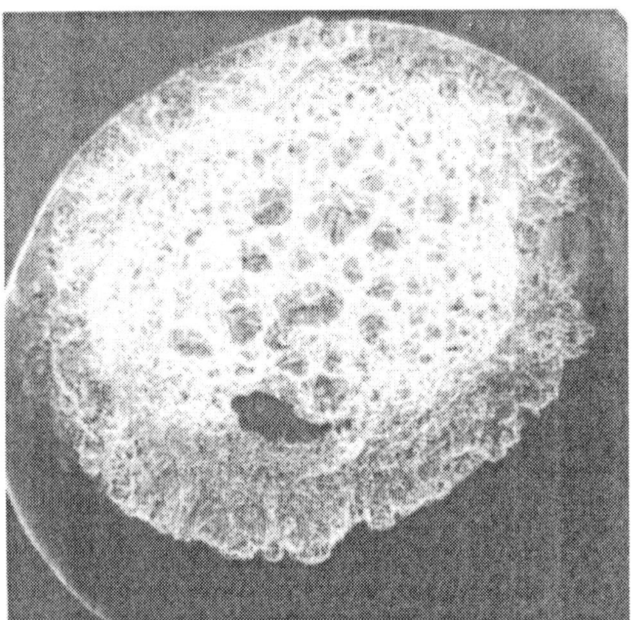

Fig. 14.21 Intermetallic evolution after storage test with evident Kirkendall voids

the resistivity of pure metals is 2.3 μΩ · cm for Au and 3.7 μΩ · cm for Al. From a mechanical point of view, the tensile strength of intermetallic compounds is higher than that of pure metals and hence the formation of intermetallic phases by itself does not necessarily have a detrimental effect on bonds. The bond strength degradation and the eventual failure is caused by the Kirkendall voiding phenomena: due to the different diffusion rate of the Au and Al through the intermetallic phases, vacancies are generated and their coalescence leads to Kirkendall voids formation, an example of which is shown in Fig. 14.21

Kirkendall voiding takes place at the Au-rich side of the joint, in particular in Au_5Al_2 compound, but occurs also in $AuAl_2$ phase. This phase has a purple colour and its presence, often observed in failed bond, gave rise to the term "purple plague" commonly used to indicate Au/Al joint degradation. Of course not the presence of purple intermetallic phase itself but the Kirkendall voids formation is the cause of the bond degradation, and failures can occur also in the Au-rich phase without the appearance of the purple coloration (see for instance [30]).

Important Kirkendall voiding phenomena can take place just during bonding operation, if excessive diffusion is allowed to occur, and bonds of poor quality can be the result. Afterwards Kirkendall voids can grow when the joint is exposed to high temperature storage, and the degradation affects both electrical and mechanical features of the bond causing an increase of electrical resistance and a loss of mechanical strength as shown in Fig. 14.22 [31].

Another aspect to consider is the different thermal expansion coefficients for gold, aluminium and the various intermetallic compounds, which cause thermo-

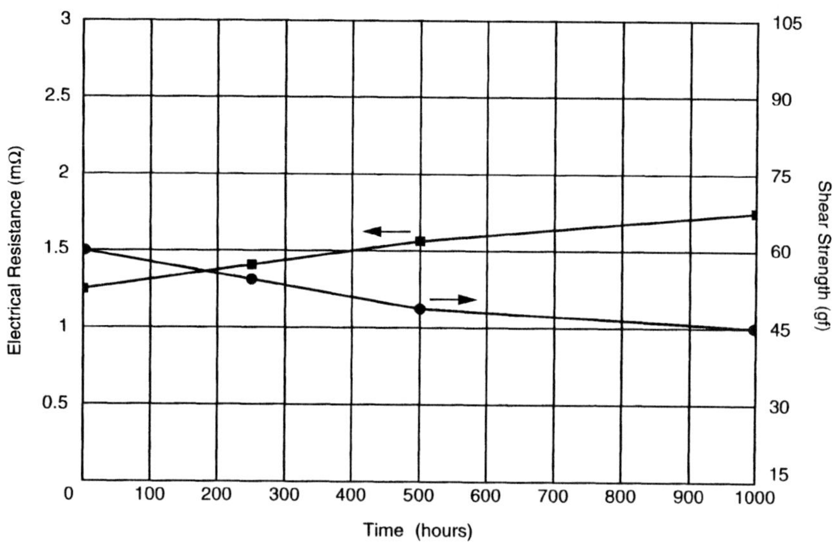

Fig. 14.22 Degradation of bond electrical resistance and bond shear strength in storage test (180 °C)

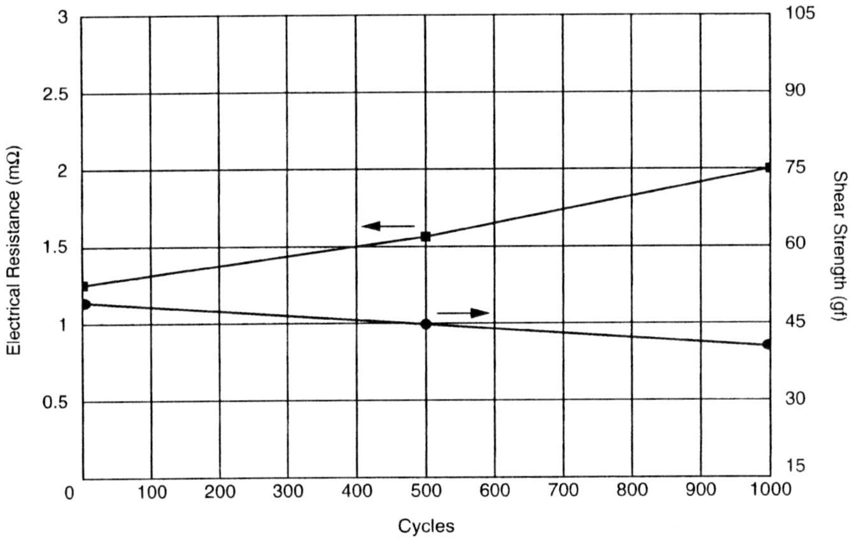

Fig. 14.23 Degradation of bond electrical resistance and bond shear strength during temperature cycling test (−65/+150 °C)

mechanical stresses at the interfaces when the joint is submitted to temperature excursion. By effect of these stresses, cracks can develop within the intermetallic phases strongly enhancing bond degradation. This circumstance is observed during temperature cycling and thermal shocks as shown in Fig. 14.23 [31].

It should be underlined that, if the bonding is correctly done, at practical operating temperature, Au/Al intermetallic degradation is not expected normally to take place during the useful life of ICs. However, different factors (as reviewed in [30]) can contribute to make the device more critical as concern this failure mode. We want to mention, just as an example, the presence of some flame retarding additives in the epoxy resins which have been found to give an evident detrimental effect on bond degradation. For these reasons, the reliability of Au/Al joint has to be carefully considered, especially for Smart Power ICs where the working temperature can be meaningfully higher than that of other ICs.

14.4.3.2
Corrosion of Aluminium

Most of the packages used for Smart Power ICs are in plastic and hence, by definition, not hermetic: water reaches the die by penetration along the interface between resin and frame or by bulk diffusion through the resin. Contamination ions can be present at die surface or, more probably, are carried in by water from resin or from frame plating. In this way, a thin film of electrolytic solution builds up on die surface and the corrosion of aluminium alloy metallization can take place.

The microenvironment present on the die surface in wet conditions is very difficult to characterise and several of its features can influence the occurrence of corrosion phenomena:

- Extension and thickness of water film on the die surface which depend on: the external environmental conditions; degree of hermeticity and hence structure of the package; gap at die-resin interface, related mainly to the adhesion properties of the resin; die power dissipation which can increase the temperature on die surface and hence locally decrease the Relative Humidity (RH);
- Amount and type of contamination ions coming from package materials or due to wafer contamination.

Furthermore, some die "back-end" features, in particular the type of Al alloy and passivation, can play an important role in corrosion phenomena.

In the last years improvements in the assembly process, the high purity reached by molding compounds and the use of silicon nitride passivation have allowed to strongly reduce the incidence of Al corrosion. This failure mode can still be encountered random for process "out of control" in consolidated package line while it can be a more structural problem for new advanced SMD packages where the resin thickness is very reduced and heavy delamination phenomena with loss of hermeticity can take place during surface mounting.

Before discussing in detail the mechanism of Al corrosion in ICs we will recall some general characteristics of aluminum/water electrochemical system. Figure 14.24 shows the potential-pH (or Pourbaix) equilibrium diagram of aluminium-water system [32]. Without explaining this diagram in depth we want to underline that aluminium oxidation reactions have a very low electrochemical potential, below that of H^+ reduction, which means high attitude to undergo electrochemical corrosion. On the other hand for a wide range of pH the Al

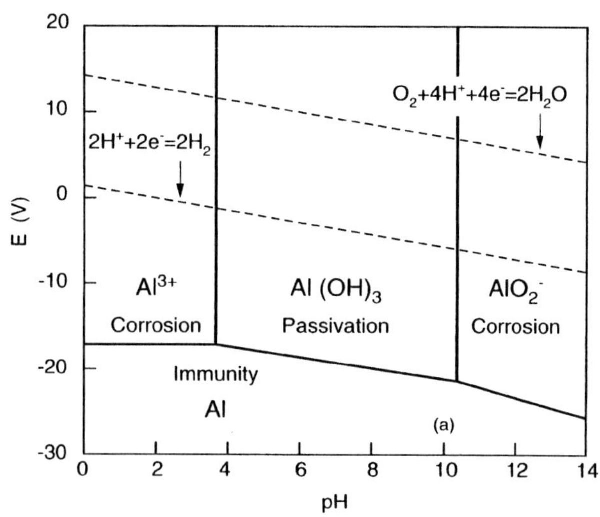

Fig. 14.24 Pourbaix diagram for system aluminium-water at 25 °C (after [32])

oxidation gives rise to a build-up of a passivation oxide film which can prevent further reactions.

Therefore aluminium corrosion can only take place when:

- The pH of solution reaches values for which passivation oxide is not stable;
- Contaminating ions able to locally destroy the film of passivation oxide are present;
- The quality of the grown oxide layer it is not good enough to prevent further oxidation reaction.

As we will see later on, all these circumstances can occur, under particular conditions, in ICs. Furthermore, we have to state that in wet conditions the corrosion of aluminium is basically of electrochemical nature, which means that an Al oxidation reaction occurs in some region of the metal and supply electrons for a reduction reaction, hydrogen discharge or similar, which takes place elsewhere in the metallic conductor. The electrical circuit is "closed" by ionic conduction in the electrolytic solution.

Coming back to the problem of the corrosion of ICs in plastic package, we have to distinguish between situation without and with applied voltage; the first case happens when the device is simply stored in wet environment; the second case occurs when, in addition, an external bias is also applied. Analysing the first situation, it should be considered that, in general, the die metallizations are overcoated with a protective passivation film which is very effective to prevent water

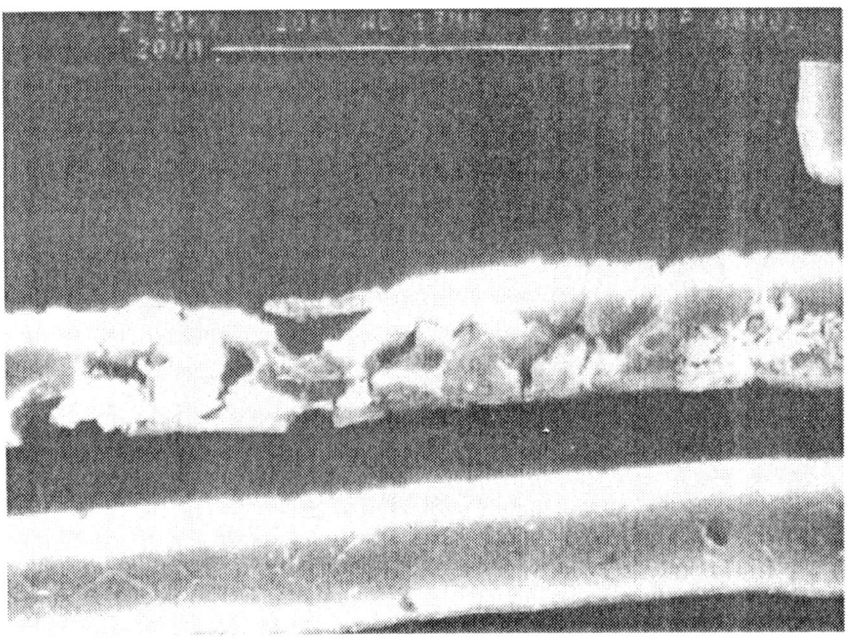

Fig. 14.25 Localised corrosion in a metal line (photo after passivation removal)

penetration, especially if it consists of silicon nitride. Therefore the corrosion starts only in relation to passivation defects, such as cracks, and appears localised; an example is shown in Fig. 14.25.

Of course, the passivation is opened in the contacting pads to allow the bonding and there the aluminium is directly exposed. However also in this case, as explained before, in general the presence of a thin layer of aluminium oxide prevents the direct contact of the metal with the water film. This oxide layer can lose its protective properties in very basic/acid solutions, where it is not stable (see Fig. 14.24), or in presence of impurity able to attack it. For instance chlorine ions are very dangerous as they can locally destroy passivation layer through reactions like:

$$Al(OH)_3 + 2Cl^- \rightarrow Al(OH)_2Cl_2^- + OH^- \qquad (14.17)$$

or similar. As the bare metal is exposed, electrochemical reactions occur with the aluminium which undergoes an oxidation and another species, typically hydrogen ions, which is reduced according to the scheme:

$$\begin{array}{ll} 2Al \rightarrow 2Al^{3+} + 6e & \text{anodic reaction} \\ 6H^+ + 6e \rightarrow 3H_2 & \text{cathodic reaction} \end{array} \qquad (14.18)$$

a precipitation of Al ions takes place subsequently according to the reaction:

$$Al^{3+} + 3H_2O \rightarrow Al(OH)_3 + 3H^+ \qquad (14.19)$$

If gold wires are used, the cathodic reactions take place at gold surface due to the lower overvoltage required for H^+ reduction and the overall reaction is enhanced. For this reason, the corrosion phenomena of aluminium pad are accelerated by the presence of Au/Al couple and in this sense we refer to galvanic corrosion. However, corrosion can take place also without any galvanic couple; in this case, the small areas of bare aluminium work as anode while the passivated metal acts as cathode.

To complete the analysis of corrosion phenomena in unbiased conditions, we want to underline that corrosion phenomena were also observed without the presence of aggressive impurities or of abnormal pH values (see for instance [33]); this takes place if, for some reason, the growth of the passivation layer of $Al(OH)_3$ is defective and gives rise to non-compact or porous films. In this case, the oxide film is not able to prevent further aluminium oxidation and this reaction can continue until eventually extending to all the aluminium thickness. The reaction takes place preferentially at the grain boundary and, at failure analysis, an intergranular abnormal oxidation appears with the typical aspect as shown in Fig. 14.26; for this reason, this failure mode is often referred to as aluminium oxidation.

These phenomena can be enhanced by some compounds present in the moulding resin but depend mainly on the initial state of the aluminium surface which is normally covered by a thin layer of native Al_2O_3. If this layer is of poor quality due to presence, for instance, of contamination the following growth of $Al(OH)_3$ can result anomalous. This type of failure mode is strongly reduced for Al/Si alloy metallization, as the silicon segregates at the grain boundary inhibiting the phenomenon [33].

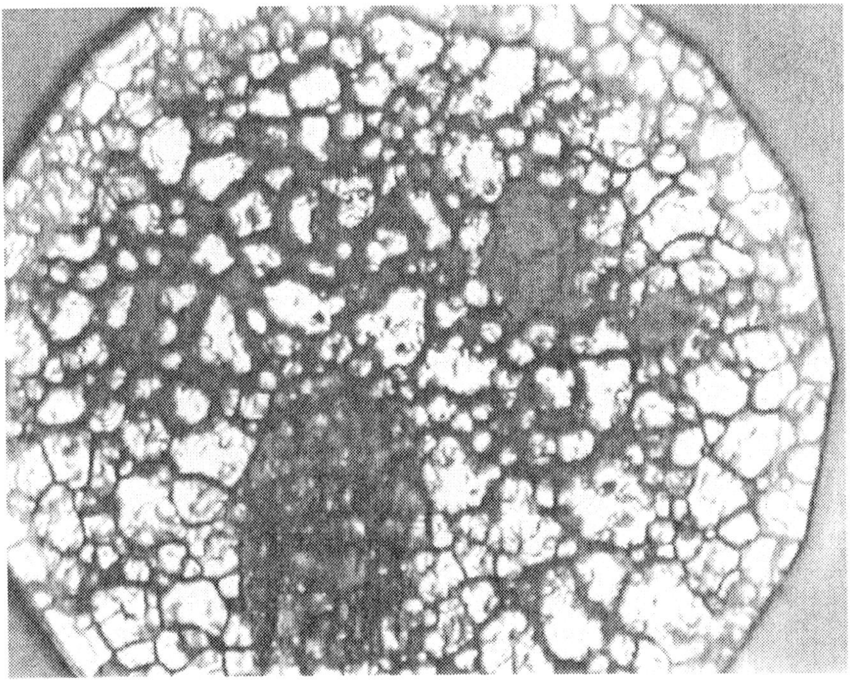

Fig. 14.26 Abnormal aluminium oxidation after pressure pot test

Finally, we want to mention that corrosion was sometimes observed also in aluminium wires, especially in presence of particular metal impurity in the aluminium alloy.

Analysing now corrosion phenomena with an external bias applied, it should be considered that two types of phenomena can take place: anodic and cathodic corrosion (see [34]). The first phenomenon occurs on positively biased metal lines; in this case aluminium oxidation reaction normally takes place with the growth of a protective oxide. Corrosion occurs if the electric field is high enough to cause oxide film breakdown or if chlorine ions able to attack the oxide film are present. The cathodic corrosion happens on the metal with negative bias where a hydrogen reduction reaction occurs. This reaction takes place mainly at grain boundary increasing the pH locally and causing the intergrain corrosion of the aluminium.

Figure 14.27 shows the result of a THB (Temperature Humidity Bias) experiment, carried out on a dedicated test pattern consisting of long parallel stripes with different types of passivation; a fixed voltage difference between the two stripes was kept during the test. On the stripes without any passivation layer the corrosion is cathodic; in the case of Phosphorus-doped Silicon Glass (PSG) protection, cathodic corrosion is still dominant but the time to failures increase due to the presence of the passivation. As regards nitride passivation the failure

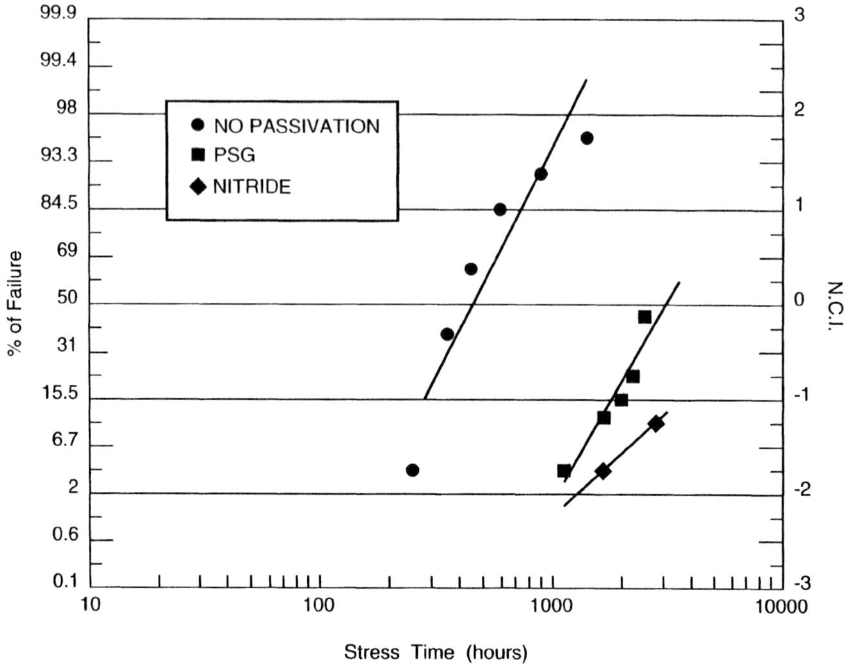

Fig. 14.27 Failure distributions (Log-Normal chart) in THB test on a test pattern with different passivations

takes place on the positively biased metal but near the pad and not where the two stripes are faced.

This example shows the higher effectiveness of nitride to prevent water and contaminant penetration. Furthermore, it should be noticed that the phosphorus of PSG can react to some extent with water, generating phosphoric acid, hence high content of phosphorus in PSG can cause heavy corrosion phenomena [34] and has to be avoided at least in the final layer.

In order to analyse corrosion phenomena with and without applied bias, the basic tests (see par. 14.3.2) are respectively temperature humidity bias and pressure pot. Different models have been proposed in the past years to correlate the results in different stress conditions and to predict the life time on the field. We want mention here two of the most common:

The Reich-Hakim model [35]:

$$\mathrm{MTF} = e^{[A + B(T + RH)]} \qquad (14.20)$$

where:

MTF is the Median Time to Failure;
T is the temperature (°C);
RH is the relative humidity.

The Peck model [36]:

$$MTF = A \cdot (RH)^n \cdot e^{Ea/KT} \quad (14.21)$$

where:
MTF is the Median Time to Failure;
T is the temperature (°K);
RH is the relative humidity;
k is the Boltzmann constant.

Each model can work rather well in a restricted range of conditions but none of them seems to have a general validity.

14.5 New Approaches for Reliability of Smart Power ICs

In the second and third paragraph we have focused the attention on the kind of tests to be performed for the reliability assurance of Smart Power ICs. In the fourth, we have reviewed the main failure mechanisms. In all these paragraphs some limits to the present approach have been underlined.

In this last paragraph analysing the future strategies we will answer to those limits and focus on another important issue dominating the reliability qualification scenario, which is fundamentally the reduction of the qualification cycle time.

As mentioned before, we have on one side very complex Smart Power ICs, designed in more and more sophiscated technologies and this means long design cycle time; on the other products that have a short life on the market, such as for the computer and industrial segment. Both these two aspects force the qualification cycle time to be reduced. To face this problem we are working on two new approaches: similarity by using generic data and process conformity. The key tool on which these two new approaches are based is the formalization, even if qualitative, of the knowledge gained by the great experience in the reliability field.

Similarity by using generic data is an approach that allows as a principle to perform qualification by extension of products belonging to the same technological family, which is key tool for a "standard" implementation of this approach. A technological family can be defined as the set of those products with the same functionality (analog, digital) designed by using the same structure both at package and at die (standard cells, dedicated cells) level. For the Smart Power ICs scenario, in which we are mainly dealing with analog devices designed with dedicated cells, the technological family concept itself does not allow to adopt the similarity approach.

We have developed a new concept for Smart Power ICs similarity: *Elementarization.*

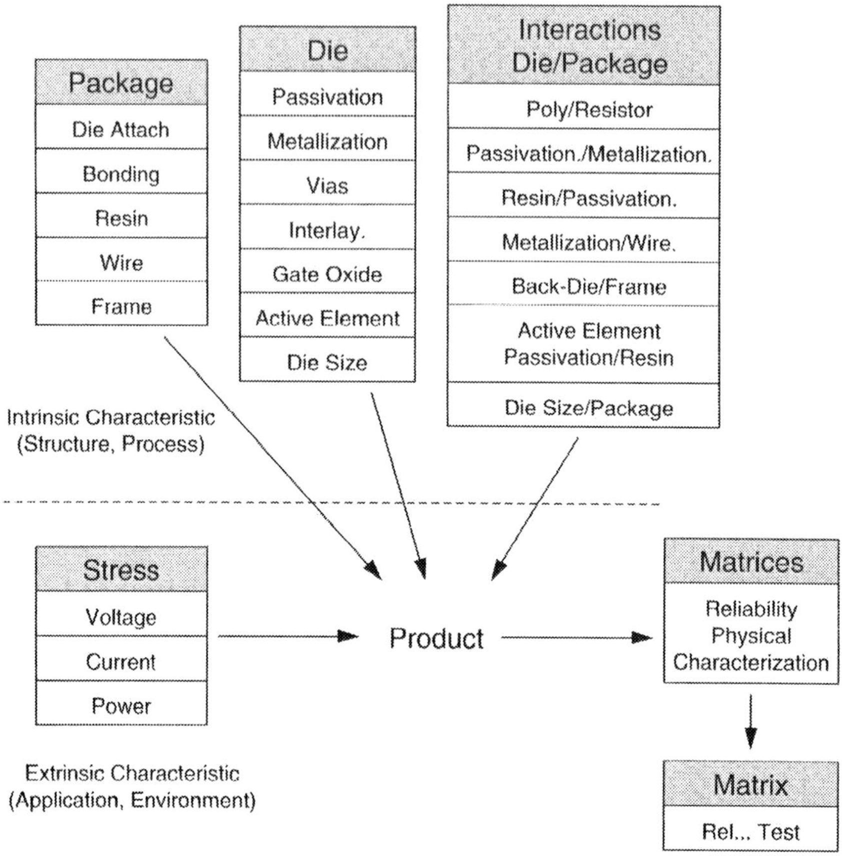

Fig. 14.28 IC Elementarization

This innovative approach bases the qualification of products on three main steps:

1) The product has to be "solved" (see Fig. 14.28) into elements (elementary structures): package elements (die attach, bonding, resin, etc.), die elements (passivation, metallization, etc.), interaction elements (die size vs package, for example).
2) For each of these elements the reliability knowledge/data available in the historical database, collecting all the information from qualification exercise, are represented in a matrix considering the number of qualifications already performed.
3) The reliability plan will include only those stress tests dedicated to the study of those elements that have not already been qualified by at least three different qualification exercises in the same class of stress.

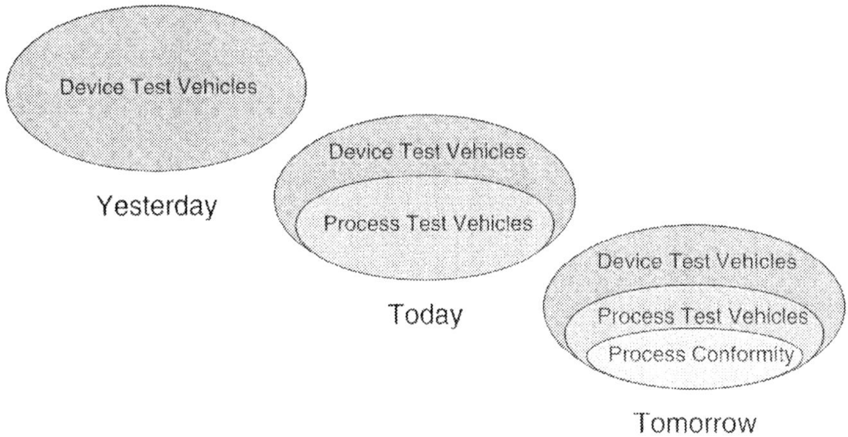

Fig. 14.29 Reliabilty tools evolution

In synthesis, the qualification of a Smart Power IC by similarity through the elementarization approach will mean the qualification of all its elements.

Process conformity. This new approach is more oriented towards process reliability qualification; in particular, as concerns process reliability monitoring and process change and transfer qualification. The idea is to analyse process data (such as: process flow; wafer parametric testing data; product yield) from a reliability point of view with the main aim to have a fast detection of reliability risks activating quicker corrective actions.

With reference to the existing reliability tools it is possible to quantify the reduction of the reliability evaluation cycle time as follows (see also Fig. 14.29): by means of stress tests on products it needs months, by means of Wafer Level Reliability (WLR) and of test on process test vehicle it needs weeks, by process conformity days.

The basic idea of process conformity, to exploit existing process data to make a reliability analysis, is in principle very simple. The difficulty, in the practise, lies on the fact that most of reliability models are empirical ones. They are defined to allow an extrapolation based on data derived from stress tests. The relationship between reliability performances and "process features" is not represented in those models. The challenge is right to formalise the impact of process features into reliability models.

The way to study the impact of "process features" on reliability performance has been planned according to the corner lot approach. By intentionally "changing something" it is possible to study its effect. The matrix of experiments has to be defined taking into account the natural variation of process parameters and hence changing in the defined range only those process parameters that can affect reliability. At the end this experimental exercise it should be possible to assess the reliability spec. limits on the basis of which process change and transfer can be analysed and the process monitored.

References

1. Mauri P (1992) Database Structure and Discovery Tools for Integrated Circuit Reliability Evaluation, Eigth International Conference on Data Engeneering, IEEE
2. Borri FR, Lonzi L (1993) Reliability Approach for Smart Power ICs, ESREF '93, Bordeaux
3. Ghio E, Cognetti C, D'Espinosa G, Fioravanti G, Lonzi L (1989) Power Devices Reliability, in Microelectronic Reliability, Vol 2, Artech House
4. Ghio E, Borri FR, Boccaletti G, D'Espinosa G, Fioravanti G (1989) Accelerated Tests, in Microelectronic Reliability, Vol 2, Artech House
5. Final Report of "International Wafer Level Reliability Workshop", (1992) IEEE
6. Hu C (1993) Simulating Hot Carriers Effects on Circuit Performance, IRPS Tutorial Note,
7. Boella G, Mauri P, Mussio P, Piccoli M (1990) Design of an Automatic Assistant for Failure Analysis of Linear Integrated Circuit, Proceedings of Expert System and Their Application, Tenth International Workshop, Avignon
8. Takeda E, Izawa R, Umeda K, Magai R (1991) AC Hot Carriers Effects in Scaled Down MOS Devices, IEEE IRPS
9. Hillen MW, Verwey JF (1986) Mobile Ions in SiO_2 Layers on Si, in Instabilty in Semiconductor Devices, Vol 1, Barbottin G and Vapaille A editors, Elsevier Science Publisher B. V. (North Holland)
10. Muller RS, Kamins TI (1986) Devices For Electronics Integrated Circuit, John Wiley & Sons,
11. Lycoudes NE, Childers CC (1980) Semiconductor Instability Failure Mechanisms Review, IEEE Transaction of Reliability, Vol R-29
12. Schroen W (1973) Failure Analysis of Surface Inversion, IEEE IRPS
13. Takeda E, Suzuki N (1983) An Empirical Model for Device Degradation Due to Hot-Carriers Injection, IEEE Electron Device Letters, Vol EDL N 4
14. Hu C et al. (1995) Hot-Electron Induced MOSFET Degradation: Model, Monitor and Improvement, IEEE Transaction on Electron Devices, Vol ED-32 N 2
15. Crook DL (1979) Method of Determining Reliabilty Screens for Time Dependent Dielectric Breakdown, IEEE IRPS
16. Lee JC, Chen HC, Hu C (1988) Modelling and Characterization of Gate Oxide Reliability, IEEE Transaction on Electron Devices, Vol 35
17. Tobias PA, Trinidade DC (1986) Applied Reliability, Van Nostrand Reinhold Company
18. Villa FF (1986) Improved Second Breakdown of Integrated Bipolar Power Transistors, IEEE Transaction Electron Devices, Vol ED-33
19. Carlson PJ (1986) Power MOSFET Ruggedness Testing and Performance, PCIM
20. Deng AC, Chua LO (1989) CMOS Latchup Theory, International Journal of Circuit Theory and Application, Vol 17
21. Scorzoni A, Neri B, Caprile C, Fantini F (1991) Electromigration in Thin Films Interconnection lines: Models, Methods and Results, Materials Science Reports Vol 7, n 4, 5
22. Schreiber HU (1981) Activation Energies for Different Electromigration Mechanisms in Aluminium, Solid State Electronics, Vol 24
23. Antognetti P, Bisio GR, Curatelli F, Palara S (1980) Three-Dimensional Transient Thermal Simulation: Application to Delayed Short Circuit Protection in Power ICs, IEEE Journal of Solid State Circuits, Vol SC-15
24. Lonzi L, Piccoli M, Speroni F (1994) Failure Mechanisms in Smart Power ICs, ESREF '94, Glasgow
25. Fredriksson G (1985) How Pulsed Operation Affects Aluminium Bond Wires, Quality and Reliability Engineering International, Vol 1
26. Dunn CF, McPherson JW (1990) Temperature-Cycling Acceleration Factors for Aluminium Metallization Failure in VLSI Applications, IEEE IRPS
27. Hansen M, Anderko K (1958) Constitution of Binary Alloys, Mc Graw Hill

28. Philofsky E (1970) Purple Plague Revisited, IEEE IRPS
29. Borri FR, D'Espinosa G (1986) Power Circuit Reliability in Power Integrated Circuit Physics and Applications, Antognetti P editor, Mc Graw Hill
30. Footner PK, Richardson PB, Yates RB (1987) Purple Plague: Eliminated or Just Forgotten?, Quality and Reliability Engineering International, Vol 3
31. Lonzi L (1982) Gold Wire Bonding on Aluminium Pad Reliability, ESSDERC '82, Munich
32. McDonald DD, Butler P (1973) The Thermodinamics of the Aluminium-Water System at Elevated Temperatures, Corrosion Science, Vol 13
33. Nishida T, Mukai K (1985) Moisture Resistance of Polyimide Multilevel Interconnection LSIs, IEEE IRPS
34. Lermer L, Eldridge JM (1982) Effects of Several Parameters on the Corrosion Rates of Al Conductors in Integrated Circuits, Journal of the Electrochemical Society
35. Reich B, Hakim EB (1972) Temperature-Humidity Acceleration Factors for Plastic Transistors and Integrated Circuits, Solid-State Technology
36. Peck DS (1986) Comprehensive Model for Humidity Testing Correlation, IEEE IRPS

Subject Index

Printing: Mercedes-Druck, Berlin
Binding: Buchbinderei Lüderitz & Bauer, Berlin

Printed in the United Kingdom
by Lightning Source UK Ltd.
107685UKS00004B/43-45

9 783540 432388